Massive MIMO Systems

Massive MIMO Systems

Special Issue Editors

Kazuki Maruta
Francisco Falcone

MDPI • Basel • Beijing • Wuhan • Barcelona • Belgrade • Manchester • Tokyo • Cluj • Tianjin

Special Issue Editors
Kazuki Maruta
Chiba University
Japan

Francisco Falcone
Public University of Navarre
Spain

Editorial Office
MDPI
St. Alban-Anlage 66
4052 Basel, Switzerland

This is a reprint of articles from the Special Issue published online in the open access journal *Electronics* (ISSN 2079-9292) (available at: https://www.mdpi.com/journal/electronics/special_issues/mimo).

For citation purposes, cite each article independently as indicated on the article page online and as indicated below:

LastName, A.A.; LastName, B.B.; LastName, C.C. Article Title. *Journal Name* **Year**, *Article Number*, Page Range.

ISBN 978-3-03936-016-1 (Hbk)
ISBN 978-3-03936-017-8 (PDF)

Contents

About the Special Issue Editors

Kazuki Maruta completed his Bachelor of Engineering and Master of Engineering degrees, as well as his PhD, at Kyushu University, Japan, in 2006, 2008 and 2016, respectively. From 2008 to 2017, he worked on the research and development of interference compensation techniques for future wireless communication systems at NTT Access Network Service Systems Laboratories. From 2017 to 2020, he was an Assistant Professor at the Graduate School of Engineering, Chiba University. He is currently a Specially Appointed Associate Professor at the Academy for Super Smart Society, Tokyo Institute of Technology. His research interests include MIMO, adaptive array signal processing, channel estimation, medium access control protocols and moving networks. He is a member of IEEE and IEICE. He won the IEICE Young Researcher's Award in 2012, the IEICE Radio Communication Systems (RCS) Active Researcher Award in 2014, the Asia-Pacific Microwave Conference (APMC) Prize in 2014, and the IEICE RCS Outstanding Researcher Award in 2018. He was also co-recipient of the IEICE Best Paper Award in 2018.

Francisco Falcone completed his degree in telecommunication engineering and his PhD in communication engineering at the Public University of Navarra (UPNA), Spain, in 1999 and 2005, respectively. From 1999 to 2000, he was a Microwave Commissioning Engineer with Siemens-Italtel, where he deployed microwave access systems. From 2000 to 2008, he was a Radio Access Engineer at Telefónica Móviles, performing radio network planning and optimization tasks in mobile network deployment. In 2009, he was a co-founding member and the Director of Tafco Metawireless, a spin-off company from UPNA. He was an Assistant Lecturer at the Electrical and Electronic Engineering Department, UPNA, from 2003 to 2009. In 2009, he became an Associate Professor with the EE Department, and was the Head of Department from 2012 to 2018. From January 2018 to May 2018, he was a Visiting Professor at the Kuwait College of Science and Technology, Kuwait. He is also affiliated with the Institute for Smart Cities (ISC), UPNA, which hosts around 140 researchers, and is the Head of the ICT section. He has more than 500 contributions in indexed international journals, book chapters and conference contributions. His research interests are related to computational electromagnetics applied to the analysis of complex electromagnetic scenarios, with a focus on the analysis, design and implementation of heterogeneous wireless networks to enable context aware environments. Prof. Falcone was a recipient of the CST 2003 and CST 2005 Best Paper Award, the PhD Award from the Colegio Oficial de Ingenieros de Telecomunicación (COIT) in 2006, the Doctoral Award UPNA, 2010, the 1st Juan Gomez Peñalver Research Award from the Royal Academy of Engineering of Spain in 2010, the XII Talgo Innovation Award 2012, the IEEE 2014 Best Paper Award, 2014, the ECSA-3 Best Paper Award, 2016, and the ECSA-4 Best Paper Award, 2017. He is a Senior Member of the IEEE.

electronics

MDPI

Editorial

Massive MIMO Systems: Present and Future

Kazuki Maruta [1,*] and Francisco Falcone [2]

[1] Graduate School of Engineering, Chiba University, 1–33 Yayoi-cho, Inage-ku, Chiba-shi 263-8522, Japan
[2] Department of Electrical, Electronic and Communication Engineering & Institute for Smart Cities, Public University of Navarre, 31006 Pamplona, Spain; francisco.falcone@unavarra.es
* Correspondence: kazuki.maruta@ieee.org

Received: 17 February 2020; Accepted: 21 February 2020; Published: 26 February 2020

1. Introduction

We are going to see the first decade since the fundamental concept of massive multiple-input multiple-output (MIMO) (also called large-scale MIMO) has emerged [1]. Massive MIMO is expected to be one of the most promising technologies towards the fifth generation mobile communications (5G) and beyond. Implementation [2,3] and trials [4,5] are actively proceeded. Especially, massive array beamforming has a good match to millimeter wave communication [6] which suffers from link budget shortfall due to its high frequency. Further, thanks to its excessive degree of freedom (DoF), massive MIMO has unlimited potentiality to further enhance system capabilities [7] and still expands various research topics with depth. It should be further discussed and believed to break limitations in wireless communications such as spectral and energy efficiencies for better support of continuously increasing mobile data traffic, as well as terminals driven by Internet of things (IoT). The key contribution of this special issue is to provide readers with new insights and facilitate plentiful discussions in this field.

2. The Present Issue

This special issue consists of nineteen papers covering wide and important topics in the field of massive MIMO systems, including both fundamental regions such as computation complexity, energy efficiency, pilot contamination, channel estimation, antenna design, non-orthogonal multiple access (NOMA) and millimeter-wave beamforming, as well as emerging topic such as machine learning incorporation. From the system model aspect, variety of scenario have also been covered such as single/multi-cell, distributed antennas, heterogeneous network, IEEE802.11ac and long term evolution (LTE) standards.

Distributed antenna systems (DAS) or base station (BS) cooperation have actively investigated since it can provide array diversity or multiplexing gain due to low spatial correlation of distributed antennas. Its extension to massive MIMO was analyzed in terms of spectral and energy efficiencies with considering hardware impairment such as phase noise [8] and analog-to-digital converter (ADC) resolution [9]. In the distributed massive MIMO structure, sounding reference signal (SRS) design and channel estimation scheme were proposed in order to mitigate the pilot contamination impact [10].

Work in [11] proposed a path loss based pilot allocation strategy and pseudo-random code pilot design. In [12], a modified heuristic pilot assignment algorithm was proposed. Its optimization criteria is to maximize the minimum uplink signal-to-interference plus noise power ratio (SINR). Efficient channel state information (CSI) estimation scheme was proposed in [13]. It exploits prior CSI of the previous timeslot having temporal correlation in the angular domain. Differential modulation unnecessitates channel estimation and is preferable especially in massive MIMO systems. In [14], incoherent detection for differential modulation was expanded to multiple symbols in the single cell scenario. For further capacity enhancement, multiplexing in the power domain, i.e., NOMA enabled by successive interference cancellation (SIC), was introduced [15].

In millimeter-wave communication, almost line-of-sight (LoS) channel or Ricean fading channel is expected. Exploiting CSI of the LoS component, spectral efficiency of equal gain transmission and combining (EGT/EGC) was analyzed in Ricean fading frequency selective fading channel with cooperative relaying scenarios [16]. Such relaying approach is also effective in heterogeneous network where small cell BSs play a role of relay the macro cell BS and user terminals. Reference [17] proposed eigenvector decomposition based hybrid beamforming in the above scenario.

In the practical viewpoint, limited statistical CSI feedback constraint was considered and machine learning based user grouping aided hybrid beamforming was proposed [18]. Further, CSI estimation elimination approach, which applies a blind adaptive array signal processing, has been proposed and its practical performance was evaluated with considering medium access control (MAC) layer overhead of IEEE802.11ac and frequency division duplex (FDD) based LTE standards [19].

Computation complexity for pre/post coding is also significant problem on massive MIMO systems. Suppose uplink transmission, iteration-based new detection algorithms were proposed. One is the extension of linear minimum mean squared error (MMSE) post coding and log-likelihood ratio (LLR) calculation [20] and another is based on the maximum likelihood (ML) detection and iterative discrete estimation approaches [21].

Focusing on energy efficiency, reference [22] proposed simplified beamforming as well as power allocation strategies for the scenario wherein unicast and multicast users are non-orthogonally multiplexed. Discontinuous reception can also contribute to improve the energy efficiency. Authors in [23] introduced an artificial intelligence (AI) approach, i.e., recurrent neural network (RNN), to adapt sleep cycles of user terminals.

In realization of large-scale antenna arrays, we should pay attention to antenna manufacturing. Reference [24] developed Bayesian compressive sensing based planar array diagnostic tool for efficient and reliable testing. New antenna structures were designed; dual-polarized diamond-ring slot antenna array exhibiting wide bandwidth [25], and leaky-wave antenna array incorporating metamaterial shield [26] which can suppress the mutual coupling.

3. Future

Now discussions towards 6G has started. Massive MIMO is still expected as a promising contributor for 6G [27–29], e.g., referred as 'ultra massive MIMO'. Its potentiality will be truly realized through relentless effort on R&D including the advance of hardware performance. Variety of massive MIMO technologies, which were widely addressed in this special issue, could be one of the most important solutions to bring a breakthrough towards beyond 5G or 6G.

Author Contributions: K.M. and F.F. worked together in the whole editorial process of the special issue, 'Massive MIMO Systems' published by journal Electronics. K.M. drafted this editorial summary. K.M. and F.F. reviewed, edited and finalized the manuscript. All authors have read and agree to the published version of the manuscript.

Acknowledgments: First of all we would like to thank all researchers who submitted articles to this special issue for their excellent contributions. We are also grateful to all reviewers who contributed evaluations of scientific merits and quality of the manuscripts and provided countless valuable suggestions to improve their quality and the overall value for the scientific community. We would like to acknowledge the editorial board of Electronics journal, who invited us to guest edit this special issue. We are also grateful to the Electronics Editorial Office staff who worked thoroughly to maintain the rigorous peer-review schedule and timely publication.

Conflicts of Interest: The authors declare no conflicts of interest.

Abbreviations

The following abbreviations are used in this manuscript:

MIMO	Multiple-Input Multiple-Output
5G	Fifth generation mobile communications
DoF	Degree of freedom
IoT	Internet of things
NOMA	Non-orthogonal multiple access
LTE	Long term evolution
DAS	Distributed antenna systems
BS	Base station
ADC	Analog-to-digital converter
SRS	Sounding reference signal
SINR	Signal-to-interference plus noise power ratio
CSI	Channel state information
SIC	Successive interference cancellation
LoS	Line-of-sight
EGT	Equal gain transmission
EGC	Equal gain combining
MAC	Medium access control
FDD	Frequency division duplex
MMSE	Minimum mean squared error
LLR	Log-likelihood ratio
ML	Maximum likelihood
AI	Artificial intelligence
RNN	Recurrent neural network

References

1. Marzetta, T.L. Noncooperative Cellular Wireless with Unlimited Numbers of Base Station Antennas. *IEEE Trans. Wirel. Commun.* **2010**, *9*, 3590–3600. [CrossRef]
2. Malkowsky, S.; Vieira, J.; Liu, L.; Harris, P.; Nieman, K.; Kundargi, N.; Wong, I.C.; Tufvesson, F.; Öwall, V.; Edfors, O. The World's First Real-Time Testbed for Massive MIMO: Design, Implementation, and Validation. *IEEE Access* **2017**, *5*, 9073–9088. [CrossRef]
3. Wirth, T.; Mehlhose, M.; Thiele, L.; Haustein, T. Proof-of-concept of flexible massive MIMO beamforming at 2.4 GHz. In Proceedings of the 2017 51st Asilomar Conference on Signals, Systems, and Computers, Pacific Grove, CA, USA, 29 October–1 November 2017; pp. 1061–1065. [CrossRef]
4. Obara, T.; Okuyama, T.; Inoue, Y.; Aoki, Y.; Suyama, S.; Lee, J.; Okumura, Y. Experimental Trial of 5G Super Wideband Wireless Systems Using Massive MIMO Beamforming and Beam Tracking Control in 28GHz Band. *IEICE Trans. Commun.* **2017**, *100*, 1256–1268. [CrossRef]
5. Sakai, M.; Kamohara, K.; Iura, H.; Nishimoto, H.; Ishioka, K.; Murata, Y.; Yamamoto, M.; Okazaki, A.; Nonaka, N.; Suyama, S.; et al. Experimental Field Trials on MU-MIMO Transmission for High SHF Wide-band Massive MIMO in 5G. *IEEE Trans. Wirel. Commun.* **2020**. [CrossRef]
6. Wu, W.; Liu, D. Hybrid BD-GMD Precoding for Multiuser Millimeter-Wave Massive MIMO Systems. *IEICE Trans. Commun.* **2018**. [CrossRef]
7. Maruta, K.; Ahn, C.J. Uplink Interference Suppression by Semi-Blind Adaptive Array With Decision Feedback Channel Estimation on Multicell Massive MIMO Systems. *IEEE Trans. Commun.* **2018**, *66*, 6123–6134. [CrossRef]
8. Lv, Q.; Li, J.; Zhu, P.; Wang, D.; You, X. Downlink Spectral Efficiency Analysis in Distributed Massive MIMO with Phase Noise. *Electronics* **2018**, *7*, 317. [CrossRef]
9. Li, J.; Lv, Q.; Yang, J.; Zhu, P.; You, X. Spectral and Energy Efficiency of Distributed Massive MIMO with Low-Resolution ADC. *Electronics* **2018**, *7*, 391. [CrossRef]
10. Yu, S.; Lee, J.W. Channel Sounding for Multi-User Massive MIMO in Distributed Antenna System Environment. *Electronics* **2019**, *8*, 36. [CrossRef]

11. Saraereh, O.A.; Khan, I.; Lee, B.M.; Tahat, A. Efficient Pilot Decontamination Schemes in 5G Massive MIMO Systems. *Electronics* **2019**, *8*, 55. [CrossRef]
12. Al-hubaishi, A.S.; Noordin, N.; Sali, A.; Subramaniam, S.; Mohammed Mansoor, A. An Efficient Pilot Assignment Scheme for Addressing Pilot Contamination in Multicell Massive MIMO Systems. *Electronics* **2019**, *8*, 372. [CrossRef]
13. Lu, W.; Wang, Y.; Wen, X.; Peng, S.; Zhong, L. Downlink Channel Estimation in Massive Multiple-Input Multiple-Output with Correlated Sparsity by Overcomplete Dictionary and Bayesian Inference. *Electronics* **2019**, *8*, 473. [CrossRef]
14. Dao, H.T.; Kim, S. Multiple-Symbol Non-Coherent Detection for Differential QAM Modulation in Uplink Massive MIMO Systems. *Electronics* **2019**, *8*, 693. [CrossRef]
15. Ha, J.G.; Ro, J.H.; Song, H.K. Throughput Enhancement in Downlink MU-MIMO Using Multiple Dimensions. *Electronics* **2019**, *8*, 758. [CrossRef]
16. Yan, Q.; Sun, Y.; Yue, D.W. LOS-Based Equal Gain Transmission and Combining in General Frequency-Selective Ricean Massive MIMO Channels. *Electronics* **2019**, *8*, 79. [CrossRef]
17. Hefnawi, M. Hybrid Beamforming for Millimeter-Wave Heterogeneous Networks. *Electronics* **2019**, *8*, 133. [CrossRef]
18. Khammari, H.; Ahmed, I.; Bhatti, G.; Alajmi, M. Spatio-Radio Resource Management and Hybrid Beamforming for Limited Feedback Massive MIMO Systems. *Electronics* **2019**, *8*, 1061. [CrossRef]
19. Muramatsu, F.; Nishimori, K.; Taniguchi, R.; Hiraguri, T.; Mitsui, T. Evaluation of Multi-Beam Massive MIMO Considering MAC Layer Using IEEE802.11ac and FDD-LTE. *Electronics* **2019**, *8*, 225. [CrossRef]
20. Khan, I.; Zafar, M.; Ashraf, M.; Kim, S. Computationally Efficient Channel Estimation in 5G Massive Multiple-Input Multiple-Output Systems. *Electronics* **2018**, *7*, 382. [CrossRef]
21. Feng, H.; Zhao, X.; Li, Z.; Xing, S. A Novel Iterative Discrete Estimation Algorithm for Low-Complexity Signal Detection in Uplink Massive MIMO Systems. *Electronics* **2019**, *8*, 980. [CrossRef]
22. Wang, W.; Huang, Y.; You, L.; Xiong, J.; Li, J.; Gao, X. Energy Efficiency Optimization for Massive MIMO Non-Orthogonal Unicast and Multicast Transmission with Statistical CSI. *Electronics* **2019**, *8*, 857. [CrossRef]
23. Memon, M.L.; Maheshwari, M.K.; Saxena, N.; Roy, A.; Shin, D.R. Artificial Intelligence-Based Discontinuous Reception for Energy Saving in 5G Networks. *Electronics* **2019**, *8*, 778. [CrossRef]
24. Famoriji, O.; Zhang, Z.; Fadamiro, A.; Zakariyya, R.; Lin, F. Planar Array Diagnostic Tool for Millimeter-Wave Wireless Communication Systems. *Electronics* **2018**, *7*, 383. [CrossRef]
25. Ojaroudi Parchin, N.; Jahanbakhsh Basherlou, H.; Alibakhshikenari, M.; Ojaroudi Parchin, Y.; Al-Yasir, Y.I.A.; Abd-Alhameed, R.A.; Limiti, E. Mobile-Phone Antenna Array with Diamond-Ring Slot Elements for 5G Massive MIMO Systems. *Electronics* **2019**, *8*, 521. [CrossRef]
26. Alibakhshikenari, M.; Virdee, B.S.; See, C.H.; Abd-Alhameed, R.A.; Falcone, F.; Limiti, E. High-Isolation Leaky-Wave Array Antenna Based on CRLH-Metamaterial Implemented on SIW with ±30° Frequency Beam-Scanning Capability at Millimetre-Waves. *Electronics* **2019**, *8*, 642. [CrossRef]
27. Yang, P.; Xiao, Y.; Xiao, M.; Li, S. 6G Wireless Communications: Vision and Potential Techniques. *IEEE Netw.* **2019**, *33*, 70–75. [CrossRef]
28. Bi, Q. Ten Trends in the Cellular Industry and an Outlook on 6G. *IEEE Commun. Mag.* **2019**, *57*, 31–36.
29. Zhang, Z.; Xiao, Y.; Ma, Z.; Xiao, M.; Ding, Z.; Lei, X.; Karagiannidis, G.K.; Fan, P. 6G Wireless Networks: Vision, Requirements, Architecture, and Key Technologies. *IEEE Veh. Technol. Mag.* **2019**, *14*, 28–41.

electronics

MDPI

Article

Downlink Spectral Efficiency Analysis in Distributed Massive MIMO with Phase Noise

Qian Lv, Jiamin Li *, Pengcheng Zhu, Dongming Wang and Xiaohu You

National Mobile Communications Research Laboratory, Southeast University, Nanjing 210096, China; seulvqian@seu.edu.cn (Q.L.); p.zhu@seu.edu.cn (P.Z.); wangdm@seu.edu.cn (D.W.); xhyu@seu.edu.cn (X.Y.)
* Correspondence: lijiamin@seu.edu.cn; Tel.: +86-025-5209-1635

Received: 25 October 2018 ; Accepted:10 November 2018 ; Published: 12 November 2018

Abstract: To achieve the advantages provided by massive multiple-input multiple-output (MIMO), a large number of antennas need to be deployed at the base station. However, for the reason of cost, inexpensive hardwares are employed in the realistic scenario, which makes the system distorted by hardware impairments. Hence, in this paper, we analyze the downlink spectral efficiency in distributed massive MIMO with phase noise and amplified thermal noise. We provide an effective channel model considering large-scale fading, small-scale fast fading and phase noise. Based on the model, the estimated channel state information (CSI) is obtained during the pilot phase. Under the imperfect CSI, the closed-form expressions of downlink achievable rates with maximum ratio transmission (MRT) and zero-forcing (ZF) precoders in distributed massive MIMO are derived. Furthermore, we also give the user ultimate achievable rates when the number of antennas tends to infinity with both precoders. Based on these expressions, we analyze the impacts of phase noise on the spectral efficiency. It can be concluded that the same limit rate is achieved with both precoders when phase noise is present, and phase noise limits the spectral efficiency. Numerical results show that ZF outdoes MRT precoder in spectral efficiency and ZF precoder is more affected by phase noise.

Keywords: distributed massive MIMO; phase noise; amplified thermal noise; spectral efficiency

---

1. Introduction

Massive multiple-input multiple-output (MIMO) is becoming a promising technology to provide significant gains [1–6]. Since it was first proposed, massive MIMO has been studied extensively. The main feature of massive MIMO is that hundreds (or even thousands) of antennas are employed at each base station, simultaneously serving tens of users in the same time-frequency resource, which offers big advantages compared to conventional MIMO. Firstly, it can bring unprecedented spatial degrees-of-freedom, which enables the improvement of spectral efficiency and energy efficiency even with simple linear receivers or precoders [7]. In addition, user channels in massive MIMO systems are nearly orthogonal and fast fading, intra-cell interference can be averaged out. Massive MIMO can be divided into two categories: one is co-located massive MIMO and the other is distributed massive MIMO [8]. The latter has promising advantages of increasing energy efficiency, system coverage and spectral efficiency, which results from the increase in macro-diversity gain and the reduction in access distance [9–15]. Considering these advantages, we analyze the spectral efficiency of distributed massive MIMO in this paper. Notably, due to the different access distance between each user and all remote antenna units (RAUs), the channel vectors are non-isotropic, which makes the analysis of performance in distributed massive MIMO more difficult and more complex.

In practical communication systems, inevitable hardware impairments occur and cannot be eliminated even after applying calibration and compensation techniques [16,17]. These impairments can be divided into two categories: multiplicative distortion and additive distortion. Phase noise introduced by the local oscillators of transceivers is the multiplicative distortion. It will cause random rotations of the transmitted data symbols, which degrades the system performance. Furthermore, phase noise makes the estimated channel state information (CSI) more inaccurate and it introduces a phenomenon called channel aging which means the estimated CSI obtained during pilot phase is different from that used for downlink transmission. It is pointed out in [8] that the deployment cost and circuit power consumption of massive MIMO scale linearly with the number of antennas. Therefore inexpensive but hardware-constraint hardware may be deployed for the reason of cost, which makes the hardware impairments more severe in massive MIMO.

Analyzing the spectral efficiency is a fundamental method to evaluate the impacts of phase noise. The impacts of phase noise for uplink transmission have been studied in [18–21] and for downlink transmission were investigated in [22–24]. The impacts of phase noise on physical layer security for downlink massive MIMO were investigate in [22]. The achievable rate was derived in [23] considering the frequency-selective multipath fading channel. The capacity of downlink transmission with linear precoders was analyzed in [24] but it assumed that the number of antennas and users was asymptotically large and it only considered a co-located MIMO system.

Herein, considering a distributed massive MIMO with phase noise and amplified thermal noise, we analyzed the downlink spectral efficiency for any number of antennas and users. Followings are the key contributions of this paper:

1. In distributed massive MIMO, the channel vectors are non-isotropic and the correlation between channel vectors and intended beams for each user are destroyed by phase noise. Hence, we first obtain the distributions of the desired signal and interference powers, which is challenging and complex.
2. Considering both zero-forcing (ZF) and maximum ratio transmission (MRT) precoders, we obtain the closed-form expressions of the downlink ergodic achievable rates with imperfect CSI and hardware impairments in distributed massive MIMO. These closed-form expressions are accurate for any number of antennas and users in both distributed massive MIMO and co-locate massive MIMO. Furthermore, they are derived under imperfect CSI which is more realistic, and they enable the analysis of performance degradation caused by phase noise.
3. The ultimate achievable rate per user is obtained when the number of antennas per remote antenna unit (RAU) goes infinity. It can be used to investigate the asymptotic performance of distributed massive MIMO with hardware impairments.
4. The theoretical results are verified by Monte Carlo simulations, and we have a deep insight into the impacts of phase noise.

The rest of this paper is organized as follows. System model including system configuration, a model describing phase noise and an effective channel model is introduced in Section 2. We obtain the estimated CSI during the uplink pilot training phase and analyze the spectral efficiency with linear precoders in Section 3. Numerical results are given in Section 4. A conclusion is provided in Section 5.

Notation: Column vectors $\mathbf{x}$ and matrices $\mathbf{X}$ are denoted by bold letters in lower case and in upper case, respectively. $\mathbf{I}_N$ is a $N \times N$ identity matrix. $(\cdot)^{\mathrm{H}}$ and $(\cdot)^{\mathrm{T}}$ are the conjugate transpose and transpose operator, respectively. Scalars x are denoted by italic letters. $|x|$ represents the absolute value of x and $\|\mathbf{X}\|$ denotes the spectral norm of $\mathbf{X}$. $\mathbb{E}[\cdot]$ and $\mathrm{var}(\cdot)$ represent the expectation operator and variance operator, respectively. $\mathcal{CN}(0, \sigma^2)$ represents circularly symmetric complex Gaussian distribution with mean zero and variance σ^2. $\Gamma(k, \theta)$ means Gamma distribution with shape parameter k and scale parameter θ. Similarly, Nakagami(m, Ω) means Nakagami distribution with shape parameter m and controlling spread parameter Ω.

2. System Model

Considering a distributed massive MIMO system, we first describe the system configuration and give the conventional channel model. Next, we present a model describing phase noise and give an effective channel model incorporating the impacts of phase noise.

We consider the downlink transmission of a single-cell multi-user distributed massive MIMO system comprising M RAUs and K single-antenna users as in Figure 1. Each RAU is equipped with an array of N antennas. All users and RAUs are randomly distributed in the cell.

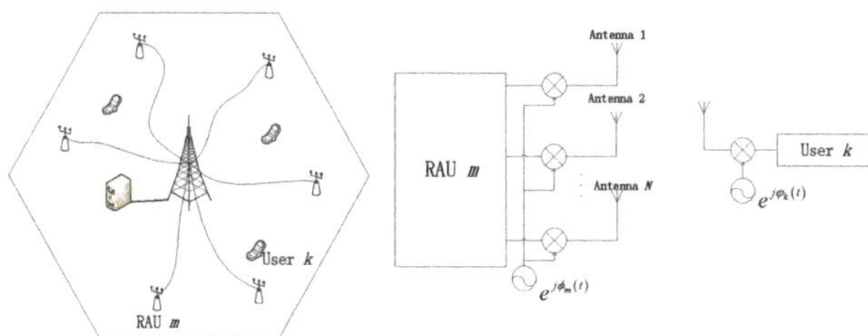

Figure 1. System Model.

Frequency-flat fading channels are assumed and the system runs in time-division duplex (TDD) protocol. The channel vector between all RAUs and the k-th user is given by

$$\tilde{\mathbf{g}}_k \triangleq \left[\tilde{g}_k^1 \cdots \tilde{g}_k^{MN}\right] = \mathbf{\Lambda}_k^{1/2}\mathbf{h}_k, \tag{1}$$

where $\mathbf{\Lambda}_k = \mathbb{E}\left[\tilde{\mathbf{g}}_k\tilde{\mathbf{g}}_k^{\mathrm{H}}\right] = \mathrm{diag}(\lambda_{1,k} \cdots \lambda_{M,k}) \otimes \mathbf{I}_N$ is the covariance matrix, $\lambda_{m,k} \triangleq cd_{m,k}^{-\alpha}$ denotes the path loss between the m-th RAU and the k-th user, $d_{m,k}$ is the corresponding distance, α is the path loss exponent, c is the median of the mean path gain at a reference distance $d_{m,k} = 1$ km, and $\mathbf{h}_k \sim \mathcal{CN}(0, \mathbf{I}_{MN})$ is the small-scale fast fading vector.

In this paper, we consider a more realistic scenario where the antennas deployed at each RAU are inexpensive and hardware-constrained. Specifically, each antenna experiences phase noise which distorts communication. The phase noise means the multiplicative phase drift in the signal, which comes from the local oscillators (LOs) of the RAUs and users. We assume that the LOs are free-running without a phase-locked loop (PLL), and then the phase noise is commonly modeled as a discrete-time independent Wiener process [8,25]. Mathematically, the phase noises at the LOs of the n-th antenna and the k-th user are denoted as

$$\phi_n(t) \sim \mathcal{N}\left(\phi_n(t-1), \sigma_{\phi,n}^2\right), \tag{2}$$

$$\varphi_k(t) \sim \mathcal{N}\left(\varphi_k(t-1), \sigma_{\varphi,k}^2\right), \tag{3}$$

which equal the previous realization $\phi_n(t-1)$ and $\varphi_k(t-1)$ plus an independent zero-mean Gaussian random increment with variances $\sigma_{\phi,n}^2$ and $\sigma_{\varphi,k}^2$. The variances are dependent on the carrier frequency and symbol time [25].

The phase noise can be independent or correlated between antennas of each RAU. In our analysis, we have assumed that the phase noise correlated between antennas of one RAU and independent among RAUs. Then by expressing the total phase noise as a multiplicative factor, we can rewrite the channel vector as

$$\mathbf{g}_k(t) = \mathbf{\Theta}_k(t)\,\tilde{\mathbf{g}}_k, \tag{4}$$

where $\mathbf{\Theta}_k(t) \triangleq \mathrm{diag}\left(e^{j\theta_k^1(t)}, \cdots, e^{j\theta_k^{MN}(t)}\right) = e^{j\varphi_k(t)}\mathbf{\Phi}(t) \in \mathbb{C}^{MN\times MN}$ is the total phase noise, wherein $\mathbf{\Phi}(t) \triangleq \mathrm{diag}(e^{j\phi_1(t)}, \cdots, e^{j\phi_M(t)}) \otimes \mathbf{I}_N$ is the phase noise induced by all RAUs, and similarly, $e^{j\varphi_k(t)}$ corresponds to the phase drift pruduced by the k-th user. Notably, because of the presence of phase noise, the effective channel becomes time-dependent.

Remark 1. *The conventional channel model without phase noise is obtained when* $\sigma_{\phi,n}^2 = \sigma_{\varphi,k}^2 = 0, \forall n, k$.

3. Downlink Spectral Efficiency Analysis

In this section, firstly, based on the effective channel model given above, we assume pilot sequence aided transmission is employed and give the channel estimation. Next, since the channel vectors are non-isotropic in distributed massive MIMO and the correlation between channel vectors and intended beams for each user is destroyed by phase noise, we give the approximated distribution of desired signal and interference powers. After that, we derive the closed-form expressions of the ergodic achievable downlink rates with both MRT and ZF precoders.

3.1. Channel Estimation

As mentioned before, the transmission protocol is assumed as TDD. Each coherence block occupying T channel uses is split into two parts: one for uplink pilot symbols and the other for downlink data symbols. In order to guarantee that the pilot symbols of K users are orthogonal to each other, it's necessary to allocate $\tau \geq K$ symbols for pilot transmission. Then the remaining $T - \tau$ channel uses are used for downlink data transmission.

During the pilot training phase, the pilot sequence $\mathbf{x}_k \triangleq [x_k(1), \cdots, x_k(\tau)]^{\mathrm{T}}$ is assigned to user k. Incorporating the hardware impairments, the received pilot vector $\mathbf{y}_{\mathrm{P}}$ at the base station at time $t \in [0, \tau]$ is given as

$$\mathbf{y}_{\mathrm{P}}(t) = \sum_{k=1}^{K} \mathbf{g}_k(t)\, x_k(t) + \mathbf{n}^{\mathrm{BS}}(t), \tag{5}$$

where $\mathbf{n}^{\mathrm{BS}}(t) \sim \mathcal{CN}\left(0, \xi^{\mathrm{BS}}\mathbf{I}_{MN}\right)$ is the amplified thermal noise at time slot t, and its variance ξ^{BS} is larger than the variance σ^2 of thermal noise. This is because of the effects of low noise amplifiers, mixers and other components.

Let $\mathbf{\Psi} \triangleq \left[\mathbf{y}_{\mathrm{P}}^{\mathrm{T}}(1), \cdots, \mathbf{y}_{\mathrm{P}}^{\mathrm{T}}(\tau)\right]^{\mathrm{T}} \in \mathbb{C}^{\tau MN \times 1}$. Motivated by [8,26], the Linear Minimum Mean Square Error (LMMSE) estimation of the channel of the k-th user obtained by pilot training is given by

$$\hat{\mathbf{g}}_k(t) = \mathbf{\Lambda}_k \mathbf{H}_k(t) \mathbf{\Sigma}^{-1} \mathbf{\Psi}, \tag{6}$$

where

$$\begin{aligned}
\mathbf{H}_k(t) &= \left[\mathbf{H}_{k,1}(t), \cdots, \mathbf{H}_{k,\tau}(t)\right], \\
\mathbf{\Sigma} &\triangleq \textstyle\sum_{j=1}^{K} \mathbf{B}_j + \xi^{\mathrm{BS}} \mathbf{I}_{\tau MN}, \\
\mathbf{H}_{k,i}(t) &= x_k^*(i) \mathbf{D}_{k,i}(t), \\
\mathbf{D}_{k,i}(t) &= \operatorname{diag}\left(e^{-\frac{\sigma_{\phi,1}^2 + \sigma_{\varphi,k}^2}{2}|t-i|}, \cdots, e^{-\frac{\sigma_{\phi,M}^2 + \sigma_{\varphi,k}^2}{2}|t-i|}\right) \otimes \mathbf{I}_N, \\
\left[\mathbf{B}_j\right]_{u,v} &\triangleq \mathbf{\Lambda}_j x_j(u) x_j^*(v) \operatorname{diag}\left(e^{-\frac{\sigma_{\phi,1}^2 + \sigma_{\varphi,j}^2}{2}|u-v|}, \cdots, e^{-\frac{\sigma_{\phi,M}^2 + \sigma_{\varphi,j}^2}{2}|u-v|}\right) \otimes \mathbf{I}_N.
\end{aligned}$$

The pilot sequences can be designed in different ways. Without loss of generation, in this paper we assume that the number of pilot symbols is equal to that of users, i.e., $\tau = K$. More specifically, we assume that the set of orthogonal pilot sequences $\mathbf{X}_{\mathrm{P}} \triangleq [\mathbf{x}_1, \cdots, \mathbf{x}_K]$ is a diagonal matrix and each element of it is $\sqrt{\rho_{\mathrm{P}}}$, wherein ρ_{P} is the average transmit power of pilot symbols. This is equivalent to the assumption made in [18,20].

Under these assumptions, we give a definition of

$$\beta_{m,k}(t) = \frac{e^{-\frac{\sigma_{\varphi,k}^2+\sigma_{\phi,m}^2}{2}|t-k|}\sqrt{\rho_{\rm p}}\lambda_{m,k}}{\sqrt{\rho_{\rm p}\lambda_{m,k}+\xi^{\rm BS}}}, \tag{7}$$

then we can rewrite the LMMSE estimation $\hat{\mathbf{g}}_k(t)$ in (6) as

$$\begin{aligned}\hat{\mathbf{g}}_k(t) &= \mathbf{\Lambda}_k\mathbf{H}_k(t)\mathbf{\Sigma}^{-1/2}\hat{\mathbf{h}}_k \\ &= \left(\beta_{1,k}(t)\hat{\mathbf{h}}_{1,k}^{\rm T},\cdots,\beta_{M,k}(t)\hat{\mathbf{h}}_{M,k}^{\rm T}\right)^{\rm T},\end{aligned} \tag{8}$$

where $\beta_{m,k}(t)$ is the equivalent large-scale fading part from user k to RAU m and $\hat{\mathbf{h}}_k = [\hat{\mathbf{h}}_{1,k}^{\rm T},\cdots,\hat{\mathbf{h}}_{M,k}^{\rm T}]^{\rm T} = \mathbf{\Sigma}^{-1/2}\mathbf{\Psi} \sim \mathcal{CN}(0,\mathbf{I}_{MN})$ represents the equivalent small-scale fast fading part.

Because of the orthogonality principle of LMMSE estimation theory, the channel vector $\mathbf{g}_k(t)$ can be decomposed as

$$\mathbf{g}_k(t) = \hat{\mathbf{g}}_k(t) + \mathbf{e}_k(t), \tag{9}$$

where $\mathbf{e}_k(t)$ is the uncorrelated and statistically independent of $\hat{\mathbf{g}}_k(t)$ estimation error.

During the pilot transmission phase, we obtain the estimated channel showing in (8). In our analysis, it is assumed that the beamforming vector is designed by using the estimated CSI once during the pilot transmission phase and then is applied for the entire duration of the downlink transmission phase.

3.2. Downlink Signal Model

For downlink transmission, the received signal of user k at time $t \in [\tau+1, T]$ is given as

$$r_k(t) = \sqrt{\rho_{\rm dl}}\tilde{\mathbf{g}}_k^{\rm H}\Theta_k^*(t)\,\mathbf{x} + n^{\rm UE}(t), \tag{10}$$

where $\rho_{\rm dl}$ is the downlink transmission power, $n^{\rm UE}(t) \sim \mathcal{CN}(0,\xi^{\rm UE})$ is the amplified thermal noise of users at time slot t, $\xi^{\rm UE}$ is the variance of the noise, and $\mathbf{x} \in \mathbb{C}^{MN\times 1}$ is the signal vector transmitted by all M RAUs. Specifically, $\mathbf{x}$ can be given by

$$\mathbf{x} = \sum\nolimits_{l=1}^{K}\mathbf{w}_l s_l, \tag{11}$$

where $s_l \sim \mathcal{CN}(0,1)$ is the transmitted data symbol assigned for user l, and $\mathbf{w}_l$ is the beamforming vector designed at time slot τ. MRT and ZF precoders are considered in our analysis. Mathematically, these two linear precoders can be defined as

$$\mathbf{w}_l = \begin{cases}\frac{\hat{\mathbf{g}}_l(\tau)}{\|\hat{\mathbf{g}}_l(\tau)\|} & \text{MRT},\\ \frac{\mathbf{a}_l(\tau)}{\|\mathbf{a}_l(\tau)\|} & \text{ZF},\end{cases} \tag{12}$$

where $\mathbf{a}_l(\tau)$ is the l-th column of $\hat{\mathbf{G}}(\tau)\left(\hat{\mathbf{G}}^{\rm H}(\tau)\hat{\mathbf{G}}(\tau)\right)^{-1}$, wherein $\hat{\mathbf{G}}(\tau) = [\hat{\mathbf{g}}_1(\tau),\cdots,\hat{\mathbf{g}}_K(\tau)]$. Considering (4), we can rewrite (10) as

$$r_k(t) = \sqrt{\rho_{\rm dl}}\mathbf{g}_k^{\rm H}(\tau)\,\tilde{\Theta}_k(t)\,\mathbf{x} + n^{\rm UE}(t), \tag{13}$$

where

$$\tilde{\Theta}_k(t) \triangleq \operatorname{diag}\left(e^{-j\left(\theta_k^1(t)-\theta_k^1(\tau)\right)},\cdots,e^{-j\left(\theta_k^{MN}(t)-\theta_k^{MN}(\tau)\right)}\right). \tag{14}$$

It is assumed that users have the statistical properties of the channel and they don't carry out channel estimation. So only statical CSI can be used by downlink users to detect the signal. Motivated by [27], we rewrite the received data as

$$r_k(t) = \sqrt{\rho_{\text{dl}}}\mathbb{E}\left[\mathbf{g}_k^{\text{H}}(\tau)\,\tilde{\Theta}_k(t)\,\mathbf{w}_k\right]s_k + n', \tag{15}$$

where

$$n' = \sqrt{\rho_{\text{dl}}}\left(\mathbf{g}_k^{\text{H}}(\tau)\,\tilde{\Theta}_k(t)\,\mathbf{w}_k - \mathbb{E}\left[\mathbf{g}_k^{\text{H}}(\tau)\,\tilde{\Theta}_k(t)\,\mathbf{w}_k\right]\right)s_k + \sqrt{\rho_{\text{dl}}}\sum_{i\neq k}^{K}\mathbf{g}_k^{\text{H}}(\tau)\,\tilde{\Theta}_k(t)\,\mathbf{w}_i s_i + n^{\text{UE}}(t).$$

Suppose $\mathbb{E}\left[\mathbf{g}_k^{\text{H}}(\tau)\,\tilde{\Theta}_k(t)\,\mathbf{w}_k\right]s_k$ is the only signal needed at user k, and treating n' as unrelated Gaussian distributed additive noise [28,29], the achievable downlink rate of user k is denoted as

$$R_k = \log_2\left(1 + \frac{\rho_{\text{dl}}\left|\mathbb{E}\left[\mathbf{g}_k^{\text{H}}(\tau)\,\tilde{\Theta}_k(t)\,\mathbf{w}_k\right]\right|^2}{A(t) + B(t) + \xi^{\text{UE}}}\right), \tag{16}$$

where

$$A(t) = \rho_{\text{dl}}\text{var}\left(\mathbf{g}_k^{\text{H}}(\tau)\,\tilde{\Theta}_k(t)\,\mathbf{w}_k\right),$$
$$B(t) = \sum\nolimits_{i\neq k}^{K}\rho_{\text{dl}}\mathbb{E}\left[\left|\mathbf{g}_k^{\text{H}}(\tau)\,\tilde{\Theta}_k(t)\,\mathbf{w}_i\right|^2\right].$$

3.3. Downlink Achievable Rates

From (16) we can find that the correlation between $\mathbf{g}_k$ and $\mathbf{w}_k$ are destroyed by phase noise and the powers of non-isotropic channel vectors projected onto the precoder subspace are necessary to obtain the closed-form expressions. Hence we first present some preliminary lemmas which help us to obtain the approximated and isotropic results.

Lemma 1 ([28]). *For an isotropic random vector* $\mathbf{x} \in \mathbb{C}^{N\times 1}$ *whose elements are independent and all distributed as* $\mathcal{CN}(0,\sigma^2)$*, then the distribution of* $\mathbf{x}^{\text{H}}\mathbf{x}$ *is* $\Gamma(N,\sigma^2)$.

The strength of the estimated channel from user k to all RAUs is

$$\hat{\mathbf{g}}_k^{\text{H}}(t)\hat{\mathbf{g}}_k(t) = \sum\nolimits_{m=1}^{M}\beta_{m,k}^2(t)\hat{\mathbf{h}}_{m,k}^{\text{H}}\hat{\mathbf{h}}_{m,k}, \tag{17}$$

According to Lemma 1, $\beta_{m,k}^2(t)\hat{\mathbf{h}}_{m,k}^{\text{H}}\hat{\mathbf{h}}_{m,k}$ is distributed as $\Gamma(N,\beta_{m,k}^2(t))$. Hence (17) is the sum of M non-identically distributed but independent items. To obtain its distribution, Lemma 2 stated bellow can be used.

Lemma 2 ([15]). *If* $\{x_i\}$ *are a set of random variables and independent of each other, each term is distributed as* $\Gamma(\chi_i,\theta_i)$*. Then the distribution of the sum* $\sum_i x_i$ *can be approximated as* $\sum_i X_i \sim \Gamma(\chi,\theta)$ *wherein*

$$\chi = \frac{(\sum_i \chi_i\theta_i)^2}{\sum_i \chi_i{\theta_i}^2},\ \theta = \frac{\sum_i \chi_i{\theta_i}^2}{\sum_i \chi_i\theta_i}. \tag{18}$$

Remark 2. *From Lemma 2, the distribution of* $\hat{\mathbf{g}}_k^{\mathrm{H}}(t)\hat{\mathbf{g}}_k(t)$ *can be approximated as* $\Gamma(\chi_k(t), \theta_k(t))$, *wherein*

$$\chi_k(t) = N\frac{\left(\sum_{m=1}^{M}\beta_{m,k}^2(t)\right)^2}{\sum_{m=1}^{M}\beta_{m,k}^4(t)}, \tag{19}$$

$$\theta_k(t) = \frac{\sum_{m=1}^{M}\beta_{m,k}^4(t)}{\sum_{m=1}^{M}\beta_{m,k}^2(t)}. \tag{20}$$

Similarly, we can also give the distribution of $\mathbf{e}_k^{\mathrm{H}}(t)\mathbf{e}_k(t)$ as $\Gamma(\chi_{\mathrm{e}}(t), \theta_{\mathrm{e}}(t))$, wherein

$$\chi_{\mathrm{e}}(t) = N\frac{\left(\sum_{m=1}^{M}\eta_{m,k}^2(t)\right)^2}{\sum_{m=1}^{M}\eta_{m,k}^4(t)}, \tag{21}$$

$$\theta_{\mathrm{e}}(t) = \frac{\sum_{m=1}^{M}\eta_{m,k}^4(t)}{\sum_{m=1}^{M}\eta_{m,k}^2(t)}, \tag{22}$$

where $\eta_{m,k}^2(t) = \lambda_{m,k} - \beta_{m,k}^2(t)$.

Based on the lemmas and analysis above, we give the following lemma about the projection principle of non-isotropic vectors.

Lemma 3 ([28]). *When we project an MN-dimensional non-isotropic estimated channel vector* $\hat{\mathbf{g}}_k \in \mathbb{C}^{MN\times 1}$ *onto a p-dimensional subspace, we can give the approximated distribution of the projection power as* $\Gamma(p\chi_k/(MN), \theta_k)$.

Remark 3. *The dimension p can be given by* $p = MN$ *with MRT precoder and* $p = MN - K + 1$ *with ZF precoder, respectively, and for any independent beam, we can have* $p = 1$ *[30,31].*

When MRT and ZF precoders are employed, based on the analysis above, we can give the distribution of the signal power at user k and the distribution of the interference power at user k

$$\left|\hat{\mathbf{g}}_k^{\mathrm{H}}(\tau)\frac{\hat{\mathbf{g}}_k(\tau)}{\|\hat{\mathbf{g}}_k(\tau)\|}\right|^2 \sim \Gamma\left(\chi_k(\tau), \theta_k(\tau)\right), \tag{23}$$

$$\left|\hat{\mathbf{g}}_k^{\mathrm{H}}(\tau)\frac{\mathbf{a}_k(\tau)}{\|\mathbf{a}_k(\tau)\|}\right|^2 \sim \Gamma\left(\tfrac{MN-K+1}{MN}\chi_k(\tau), \theta_k(\tau)\right), \tag{24}$$

$$\left|\mathbf{e}_k^{\mathrm{H}}(\tau)\mathbf{w}_i\right|^2 \sim \Gamma\left(\frac{1}{MN}\chi_{\mathrm{e}}(\tau), \theta_{\mathrm{e}}(\tau)\right). \tag{25}$$

Notably, $\mathbf{w}_i$ in (25) can be either MRT precoder or ZF precoder and the equation still holds when $i = k$, due to the independence of $\mathbf{e}_k(t)$ and $\mathbf{w}_i$.

Based on the analysis above, we can give the approximated distribution of $\left|\hat{\mathbf{g}}_k^{\mathrm{H}}(\tau)\,\tilde{\Theta}_k(t)\,\mathbf{w}_k\right|^2$ as

$$\left|\hat{\mathbf{g}}_k^{\mathrm{H}}(\tau)\,\tilde{\Theta}_k(t)\,\frac{\hat{\mathbf{g}}_k(\tau)}{\|\hat{\mathbf{g}}_k(\tau)\|}\right|^2 \sim \Gamma\left(\chi_k(\tau), \theta_k(\tau)\right), \tag{26}$$

$$\left|\hat{\mathbf{g}}_k^{\mathrm{H}}(\tau)\,\tilde{\Theta}_k(t)\,\frac{\mathbf{a}_k(t)}{\|\mathbf{a}_k(t)\|}\right|^2 \sim \Gamma\left(\tfrac{MN-K+1}{MN}\chi_k(t), \theta_k(t)\right). \tag{27}$$

Figure 2 verifies the accuracy of the approximation in (26) and (27). It illustrates the cumulative distribution function (CDF) curves of $\left|\hat{\mathbf{g}}_k^{\mathrm{H}}(\tau)\,\tilde{\Theta}_k(t)\,\mathbf{w}_k\right|^2$ with MRT precoder. The phase noise variance is set as $\sigma_\phi^2 = \sigma_\varphi^2 = 10^{-2}$. It can be seen that although the random variable $\tilde{\Theta}_k(t)$ will destroy the correlation of $\hat{\mathbf{g}}_k$ and $\mathbf{w}_k$, the approximation is exactly accurate when the variance of the phase noise is 10^{-2}, and it will be more accurate when the variance of the phase noise is lower 10^{-2}.

For ZF precoder, we can get the same conclusion. Meanwhile, the phase noise variance is generally $\sigma_{\phi}^2 = \sigma_{\varphi}^2 = 1.58 \times 10^{-4}$ [8,20,26]. Hence, it's reasonable to use (26) and (27) to analyze the downlink spectral efficiency.

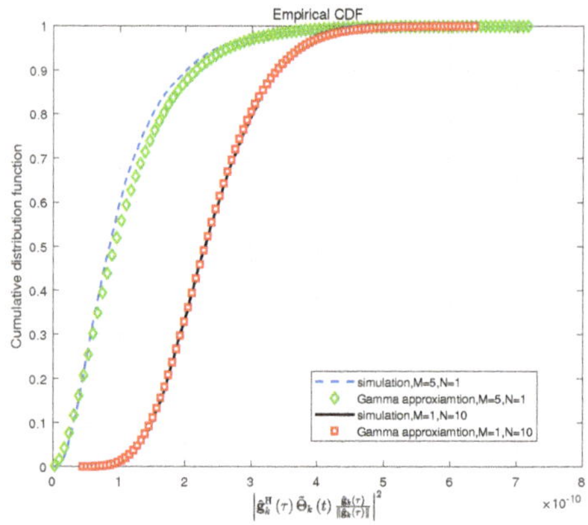

Figure 2. Cumulative distribution function of signal power with MRT precoder under different M and N.

Based on the lemmas above, the spectral efficiency for downlink transmission with both MRT and ZF precoders under hardware impairments is analyzed. The theorems and corollary stated below give the closed-form expressions of the downlink achievable rates and system asymptotic performance.

Theorem 1. *When MRT precoder is used, the closed-form expression of the downlink achievable rate under hardware impairments is given by*

$$R^{\text{mrt}}(t) = \frac{D^{\text{mrt}}(t)}{A^{\text{mrt}}(t) + B^{\text{mrt}}(t) + \xi^{\text{UE}}/\rho_{\text{dl}}}, \tag{28}$$

where

$$\begin{aligned}
D^{\text{mrt}}(t) &= \left(\frac{\Gamma\left(\chi_k'(t) + \frac{1}{2}\right)}{\Gamma\left(\chi_k'(t)\right)}\right)^2 \theta_k'(t), \\
\chi_k'(t) &= N\frac{\left(\sum_{m=1}^{M} e^{-\left(\sigma_{\varphi,k}^2+\sigma_{\phi,m}^2\right)|t-\tau|}\beta_{m,k}^2(t)\right)^2}{\sum_{m=1}^{M} e^{-2\left(\sigma_{\varphi,k}^2+\sigma_{\phi,m}^2\right)|t-\tau|}\beta_{m,k}^4(t)}, \\
\theta_k'(t) &= \frac{\sum_{m=1}^{M} e^{-2\left(\sigma_{\varphi,k}^2+\sigma_{\phi,m}^2\right)|t-\tau|}\beta_{m,k}^4(t)}{\sum_{m=1}^{M} e^{-\left(\sigma_{\varphi,k}^2+\sigma_{\phi,m}^2\right)|t-\tau|}\beta_{m,k}^2(t)}, \\
A^{\text{mrt}}(t) &= N\sum_{m=1}^{M}\beta_{m,k}^2(\tau) + \frac{1}{M}\sum_{m=1}^{M}\eta_{m,k}^2(\tau) - D^{\text{mrt}}(t), \\
B^{\text{mrt}}(t) &= \sum_{i\neq k}^{K}\frac{N\sum_{m=1}^{M}\beta_{m,i}^2(\tau)\lambda_{m,k}}{\theta_i(\tau)(\chi_i(\tau)-1)}.
\end{aligned}$$

Proof of Theorem 1. Please refer to Appendix A. □

Theorem 2. *When ZF precoder is used, the closed-form expression of the downlink achievable rate under hardware impairments is given by*

$$R^{\mathrm{zf}}(t) = \frac{D^{\mathrm{zf}}(t)}{A^{\mathrm{zf}}(t) + B^{\mathrm{zf}}(t) + \xi^{\mathrm{UE}}/\rho_{\mathrm{dl}}}, \tag{29}$$

where

$$\begin{aligned}
D^{\mathrm{zf}}(t) &= \left(\frac{\Gamma\left(\kappa(t)+\frac{1}{2}\right)}{\Gamma\left(\kappa(t)\right)}\right)^2 \theta_k'(t),\\
A^{\mathrm{zf}}(t) &= \frac{MN-K+1}{M}\sum_{m=1}^{M}\beta_{m,k}^2(\tau) - D^{\mathrm{zf}}(t),\\
B^{\mathrm{zf}}(t) &= \frac{K}{M}\sum_{m=1}^{M}\eta_{m,k}^2(\tau),\\
\kappa(t) &= \frac{MN-K+1}{MN}\chi_k'(t).
\end{aligned}$$

Proof of Theorem 2. Please refer to Appendix B. □

Then, in order to study the effects of phase noise further, we investigate a case where the number of antennas employed at each RAU goes infinity and the number of RAUs and users is fixed. The asymptotic performance provided in Corollary 1 is obtained based on (28) and (29).

Corollary 1. *Let $N \to \infty$, the ultimate rate of user k with both MRT and ZF precoders is given by*

$$R_k^{\infty}(t) = \frac{\sum_{m=1}^{M} e^{-\left(\sigma_{\varphi,k}^2+\sigma_{\phi,m}^2\right)|t-\tau|}\beta_{m,k}^2(\tau)}{\sum_{m=1}^{M}\beta_{m,k}^2(\tau) - \sum_{m=1}^{M} e^{-\left(\sigma_{\varphi,k}^2+\sigma_{\phi,m}^2\right)|t-\tau|}\beta_{m,k}^2(\tau)}, \tag{30}$$

Proof. Since the proof is similar for both precoders, we only provide the proof for MRT precoder. It can be seen that $\chi_k'(t) \to \infty$ when the number of antennas $N \to \infty$. Therefore we can have $\lim_{N\to\infty}\left(\left(\frac{\Gamma\left(\chi_k'(t)+\frac{1}{2}\right)}{\Gamma\left(\chi_k'(t)\right)}\right)^2 - \chi_k'(t)\right) = 0$ [29]. Then the limiting rate of user k can be obtained directly by dividing the denominator and numerator of (28) by N. From Corollary 1 we can see that the ultimate rate without phase noise will be unlimited when N tends to infinity, which means that phase noise limits the downlink spectral efficiency. □

4. Numerical Results

In this section, a series of Monte Carlo simulations is used to verify the theoretical results obtained in Section 3. A circular single-cell massive MIMO system is considered. All of the RAUs and users are randomly distributed in the cell and the minimum access distance between RAUs and users is set as $r_0 = 30$ m. The channels are generated by (4), and other simulation parameters are presented in Table 1.

Table 1. Basic simulation parameters.

Number of RAUs M	5
Number of users K	2
Cell radius R	1 km
Path loss exponent α	3.7
Power of uplink pilot symbol ρ_{p}	1 Watt
downlink transmit power ρ_{dl}	10 Watts
variance of thermal noise is assumed as σ^2	−174 dBm

Figure 3 illustrates the theoretical and simulated spectral efficiency with MRT and ZF precoders versus the number of antennas per RAU. The spectral efficiency is the average rate between users. We assume that the variances of phase noise and amplified thermal noise are $\sigma^2_{\phi,m} = \sigma^2_{\varphi,k} = 1.58 \times 10^{-4}\,\forall m,k$ and $\xi^{\mathrm{UE}} = \xi^{\mathrm{BS}} = 1.58\sigma^2$. t is set as $\tau + 1$. It can be seen that the closed-form expressions in (28), (29) and the simulation results in (16) match well with each other. For both precoders, the spectral efficiency increases and gets more and more close to the limiting average rate with the increasing of N. When $N = 100$, the system achieves 80% of the ultimate rate with ZF precoder and 76% with MRT precoder. Furthermore, it can be seen that ZF precoder achieves better performance than MRT.

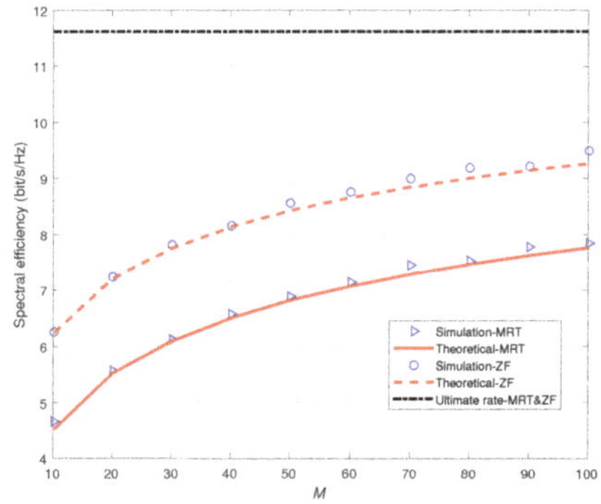

Figure 3. Spectral efficiency against M with MRT and ZF precoders.

Next, we investigate the effects of phase noise. Figure 4 illustrates the theoretical spectral efficiency with MRT and ZF precoders against the variance of phase noise. Notably, the variance of phase noise reflects the strength of phase noise. It is assumed that the number of antennas $N = 50$ and other system parameters have the same value as Figure 3. Figure 4 reveals that the spectral efficiency decreases monotonically with the variance of phase noise increasing. In addition, phase noise have a greater impact on ZF precoder. This results form the fact that ZF precoder is more sensitive to CSI. It can be noted that as the variance increases, the performance gap between MRT and ZF precoders becomes smaller. This is because when the phase noise is severe, the loss caused by unknown CSI at user side dominates rather than the interference between users.

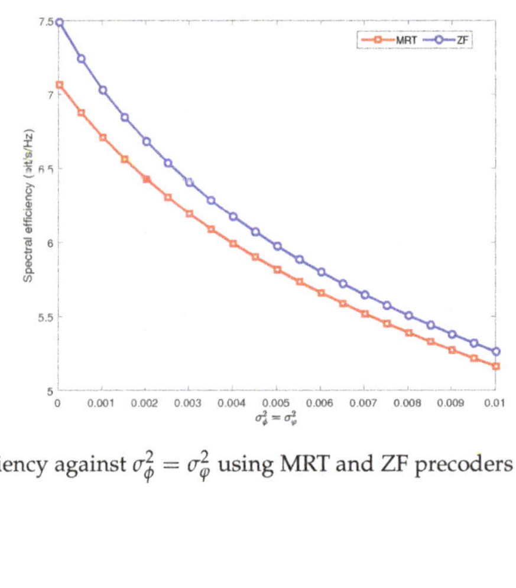

Figure 4. Spectral efficiency against $\sigma^2_{\phi} = \sigma^2_{\varphi}$ using MRT and ZF precoders with $\sigma^2_{\phi,m} = \sigma^2_{\varphi,k} = \sigma^2_{\phi} = \sigma^2_{\varphi}, \forall m,k$.

Figure 5 illustrates the instantaneous spectral efficiency during the downlink transmission phase. The number of antennas employed at each RAU is assumed as $N = 40$ and the variance of phase noise is set as $\sigma^2_{\phi,m} = \sigma^2_{\varphi,k} = 1.58 \times 10^{-4}\,\forall m,k$. In addition, the coherence time of channel is set as $T = 200$. As shown in Figure 5, the spectral efficiency degrades as t increases. This is because the uncertainty of the phase drift between downlink transmission phase and pilot training phase increases with the growing of t. Figure 5 reveals that it's improper to use the estimated CSI obtained during the pilot phase for the whole data transmission phase.

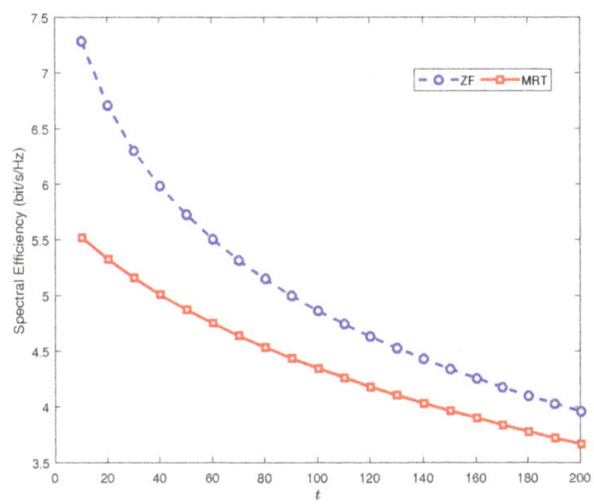

Figure 5. Spectral efficiency against t with MRT and ZF precoders.

5. Conclusions

In this paper, we analyzed downlink spectral efficiency with hardware impairments in distributed massive MIMO. Initially, employing pilot symbol assisted transmission, we obtained the estimated CSI in a more realistic scenario where transmission is distorted by phase noise and amplified thermal noise. Next, we used the imperfect CSI to derive the closed-form expressions for downlink achievable rates with MRT and ZF precoders. In addition, we obtained the ultimate rate when $N \to \infty$. It can be seen that the rate performance was limited by phase noise. Then, numerical results proved that the theoretical analysis was accurate. Furthermore, they also revealed that ZF can achieve larger spectral efficiency than MRT precoder, and hardware impairments had a greater impact on ZF precoder. Finally, spectral efficiency degraded with the increasing of the variance of phase noise and downlink transmission time.

In the future work, we intend to extend our research considering a more effective phase noise model which could lead finer precoding strategies to improve the theoretical rates.

Author Contributions: Formal analysis, Q.L.; Supervision, J.L., P.Z., D.W. and X.Y.; Validation, Q.L.; Writing—original draft, Q.L.; Writing—review & editing, Q.L. and J.L.

Funding: This work was supported in part by National Natural Science Foundation of China (NSFC) (Grant NO. 61501113, 61571120, 61871122), Jiangsu Provincial Natural Science Foundation (Grant No.BK20150630, BK20180011), six talent peaks project in Jiangsu province, and National Key Special Program No.2018ZX03001008-002.

Conflicts of Interest: The authors declare no conflict of interest.

Appendix A. Proof of Theorem 1

When MRT precoder is chosen, the following three terms $\left|\mathbb{E}\left[\mathbf{g}_k^{\mathrm{H}}(\tau)\,\tilde{\Theta}_k(t)\,\mathbf{w}_k\right]\right|^2$, $A(t)$ and $B(t)$ need to be calculated showing in (16).

For the term $\left|\mathbb{E}\left[\mathbf{g}_k^{\mathrm{H}}(\tau)\,\tilde{\Theta}_k(t)\,\mathbf{w}_k\right]\right|^2$, we can obtain

$$\begin{aligned}
&\left|\mathbb{E}\left[\mathbf{g}_k^{\mathrm{H}}(\tau)\,\tilde{\Theta}_k(t)\,\mathbf{w}_k\right]\right|^2 \\
&\overset{(a)}{=}\left|\mathbb{E}\left[\mathbf{g}_k^{\mathrm{H}}(\tau)\,\mathbb{E}\left[\tilde{\Theta}_k(t)\right]\mathbf{w}_k\right]\right|^2 \\
&\overset{(b)}{=}\left|\mathbb{E}\left[\left|\hat{\mathbf{g}}_k^{\mathrm{H}}(\tau)\,\mathbf{D}_{k,\tau}(t)\mathbf{w}_k\right|\right]\right|^2 \\
&\overset{(c)}{=}\left(\frac{\Gamma\left(\chi_k'(\tau)+\frac{1}{2}\right)}{\Gamma\left(\chi_k'(\tau)\right)}\right)^2\theta_k'(\tau),
\end{aligned} \tag{A1}$$

where (a) is obtained because $\tilde{\Theta}_k(t)$ is independent of $\mathbf{g}_k(\tau)$ and $\mathbf{w}_k$. By exploiting the fact that $\mathbb{E}\left[e^{-j\left(\theta_k^m(t)-\theta_k^m(\tau)\right)}\right] = e^{-\frac{\sigma_{\varphi,k}^2+\sigma_{\phi,m}^2}{2}|t-\tau|}$ and $\mathbf{e}_k(t)$ is independent of $\mathbf{w}_k$, we can get (b), and (c) results from Lemma 2, Lemma 3 and the relationship between Gamma distribution and Nakagami distribution.

For the term $A(t)$, we obtain

$$\begin{aligned}
&\operatorname{var}\left(\mathbf{g}_k^{\mathrm{H}}(\tau)\,\tilde{\Theta}_k(t)\,\mathbf{w}_k\right) \\
&=\mathbb{E}\left[\left|\hat{\mathbf{g}}_k^{\mathrm{H}}(\tau)\,\tilde{\Theta}_k(t)\,\mathbf{w}_k\right|^2\right]+\mathbb{E}\left[\left|\mathbf{e}_k^{\mathrm{H}}(\tau)\,\tilde{\Theta}_k(t)\,\mathbf{w}_k\right|^2\right] \\
&-\left|\mathbb{E}\left[\mathbf{g}_k^{\mathrm{H}}(\tau)\,\tilde{\Theta}_k(t)\,\mathbf{w}_k\right]\right|^2.
\end{aligned} \tag{A2}$$

The first term of (A2) can be calculated as

$$\mathbb{E}\left[\left|\hat{\mathbf{g}}_k^{\mathrm{H}}(\tau)\,\tilde{\Theta}_k(t)\,\frac{\hat{\mathbf{g}}_k(\tau)}{\|\hat{\mathbf{g}}_k(\tau)\|}\right|^2\right]\overset{(a)}{=}N\sum\nolimits_{m=1}^{M}\beta_{m,k}^2(\tau), \tag{A3}$$

where (a) results from (26).

Next the second term can be given by

$$\mathbb{E}\left[\left|\mathbf{e}_k^{\mathrm{H}}(\tau)\,\tilde{\Theta}_k(t)\,\frac{\hat{\mathbf{g}}_k(\tau)}{\|\hat{\mathbf{g}}_k(\tau)\|}\right|^2\right]\overset{(a)}{=}\frac{1}{M}\sum\nolimits_{m=1}^{M}\eta_{m,k}^2(\tau), \tag{A4}$$

where (a) can be obtained by exploiting the fact that $\|\mathbf{e}_k^{\mathrm{H}}(\tau)\,\tilde{\Theta}_k(t)\,\|^2\sim\Gamma(\chi_{\mathrm{e}}(\tau),\theta_{\mathrm{e}}(\tau))$ and (25).

For the last term $B(t)$, we first calculate

$$\begin{aligned}
\mathbb{E}\left[\frac{1}{\|\hat{\mathbf{g}}_i\|^2}\right]&\overset{(a)}{=}\int_0^\infty\frac{1}{x}x^{\chi_i-1}\frac{e^{-x/\theta_i}}{\theta_i^{\chi_i}\Gamma(\chi_i)}dx \\
&=\frac{1}{\theta_i^{\chi_i}\Gamma(\chi_i)}\theta_i^{\chi_i-1}\Gamma(\chi_i-1) \\
&=\frac{1}{\theta_i(\chi_i-1)},
\end{aligned} \tag{A5}$$

where (a) results from Remark 2 and we omit (t) in (A5). Based on (A5), we have

$$\begin{aligned}
&\mathbb{E}\left[\left|\mathbf{g}_k^{\mathrm{H}}(\tau)\,\tilde{\Theta}_k(t)\,\mathbf{w}_i\right|^2\right] \\
&\overset{(a)}{\asymp} \mathbb{E}\left[\left|\mathbf{g}_k^{\mathrm{H}}(\tau)\,\tilde{\Theta}_k(t)\hat{\mathbf{g}}_i(\tau)\right|^2\right]\mathbb{E}\left[\frac{1}{\|\hat{\mathbf{g}}_i(\tau)\|^2}\right] \\
&= \frac{\mathbb{E}\left[\sum_{m=1}^{M}\beta_{m,i}^2(\tau)\lambda_{m,k}\hat{\mathbf{h}}_{m,k}^{\mathrm{H}}\hat{\mathbf{h}}_{m,k}\right]}{\theta_i(\tau)(\chi_i(\tau)-1)} \\
&= \frac{N\sum_{m=1}^{M}\beta_{m,i}^2(\tau)\lambda_{m,k}}{\theta_i(\tau)(\chi_i(\tau)-1)},
\end{aligned} \tag{A6}$$

where $x \asymp y$ means $\lim_{N\to\infty}(x-y)=0$, (a) results from Lemma 4 (ii) of [32].

Finally, combining (A1)–(A6) concludes the proof.

Appendix B. Proof of the Theorem 2

To derive the closed-form expression of (16) with ZF precoder, the following three terms $\left|\mathbb{E}\left[\mathbf{g}_k^{\mathrm{H}}(\tau)\,\tilde{\Theta}_k(t)\,\mathbf{w}_k\right]\right|^2$, $A(t)$ and $B(t)$ need to be calculated, wherein $\mathbf{w}_k = \frac{\mathbf{a}_k(\tau)}{\|\mathbf{a}_k(\tau)\|}$.

For the term $\left|\mathbb{E}\left[\mathbf{g}_k^{\mathrm{H}}(\tau)\,\tilde{\Theta}_k(t)\,\mathbf{w}_k\right]\right|^2$, we obtain

$$\begin{aligned}
&\left|\mathbb{E}\left[\mathbf{g}_k^{\mathrm{H}}(\tau)\,\tilde{\Theta}_k(t)\mathbf{w}_k\right]\right|^2 \\
&\overset{(a)}{=} \left|\mathbb{E}\left[\mathbf{g}_k^{\mathrm{H}}(\tau)\,\mathbb{E}\left[\tilde{\Theta}_k(t)\right]\mathbf{w}_k\right]\right|^2 \\
&\overset{(b)}{=} \left|\mathbb{E}\left[\hat{\mathbf{g}}_k^{\mathrm{H}}(\tau)\mathbf{D}_{k,\tau}(t)\frac{\mathbf{a}_k(\tau)}{\|\mathbf{a}_k(\tau)\|}\right]\right|^2 \\
&\overset{(c)}{=} \left(\frac{\Gamma\left(\kappa(t)+\frac{1}{2}\right)}{\Gamma(\kappa(t))}\right)^2\theta_k'(t),
\end{aligned} \tag{A7}$$

where (a) is obtained because $\tilde{\Theta}_k(t)$ is independent of $\mathbf{g}(\tau)$ and $\mathbf{w}_k$. (b) is obtained due to the independence of $\mathbf{e}_k(\tau)$ and $\mathbf{w}_k$. (c) results from Lemma 2, Lemma 3 and the relationship between Gamma distribution and Nakagami distribution.

Similar to the proof of Theorem 1, to get the closed-form expression of $A(t)$, we need to calculate the following two terms

$$\mathbb{E}\left[\left|\hat{\mathbf{g}}_k^{\mathrm{H}}(\tau)\,\tilde{\Theta}_k(t)\,\frac{\mathbf{a}_k(\tau)}{\|\mathbf{a}_k(\tau)\|}\right|^2\right], \tag{A8}$$

$$\mathbb{E}\left[\left|\mathbf{e}_k^{\mathrm{H}}(\tau)\,\tilde{\Theta}_k(t)\,\frac{\mathbf{a}_k(\tau)}{\|\mathbf{a}_k(\tau)\|}\right|^2\right]. \tag{A9}$$

The first term (A8) can be given by

$$\mathbb{E}\left[\left|\hat{\mathbf{g}}_k^{\mathrm{H}}(\tau)\,\tilde{\Theta}_k(t)\,\frac{\mathbf{a}_k(\tau)}{\|\mathbf{a}_k(\tau)\|}\right|^2\right] = \frac{MN-K+1}{M}\sum_{m=1}^{M}\beta_{m,k}^2(\tau), \tag{A10}$$

which results from (27).

Next the second term can be given by

$$\mathbb{E}\left[\left|\mathbf{e}_k^{\mathrm{H}}(\tau)\,\tilde{\Theta}_k(t)\,\frac{\mathbf{a}_k(\tau)}{\|\mathbf{a}_k(\tau)\|}\right|^2\right] = \frac{1}{M}\sum\nolimits_{m=1}^{M}\eta_{m,k}^2(\tau), \tag{A11}$$

which results from (25) and the fact that $\|\mathbf{e}_k^{\mathrm{H}}(\tau)\,\tilde{\Theta}_k(t)\|^2 \sim \Gamma(\chi_{\mathrm{e}}(\tau), \theta_{\mathrm{e}}(\tau))$.

For the term $B(t)$, we can have

$$\begin{aligned} &\mathbb{E}\left[\left|\mathbf{g}_k^{\mathrm{H}}(\tau)\,\tilde{\Theta}_k(t)\,\mathbf{w}_i\right|^2\right] \\ &\overset{(a)}{=} \mathbb{E}\left[\left|\mathbf{e}_k^{\mathrm{H}}(\tau)\,\tilde{\Theta}_k(t)\,\mathbf{w}_i\right|^2\right] \\ &= \frac{1}{M}\sum\nolimits_{m=1}^{M}\eta_{m,k}^2(\tau), \end{aligned} \tag{A12}$$

where (a) results from the property of ZF precoder.

Substituting (A7) and (A10)–(A12) into (16) completes the proof.

References

1. Wang, J.; Zhu, H.; Gomes, N.J.; Wang, J. Frequency Reuse of Beam Allocation for Multiuser Massive MIMO Systems. *IEEE Trans. Wirel. Commun.* **2018**, *17*, 2346–2359. [CrossRef]
2. Wang, C.X.; Haider, F.; Gao, X.; You, X.H.; Yang, Y.; Yuan, D.; Aggoune, H.; Haas, H.; Fletcher, S.; Hepsaydir, E. Cellular architecture and key technologies for 5G wireless communication networks. *IEEE Commun. Mag.* **2014**, *52*, 122–130. [CrossRef]
3. Jiang, F.; Chen, J.; Swindlehurst, A.L.; López-Salcedo, J.A. Massive MIMO for Wireless Sensing with a Coherent Multiple Access Channel. *IEEE Trans. Signal Process.* **2015**, *63*, 3005–3017. [CrossRef]
4. Ciuonzo, D.; Rossi, P.S.; Dey, S. Massive MIMO Channel-Aware Decision Fusion. *IEEE Trans. Signal Process.* **2015**, *63*, 604–619. [CrossRef]
5. Ciuonzo, D.; Rossi, P.S.; Dey, S. Massive MIMO meets decision fusion: Decode-and-fuse vs. decode-then-fuse. In Proceedings of the 2014 IEEE 8th Sensor Array and Multichannel Signal Processing Workshop (SAM), A Coruna, Spain, 22–25 June 2014; pp. 265–268. [CrossRef]
6. Shirazinia, A.; Dey, S.; Ciuonzo, D.; Rossi, P.S. Massive MIMO for Decentralized Estimation of a Correlated Source. *IEEE Trans. Signal Process.* **2016**, *64*, 2499–2512. [CrossRef]
7. Cao, J.; Wang, D.; Li, J.; Sun, Q.; Hu, Y. Uplink spectral efficiency analysis of multi-cell multi-user massive MIMO over correlated Ricean channel. *Sci. China Inf. Sci.* **2018**, *61*, 082305. [CrossRef]
8. Björnson, E.; Matthaiou, M.; Debbah, M. Massive MIMO with non-ideal arbitrary arrays: Hardware scaling laws and circuit-aware design. *IEEE Trans. Wirel. Commun.* **2015**, *14*, 4353–4368. [CrossRef]
9. Dai, L. An Uplink Capacity Analysis of the Distributed Antenna System (DAS): From Cellular DAS to DAS with Virtual Cells. *IEEE Trans. Wirel. Commun.* **2014**, *13*, 2717–2731.
10. Liu, Z.; Dai, L. On the Scaling Behavior of Average Ergodic Capacity of Distributed MIMO Systems. In Proceedings of the 2015 IEEE Global Communications Conference (GLOBECOM), San Diego, CA, USA, 6–10 December 2015; pp. 1–6. [CrossRef]
11. Wang, J.; Dai, L. Asymptotic Rate Analysis of Downlink Multi-User Systems with Co-Located and Distributed Antennas. *IEEE Trans. Wirel. Commun.* **2015**, *14*, 3046–3058. [CrossRef]
12. Wang, J.; Dai, L. Downlink Rate Analysis for Virtual-Cell Based Large-Scale Distributed Antenna Systems. *IEEE Trans. Wirel. Commun.* **2016**, *15*, 1998–2011. [CrossRef]
13. Khadka, A.; Adachi, K.; Sun, S.; Wang, J.; Zhu, H.; Wang, J. Cooperative Transmission Strategy over Users' Mobility for Downlink Distributed Antenna Systems. In Proceedings of the GLOBECOM 2017–2017 IEEE Global Communications Conference, Singapore, 4–8 December 2017; pp. 1–6. [CrossRef]
14. Zhu, H.; Wang, J. Radio Resource Allocation in Multiuser Distributed Antenna Systems. *IEEE J. Sel. Areas Commun.* **2013**, *31*, 2058–2066. [CrossRef]

15. Lv, Q.; Li, J.; Zhu, P.; You, X. Spectral Efficiency Analysis for Bidirectional Dynamic Network with Massive MIMO Under Imperfect CSI. *IEEE Access* **2018**, *6*, 43660–43671. [CrossRef]
16. Studer, C.; Wenk, M.; Burg, A. MIMO transmission with residual transmit-RF impairments. In Proceedings of the 2010 International ITG Workshop on Smart Antennas (WSA), Bremen, Germany, 23–24 February 2010; pp. 189–196. [CrossRef]
17. Papazafeiropoulos, A.; Clerckx, B.; Ratnarajah, T. Mitigation of phase noise in massive MIMO systems: A rate-splitting approach. In Proceedings of the 2017 IEEE International Conference on Communications (ICC), Paris, France, 21–25 May 2017; pp. 1–7. [CrossRef]
18. Pitarokoilis, A.; Mohammed, S.K.; Larsson, E.G. Uplink Performance of Time-Reversal MRC in Massive MIMO Systems Subject to Phase Noise. *IEEE Trans. Wirel. Commun.* **2015**, *14*, 711–723. [CrossRef]
19. Pitarokoilis, A.; Mohammed, S.K.; Larsson, E.G. Achievable rates of ZF receivers in massive MIMO with phase noise impairments. In Proceedings of the 2013 Asilomar Conference on Signals, Systems and Computers, Pacific Grove, CA, USA, 3–6 November 2013; pp. 1004–1008. [CrossRef]
20. Li, J.; Wang, D.; Zhu, P.; You, X. Uplink Spectral Efficiency Analysis of Distributed Massive MIMO with Channel Impairments. *IEEE Access* **2017**, *5*, 5020–5030. [CrossRef]
21. Fang, Y.; Li, X.; Qiu, L. Asymptotic Equivalent Performance of Uplink Massive MIMO Systems with Phase Noise. In Proceedings of the 2018 IEEE International Conference on Communications (ICC), Kansas City, MO, USA, 20–24 May 2018; pp. 1–6. [CrossRef]
22. Zhu, J.; Schober, R.; Bhargava, V.K. Physical layer security for massive MIMO systems impaired by phase noise. In Proceedings of the 2016 IEEE 17th International Workshop on Signal Processing Advances in Wireless Communications (SPAWC), Edinburgh, UK, 3–6 July 2016; pp. 1–5. [CrossRef]
23. Corvaja, R.; Armada, A.G. Phase Noise Degradation in Massive MIMO Downlink with Zero-Forcing and Maximum Ratio Transmission Precoding. *IEEE Trans. Veh. Technol.* **2016**, *65*, 8052–8059. [CrossRef]
24. Krishnan, R.; Khanzadi, M.R.; Krishnan, N.; Wu, Y.; i Amat, A.G.; Eriksson, T.; Schober, R. Linear Massive MIMO Precoders in the Presence of Phase Noise - A Large-Scale Analysis. *IEEE Trans. Veh. Technol.* **2016**, *65*, 3057–3071. [CrossRef]
25. Petrovic, D.; Rave, W.; Fettweis, G. Effects of Phase Noise on OFDM Systems with and without PLL: Characterization and Compensation. *IEEE Trans. Commun.* **2007**, *55*, 1607–1616. [CrossRef]
26. Papazafeiropoulos, A.; Clerckx, B.; Ratnarajah, T. Rate-splitting to mitigate residual transceiver hardware impairments in massive MIMO systems. *IEEE Trans. Veh. Technol.* **2017**, *66*, 8196–8211. [CrossRef]
27. Li, J.; Wang, D.; Zhu, P.; You, X. Spectral efficiency analysis of single-cell multi-user large-scale distributed antenna system. *IET Commun.* **2014**, *8*, 2213–2221. [CrossRef]
28. Li, J.; Wang, D.; Zhu, P.; Wang, J.; You, X. Downlink Spectral Efficiency of Distributed Massive MIMO Systems with Linear Beamforming Under Pilot Contamination. *IEEE Trans. Veh. Technol.* **2018**, *67*, 1130–1145. [CrossRef]
29. Jose, J.; Ashikhmin, A.; Marzetta, T.L.; Vishwanath, S. Pilot contamination and precoding in multi-cell TDD systems. *IEEE Trans. Wirel. Commun.* **2011**, *10*, 2640–2651. [CrossRef]
30. Hosseini, K.; Yu, W.; Adve, R.S. Large-scale MIMO versus network MIMO for multicell interference mitigation. In Proceedings of the 2014 IEEE 15th International Workshop on Signal Processing Advances in Wireless Communications (SPAWC), Toronto, ON, Canada, 22–25 June 2014; pp. 70–74. [CrossRef]
31. Li, J.; Wang, D.; Zhu, P.; You, X. Benefits of Beamforming Training Scheme in Distributed Large-Scale MIMO Systems. *IEEE Access* **2018**, *6*, 7432–7444. [CrossRef]
32. Hoydis, J.; ten Brink, S.; Debbah, M. Massive MIMO in the UL/DL of Cellular Networks: How Many Antennas Do We Need? *IEEE J. Sel. Areas Commun.* **2013**, *31*, 160–171. [CrossRef]

electronics

MDPI

Article

Spectral and Energy Efficiency of Distributed Massive MIMO with Low-Resolution ADC

Jiamin Li, Qian Lv *, Jing Yang, Pengcheng Zhu and Xiaohu You

National Mobile Communications Research Laboratory, Southeast University, Nanjing 210096, China; lijiamin@seu.edu.cn (J.L.); jingyang@seu.edu.cn (J.Y.); p.zhu@seu.edu.cn (P.Z.); xhyu@seu.edu.cn (X.Y.)

* Correspondence: seulvqian@seu.edu.cn; Tel.: +25-5209-1635

Received: 15 November 2018; Accepted: 3 December 2018; Published: 4 December 2018

Abstract: In this paper, considering a more realistic case where the low-resolution analog-to-digital convertors (ADCs) are employed at receiver antennas, we investigate the spectral and energy efficiency in multi-cell multi-user distributed massive multi-input multi-output (MIMO) systems with two linear receivers. An additive quantization noise model is provided first to study the effects of quantization noise. Using the model provided, the closed-form expressions for the uplink achievable rates with a zero-forcing (ZF) receiver and a maximum ratio combination (MRC) receiver under quantization noise and pilot contamination are derived. Furthermore, the asymptotic achievable rates are also given when the number of quantization bits, the per user transmit power, and the number of antennas per remote antenna unit (RAU) go to infinity, respectively. Numerical results prove that the theoretical analysis is accurate and show that quantization noise degrades the performance in spectral efficiency, but the growth in the number of antennas can compensate for the degradation. Furthermore, low-resolution ADCs with 3 or 4 bits outperform perfect ADCs in energy efficiency. Numerical results imply that it is preferable to use low-resolution ADCs in distributed massive MIMO systems.

Keywords: distributed massive MIMO; energy efficiency; spectral efficiency; pilot contamination; quantization noise

---

1. Introduction

Massive multi-input multi-output (MIMO) systems are an essential technology for the fifth generation (5G) mobile networks because they can significantly improve spectral efficiency and energy efficiency [1–6]. In massive MIMO systems, a relatively small number of users are served by hundreds or thousands of antennas employed at base stations in the same time-frequency resource. The huge number of antennas provides a high number of degrees-of-freedom, which favors low-complexity receivers, such as maximum ratio combination (MRC) and zero-forcing (ZF), and beamforming, such as ZF and maximum ratio transmission (MRT) [2,7,8]. Therefore, we consider MRC and ZF receivers for uplink transmission in this paper. There are two categories for massive MIMO: co-located massive MIMO and distributed massive MIMO [9]. Compared to co-located massive MIMO, distributed massive MIMO has advantages of increasing spectral efficiency, energy efficiency, and system coverage due to the reduced access distance [8,10–12]. Hence, a distributed massive MIMO system is considered in this paper.

Although massive MIMO systems have significant performance gains, they also face new challenges: high total power consumption, expensive hardware, and mass data processing [13]. Specifically, each antenna is equipped with a radio frequency (RF) chain, including an analog-to-digital converter (ADC) unit in massive MIMO systems. However, with the increase in the antenna number, the hardware complexity and the power consumption of ADCs increase exponentially with the number of quantization bits [14]. Therefore, one promising solution is to employ low-resolution ADCs in massive MIMO systems. The study of low-resolution ADC in MIMO or massive MIMO systems has caused widespread concern.

Spectral and energy efficiency are two fundamental metrics to analyze the impacts of low-resolution ADCs. Spectral efficiency was investigated in [15–17]. The performance of 1-bit resolution ADC in MIMO systems was studied in [15] considering the nonlinear characteristics of a quantizer. In massive MIMO systems with low-resolution ADCs, the uplink achievable rate using the common MRC receivers has been investigated in [16], and the uplink achievable rate using the common ZF receivers was studied in [17]. However, these two papers made an assumption that the base station had perfect channel state information (CSI), and they only considered a single-cell massive MIMO system. In fact, the CSI is not available at the base station. On the other hand, energy efficiency was studied in [13,18–20]. The optimal number of quantized bits and antenna selection were considered to maximize the energy efficiency of general MIMO with low-resolution ADCs in [13]. It was pointed out in [18] that very low bit resolution is not preferable from the perspective of energy efficiency. A function about energy efficiency and the number of quantized bits was obtained in [19].

The previous papers mainly studied a single-cell system, made an assumption that the base station had perfect CSI, and did not analyze spectral efficiency and energy efficiency simultaneously. Hence, in this paper, a multi-cell multi-user massive MIMO system with low-resolution ADCs is considered, and we assume that the base stations estimate CSI during the uplink pilot transmission phase. Furthermore, the uplink spectral and energy efficiency are both analyzed. Here are the key contributions of this paper:

1. A joint uplink signal model is provided and it enables us to study the effects of pilot contamination and quantization noise simultaneously.
2. Under imperfect CSI and considering MRC and ZF receivers, we derive the closed-form expressions for the uplink achievable rates. The asymptotic performance with quantization bits, the number of antennas per RAU, and per user transmit power are also obtained.
3. The theoretical results are verified by performing Monte Carlo simulations, and we obtain deep insight into the impacts of quantization noise on the uplink spectral efficiency and energy efficiency in distributed massive MIMO systems.

2. System Model

We consider a distributed massive MIMO system. There are L adjacent cells, and each cell consists of M remote antennas units (RAUs) and K single-antenna users. Each RAU is equipped with an array of N antennas. Each antenna is equipped with a low-resolution ADC, which means system performance will be degraded by quantization noise. RAUs in the same cell transmit or receive signals simultaneously while the beamforming design and signal processing are performed in a baseband processing unit. An example is given in Figure 1. There are $L = 7$ adjacent cells, and in cell-1, there are $K = 6$ users and $M = 6$ RAUs.

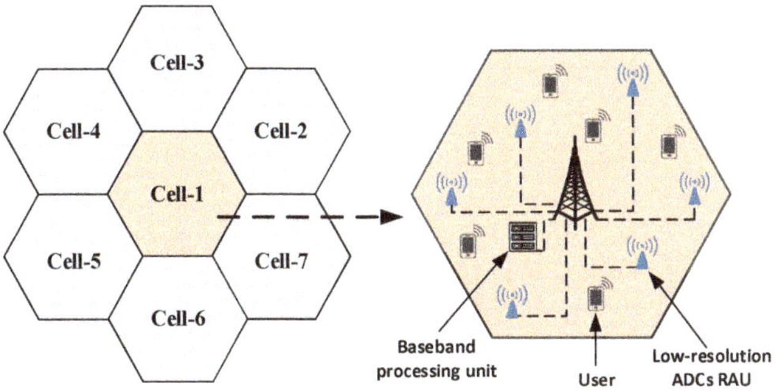

Figure 1. System configuration.

2.1. Quantization Noise Model

For uplink transmission, the signal vector received by all RAUs in cell l can be given by

$$\mathbf{y}_l = \sqrt{p_\mathrm{u}}\sum\nolimits_{i=1}^{L}\mathbf{G}_{l,i}\mathbf{x}_i + \mathbf{n}_l, \tag{1}$$

where $\mathbf{x}_i$ is the $K \times 1$ signal vector transmitted by the K users in cell i, p_u is the uplink transmitted power, and $\mathbf{n}_l \sim \mathcal{CN}(0, \mathbf{I}_{MN})$ is the additive white Gaussian noise, $\mathbf{G}_{l,i} = [\mathbf{g}_{l,i,1}, \ldots, \mathbf{g}_{l,i,K}]$ is the $MN \times K$ channel matrix from M RAUs in cell l to K users in cell i, wherein

$$\mathbf{g}_{l,i,k} = \left[\sqrt{\lambda_{l,1,i,k}}\mathbf{h}_{l,1,i,k}^\mathrm{T}, \cdots, \sqrt{\lambda_{l,M,i,k}}\mathbf{h}_{l,M,i,k}^\mathrm{T}\right]^\mathrm{T} \tag{2}$$

where $\lambda_{l,m,i,k}$ is the path loss between the k-th user in the i-th cell and the m-th RAU in the l-th cell, which is dependent on the corresponding distance, and $\mathbf{h}_{l,m,i,k} \sim \mathcal{CN}(0, \mathbf{I}_N)$ denotes the small scale fading.

This paper assumes that the CSI is unknown to the base station, and pilot training is performed. Motivated by [21], based on the minimum mean square error (MMSE) channel estimation, the equivalent estimated channel can be given by

$$\hat{\mathbf{g}}_{i,l,k} = \left[\sqrt{\beta_{i,1,l,k}}\hat{\mathbf{h}}_{i,k,1}^\mathrm{T}, \cdots, \sqrt{\beta_{i,M,l,k}}\hat{\mathbf{h}}_{i,k,M}^\mathrm{T}\right]^\mathrm{T} \tag{3}$$

where

$$\beta_{i,m,l,k} = \frac{\lambda_{i,m,l,k}^2}{\sum_{j=1}^{L}\lambda_{i,m,j,k} + 1/(\tau p_\mathrm{u})}. \tag{4}$$

τ denotes the length of pilot sequences, $\beta_{i,m,l,k}$ denotes the equivalent path loss between the k-th user in the l-th cell and the m-th RAU in the i-th cell, and $\hat{\mathbf{h}}_{i,k} \triangleq [\hat{\mathbf{h}}_{i,k,1}^\mathrm{T}, \cdots, \hat{\mathbf{h}}_{i,k,M}^\mathrm{T}]^\mathrm{T} \sim \mathcal{CN}(0, \mathbf{I}_{MN})$ represents the equivalent small scale fading part of the estimated channel. Because of the orthogonality principle of MMSE estimation theory, $\mathbf{g}_{i,l,k}$ can be decomposed as

$$\mathbf{g}_{i,l,k} = \hat{\mathbf{g}}_{i,l,k} + \tilde{\mathbf{g}}_{i,l,k} \tag{5}$$

where $\tilde{\mathbf{g}}_{i,l,k} \sim \mathcal{CN}\left(0, \mathrm{diag}\left(\eta_{i,1,l,k}, \cdots, \eta_{i,M,l,k}\right) \otimes \mathbf{I}_N\right)$ is the uncorrelated and statistically independent of $\hat{\mathbf{g}}_{i,l,k}$ estimation error, and $\eta_{i,m,l,k} \triangleq \lambda_{i,m,l,k} - \beta_{i,m,l,k}$.

After the received analog signals pass through the low-resolution ADCs, the quantized digital signal vector can be obtained as

$$\mathbf{y}_{l,\mathrm{q}} = \mathcal{Q}(\mathbf{y}_l) = \mathcal{Q}\left(\sqrt{p_\mathrm{u}}\sum\nolimits_{i=1}^{L}\mathbf{G}_{l,i}\mathbf{x}_i + \mathbf{n}_l\right) \tag{6}$$

where $\mathcal{Q}(.)$ represents the quantization function. Assuming that the gain of automatic gain control is appropriately set, the additive quantization noise model (AQNM) can be employed to reformulate the quantized signal vector as

$$\mathbf{y}_{l,q} = \alpha \mathbf{y}_l + \mathbf{n}_{l,q} = \alpha\sqrt{p_{\mathrm{u}}} \sum_{i=1}^{L} \mathbf{G}_{l,i}\mathbf{x}_i + \alpha \mathbf{n}_l + \mathbf{n}_{l,\mathrm{q}} \tag{7}$$

where $\alpha = 1 - \rho$, ρ is the inverse of the signal-to-quantization-noise ratio, and $\mathbf{n}_{l,\mathrm{q}}$ denotes the additive uncorrelated quantization noise vector, which is Gaussian-distributed. The parameter ρ is a constant dependent on the number of quantization bits b. According to [16], the covariance matrix of quantization noise $\mathbf{n}_{l,\mathrm{q}}$ for a fixed channel realization can be denoted as

$$\mathbf{R}_{\mathbf{n}_{l,\mathrm{q}}} = \alpha(1-\alpha)\mathrm{diag}\left(p_{\mathrm{u}} \sum_{i=1}^{L} \mathbf{G}_{l,i}\mathbf{G}_{l,i}^{\mathrm{H}} + \mathbf{I}\right). \tag{8}$$

2.2. The Energy Efficiency Model

(1) Achievable uplink rates: In the uplink transmission phase, the quantized signal processed by the linear detector of user k in cell l is presented as

$$\begin{aligned} r_{l,k} &= \mathbf{a}_{l,k}^{\mathrm{H}}\mathbf{y}_{l,\mathrm{q}} \\ &= \alpha\sqrt{p_{\mathrm{u}}} \sum_{i=1}^{L} \sum_{j=1}^{K} \mathbf{a}_{l,k}^{\mathrm{H}}\hat{\mathbf{g}}_{l,i,j}x_{i,j} + \alpha\sqrt{p_{\mathrm{u}}} \sum_{i=1}^{L} \sum_{j=1}^{K} \mathbf{a}_{l,k}^{\mathrm{H}}\tilde{\mathbf{g}}_{l,i,j}x_{i,j} + \alpha\mathbf{a}_{l,k}^{\mathrm{H}}\mathbf{n}_l + \alpha\mathbf{a}_{l,k}^{\mathrm{H}}\mathbf{n}_{l,\mathrm{q}} \end{aligned} \tag{9}$$

where $\mathbf{a}_{l,k}$ is the linear receiver vector in cell l for user k, and $x_{i,j} \sim \mathcal{CN}(0,1)$ is the j-th column of $\mathbf{x}_i$. In this paper, we focus on two linear receivers, namely MRC and ZF. Mathematically, $\mathbf{a}_{l,k}$ can be given by

$$\mathbf{a}_{l,k} = \begin{cases} \hat{\mathbf{g}}_{l,l,k}, & \text{for MRC} \\ \mathbf{f}_{l,l,k}, & \text{for ZF} \end{cases} \tag{10}$$

where $\mathbf{f}_{l,l,k}$ is the k-th column of $\hat{\mathbf{G}}_{l,l}\left(\hat{\mathbf{G}}_{l,l}^{\mathrm{H}}\hat{\mathbf{G}}_{l,l}\right)^{-1}$, and $\hat{\mathbf{G}}_{l,l} = [\hat{\mathbf{g}}_{l,l,1}, \cdots, \hat{\mathbf{g}}_{l,l,K}]$.

Motivated by [22,23], treating the interference as worst-case unrelated additive noise, the lower bound of the achievable uplink rate of the k-th user in the l-th cell can be given by

$$R_{l,k}(\mathbf{p}) = \mathbb{E}\left[\log_2\left(1 + \frac{p_{\mathrm{u}}\alpha^2|\mathbf{a}_{l,k}^{\mathrm{H}}\hat{\mathbf{g}}_{l,l,k}|^2}{\mathbb{E}\left[\mathbf{a}_{l,k}^{\mathrm{H}}\left(p_{\mathrm{u}}\alpha^2\sum_{(i,j)\neq(l,k)}\hat{\mathbf{g}}_{l,i,j}\hat{\mathbf{g}}_{l,i,j}^{\mathrm{H}} + p_{\mathrm{u}}\alpha^2\sum_{(i,j)}\tilde{\mathbf{g}}_{l,i,j}\tilde{\mathbf{g}}_{l,i,j}^{\mathrm{H}} + \alpha^2 + \mathbf{R}_{\mathbf{n}_{l,\mathrm{q}}}\right)\mathbf{a}_{l,k}|\hat{\mathbf{G}}_{l,l}\right]}\right)\right] \tag{11}$$

$$\overset{(a)}{=} \mathbb{E}\left[\log_2\left(1 + \frac{p_{\mathrm{u}}\alpha^2|\mathbf{a}_{l,k}^{\mathrm{H}}\hat{\mathbf{g}}_{l,l,k}|^2}{\mathbb{E}\left[\mathcal{I}_{l,k} + p_{\mathrm{u}}\alpha^2\sum_{i\neq l}|\mathbf{a}_{l,k}^{\mathrm{H}}\hat{\mathbf{g}}_{l,i,k}|^2 + \alpha^2\|\mathbf{a}_{l,k}^{\mathrm{H}}\|^2\right]}\right)\right] \tag{12}$$

where $\mathbf{p}$ is the transmitted power vector of K users. Since the denominator of Equation (11) is a conditional expectation operator and the estimated error vector and estimated channel vector are independent, $\mathcal{I}_{l,k}$ is given as

$$\mathcal{I}_{l,k} = p_{\mathrm{u}}\alpha^2 \sum_{i=1}^{L} \sum_{j\neq k} \mathbf{a}_{l,k}^{\mathrm{H}}\mathbb{E}\left[\hat{\mathbf{g}}_{l,i,j}\hat{\mathbf{g}}_{l,i,j}^{\mathrm{H}}\right]\mathbf{a}_{l,k} + p_{\mathrm{u}}\alpha^2 \sum_{(i,j)} \mathbf{a}_{l,k}^{\mathrm{H}}\mathbb{E}\left[\tilde{\mathbf{g}}_{l,i,j}\tilde{\mathbf{g}}_{l,i,j}^{\mathrm{H}}\right]\mathbf{a}_{l,k} + \mathbf{a}_{l,k}^{\mathrm{H}}\mathbb{E}\left[\mathbf{R}_{\mathbf{n}_{l,\mathrm{q}}}\right]\mathbf{a}_{l,k} \tag{13}$$

wherein

$$\begin{aligned}
&\mathbb{E}\left[\hat{\mathbf{g}}_{l,i,j}\hat{\mathbf{g}}_{l,i,j}^{\mathrm{H}}\right] = \mathrm{diag}\left(\beta_{l,1,i,j}\mathbf{I}_N,\cdots,\beta_{l,M,i,j}\mathbf{I}_N\right)\\
&\mathbb{E}\left[\tilde{\mathbf{g}}_{l,i,j}\tilde{\mathbf{g}}_{l,i,j}^{\mathrm{H}}\right] = \mathrm{diag}\left(\eta_{l,1,i,j}\mathbf{I}_N,\cdots,\eta_{l,M,i,j}\mathbf{I}_N\right)\\
&\mathbb{E}\left[\mathbf{R}_{\mathbf{n}_{l,q}}\right] = \mathrm{diag}\left(p_{\mathrm{u}}\hat{\mathbf{G}}_{l,l}\hat{\mathbf{G}}_{l,l}^{\mathrm{H}}\right) + \mathbf{D}_{\mathrm{R}}\\
&\mathbf{D}_{\mathrm{R}} = \mathrm{diag}\left(\left(p_{\mathrm{u}}\sum_{j=1}^{K}\eta_{l,1,l,j} + p_{\mathrm{u}}\sum_{i\neq l}\sum_{j=1}^{K}\lambda_{l,1,i,j}+1\right)\mathbf{I}_N,\cdots,\left(p_{\mathrm{u}}\sum_{j=1}^{K}\eta_{l,M,l,j} + p_{\mathrm{u}}\sum_{i\neq l}\sum_{j=1}^{K}\lambda_{l,M,i,j}+1\right)\mathbf{I}_N\right).\\
&\eta_{l,m,i,j} = \eta_{l,m,i,j} - \beta_{l.m,i,j}.
\end{aligned} \tag{14}$$

(2) Power consumption model: According to [24–26], for cell l, the total power consumption model can be given by

$$P_l = P_{\mathrm{TC}} + P_{\mathrm{LP}} + P_{\mathrm{T}} + P_{\mathrm{BH}}. \tag{15}$$

The first term P_{TC} is the power consumption of transceiver chains, which can be given by

$$P_{\mathrm{TC}} = M(NP_{\mathrm{BS}} + \rho P_{\mathrm{SYN}}) + (1-\rho)P_{\mathrm{SYN}} + KP_{\mathrm{UE}} + MP_{\mathrm{ADC}} \tag{16}$$

where P_{BS} and P_{UE} are the power consumption of running the circuit components employed at the base station and users, P_{SYN} are the power consumed by the local oscillator, and $P_{\mathrm{ADC}} = a_0N2^b + a_1$ are the power consumed by ADC, wherein a_0 and a_1 are constant parameters, $\rho = 1$ for the distributed antenna system (DAS), and $\rho = 0$ for the co-located antenna system (CAS). This results from the assumption that antennas at the same RAU are connected to a common oscillator, while oscillators at different RAUs are different in the DAS, and all antennas are connected to a single oscillator in the CAS.

The second term P_{LP} is the power consumption of the MRC/ZF receiver at the base station, which can be given by

$$P_{\mathrm{LP}} = B\tfrac{T-\tau}{T}\tfrac{2MNK}{L_{\mathrm{BS}}} + \frac{B}{T}\left(\tfrac{3MNK}{L_{\mathrm{BS}}}(1-d) + d\left(\tfrac{K^3}{3L_{\mathrm{BS}}} + \tfrac{MNK(3K+1)}{L_{\mathrm{BS}}}\right)\right) \tag{17}$$

where $d = 0$ for MRC while $d = 1$ for ZF, B is the bandwidth, T denotes the symbols for uplink transmission, and L_{BS} is the computational efficiency of arithmetic complex-valued operations for a Joule.

The third term P_{T} is transmit power, which can be represented as

$$P_{\mathrm{T}} = \tfrac{T-\tau}{T}\tfrac{K}{\xi}p_{\mathrm{u}} \tag{18}$$

where ξ is the amplified efficiency.

For the last term, P_{BH} is the power consumed of backhaul in the DAS, while it can be neglected in the CAS. Specifically, P_{BH} in the DAS can be given by

$$P_{\mathrm{BH}} = M\left(P_0 + BP_{\mathrm{BT}}\textstyle\sum_{k=1}^{K}R_{l,k}(\mathbf{p})\right) \tag{19}$$

where P_0 and P_{BT} are the fixed and traffic-dependent power consumption at each backhaul, respectively.

(3) Global energy efficiency model: Based on the above analysis, the total power consumption for all cells can be given by

$$P_{\mathrm{Total}}(\mathbf{p}) = LP_{\mathrm{IND}} + \tfrac{T-\tau}{\xi T}LKp_{\mathrm{u}} + P_{\mathrm{BT}}MB\textstyle\sum_{l=1}^{L}\sum_{k=1}^{K}R_{l,k}(\mathbf{p}) \tag{20}$$

where P_{IND} is the power consumption independent of $\mathbf{p}$ and can be given by

$$P_{\mathrm{IND}} = P_{\mathrm{TC}} + P_{\mathrm{LP}} + MP_0. \tag{21}$$

According to [24], the global energy efficiency is defined as the ratio of the achievable sum rate to the total power consumption in Watts. Mathematically, it can be defined as

$$\varphi(\mathbf{p}) = \frac{B\sum_{l=1}^{L}\sum_{k=1}^{K} R_{l,k}(\mathbf{p})}{P_{\mathrm{Total}}}. \tag{22}$$

3. Energy Efficiency Analysis

From Equations (22) and (A4), we can see that it is difficult to directly calculate Equation (A4) to analyze the energy efficiency. Therefore, we first derive the closed-form expressions of uplink achievable rates. The results are shown in the following theorems.

Theorem 1. *Using MRC receiver with low-resolution ADCs and pilot contamination, the closed-form expression for the uplink achievable rate of the k-th user in the l-th cell is given by*

$$R_{l,k}^{mrc} = log_2\left(1 + \frac{p_{\mathrm{u}}\alpha\left[\left(N\sum_{m=1}^{M}\beta_{l,m,l,k}\right)^2 + N\sum_{m=1}^{M}\beta_{l,m,l,k}^2\right]}{p_{\mathrm{u}}\alpha\mathbf{\Omega}_{l,k} + p_{\mathrm{u}}\alpha\Xi_{l,k} + (1-\alpha)\mathbf{\Phi}_{l,k} + \alpha\sum_{m=1}^{M}\beta_{l,m,l,k}^2}\right) \tag{23}$$

where

$$\begin{aligned}
\mathbf{\Omega}_{l,k} &= N\sum_{m=1}^{M}\sum_{i=1}^{L}\sum_{j=1}^{K}\beta_{l,m,l,k}\eta_{l,m,i,j}\\
\Xi_{l,k} &= N\sum_{m=1}^{M}\beta_{l,m,l,k}\left(\sum_{i=1}^{L}\sum_{j\neq k}\beta_{l,m,i,j} + \sum_{i\neq l}\beta_{l,m,i,k}\right) + \sum_{i\neq l}\left(N\sum_{m=1}^{M}\beta_{l,m,l,k}^{1/2}\beta_{l,m,i,k}^{1/2}\right)^2\\
\mathbf{\Phi}_{l,k} &= p_{\mathrm{u}}\left(\left(N\sum_{m=1}^{M}\beta_{l,m,l,k}\right)^2 + N\sum_{m=1}^{M}\beta_{l,m,l,k}^2 + N\sum_{j\neq k}\sum_{m=1}^{M}\beta_{l,m,l,k}\beta_{l,m,l,j}\right) + \Upsilon_{l,k}\\
\Upsilon_{l,k} &= N\sum_{m=1}^{M}\beta_{l,m,l,k}\left(1 + p_{\mathrm{u}}\sum_{j=1}^{K}\eta_{l,m,l,j} + p_{\mathrm{u}}\sum_{i\neq l}\sum_{j=1}^{K}\lambda_{l,m,i,j}\right).
\end{aligned} \tag{24}$$

Proof. The proof is given in Appendix B. □

Theorem 2. *Using ZF receiver with low-resolution ADCs and pilot contamination, the closed-form expression for the uplink achievable rate of user k in the l-th cell is given by*

$$R_{l,k}^{zf} = log_2\left(1 + \frac{p_{\mathrm{u}}\alpha\zeta\sum_{m=1}^{M}\beta_{l,m,l,k}}{\alpha p_{\mathrm{u}}\sum_{m=1}^{M}\left(\sum_{i\neq l}\sum_{j\neq k}\beta_{l,m,i,j} + \sum_{i=1}^{L}\sum_{j=1}^{K}\eta_{l,m,i,j} + \zeta\sum_{l\neq l}\beta_{l,m,i,k}\right) + (1-\alpha)\left(\zeta p_{\mathrm{u}}\sum_{m=1}^{M}\beta_{l,m,l,k} + \Delta_{l,k}\right) + \alpha M}\right) \tag{25}$$

where $\zeta = MN - K + 1$, *and*

$$\Delta_{l,k} = \sum_{m=1}^{M}\left(1 + p_{\mathrm{u}}\sum_{j=1}^{K}\eta_{l,m,l,j} + p_{\mathrm{u}}\sum_{i\neq l}\sum_{j=1}^{K}\lambda_{l,m,i,j}\right). \tag{26}$$

Proof. The proof is given in Appendix C. □

From Equations (23) and (25), it can be concluded that the quantization noise influences both the numerator and the denominator of Equation (A4). This means that the quantization noise is unlike the additive noise, which only affects the denominator.

Based on the theorems above, we analyze the asymptotic performance with quantization bits, per user transmit power, and the number of antennas per RAU, respectively. The results are given below.

Case 1: With a fixed transmitted power per user p_{u} and a total number of antennas per cell MN, when the number of quantization bits $b \rightarrow \infty$, the inverse of the signal-to-quantization-noise ratio ρ tends toward zero, which means that α in Equations (23) and (25) tends toward 1. The following results can then be obtained in this case by replacing the α in Equations (23) and (25) with 1

$$\tilde{R}_{l,k}^{\mathrm{mrc}} = \log_2\left(1 + \frac{p_{\mathrm{u}}\left[\left(N\sum_{m=1}^{M}\beta_{l,m,l,k}\right)^2 + N\sum_{m=1}^{M}\beta_{l,m,l,k}^2\right]}{p_{\mathrm{u}}\mathbf{\Omega}_{l,k} + p_{\mathrm{u}}\Xi_{l,k} + \sum_{m=1}^{M}\beta_{l,m,l,k}^2}\right) \tag{27}$$

$$\tilde{R}_{l,k}^{\mathrm{zf}} = \log_2\left(1 + \frac{p_{\mathrm{u}}\zeta\sum_{m=1}^{M}\beta_{l,m,l,k}}{p_{\mathrm{u}}\sum\limits_{m=1}^{M}\left(\sum\limits_{i\neq l}\sum\limits_{j\neq k}\beta_{l,m,i,j} + \sum\limits_{i=1}^{L}\sum\limits_{j=1}^{K}\eta_{l,m,i,j} + \zeta\sum\limits_{i\neq l}\beta_{l,m,i,k}\right) + M}\right). \tag{28}$$

Case 1 shows the achievable uplink rates without considering the quantization noise caused by ADC. It can be seen that the spectral efficiency is only limited by pilot contamination and channel estimation error. Moreover, since the power consumption of an ADC $P_{\mathrm{ADC}} = a_0 N 2^b + a_1$ is an exponential function of b, P_{ADC} tends toward infinity when $b \rightarrow \infty$. As shown in Equation (20), the total power consumption also goes to infinity. Hence, the limited achievable rates and unlimited power consumption lead to the fact that the global energy efficiency tends toward zero, that is $\phi(\mathbf{p}) \rightarrow 0$ when $b \rightarrow \infty$, while p_{u} and MN are fixed.

Case 2: With a fixed number of quantization bits b and antennas per cell MN, when $p_{\mathrm{u}} \rightarrow \infty$, the ultimate rates of user k in cell l with both receivers can be directly obtained by dividing the dominators and numerators of Equations (23) and (25) by p_{u}, which are given by

$$\tilde{R}_{l,k}^{\mathrm{mrc}} = \log_2\left(1 + \frac{\alpha\left[\left(N\sum_{m=1}^{M}\beta_{l,m,l,k}\right)^2 + N\sum_{m=1}^{M}\beta_{l,m,l,k}^2\right]}{\alpha\mathbf{\Omega}_{l,k} + \alpha\Xi_{l,k} + (1-\alpha)\mathbf{\Phi}'_{l,k}}\right) \tag{29}$$

$$\tilde{R}_{l,k}^{\mathrm{zf}} = \log_2\left(1 + \frac{\alpha\zeta\sum_{m=1}^{M}\beta_{l,m,l,k}}{\alpha\sum\limits_{m=1}^{M}\left(\sum\limits_{i\neq l}\sum\limits_{j\neq k}\beta_{l,m,i,j} + \sum\limits_{i=1}^{L}\sum\limits_{j=1}^{K}\eta_{l,m,i,j} + \zeta\sum\limits_{i\neq l}\beta_{l,m,i,k}\right) + (1-\alpha)\left(\zeta\sum\limits_{m=1}^{M}\beta_{l,m,l,k} + \Delta'_{l,k}\right)}\right) \tag{30}$$

where

$$\mathbf{\Phi}_{l,k} = \left(\left(N\sum_{m=1}^{M}\beta_{l,m,l,k}\right)^2 + N\sum_{m=1}^{M}\beta_{l,m,l,k}^2 + N\sum_{j\neq k}\sum_{m=1}^{M}\beta_{l,m,l,k}\beta_{l,m,l,j}\right) + \Upsilon'_{l,k}$$
$$\Upsilon'_{l,k} = N\sum_{m=1}^{M}\beta_{l,m,l,k}\left(\sum_{j=1}^{K}\eta_{l,m,l,j} + \sum_{i\neq l}\sum_{j=1}^{K}\lambda_{l,m,i,j}\right)$$
$$\Delta'_{l,k} = \sum_{m=1}^{M}\left(\sum_{j=1}^{K}\eta_{l,m,l,j} + \sum_{i\neq l}\sum_{j=1}^{K}\lambda_{l,m,i,j}\right).$$

Case 2 indicates that, as p_{u} grows indefinitely, the achievable uplink rates approach certain values dependent on the resolution of ADC. This observation shows that the performance degradation due to low-resolution ADCs cannot be compensated by increasing the transmit power. Furthermore, it can be seen that, as the transmit power increases, the system energy efficiency tends toward zero. This is because the total power consumption presented in Equation (20) tends toward infinity as p_{u} increases indefinitely, but the unlimited growth in the transmit power cannot improve the achievable uplink rates indefinitely.

Case 3: With a fixed number of quantization bits b, the number of RAUs per cell M and the transmitted power per user p_{u}, when $N \rightarrow \infty$, the limiting rates of user k in cell l with both receivers can be directly obtained by dividing the dominators and numerators of Equations (23) and (25) by N^2, which are given by

$$R_{l,k}^{\mathrm{mrc}} = \log_2\left(1+\frac{\alpha\left(\sum_{m=1}^{M}\beta_{l,m,l,k}\right)^2}{\alpha\sum_{i\neq l}\left(\sum_{m=1}^{M}\beta_{l,m,l,k}^{1/2}\beta_{l,m,i,k}^{1/2}\right)^2+(1-\alpha)\left(\sum_{m=1}^{M}\beta_{l,m,l,k}\right)^2}\right) \quad (31)$$

$$R_{l,k}^{\mathrm{zf}} = \log_2\left(1+\frac{\alpha\sum_{m=1}^{M}\beta_{l,m,l,k}}{\alpha\sum_{i\neq l}\sum_{m=1}^{M}\beta_{l,m,i,k}+(1-\alpha)\sum_{m=1}^{M}\beta_{l,m,l,k}}\right). \quad (32)$$

Case 3 shows that, when the number of antennas per RAU grows without bound, the impacts of quantization noise vanish. However, the achievable rates with both receivers tend toward certain and limited values as N goes infinity. This results from the presence of pilot contamination. Furthermore, since the power consumption of transceiver chains and linear processing at the base station are proportional to N, they tend toward infinity when $N \to \infty$. As shown in Equation (20), the total power consumption also goes to infinity. Hence, the limited achievable rates and unlimited power consumption lead to the fact that the global energy efficiency tends toward zero, that is $\phi(\mathbf{p}) \to 0$ when $N \to \infty$ while b, M, and p_{u} are fixed.

4. Numerical Results

In this section, we verify the accuracy of the theoretical results in Section 3 by a series of Monte Carlo simulations. A multi-cell distributed massive MIMO system is considered, which consists of $L = 7$ cells, $M = 7$ RAUs per cell, $K = 6$ users per cell, and the cell radius D is normalized to 1. In each cell, all users are uniformly distributed, while RAUs have fixed locations with radiuses $r_1 = 0, r_2 = \cdots = r_7 = (3-\sqrt{3})/2$ and angles $\theta_1 = 0, \theta_2 = \pi/6, \theta_3 = \pi/2, \theta_4 = 5\pi/6, \theta_5 = 7\pi/6, \theta_6 = 3\pi/2, and\ \theta_7 = 11\pi/6$. The path loss between the k-th user in the i-th cell and the m-th RAU in the l-th cell $\lambda_{l,m,i,k}$ is modeled as $\lambda_{l,m,i,k} = d_{l,m,i,k}^{-\iota}$, where $d_{l,m,i,k}$ is the corresponding distance, and ι assumed as $\iota = 3.7$ is the path loss exponent. Moreover, the length of pilot sequences is $\tau = K$. The coherence time of the channel is assumed as $T = 196$ symbols, and the power consumption parameters are given in Table 1.

Table 1. Power consumption parameters.

Parameters	Values
Transmitted power per user p_{u}	0.02 Watts
Power consumption per antenna at base station P_{BS}	0.1 Watts
Power consumption per antenna at users P_{UE}	0.1 Watts
Power consumption per local oscillator P_{SYN}	1 Watts
Computational efficiency at base stations L_{BS}	12.8 Gflops/Watt
Power amplified efficiency ξ	0.4
Fixed power consumption per backhaul P_0	0.825 Watts
Traffic-dependent backhaul power P_{BT}	0.25 Watts/(Gbit/s)
Parameters of ADC $a_0 \& a_1$	10^{-3} & 0.02

We first prove the accuracy of the theoretical results given in Theorems 1 and 2. Figure 2 illustrates the uplink spectral efficiency per cell versus the number of quantization bits with different numbers of antennas per RAU. It can be seen that the closed-form expressions and simulation results match well with each other using both MRC and ZF receivers. As the number of antennas per RAU increases, the uplink spectral efficiency grows obviously for both receivers. Furthermore, for both receivers, the uplink spectral efficiency increases rapidly with the increase in quantization bits b when b is small, while the growth of b cannot improve the spectral efficiency further when b is large. It can be concluded that low-resolution ADCs are acceptable in massive MIMO systems, and employing a large number of antennas at each RAU can compensate for the performance degradation. In the following, the closed-form expressions will be used for numerical work.

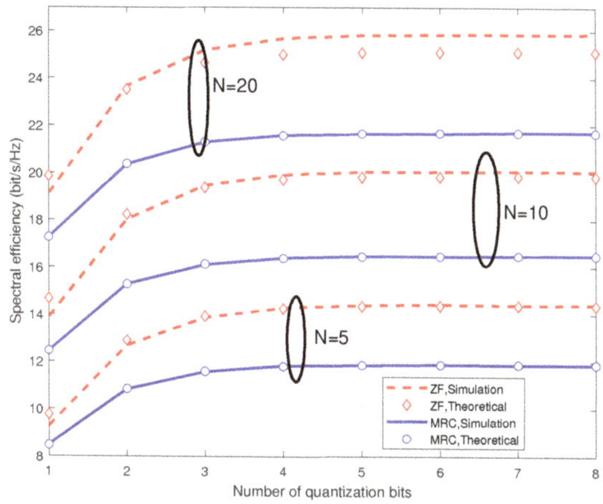

Figure 2. Spectral efficiency versus the number of quantization bits with different numbers of antennas per RAU.

Next, Figure 3 illustrates the energy efficiency versus spectral efficiency with different numbers of quantization bits and of antennas per RAU. It can be seen that, as the number of antennas and the number of quantization bits increase, energy efficiency increases first and then decreases. This is because the power consumption and spectral efficiency both increase with the increase in antennas and quantization bits, but the improvement of spectral efficiency dominates first, and the power consumption then dominates. The results illustrate that we cannot improve the spectral efficiency and energy efficiency simultaneously without bound, and there needs to be a tradeoff between them, which was investigated in [27–29]. Moreover, it can be seen in Figure 3b that $b = 3$ or $b = 4$ are preferable under the system configuration mentioned above. If b increases further, the spectral efficiency can be slightly improved while the energy efficiency decreases rapidly. It should be noted that the optimal number of bits is dependent on system configuration and system parameters. Figure 3 also indicates that low-resolution ADCs (b = 3 or 4 bits in our simulation results) are preferable in distributed massive MIMO systems.

Finally, the energy efficiency against spectral efficiency with different numbers of quantization bits and transmitted power per user is presented in Figure 4. The same conclusion about the relationship between energy efficiency and spectral efficiency with different b can be obtained from Figure 4. As for the transmitted power per user, it can be seen that, with its increase, the energy efficiency increases first and then decreases. This results from the fact that the power consumption linearly increases with the growth of transmitted power, but the spectral efficiency increases first and then tends toward a certain value.

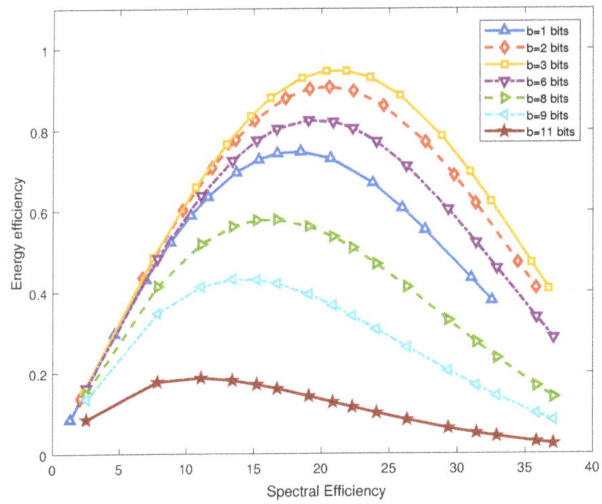

(a)

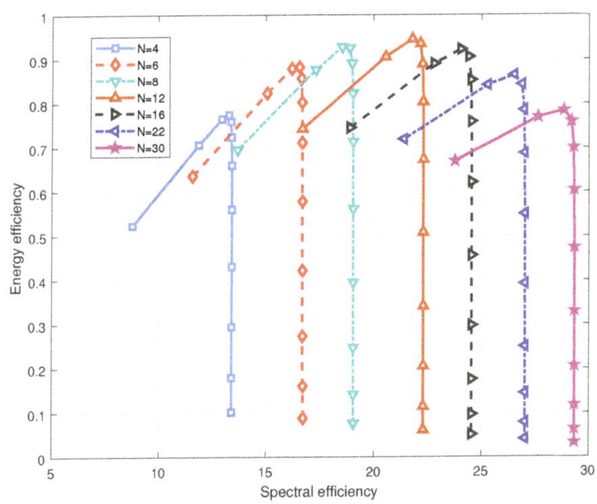

(b)

Figure 3. Energy efficiency versus spectral efficiency with different numbers of quantization bits b and different numbers of antennas N per RAU. (**a**) Each line corresponds to different numbers of quantization bits with b = [1, 2, 3, 6, 8, 9, 11] and the points on each line correspond to different numbers of antennas per RAU with N= [1:1:6, 8, 10, 12, 15, 20, 30, 40, 50, 80, 100]. (**b**) Each line corresponds to different numbers of antennas per RAU with N= [4, 6, 8, 12, 16, 22, 30] and the points on each line correspond to different numbers of quantization bits with b = [1:1:12].

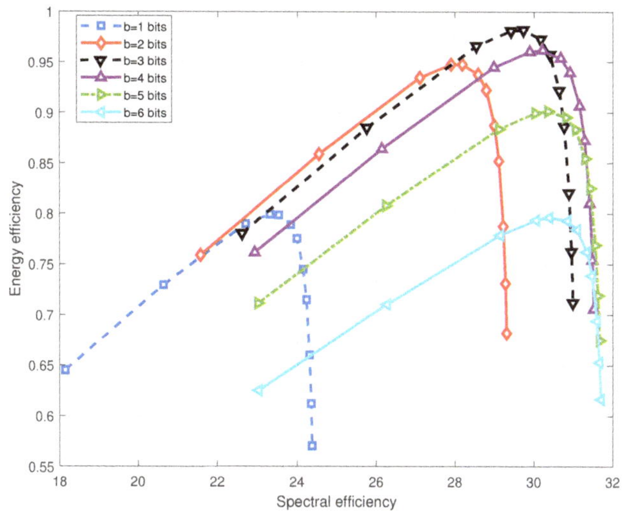

Figure 4. Energy efficiency versus spectral efficiency with the number of quantization bits b = [1:1:6] bits and transmitted power per user $p_u = [0.01, 0.02, 0.05, 0.08, 0.1, 0.15, 0.2, 0.3, 0.4, 0.6, 0.8, 1]$ W, where each line corresponds to different b and the points on each line correspond to different p_u.

5. Conclusions

In this paper, we analyzed the uplink spectral and energy efficiency simultaneously in distributed massive MIMO systems with low-resolution ADCs. Furthermore, this paper considered a more realistic scenario where the base station did not have CSI and it obtained the estimated CSI during the pilot phase. In this case, the pilot contamination presents and degrades the system performance. We first gave an additive quantization noise model and got the estimated CSI with pilot contamination. Under the imperfect CSI, we derived the closed-form expressions for achievable uplink rates using MRC and ZF receivers. Furthermore, we obtained the asymptotic performance with the number of quantization bits, the per user transmit power, and the per RAU antenna number, respectively. The theorems are verified by simulation. It can be noted that the increase in antennas can compensate for the spectral efficiency degradation caused by quantization noise. Furthermore, the energy efficiency with low-resolution ADCs are better than that with perfect ADCs. Numerical results imply that it is preferable to use low-resolution ADCs in distributed massive MIMO systems.

We intend to extend our research considering the tradeoff between spectral efficiency and energy efficiency, which involves multi-objective optimization. Furthermore, in order to make the system more energy-efficient, we plan to extend our research considering RAU selection.

Author Contributions: Formal analysis, J.L.; Methodology, J.L.; Supervision, P.Z. and X.Y.; Validation, J.L. and J.Y.; Writing original draft, J.L. and Q.L.; Writing review & editing, J.L. and Q.L.

Funding: This work was supported in part by National Natural Science Foundation of China (NSFC) (Grant No. 61501113, 61571120) and the Jiangsu Provincial Natural Science Foundation (Grant No. BK20150630, BK20180011).

Conflicts of Interest: The authors declare no conflict of interest.

Appendix A. Lemmas for Proof

In order to derive the closed-form expressions with both receivers, we provide the following preliminary lemmas first.

Lemma A1 ([22]). *Suppose $\{X_i\}$ are independent Gamma distributed random variables, i.e., $\{X_i\} \sim \Gamma(k_i, \theta_i)$. Then the first two moments of the sum $\sum_i X_i$ can be given by*

$$\mathbb{E}\left[\sum_i X_i\right] = \sum_i k_i\theta_i, \tag{A1}$$

$$\mathbb{E}\left[\left(\sum_i X_i\right)^2\right] = \sum_i k_i\theta_i^2 + \left(\sum_i k_i\theta_i\right)^2. \tag{A2}$$

Lemma A2 ([21]). *For the p-dimensional non-isotropic channel vector $\mathbf{x}$ whose strength is distributed as $\mathbf{x}^{\mathrm{H}}\mathbf{x} \sim \Gamma(k, \theta)$, when projected onto a s-dimensional subspace, the distribution of the projection power can be approximated as $\Gamma(sk/p, \theta)$.*

Lemma A3 ([30]). *If $\mathbf{x}$ is an $N \times 1$ isotropic random vector and $\mathbf{A}$ is a constant matrix. Then we can have*

$$\mathbb{E}_{\mathbf{x}}\left[\mathbf{x}^{\mathrm{H}}\mathbf{A}\mathbf{x}\right] = \frac{\mathrm{tr}(\mathbf{A})}{N}. \tag{A3}$$

Appendix B. Proof of Theorem 1

From Lemma 4 of [31], we can obtain the approximation of Equation (11) as follows:

$$R_{l,k}(\mathbf{p}) \approx \log_2\left(1 + \frac{p_{\mathrm{u}}\alpha^2\mathbb{E}\left[|\mathbf{a}_{l,k}^{\mathrm{H}}\hat{\mathbf{g}}_{l,l,k}|^2\right]}{\mathbb{E}\left[\mathcal{I}_{l,k} + p_{\mathrm{u}}\alpha^2\sum_{i\neq l}|\mathbf{a}_{l,k}^{\mathrm{H}}\hat{\mathbf{g}}_{l,i,k}|^2 + \alpha^2\|\mathbf{a}_{l,k}^{\mathrm{H}}\|^2\right]}\right). \tag{A4}$$

Consider the MRC receiver, it can be seen from Equation (A4) that the following terms need to be simplified:

$$\mathbb{E}\left[\|\hat{\mathbf{g}}_{l,l,k}\|^4\right] = \hat{k}_{l,l,k}\hat{\theta}_{l,l,k}^2 + (\hat{k}_{l,l,k}\hat{\theta}_{l,l,k})^2 \tag{A5}$$

where

$$\hat{k}_{l,l,k} = \frac{N(\sum_{m=1}^{M}\beta_{l,m,l,k})^2}{\sum_{m=1}^{M}\beta_{l,m,l,k}^2} \qquad \hat{\theta}_{l,l,k} = \frac{\sum_{m=1}^{M}\beta_{l,m,l,k}^2}{\sum_{m=1}^{M}\beta_{l,m,l,k}}. \tag{A6}$$

This can be obtained by exploiting the fact that $\hat{\mathbf{g}}_{l,l,k}^{\mathrm{H}}\hat{\mathbf{g}}_{l,l,k} \sim \Gamma(\hat{k}_{l,l,k}, \hat{\theta}_{l,l,k})$ and Lemma A1.

Due to the independence between $\hat{\mathbf{g}}_{l,l,k}$ and $\hat{\mathbf{g}}_{l,i,j}$ when $j \neq k$, we have

$$\mathbb{E}\left[\hat{\mathbf{g}}_{l,l,k}^{\mathrm{H}}\mathbf{E}\left[\hat{\mathbf{g}}_{l,i,j}\hat{\mathbf{g}}_{l,i,j}^{\mathrm{H}}\right]\hat{\mathbf{g}}_{l,l,k}\right] \overset{(a)}{=} N\sum_{m=1}^{M}\beta_{l,m,l,k}\beta_{l,m,i,j} \tag{A7}$$

where (a) results from the fact that the channel strength is Gamma-distributed.

Because of the pilot contamination, $\hat{\mathbf{g}}_{l,l,k}$ and $\hat{\mathbf{g}}_{l,i,k}$ are dependent, we have

$$\mathbb{E}\left[|\hat{\mathbf{g}}_{l,l,k}^{\mathrm{H}}\hat{\mathbf{g}}_{l,i,k}|^2\right] \overset{(a)}{=} \left(N\sum_{m=1}^{M}\beta_{l,m,l,k}^{1/2}\beta_{l,m,i,k}^{1/2}\right)^2 + N\sum_{m=1}^{M}\beta_{l,m,l,k}\beta_{l,m,i,k} \tag{A8}$$

where (a) is obtained due to the fact that the channel strength is Gamma-distributed and due to Lemma A1.

Using the fact that $\hat{\mathbf{g}}_{l,l,k}$ and $\tilde{\mathbf{g}}_{l,i,j}$ are independent, we have

$$\sum_{(i,j)}\mathbb{E}\left[\hat{\mathbf{g}}_{l,l,k}^{\mathrm{H}}\mathbb{E}\left[\tilde{\mathbf{g}}_{l,i,j}\tilde{\mathbf{g}}_{l,i,j}^{\mathrm{H}}\right]\hat{\mathbf{g}}_{l,l,k}\right] = N\sum_{m=1}^{M}\sum_{i=1}^{L}\sum_{j=1}^{K}\beta_{l,m,l,k}\eta_{l,m,i,j}. \tag{A9}$$

For the last term, we first calculate

$$\begin{aligned}
&\mathbb{E}\left[\hat{\mathbf{g}}_{l,l,k}^{\mathrm{H}}\mathrm{diag}\left(p_{\mathrm{u}}\hat{\mathbf{G}}_{l,l}\hat{\mathbf{G}}_{l,l}^{\mathrm{H}}\right)\hat{\mathbf{g}}_{l,l,k}\right] \\
&= p_{\mathrm{u}}\mathbb{E}\left[\hat{\mathbf{g}}_{l,l,k}^{\mathrm{H}}\hat{\mathbf{g}}_{l,l,k}\hat{\mathbf{g}}_{l,l,k}^{\mathrm{H}}\hat{\mathbf{g}}_{l,l,k}\right] + p_{\mathrm{u}}\sum\nolimits_{j\neq k}\mathbb{E}\left[\hat{\mathbf{g}}_{l,l,k}^{\mathrm{H}}\hat{\mathbf{g}}_{l,l,j}\hat{\mathbf{g}}_{l,l,j}^{\mathrm{H}}\hat{\mathbf{g}}_{l,l,k}\right] \\
&= p_{\mathrm{u}}\left(\hat{k}_{l,l,k}\hat{\theta}_{l,l,k}^{2} + (\hat{k}_{l,l,k}\hat{\theta}_{l,l,k})^{2} + N\sum\nolimits_{j\neq k}\sum\nolimits_{m=1}^{M}\beta_{l,m,l,k}\beta_{l,m,l,j}\right).
\end{aligned} \tag{A10}$$

Then we can obtain

$$\begin{aligned}
&\mathbb{E}\left[\hat{\mathbf{g}}_{l,l,k}^{\mathrm{H}}\mathbb{E}\left[\mathrm{diag}\left(p_{\mathrm{u}}\sum_{i=1}^{L}\mathbf{G}_{l,i}\mathbf{G}_{l,i}^{\mathrm{H}}+\mathbf{I}\right)\right]\hat{\mathbf{g}}_{l,l,k}\right] \\
&= \mathbb{E}\left[\hat{\mathbf{g}}_{l,l,k}^{\mathrm{H}}\mathrm{diag}\left(p_{\mathrm{u}}\hat{\mathbf{G}}_{l,l}\hat{\mathbf{G}}_{l,l}^{\mathrm{H}}\right)\hat{\mathbf{g}}_{l,l,k}\right] + \mathbb{E}\left[\sum_{n=1}^{MN}\left|\hat{g}_{l,n,l,k}\right|^{2}\mathbf{D}_{\mathrm{R}}\right] \\
&= p_{\mathrm{u}}\left(\hat{k}_{l,l,k}\hat{\theta}_{l,l,k}^{2} + (\hat{k}_{l,l,k}\hat{\theta}_{l,l,k})^{2} + N\sum\nolimits_{j\neq k}\sum\nolimits_{m=1}^{M}\beta_{l,m,l,k}\beta_{l,m,l,j}\right) + \Upsilon_{l,k}.
\end{aligned} \tag{A11}$$

Substituting Equations (A5), (A7)–(A9), and (A11) into Equation (A4) yields the closed-form expression expressed by Equation (23). This completes the proof.

Appendix C. Proof of the Theorem 2

Consider a ZF receiver, similar to the proof of Theorem 1. The following terms need to be calculated.

For the term $\mathbb{E}\left[\frac{1}{\|\mathbf{a}_{l,k}^{\mathrm{H}}\|^2}\right]$, we have

$$\frac{1}{\|\mathbf{a}_{l,k}^{\mathrm{H}}\|^2} = \left|\frac{\mathbf{a}_{l,k}^{\mathrm{H}}}{\|\mathbf{a}_{l,k}^{\mathrm{H}}\|^2}\hat{\mathbf{g}}_{l,l,k}\right| \sim \Gamma\left(\tfrac{MN-K+1}{MN}\hat{k}_{l,l,k},\hat{\theta}_{l,l,k}\right), \tag{A12}$$

which results form Lemma A2 and from the fact that, from the perspective of each user, an intended beam lies in a subspace of dimension $s = MN - K + 1$ with ZF receivers. Thus,

$$\mathbb{E}\left[\frac{1}{\|\mathbf{a}_{l,k}^{\mathrm{H}}\|^2}\right] = \tfrac{MN-K+1}{MN}\hat{k}_{l,l,k}\hat{\theta}_{l,l,k}. \tag{A13}$$

Next, due to the independence between $\mathbf{a}_{l,k}$ and $\hat{\mathbf{g}}_{l,i,j}$, we have

$$\begin{aligned}
&\sum\nolimits_{i\neq l}\sum\nolimits_{j\neq k}\mathbb{E}\left[\mathbf{a}_{l,k}^{\mathrm{H}}\mathbb{E}\left[\hat{\mathbf{g}}_{l,i,j}\hat{\mathbf{g}}_{l,i,j}^{\mathrm{H}}\right]\mathbf{a}_{l,k}\right] \\
&\overset{(a)}{=} \sum\nolimits_{i\neq l}\sum\nolimits_{j\neq k}\mathbb{E}\left[\mathbb{E}\left[\mathbf{a}_{l,k}^{\mathrm{H}}\hat{\mathbf{g}}_{l,i,j}\hat{\mathbf{g}}_{l,i,j}^{\mathrm{H}}\mathbf{a}_{l,k}\right]\right] \\
&\overset{(b)}{=} \frac{1}{MN}\mathbb{E}\left[\hat{\mathbf{g}}_{l,i,j}^{\mathrm{H}}\hat{\mathbf{g}}_{l,i,j}\right] \\
&\overset{(c)}{=} \tfrac{1}{M}\sum\nolimits_{i\neq l}\sum\nolimits_{j\neq k}\sum\nolimits_{m=1}^{M}\beta_{l,m,i,j}.
\end{aligned} \tag{A14}$$

where (a) results from the fact that $\mathbf{a}_{l,k}$ and $\hat{\mathbf{g}}_{l,i,j}$ are independent, (b) results from Lemma A3, and (c) results from the fact that $\hat{\mathbf{g}}_{l,i,j}^{\mathrm{H}}\hat{\mathbf{g}}_{l,i,j} \sim \Gamma(\hat{k}_{l,i,j},\hat{\theta}_{l,i,j})$.

Similarly, we have

$$\sum\nolimits_{(i,j)}\mathbb{E}\left[\frac{\mathbf{a}_{l,k}^{\mathrm{H}}}{\|\mathbf{a}_{l,k}^{\mathrm{H}}\|^2}\mathbb{E}\left[\tilde{\mathbf{g}}_{l,i,j}\tilde{\mathbf{g}}_{l,i,j}^{\mathrm{H}}\right]\frac{\mathbf{a}_{l,k}}{\|\mathbf{a}_{l,k}^{\mathrm{H}}\|^2}\right] = \tfrac{1}{M}\sum\nolimits_{i=1}^{L}\sum\nolimits_{j=1}^{K}\sum\nolimits_{m=1}^{M}\eta_{l,m,i,j}. \tag{A15}$$

Due to the pilot contamination, $\mathbf{a}_{l,k}$ and $\hat{\mathbf{g}}_{l,i,k}$ are dependent, we have

$$\sum_{i\neq l}\mathbb{E}\left[\left|\frac{\mathbf{a}_{l,k}^{\mathrm{H}}}{\|\mathbf{a}_{l,k}^{\mathrm{H}}\|}\hat{\mathbf{g}}_{l,i,k}\right|^{2}\right]=\frac{MN-K+1}{M}\sum_{i\neq l}\sum_{m=1}^{M}\beta_{l,m,i,k}. \tag{A16}$$

For the last term, we first calculate

$$\begin{aligned}&\mathbb{E}\left[\frac{\mathbf{a}_{l,k}^{\mathrm{H}}}{\|\mathbf{a}_{l,k}^{\mathrm{H}}\|}\mathrm{diag}\left(p_{\mathrm{u}}\hat{\mathbf{G}}_{l,l}\hat{\mathbf{G}}_{l,l}^{\mathrm{H}}\right)\frac{\mathbf{a}_{l,k}}{\|\mathbf{a}_{l,k}^{\mathrm{H}}\|}\right]\\&=p_{\mathrm{u}}\mathbb{E}\left[\frac{\mathbf{a}_{l,k}^{\mathrm{H}}}{\|\mathbf{a}_{l,k}^{\mathrm{H}}\|}\hat{\mathbf{g}}_{l,l,k}\hat{\mathbf{g}}_{l,l,k}^{\mathrm{H}}\frac{\mathbf{a}_{l,k}}{\|\mathbf{a}_{l,k}^{\mathrm{H}}\|}\right]\\&=\frac{MN-K+1}{MN}p_{\mathrm{u}}\hat{k}_{l,l,k}\hat{\theta}_{l,l,k}.\end{aligned} \tag{A17}$$

Then we can obtain

$$\begin{aligned}&\mathbb{E}\left[\frac{\mathbf{a}_{l,k}^{\mathrm{H}}}{\|\mathbf{a}_{l,k}^{\mathrm{H}}\|}\mathbb{E}\left[\mathrm{diag}\left(p_{\mathrm{u}}\sum_{i=1}^{L}\mathbf{G}_{l,i}\mathbf{G}_{l,i}^{\mathrm{H}}+\mathbf{I}\right)\right]\frac{\mathbf{a}_{l,k}}{\|\mathbf{a}_{l,k}^{\mathrm{H}}\|}\right]\\&=\mathbb{E}\left[\frac{\mathbf{a}_{l,k}^{\mathrm{H}}}{\|\mathbf{a}_{l,k}^{\mathrm{H}}\|}\mathrm{diag}\left(p_{\mathrm{u}}\hat{\mathbf{G}}_{l,l}\hat{\mathbf{G}}_{l,l}^{\mathrm{H}}\right)\frac{\mathbf{a}_{l,k}}{\|\mathbf{a}_{l,k}^{\mathrm{H}}\|}\right]+\mathbb{E}\left[\frac{\mathbf{a}_{l,k}^{\mathrm{H}}}{\|\mathbf{a}_{l,k}^{\mathrm{H}}\|}\mathbf{D}_{\mathrm{R}}\frac{\mathbf{a}_{l,k}}{\|\mathbf{a}_{l,k}^{\mathrm{H}}\|}\right]\\&\overset{(a)}{=}\frac{MN-K+1}{MN}p_{\mathrm{u}}\hat{k}_{l,l,k}\hat{\theta}_{l,l,k}+\frac{1}{M}\Delta_{l,k}\end{aligned} \tag{A18}$$

where (a) results form Lemma A3.

Substituting Equations (A13)–(A16), and (A18) into Equation (A4) yields the closed-form expression expressed by Equation (25). This completes the proof.

References

1. Wang, J.; Zhu, H.; Gomes, N.J.; Wang, J. Frequency Reuse of Beam Allocation for Multiuser Massive MIMO Systems. *IEEE Trans. Wirel. Commun.* **2018**, *17*, 2346–2359. [CrossRef]
2. Marzetta, T.L. Noncooperative Cellular Wireless with Unlimited Numbers of Base Station Antennas. *IEEE Trans. Wirel. Commun.* **2010**, *9*, 3590–3600. [CrossRef]
3. Larsson, E.G.; Edfors, O.; Tufvesson, F.; Marzetta, T.L. Massive MIMO for next generation wireless systems. *IEEE Commun. Mag.* **2014**, *52*, 186–195. [CrossRef]
4. Lu, L.; Li, G.Y.; Swindlehurst, A.L.; Ashikhmin, A.; Zhang, R. An Overview of Massive MIMO: Benefits and Challenges. *IEEE J. Sel. Top. Signal Process.* **2014**, *8*, 742–758. [CrossRef]
5. Wang, D.; Zhang, Y.; Wei, H.; You, X.; Gao, X.; Wang, J. An overview of transmission theory and techniques of large-scale antenna systems for 5G wireless communications. *Sci. China Inf. Sci.* **2016**, *59*, 081301. [CrossRef]
6. Zhang, J.; Wen, C.; Jin, S.; Gao, X.; Wong, K. On Capacity of Large-Scale MIMO Multiple Access Channels with Distributed Sets of Correlated Antennas. *IEEE J. Sel. Areas Commun.* **2013**, *31*, 133–148. [CrossRef]
7. Rusek, F.; Persson, D.; Lau, B.K.; Larsson, E.G.; Marzetta, T.L.; Edfors, O.; Tufvesson, F. Scaling Up MIMO: Opportunities and Challenges with Very Large Arrays. *IEEE Signal Process. Mag.* **2013**, *30*, 40–60. [CrossRef]
8. Li, J.; Wang, D.; Zhu, P.; Wang, J.; You, X. Downlink Spectral Efficiency of Distributed Massive MIMO Systems With Linear Beamforming Under Pilot Contamination. *IEEE Trans. Veh. Technol.* **2018**, *67*, 1130–1145. [CrossRef]
9. Björnson, E.; Matthaiou, M.; Debbah, M. Massive MIMO with non-ideal arbitrary arrays: Hardware scaling laws and circuit-aware design. *IEEE Trans. Wirel. Commun.* **2015**, *14*, 4353–4368. [CrossRef]
10. Zhu, H. Performance Comparison Between Distributed Antenna and Microcellular Systems. *IEEE J. Sel. Areas Commun.* **2011**, *29*, 1151–1163. [CrossRef]
11. You, X.H.; Wang, D.M.; Sheng, B.; Gao, X.Q.; Zhao, X.S.; Chen, M. Cooperative distributed antenna systems for mobile communications. *IEEE Wirel. Commun.* **2010**, *17*, 35. [CrossRef]
12. Dai, L. A Comparative Study on Uplink Sum Capacity with Co-Located and Distributed Antennas. *IEEE J. Sel. Areas Commun.* **2011**, *29*, 1200–1213. [CrossRef]

13. Bai, Q.; Nossek, J.A. Energy efficiency maximization for 5G multi-antenna receivers. *Trans. Emerg. Telecommun. Technol.* **2015**, *26*, 3–14. [CrossRef]
14. Walden, R.H. Analog-to-digital converter survey and analysis. *IEEE J. Sel. Areas Commun.* **1999**, *17*, 539–550. [CrossRef]
15. Singh, J.; Dabeer, O.; Madhow, U. On the limits of communication with low-precision analog-to-digital conversion at the receiver. *IEEE Trans. Commun.* **2009**, *57*, 3629–3639. [CrossRef]
16. Fan, L.; Jin, S.; Wen, C.; Zhang, H. Uplink Achievable Rate for Massive MIMO Systems With Low-Resolution ADC. *IEEE Commun. Lett.* **2015**, *19*, 2186–2189. [CrossRef]
17. Qiao, D.; Tan, W.; Zhao, Y.; Wen, C.; Jin, S. Spectral efficiency for massive MIMO zero-forcing receiver with low-resolution ADC. In Proceedings of the 2016 8th International Conference on Wireless Communications Signal Processing (WCSP), Yangzhou, China, 13–15 October 2016; pp. 1–6. [CrossRef]
18. Verenzuela, D.; Bjrnson, E.; Matthaiou, M. Hardware design and optimal ADC resolution for uplink massive MIMO systems. In Proceedings of the 2016 IEEE Sensor Array and Multichannel Signal Processing Workshop (SAM), Rio de Janerio, Brazil, 10–13 July 2016; pp. 1–5. [CrossRef]
19. Sarajlić, M.; Liu, L.; Edfors, O. When Are Low Resolution ADCs Energy Efficient in Massive MIMO? *IEEE Access* **2017**, *5*, 14837–14853. [CrossRef]
20. Roth, K.; Nossek, J.A. Achievable Rate and Energy Efficiency of Hybrid and Digital Beamforming Receivers With Low Resolution ADC. *IEEE J. Sel. Areas Commun.* **2017**, *35*, 2056–2068. [CrossRef]
21. Li, J.; Wang, D.; Zhu, P.; You, X. Uplink Spectral Efficiency Analysis of Distributed Massive MIMO With Channel Impairments. *IEEE Access* **2017**, *5*, 5020–5030. [CrossRef]
22. Li, J.; Wang, D.; Zhu, P.; You, X. Spectral efficiency analysis of single-cell multi-user large-scale distributed antenna system. *IET Commun.* **2014**, *8*, 2213–2221. [CrossRef]
23. Ngo, H.Q.; Larsson, E.G.; Marzetta, T.L. Energy and Spectral Efficiency of Very Large Multiuser MIMO Systems. *IEEE Trans. Commun.* **2013**, *61*, 1436–1449. [CrossRef]
24. Zuo, J.; Zhang, J.; Yuen, C.; Jiang, W.; Luo, W. Energy-Efficient Downlink Transmission for Multicell Massive DAS With Pilot Contamination. *IEEE Trans. Veh. Technol.* **2017**, *66*, 1209–1221. [CrossRef]
25. Ngo, H.Q.; Tran, L.; Duong, T.Q.; Matthaiou, M.; Larsson, E.G. On the Total Energy Efficiency of Cell-Free Massive MIMO. *IEEE Trans. Green Commun. Netw.* **2018**, *2*, 25–39. [CrossRef]
26. Björnson, E.; Sanguinetti, L.; Hoydis, J.; Debbah, M. Optimal Design of Energy-Efficient Multi-User MIMO Systems: Is Massive MIMO the Answer? *IEEE Trans. Wirel. Commun.* **2015**, *14*, 3059–3075. [CrossRef]
27. Pham, Q.; Hwang, W. Fairness-Aware Spectral and Energy Efficiency in Spectrum-Sharing Wireless Networks. *IEEE Trans. Veh. Technol.* **2017**, *66*, 10207–10219. [CrossRef]
28. Deng, L.; Rui, Y.; Cheng, P.; Zhang, J.; Zhang, Q.T.; Li, M. A Unified Energy Efficiency and Spectral Efficiency Tradeoff Metric in Wireless Networks. *IEEE Commun. Lett.* **2013**, *17*, 55–58. [CrossRef]
29. Xu, L.; Yu, G.; Jiang, Y. Energy-Efficient Resource Allocation in Single-Cell OFDMA Systems: Multi-Objective Approach. *IEEE Trans. Wirel. Commun.* **2015**, *14*, 5848–5858. [CrossRef]
30. Heath, W.H., Jr.; Wu, T.; Kwon, Y.H.; Soong, A.C.K. Multiuser MIMO in Distributed Antenna Systems With Out-of-Cell Interference. *IEEE Trans. Signal Process.* **2011**, *59*, 4885–4899. [CrossRef]
31. Lim, Y.; Chae, C.; Caire, G. Performance Analysis of Massive MIMO for Cell-Boundary Users. *IEEE Trans. Wirel. Commun.* **2015**, *14*, 6827–6842. [CrossRef]

electronics

MDPI

Article

Channel Sounding for Multi-User Massive MIMO in Distributed Antenna System Environment

Seoyoung Yu † and Jeong Woo Lee *,†

School of Electrical and Electronics Engineering, Chung-Ang University, Seoul 06974, Korea; holywillow@naver.com

* Correspondence: jwlee2@cau.ac.kr; Tel.: +82-2-820-5734

† Current address: 84 Heukseok-ro, Dongjak-gu, Seoul 06974, Korea.

Received: 20 November 2018; Accepted: 21 December 2018; Published: 1 January 2019

Abstract: We propose a generation scheme for a sounding reference signal (SRS) suitable for supporting a large number of users in massive multi-input multi-output (MIMO) system with a distributed antenna system (DAS) environment. The proposed SRS can alleviate the pilot contamination problem which occurs inherently in the multi-user system due to the limited number of orthogonal sequences. The proposed SRS sequence is generated by applying a well-chosen phase rotation to the conventional LTE/LTE-A SRS sequences without requiring an increased amount of resource usage. We also propose using the correlation-aided channel estimation algorithm as a supplemental scheme to obtain more reliable and refined channel estimation. It is shown that the proposed SRS sequence and the supplemental channel estimation scheme improve significantly the channel estimation performance in multi-user massive MIMO systems.

Keywords: massive multi-input multi-output (MIMO); distributed antenna systems (DAS); sounding reference signal (SRS); channel estimation

1. Introduction

It is expected that the amount of mobile wireless traffic in 2020 will be 1000 times higher than that of 2010 [1–3]. Along with the dramatic growth in the demand for wireless communications, performance requirements for data rate, spectral efficiency and energy efficiency are also getting higher [4–6]. To meet these growing demand and performance requirements for wireless communications, massive multi-input multi-output (m-MIMO) technology was proposed as one of the key technologies for the next generation cellular networks, known as fifth generation (5G) systems [7–10]. It is known that m-MIMO systems, whose transmitter or receiver is equipped with massive number of antennas, can improve the spectral efficiency and save energy in wireless communication systems [10–12]. As a result, m-MIMO systems have recently attracted many researchers and engineers in many aspects. Multi-user m-MIMO technology, in which a base station (BS) uses a large number of antennas to serve many pieces of user equipment (UE) simultaneously on the same time-frequency resource, is one example that is actively studied to be practically adopted in 5G systems [13–15]. The distributed antenna system (DAS) has also been considered a key technology for feasible deployment of 5G systems [16–18]. In DAS configuration, there exist a digital unit (DU) and multiple radio units (RUs) in a cell, where each RU is connected to DU via fiber optic links. The DU manages RUs in a centralized manner, by which RUs can transmit and receive signals in a cooperative manner.

The key requirement for enjoying the benefit of m-MIMO technology is to obtain the accurate channel state information (CSI) for each link at the BS, or at RUs in DAS environment. In the frequency-division duplex (FDD) approach, UEs estimate downlink (DL) channels by using DL pilot signals, or sounding reference signals (SRS), transmitted from the BS [19]. The required number of DL pilots in an FDD based approach is proportional to the number of BS antennas multiplied by the number of served UEs, which complicates the adoption of such a DL channel estimation in massive MIMO environments. Thus, the m-MIMO systems typically employ the time-division duplex (TDD) approach to estimate the DL channel. In the TDD approach, UEs estimate DL channels by sending mutually orthogonal uplink (UL) pilot signals (SRS) to BS based on the DL and UL channel reciprocity within the channel coherence interval [9]. The total number of UL pilots required in such a TDD based approach is proportional to the number of served UEs irrespective of the number of BS antennas [7,9].

For a given length of sequence, say M, we may generate at most M orthogonal sequences. Conventionally, pilot sequences are mutually orthogonal, so the maximum number of pilot sequences is limited by M. If the number of UEs in simultaneous service exceeds M, we need to reuse some or all of the already generated orthogonal sequences. This results in the so-called pilot contamination problem [20] due to the violation of orthogonality between pilot sequences. The pilot contamination mainly limits the performance improvement of m-MIMO systems [7,15]. Most of the prior works that proposed to resolve this problem have considered the use of mutually orthogonal SRS sequences. They include pilot signal coordination [21], blind channel estimation with data samples [22] and cooperative multi-cell precoding in m-MIMO systems [23]. The number of mutually orthogonal SRS sequences is mainly limited by the length of the base sequence. In Long Term Evolution (LTE)/Long Term Evolution-Advanced (LTE-A) systems, the number of orthogonal SRS sequences is 16, which is not suitable for serving a large number of users in m-MIMO environments [24,25]. Using long pilot sequences may reduce the pilot contamination, but they replace data sequences and reduce channel spectral efficiency and throughput. If pilot sequences are made too long, they may occupy even the restricted band, which must be prevented. Thus, it may be desirable to generate a larger number of SRS sequences without increasing the sequence length.

For this purpose, we propose a mechanism for generating SRS sequences with a lower level of pilot contamination which is suitable for serving a large number of UEs. In the proposed mechanism, a phase rotation is applied to the base sequence without increasing the sequence length. The resultant SRS sequences may be mutually correlated and thus still incur a pilot contamination as the number of UEs grows and exceeds the length of base sequence. This results in high channel estimation error when a linear estimation based on the orthogonality of SRS sequences is used. To resolve this problem, we propose a two-step channel estimation algorithm by which the least square (LS) estimation [26] is first applied and the minimum mean squared error (MMSE) estimation [27] is additionally applied only to the group of UEs using mutually correlated SRS sequences. The proposed *correlation-aided channel estimation* shows the improved performance in the channel estimation. It is shown that the proposed SRS combined with the supplemental channel estimation algorithm guarantees lower mean squared error (MSE) in channel estimation, which alleviates the pilot contamination problem.

The rest of the paper is organized as follows. In Section 2, we introduce the system model in consideration. In Section 3, a brief review of SRS sequences in conventional LTE/LTE-A systems is provided. We introduce the generation of proposed SRS sequences and the analysis on the resultant correlation in Section 4. In Section 5, we propose the two-step channel estimation algorithm composed of an LS estimation followed by a supplemental MMSE estimation. We also formulate and analyze the MSE obtained for SRS sequences with an LS estimation. We verify the performances in various aspects by computer simulations in Section 6 and conclude this paper in Section 7.

2. System Model

Consider a cell having a DU and R RUs each of which serves K UEs as shown in Figure 1, where each RU has N_T transmit and receive antennas. We index RU by $r \in \{0, \cdots, R-1\}$, and index the UE served by RU r as $r(k)$, $k \in \{0, \cdots, K-1\}$. We consider orthogonal frequency division multiplexing (OFDM) communications with N_c subcarriers between UE and RU. We suppose the channel reciprocity, by which the DL channel from RU to UE can be estimated by using the UL pilots sent from UEs under the constraint that the time delay from the UL channel estimation to the DL transmission is less than the coherence time of the channel [7]. Then, the UL channel estimated by sending SRS sequence from each UE to RU is used as the DL channel estimation. Let $\mathbf{s}_{r(k)} = [s_{r(k)}[0] \ \cdots \ s_{r(k)}[M-1]]^T$ denote the SRS sequence of UE $r(k)$, where the superscript T denotes a transpose of a vector. We also let $\mathbf{h}_{r(k),r'}[m] \in \mathcal{C}^{N_T \times 1}$ denote the channel gain between UE $r(k)$ and RU r' over the m-th subcarrier. Then, the channel gains corresponding to M subcarriers, $\mathbf{h}_{r(k),r'}[m]$, $m = 0, \cdots, M-1$, are estimated by using an SRS sequence.

The signal received by RU r over the subcarrier m is denoted by $\mathbf{y}_r[m] \in \mathcal{C}^{N_T \times 1}$ and obtained by

$$\mathbf{y}_r[m] = \sum_{k=0}^{K-1} \mathbf{h}_{r(k),r}[m] s_{r(k)}[m] + \sum_{r'=0, r' \neq r}^{R-1} \sum_{k=0}^{K-1} \mathbf{h}_{r'(k),r}[m] s_{r'(k)}[m] + \mathbf{n}_r[m], \tag{1}$$

where $\mathbf{n}_r[m] \in \mathcal{C}^{N_T \times 1}$ is the zero-mean additive white Gaussian noise vector with covariance matrix $\sigma_n^2 \mathbf{I}_{N_T \times N_T}$. Note that $\mathbf{h}_{r(k),r'}[m] = \beta_{r(k),r'} \mathbf{g}_{r(k),r'}[m]$, where $\beta_{r(k),r'}$ represents the large scale fading while each entry of $\mathbf{g}_{r(k),r'}[m] \in \mathcal{C}^{N_T \times 1}$ denotes the small scale fading represented by an independent and identically distributed (i.i.d.) zero mean complex Gaussian random variable with unit variance. Note that the large scale fading factor $\beta^2_{r(k),r'} = d^{-a}_{r(k),r'}$, where $d_{r(k),r'}$ is a distance between UE $r(k)$ and RU r', and a is an attenuation factor.

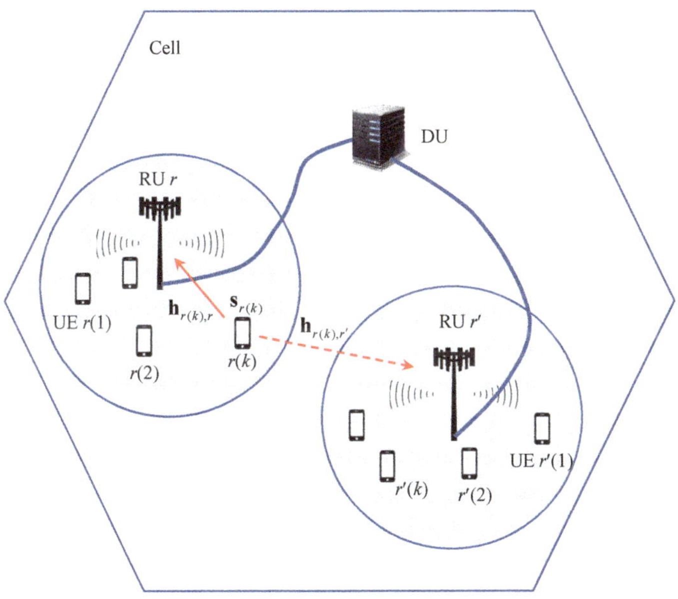

Figure 1. Multi-user m-MIMO configuration in DAS environments.

3. Conventional Channel Sounding Reference Signal in LTE/LTE-A Systems

In the following, we briefly introduce the generation of SRS in conventional LTE/LTE-A systems [24,25]. The structure of SRS symbol is illustrated in Figure 2. Basically, the SRS sequence is generated by a cyclic shift of a base sequence, which is obtained from Zadoff–Chu sequence [24,25] as presented below. Let N_{sc}^{RB} be the number of subcarriers per RB, where $N_{sc}^{RB} = 12$ in LTE/LTE-A systems, and L be the number of assigned subcarriers for SRS or sounding bandwidth. Let D be the decimation factor which is the number of SRS sequences sharing the allocated sounding bandwidth, where the length of SRS sequence is $M = L/D$. Note that L is the multiple of N_{sc}^{RB}, i.e., $L = n \cdot N_{sc}^{RB}$, $1 \leq n \leq N_{RB}^{UL}$, where N_{RB}^{UL} is the uplink system bandwidth in terms of RBs. Let L_z denote the length of Zadoff–Chu sequence used to generate the base sequence of length M, where L_z is given by the largest prime number such that $L_z < M$.

Let $x_q[m]$ denote the q-th root Zadoff–Chu sequence defined by

$$x_q[m] = \exp\left\{-j\frac{\pi q m(m+1)}{L_z}\right\}, \quad 0 \le m < L_z, \tag{2}$$

where

$$\begin{aligned} q &= \lfloor \bar{q} + 0.5 \rfloor + v(-1)^{\lfloor 2\bar{q} \rfloor}, \\ \bar{q} &= L_z(u+1)/31, \end{aligned} \tag{3}$$

with $u \in \{0, 1, \cdots, 29\}$ and $v = 0$ if $M = nN_{sc}^{RB}$, $n \le 5$ and $v = 0, 1$ if $M = nN_{sc}^{RB}$, $n \ge 6$. The base sequence $\bar{x}[m]$ is obtained by

$$\bar{x}[m] = x_q[m \mod L_z], \quad 0 \le m < M. \tag{4}$$

Note that base sequences are divided into groups, where u is the group number associated with the physical cell ID and the length of the SRS sequence, and v is the base sequence number within the group. The SRS sequence $x^{(\alpha)}[m]$ of length M is defined by applying a cyclic shift α, $\alpha \in \{0, 1 \cdots, 7\}$, to the base sequence $\bar{x}[m]$ as

$$x^{(\alpha)}[m] = e^{j2\pi\frac{\alpha}{L_c}m}\bar{x}[m], \quad 0 \le m < M, \tag{5}$$

where $L_c > 7$ to obtain distinct values of $e^{j2\pi\frac{\alpha}{L_c}}$ for different $\alpha \in \{0, 1 \cdots, 7\}$. For any α_i and α_j chosen from $\{0, \cdots, 7\}$, two sequences $\mathbf{x}^{(\alpha_i)} = [x^{(\alpha_i)}[0] \cdots x^{(\alpha_i)}[M-1]]^T$ and $\mathbf{x}^{(\alpha_j)} = [x^{(\alpha_j)}[0] \cdots x^{(\alpha_j)}[M-1]]^T$ are orthogonal if $\frac{1}{M}\mathbf{x}^{(\alpha_i)^H}\mathbf{x}^{(\alpha_j)} = \delta_{i-j}$, where $\delta_{i-j} = 1$ if $i = j$ and 0 otherwise, and the superscript H denotes a conjugate transpose of a vector. This condition reduces to $\frac{1}{M}\sum_{m=0}^{M-1} e^{j\frac{2\pi}{L_c}(\alpha_j - \alpha_i)m} = \delta_{i-j}$ by using Equation (5) and the property $|\bar{x}[m]|^2 = 1$ for all m, which is clear from Equations (2)–(4).

This condition is satisfied only if $(\alpha_j - \alpha_i)\frac{M}{L_c}$ is an even integer, or, in other words, L_c is a factor of $(\alpha_j - \alpha_i)\frac{M}{2}$, for $\alpha_i \ne \alpha_j \in \{0, \cdots, 7\}$. Consequently, L_c must be an integer that is greater than 7 and is a factor of $\frac{1}{2}M$.

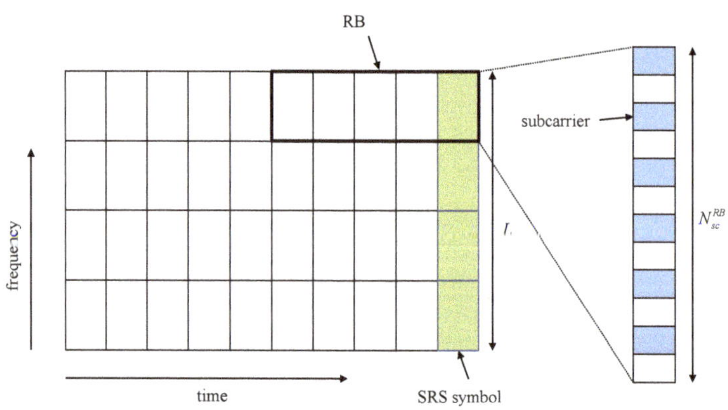

Figure 2. SRS symbol structure.

Multiple SRS sequences are defined from a single base sequence by using different values of α and the decimation factor. In LTE/LTE A, the decimation factor of two is used and the signal occupies every second subcarrier within the allocated sounding bandwidth. By using distinct SRS sequences obtained with different values of α and by using distinct sets of subcarriers as a result of decimation, multiple UEs can estimate their channel gains and can be served by RU simultaneously.

In conventional LTE/LTE-A systems using $\alpha \in \{0, \cdots, 7\}$ and the decimation factor of two, we can obtain only 16 orthogonal sequences. In the multi-user systems with higher number of UEs than 16, we need to design a larger set of SRS sequences having low cross-correlation.

4. Proposed Channel Sounding Reference Signal for Multi-User Systems

The lack of orthogonal SRS sequences may cause the pilot contamination problem. The easy way to alleviate this is using longer SRS sequences or wider sounding bandwidth. However, this approach may result in the degraded performance of channel estimation in the frequency selective environment and the lower spectral efficiency and throughput because SRS sequences replace data sequences. Moreover, if the sounding bandwidth is too wide, SRS sequences may occupy the restricted band, which must be prohibited. Thus, we aim to generate a new set of SRS sequences showing reduced pilot contamination without increasing the sounding bandwidth or sequence length. We apply phase rotation to the LTE/LTE-A SRS sequences, which is introduced in Section 3 to generate a new SRS sequence as

$$x^{(\alpha,s,p)}[m] \triangleq e^{j2\pi\frac{s}{M}m} e^{j2\pi\frac{p}{L_p}m} x^{(\alpha)}[m] = e^{j2\pi\left(\frac{s}{M}+\frac{p}{L_p}+\frac{\alpha}{L_c}\right)m}\bar{x}[m], \quad 0 \le m < M, \tag{6}$$

where $0 \le s < \frac{M}{L_c}$ and $0 \le p < L_p$, and the last equality comes from Equation (5). Note that L_c is an integer greater than 7 which can divide $\frac{1}{2}M$ as introduced in Section 3 and L_p is a prime number which is smaller than L_c. For a given SRS sequence length M, we may generate up to M orthogonal sequences. However, by using Equation (5), we can generate only eight orthogonal sequences with varying $\alpha = 0, \cdots, 7$. Thus, we use the phase rotation $e^{j2\pi\frac{s}{M}m}$, $0 \le s < \frac{M}{L_c}$, together with $e^{j2\pi\frac{\alpha}{L_c}m}$, $0 \le \alpha < L_c$, to generate M orthogonal SRS sequences without incurring pilot contamination. If the number of UEs exceeds M, we need to generate extra SRS sequences instead of reusing already generated ones. For this purpose, we apply additional phase rotation $e^{j2\pi\frac{p}{L_p}m}$, $0 \le p < L_p$, where L_p needs to be coprime with M and thus be also coprime with L_c in order to make the resultant sequences distinct from already generated M sequences. We empirically found that a prime number L_p smaller than L_c results in a good performance.

By Equations (2)–(4), we can rewrite Equation (6) as

$$x^{(\alpha,s,p)}[m] = \begin{cases} \exp\left\{j2\pi\left(\left(\frac{s}{M}+\frac{p}{L_p}+\frac{\alpha}{L_c}\right)m - \frac{qm(m+1)}{2L_z}\right)\right\}, & \text{for } 0 \le m < L_z, \\ \exp\left\{j2\pi\left(\left(\frac{s}{M}+\frac{p}{L_p}+\frac{\alpha}{L_c}\right)m - \frac{q(m-L_z)(m-L_z+1)}{2L_z}\right)\right\}, & \text{for } L_z \le m < M. \end{cases} \tag{7}$$

The correlation of SRS sequences $\mathbf{s}_{r(k)}$ and $\mathbf{s}_{r'(j)}$ is defined and expanded as

$$C_{r(k),r'(j)} \triangleq \frac{1}{M}\mathbf{s}_{r(k)}{}^{H}\mathbf{s}_{r'(j)} = \frac{1}{M}\sum_{m=0}^{M-1} s^{*}_{r(k)}[m]s_{r'(j)}[m] = \frac{1}{M}\sum_{m=0}^{M-1} e^{-j2\pi\left(\frac{s-s'}{M}+\frac{p-p'}{L_p}+\frac{\alpha-\alpha'}{L_c}\right)m}, \tag{8}$$

where $s_{r(k)}[m] = x^{(\alpha,s,p)}[m]$, $s_{r'(j)}[m] = x^{(\alpha',s',p')}[m]$ and the superscript $*$ represents the complex conjugate of a complex variable. The detailed derivation of Equation (8) is provided in Appendix A. It is clear that $C_{r(k),r'(j)}$ obtained with $s = s'$, $p = p'$ and $\alpha = \alpha'$ corresponds to the auto-correlation of $\mathbf{s}_{r(k)}$ because $\mathbf{s}_{r'(j)} = \mathbf{s}_{r(k)}$, where $C_{r(k),r(k)} = \frac{1}{M}\sum_{m=0}^{M-1} e^{j\cdot 0} = 1$. The cross-correlation $C_{r(k),r'(j)} = 0$ if $p = p'$ and either $s \neq s'$ or $\alpha \neq \alpha'$, while $C_{r(k),r'(j)} \neq 0$ if $p \neq p'$ by the following reason.

Consider $f(\phi) = \frac{1}{M}\sum_{m=0}^{M-1} e^{j2\pi\phi m}$, where $f(\phi) = 0$ if ϕM is a nonzero integer, and $f(\phi) \neq 0$, otherwise. If $p = p'$, the rightmost side of Equation (8) becomes $f\left(\frac{s'-s}{M}+\frac{\alpha'-\alpha}{L_c}\right)$. If $s \neq s'$ or $\alpha \neq \alpha'$, $\left(\frac{s'-s}{M}+\frac{\alpha'-\alpha}{L_c}\right)M$ is a nonzero integer because M is a multiple of L_c as introduced in Section 3, and thus $C_{r(k),r'(j)} = f\left(\frac{s'-s}{M}+\frac{\alpha'-\alpha}{L_c}\right) = 0$. On the other hand, if $p \neq p'$, $\left(\frac{s'-s}{M}+\frac{p'-p}{L_p}+\frac{\alpha'-\alpha}{L_c}\right)M$ cannot be a nonzero integer because $(p'-p)M$ is not divisible by L_p, and, thus $C_{r(k),r'(j)} = f(\frac{s'-s}{M} + \frac{p'-p}{L_p} + \frac{\alpha'-\alpha}{L_c}) \neq 0$, where L_p and M are coprime and $|p-p'| < L_p$.

Consequently, for each p, we can generate a set of M orthogonal SRS sequences, where $0 \leq s < \frac{M}{L_c}$ and $0 \leq \alpha < L_c$ enables the generation of M orthogonal sequences. Applying the phase rotation $e^{j2\pi \frac{p}{L_p} m}$, $0 \leq p < L_p$, in Equation (7) enables to obtain L_p sets of M orthogonal sequences. Any two sequences obtained with different values of p are mutually correlated.

Distinct SRS sequences generated with different α, s and p by Equation (7) are assigned to different UEs. The first set of M orthogonal sequences generated with $p = 0$ are assigned to the first M UEs. Then, the next set of M orthogonal sequences generated with $p = 1$ are assigned to the next M UEs. This procedure is repeated by increasing p until all UEs are assigned SRS sequences.

5. Channel Estimation

5.1. Least Square Channel Estimation

Let $\hat{\mathbf{h}}_{r(k),r}[m]$ denote the estimation of the channel gain $\mathbf{h}_{r(k),r}[m]$. We apply the LS estimation algorithm [26] to the received signal of SRS sequence for the channel estimation. In this process, we assume a block fading with length M, by which the channel is considered invariant over consecutive M subcarriers. Then, the estimation of channel gain at the m-th subcarrier in a fading block, $\hat{\mathbf{h}}_{r(k),r}[m]$, can also be denoted by $\hat{\mathbf{h}}_{r(k),r}$ and obtained as

$$\hat{\mathbf{h}}_{r(k),r}[m] = \hat{\mathbf{h}}_{r(k),r} = \frac{1}{M} \sum_{i=0}^{M-1} s^*_{r(k)}[i] \mathbf{y}_r[i]. \tag{9}$$

By using Equation (1), we can rewrite Equation (9) as

$$\begin{aligned} \hat{\mathbf{h}}_{r(k),r} &= \frac{1}{M} \sum_{i=0}^{M-1} s^*_{r(k)}[i] \left\{ \sum_{j=0}^{K-1} \mathbf{h}_{r(j),r}[i] s_{r(j)}[i] + \sum_{r'=0,r'\neq r}^{R-1} \sum_{j=0}^{K-1} \mathbf{h}_{r'(j),r}[i] s_{r'(j)}[i] + \mathbf{n}_r[i] \right\} \\ &= \frac{1}{M} \sum_{i=0}^{M-1} \left\{ \mathbf{h}_{r(k),r}[i] + \sum_{j=0,j\neq k}^{K-1} \mathbf{h}_{r(j),r}[i] s^*_{r(k)}[i] s_{r(j)}[i] + \sum_{r'=0,r'\neq r}^{R-1} \sum_{j=0}^{K-1} \mathbf{h}_{r'(j),r}[i] s^*_{r(k)}[i] s_{r'(j)}[i] + s^*_{r(k)}[i] \mathbf{n}_r[i] \right\}, \end{aligned} \tag{10}$$

where $s^*_{r(k)}[i] s_{r(k)}[i] = |s_{r(k)}[i]|^2 = 1$ is used. If the channel is actually block faded with length M, the channel gain $\mathbf{h}_{r(k),r}[m]$ is identical for all $m = 0, \cdots, M-1$ and thus, we can represent the channel gain by $\mathbf{h}_{r(k),r}$. Then, Equation (10) can be rewritten as

$$\hat{\mathbf{h}}_{r(k),r} = \mathbf{h}_{r(k),r} + \sum_{j=0,j\neq k}^{K-1} \mathbf{h}_{r(j),r} C_{r(k),r(j)} + \sum_{r'=0,r'\neq r}^{R-1} \sum_{j=0}^{K-1} \mathbf{h}_{r'(j),r} C_{r(k),r'(j)} + \frac{1}{M} \sum_{i=0}^{M-1} s^*_{r(k)}[i] \mathbf{n}_r[i]. \tag{11}$$

In case that all SRS sequences of UEs in the cell are mutually orthogonal, we have zero cross-correlation between any two SRS sequences so that Equation (11) is simplified as

$$\hat{\mathbf{h}}_{r(k),r} = \mathbf{h}_{r(k),r} + \frac{1}{M} \sum_{i=0}^{M-1} s^*_{r(k)}[i] \mathbf{n}_r[i]. \tag{12}$$

We define the normalized mean squared error (MSE) of the channel estimation between UE $r(k)$ and RU r at subcarrier m as

$$\sigma^2_{MSE,r(k)}[m] = \frac{E\|\hat{\mathbf{h}}_{r(k),r} - \mathbf{h}_{r(k),r}[m]\|^2}{\beta^2_{r(k),r} N_T}. \tag{13}$$

Then, we define the average MSE by

$$\bar{\sigma}^2_{MSE} = \frac{1}{KRM} \sum_{k=0}^{K-1} \sum_{r=0}^{R-1} \sum_{m=0}^{M-1} \sigma^2_{MSE,r(k)}[m] \tag{14}$$

and use this as the performance metric of channel estimation. Note that, if the channel is actually block faded with length M, the normalized MSE can be denoted by $\sigma^2_{MSE,r(k)}$ and Equation (14) can be simplified as $\bar{\sigma}^2_{MSE} = \frac{1}{KR} \sum_{k=0}^{K-1} \sum_{r=0}^{R-1} \sigma^2_{MSE,r(k)}$. Under the assumption of block fading with length M, the average MSE can be obtained from Equations (11), (13) and (14) as

$$\begin{aligned} \bar{\sigma}^2_{MSE} &= \frac{1}{KR} \sum_{k=0}^{K-1} \sum_{r=0}^{R-1} \frac{E\|\hat{\mathbf{h}}_{r(k),r} - \mathbf{h}_{r(k),r}\|^2}{\beta^2_{r(k),r} N_T} \\ &= \frac{1}{KR} \sum_{k=0}^{K-1} \sum_{r=0}^{R-1} \frac{1}{\beta^2_{r(k),r}} \left\{ \sum_{j=0,j\neq k}^{K-1} \beta^2_{r(j),r} |C_{r(k),r(j)}|^2 + \sum_{r'=0,r'\neq r}^{R-1} \sum_{j=0}^{K-1} \beta^2_{r'(j),r} |C_{r(k),r'(j)}|^2 + \frac{1}{M}\sigma_n^2 \right\}, \end{aligned} \tag{15}$$

where the detailed derivation is given in Appendix B.

Let us predict analytically the average MSE under the block fading environment. The large scale fading factors are assumed to be $\beta_{r(k),r} = \beta_1$ and $\beta_{r'(k),r} = \beta_2$ for all k, r and r', $r' \neq r$, which means that the large scale fading between UE and serving RU is represented by β_1 while the large scale fading between UE and other neighboring RUs are represented by β_2. We suppose $\beta_2 < \beta_1$ because UE is usually served by a nearly located RU. Then, Equation (15) can be rewritten as

$$\bar{\sigma}^2_{MSE} = \frac{1}{KR} \sum_{k=0}^{K-1} \sum_{r=0}^{R-1} \left\{ \sum_{j=0,j\neq k}^{K-1} |C_{r(k),r(j)}|^2 + \frac{\beta_2^2}{\beta_1^2} \sum_{r'=0,r'\neq r}^{R-1} \sum_{j=0}^{K-1} |C_{r(k),r'(j)}|^2 + \frac{\sigma_n^2}{\beta_1^2 M} \right\}. \tag{16}$$

We consider the reuse of M orthogonal SRS sequences to UEs repeatedly without applying the phase rotation $e^{j2\pi \frac{p}{L_p} m}$ in Equation (7). We suppose that each RU in the cell serves the equal number of UEs and distinct set of $\frac{M}{R}$ orthogonal sequences are assigned repeatedly to UEs in each RU. Then, we obtain a correlation matrix $\mathbf{C} \in \mathcal{C}^{KR \times KR}$ of SRS sequences whose (i, j)-th entry is defined by

$$\mathbf{C}_{ij} = \begin{cases} 1, & \text{if } |i-j| = kM, \quad k = 0, \cdots, \lfloor KR/M \rfloor, \\ 0, & \text{else.} \end{cases} \tag{17}$$

Every other $\frac{M}{R}$ column and row of $\mathbf{C}$ correspond to UEs in the same RU. Due to the reuse of $\frac{M}{R}$ orthogonal sequences for each RU, SRS sequences used for different RUs are always mutually orthogonal. It follows that $C_{r(k),r'(j)} = 0$ for all k, j and $r \neq r'$ and Equation (16) becomes

$$\bar{\sigma}^2_{MSE} = \frac{1}{KR} \sum_{k=0}^{K-1} \sum_{r=0}^{R-1} \sum_{j=0,j\neq k}^{K-1} |C_{r(k),r(j)}|^2 + \frac{\sigma_n^2}{\beta_1^2 M}. \tag{18}$$

From Equation (17) and the allocation rule of orthogonal sequences to UEs introduced above, we obtain

$$\frac{1}{KR} \sum_{k=0}^{K-1} \sum_{r=0}^{R-1} \sum_{j=0,j\neq k}^{K-1} |C_{r(k),r(j)}|^2 = \begin{cases} 0, & \text{if } KR \leq M, \\ \frac{1}{KR} \sum_{i=1}^{\lfloor \frac{KR}{M} \rfloor} 2(KR - iM), & \text{if } KR > M. \end{cases} \tag{19}$$

Consequently, we can rewrite Equation (18) as

$$\bar{\sigma}^2_{MSE} = \begin{cases} \frac{\sigma_n^2}{\beta_1^2 M}, & \text{if } KR \leq M, \\ \lfloor \frac{KR}{M} \rfloor \left(2 - \frac{M}{KR}\left(\lfloor \frac{KR}{M} \rfloor + 1\right)\right) + \frac{\sigma_n^2}{\beta_1^2 M}, & \text{if } KR > M. \end{cases} \tag{20}$$

Note that $\bar{\sigma}^2_{MSE}$ is independent of β_2^2 while it depends on β_1^2. Recall that β_2^2 is determined by the distance between a UE and its neighboring RU. Thus, the performance of channel estimation obtained by reusing orthogonal sequences repeatedly is not affected by how far neighboring RUs are located from UE. It is also clear from Equation (20) that the average MSE decreases as M grows. Since the number of orthogonal sequences is limited for a given sequence length, the number of UEs served simultaneously in a cell is limited. In the multi-user systems, a larger number of UEs must be accommodated, so we may need to use non-orthogonal sequences for channel sounding.

5.2. Supplemental Correlation-Aided Channel Estimation

As the number of UEs grows in the cell, it is impossible to assign orthogonal SRS sequences to all UEs even by using the proposed SRS sequences because the number of orthogonal sequences is limited to M. Thus, we propose using the supplemental channel estimation scheme to enhance the channel estimation performance even with correlated SRS sequences. Suppose DU knows SRS sequences and their cross-correlations. We rewrite Equation (11) as

$$\hat{\mathbf{h}}_{r(k),r} = \sum_{r'=0}^{R-1}\sum_{j=0}^{K-1} \mathbf{h}_{r'(j),r} C_{r(k),r'(j)} + \frac{1}{M}\sum_{i=0}^{M-1} s^*_{r(k)}[i]\mathbf{n}_r[i], \tag{21}$$

where $0 \le r < R$ and $0 \le k < K$. This can be expressed by using matrix forms as

$$\hat{\mathbf{H}} = \mathbf{HC} + \mathbf{N}, \tag{22}$$

where $\hat{\mathbf{H}} \in \mathcal{C}^{N_T \times KR}$, $\mathbf{H} \in \mathcal{C}^{N_T \times KR}$ and $\mathbf{N} \in \mathcal{C}^{N_T \times KR}$ are matrices whose columns are $\hat{\mathbf{h}}_{r(k),r}$, $\mathbf{h}_{r(k),r}$ and $\mathbf{n}_{r(k),r}$, respectively, $0 \le r < R$ and $0 \le k < K$. We let $\mathbf{C} \in \mathcal{C}^{KR \times KR}$ denote a correlation matrix whose entries are $C_{r(k),r'(j)}$, where $0 \le r, r' < R$ and $0 \le k, j < K$. We define two classes of UEs. The first one is the set of UEs that are assigned mutually orthogonal SRS sequences. UEs assigned mutually correlated SRS sequences compose the second class. We reorder $\hat{\mathbf{H}}$, $\mathbf{H}$, $\mathbf{C}$ and $\mathbf{N}$ and partition them as $\hat{\mathbf{H}} = \begin{bmatrix}\hat{\mathbf{H}}_u & \hat{\mathbf{H}}_c\end{bmatrix}$, $\mathbf{H} = \begin{bmatrix}\mathbf{H}_u & \mathbf{H}_c\end{bmatrix}$, $\mathbf{C} = \begin{bmatrix}\mathbf{I} & \mathbf{0}\\ \mathbf{0} & \mathbf{A}\end{bmatrix}$ and $\mathbf{N} = \begin{bmatrix}\mathbf{N}_u & \mathbf{N}_c\end{bmatrix}$, where submatrices with subscripts u and c correspond to the first and the second class of UEs, respectively, and $\mathbf{A}$ is a non-diagonal matrix. Then, Equation (22) can be rewritten in partitioned forms as

$$\begin{bmatrix}\hat{\mathbf{H}}_u & \hat{\mathbf{H}}_c\end{bmatrix} = \begin{bmatrix}\mathbf{H}_u & \mathbf{H}_c\end{bmatrix}\begin{bmatrix}\mathbf{I} & \mathbf{0}\\ \mathbf{0} & \mathbf{A}\end{bmatrix} + \begin{bmatrix}\mathbf{N}_u & \mathbf{N}_c\end{bmatrix}, \tag{23}$$

which implies

$$\hat{\mathbf{H}}_u = \mathbf{H}_u + \mathbf{N}_u, \tag{24}$$

$$\hat{\mathbf{H}}_c = \mathbf{H}_c\mathbf{A} + \mathbf{N}_c. \tag{25}$$

It is clear from Equation (24) that $\hat{\mathbf{h}}_{r(k),r} = \mathbf{h}_{r(k),r} + \mathbf{n}_{r(k),r}$ for the first class of UEs. So, $\hat{\mathbf{h}}_{r(k),r}$ is considered to estimate sufficiently the actual channel $\mathbf{h}_{r(k),r}$ of UE $r(k)$ in the first class. On the other hand, as can be seen from Equation (25), for the second class of UEs, $\hat{\mathbf{h}}_{r(k),r}$ includes the linear combination of other UEs' channels as well. Thus, we need a supplemental procedure to obtain the more reliable channel estimations for UEs in the second class. For this purpose, we propose the *correlation-aided channel estimation* which applies the MMSE algorithm [27] to the estimated channels for the second class of UEs as the supplemental procedure. We obtain the refined channel estimation $\tilde{\hat{\mathbf{H}}}_c$ for UEs in the second class by multiplying the MMSE nulling matrix $\mathbf{W}$ to $\hat{\mathbf{H}}_c$ as

$$\tilde{\hat{\mathbf{H}}}_c = \hat{\mathbf{H}}_c\mathbf{W}, \tag{26}$$

where

$$\mathbf{W} = \mathbf{A}^H \left(\mathbf{A}\mathbf{A}^H + \mathbf{B}^{-1}\mathbf{I}\right)^{-1} \tag{27}$$

and $\mathbf{B}$ is a diagonal matrix whose diagonal entries are $\beta^2_{r(k),r}$ corresponding to UEs in the second class. Then, after refinement of $\hat{\mathbf{H}}_c$ to $\tilde{\hat{\mathbf{H}}}_c$ by Equation (26), the channel estimation $\hat{\mathbf{H}}$ is replaced by $\hat{\mathbf{H}} = \left[\hat{\mathbf{H}}_u \quad \tilde{\hat{\mathbf{H}}}_c\right]$.

In summary, first, we apply the LS estimation to obtain $\hat{\mathbf{h}}_{r(k),r}$ for all UEs. Next, we define the first and the second classes of UEs based on the correlation matrix $\mathbf{C}$. The first class is defined by UEs whose SRS sequences are mutually orthogonal. The remaining UEs form the second class. Then, for the second class of UEs, we apply additionally the MMSE algorithm to the channel estimation obtained by LS scheme as the supplemental procedure. Finally, we use the result of LS estimation for the first class and the result of supplement estimation for the second class as the channel estimation. If the number of UEs is less than or equal to the number of mutually orthogonal SRS sequences, we only need to perform a LS estimation as introduced in Equation (9). Otherwise, we need to apply the proposed correlation-aided supplemental estimation after performing the LS estimation.

6. Numerical Results

We evaluate the performances of the proposed SRS and the proposed supplemental channel estimation algorithm in terms of MSE in the multi-user m-MIMO DAS environment. For comparison, the MSE obtained by using conventional LTE/LTE-A SRS scheme is also shown. As the proposed scheme, we consider the following three types:

(a) using repeatedly M orthogonal sequences generated by varying α and s without applying the phase rotation $e^{j2\pi \frac{p}{L_p} m}$ in Equation (7),

(b) using the phase rotation $e^{j2\pi \frac{p}{L_p} m}$ together with varying α and s for SRS generation, but not using the supplemental correlation-aided channel estimation,

(c) using the phase rotation $e^{j2\pi \frac{p}{L_p} m}$ together with varying α and s for SRS generation, and the supplemental correlation-aided channel estimation as well.

We consider the large scale fading factor between UEs and serving RU as β_1 and the large scale fading factor between UEs and other neighboring RUs as β_2, i.e., $\beta_{r(k),r} = \beta_1$ and $\beta_{r(k),r'} = \beta_2$ for all k, r and $r' \neq r$. We let $\beta_2/\beta_1 = 0.7692$, which comes from the assumption that the distance of UE from serving RU is 1.3 times shorter than the distance from other neighboring RUs, where the attenuation factor is assumed to be $a = 2$. We let each RU serve the same number of UEs. We choose $L_c = 8$ in Equations (5) and (7), and $L_p = 3$ when applying the phase rotation $e^{j2\pi \frac{p}{L_p} m}$ in types (b) and (c) of the proposed scheme. We consider a block fading channel, International Telecommunication Union Radiocommunication Sector (ITU-R) Ped A and ITU-R Veh A channels [28], where ITU-R Ped A [29] and ITU-R Veh A channels are examples of frequency selective channels. The simulation parameters used for performance evaluation are listed in Table 1.

We depict the average MSE of channel estimation with respect to the total number of UEs in the cell obtained for block fading channel, ITU-R Ped A channel and ITU-R Veh A channel in Figures 3–5, respectively. In each figure, we plot the average MSE obtained by using conventional SRS in LTE/LTE-A systems and the proposed scheme of three types.

We also plot the average MSE obtained by applying supplemental MMSE estimation to all UEs instead of applying MMSE selectively only to UEs belonging to the second class, where the phase rotation $e^{j2\pi \frac{p}{L_p} m}$ is applied and the corresponding curves are marked with 'nonselective supplemental MMSE.' In Figure 3, we include the average MSE predicted analytically by Equation (20) for the proposed scheme of type (a) over block fading channel.

Table 1. Parameters used in performance evaluation.

Parameters	Values
number of subcarriers (N_c)	1024
number of subcarriers per RB (N_{sc}^{RB})	12
SRS length (M)	48 (proposed scheme)
	24 (LTE/LTE-A with a decimation factor of 2)
sampling time	0.1 μs
average SNR (β_1^2/σ_n^2)	0 dB
channel model	block fading, ITU-R Ped A, ITU-R Veh A
number of RUs (R)	4
number of RU antennas (N_T)	8

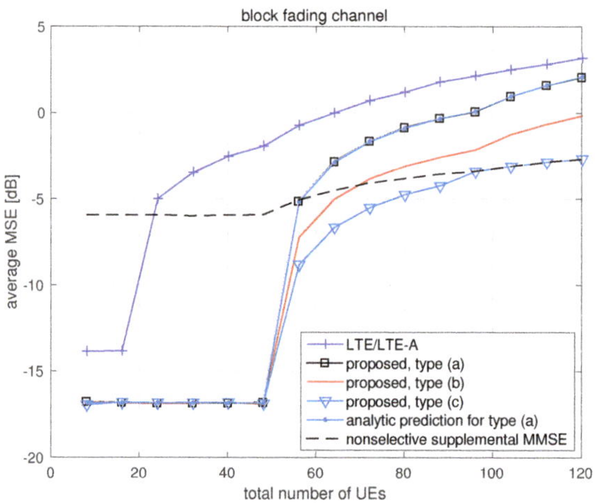

Figure 3. Average MSE of channel estimation with respect to the total number of UEs over block fading channel.

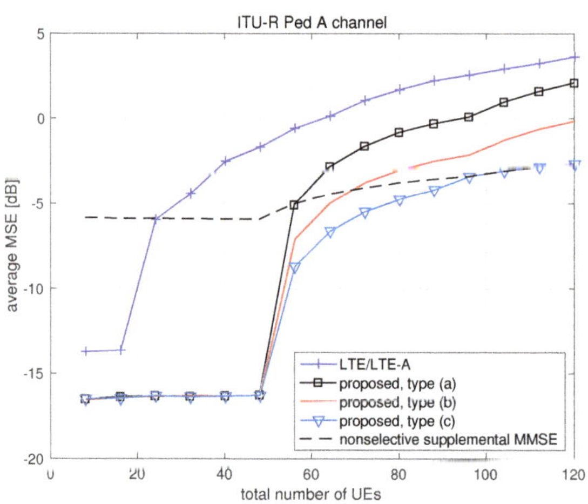

Figure 4. Average MSE of channel estimation with respect to the total number of UEs over ITU-R Ped A channel.

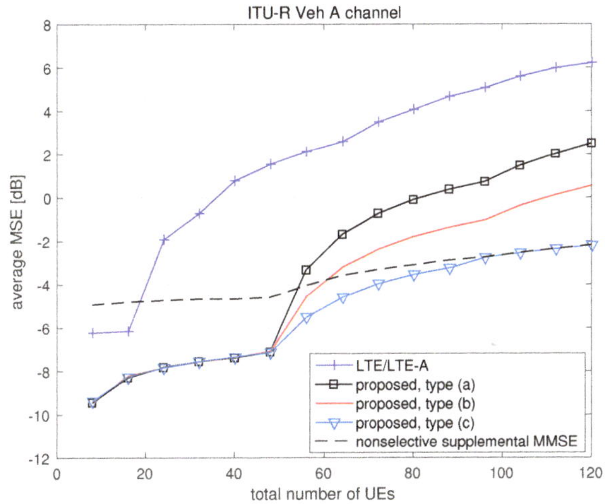

Figure 5. Average MSE of channel estimation with respect to the total number of UEs over ITU-R Veh A channel.

It is observed that the channel estimation MSE obtained by using any SRS scheme rapidly increases when the number of UEs in service exceeds the number of orthogonal SRS sequences, where the conventional LTE/LTE-A scheme and the proposed scheme generate 16 and M (=48) orthogonal SRS sequences, respectively. For more than 48 UEs, the proposed scheme also provides much lower channel estimation MSE than LTE/LTE-A scheme. Even reusing repeatedly M (=48) orthogonal SRS sequences generated in type (a) can lower significantly the channel estimation MSE for any number of UEs. It is observed that applying phase rotation $e^{j2\pi \frac{p}{L_p} m}$ to SRS generation further lowers the channel estimation MSE, and the use of supplemental selective MMSE estimation can even further improve the channel estimation performance at the cost of increasing complexity. Note that the computational complexity of supplemental selective MMSE estimation is, in general, $O(n^3)$ because it requires the matrix inversion, where n is the dimension of a square matrix $\mathbf{A}$. Considering, as a reference, the case that M orthogonal sequences are repeatedly used to generate SRS sequences, the dimension of $\mathbf{A}$ is determined from Equation (17) as $n = \min(KR, 2\max(0, KR - M))$. Applying supplemental MMSE estimation nonselectively to all UEs degrades the performance when the number of UEs is not high enough. In Figure 3, the analytic prediction of MSE for block fading channel is observed to match the numerical result very well. It is clear that the proposed scheme of type (c) shows the best channel estimation performance in terms of MSE for all numbers of UEs. The MSE gain of the proposed scheme of type (c) over LTE/LTE-A system is about 6 dB in block fading and ITU-R Ped A channels, and about 8 dB in ITU-R Veh A channel when serving 120 UEs through four RUs. The performance gain achieved by using the proposed scheme can be observed in block fading channel as well as frequency selective channel.

Conventional LTE/LTE-A systems may not effectively employ m-MIMO transmission for the service of a large number of UEs mainly due to the pilot contamination problem. However, the proposed SRS and the supplemental channel estimation can relieve the pilot contamination problem in a significant level. It is expected that the proposed scheme can be effectively adopted in multi-user m-MIMO systems.

7. Conclusions

In this paper, we proposed a generation of SRS sequences resulting in improved channel estimation performance without increasing the sequence length. The proposed SRS can easily be generated by imposing a phase rotation to the base sequence. Even though the proposed SRS sequences are mutually correlated when the number of UEs is higher than the sequence length, it lowers significantly the MSE of channel estimation and thus alleviates the pilot contamination problem. We also proposed a supplemental correlation-aided channel estimation scheme to further improve the channel estimation performance of multi-user m-MIMO technology in DAS environments. It is expected that the proposed SRS scheme and the supplemental channel estimation scheme can be effectively adopted in m-MIMO systems.

Author Contributions: Writing, J.W.L.; Software, S.Y.; Conceptualization, J.W.L.; Supervision, J.W.L.

Funding: This research was supported by the Chung-Ang University Graduate Research Scholarship in 2017 and by the Basic Science Research Program through the National Research Foundation of Korea funded by the Ministry of Education (NRF-2016R1D1A1B03933174).

Conflicts of Interest: The authors declare no conflict of interest.

Appendix A. Derivation of Equation (8)

With $s_{r(k)}[m] = x^{(\alpha,s,p)}[m]$ and $s_{r'(j)}[m] = x^{(\alpha',s',p')}[m]$, we obtain Equation (8) by using Equation (7) as

$$
\begin{aligned}
C_{r(k),r'(j)} &\triangleq \frac{1}{M}\mathbf{s}_{r(k)}{}^{H}\mathbf{s}_{r'(j)} = \frac{1}{M}\sum_{m=0}^{M-1} s^{*}_{r(k)}[m]s_{r'(j)}[m] \\
&= \frac{1}{M}\sum_{m=0}^{L_z-1} e^{-j2\pi\left(\left(\frac{s}{M}+\frac{p}{L_p}+\frac{\alpha}{L_c}\right)m-\frac{qm(m+1)}{2L_z}\right)} e^{j2\pi\left(\left(\frac{s'}{M}+\frac{p'}{L_p}+\frac{\alpha'}{L_c}\right)m-\frac{qm(m+1)}{2L_z}\right)} \\
&+ \frac{1}{M}\sum_{m=L_z}^{M-1} e^{-j2\pi\left(\left(\frac{s}{M}+\frac{p}{L_p}+\frac{\alpha}{L_c}\right)m-\frac{q(m-L_z)(m-L_z+1)}{2L_z}\right)} e^{j2\pi\left(\left(\frac{s'}{M}+\frac{p'}{L_p}+\frac{\alpha'}{L_c}\right)m-\frac{q(m-L_z)(m-L_z+1)}{2L_z}\right)} \\
&= \frac{1}{M}\sum_{m=0}^{M-1} e^{-j2\pi\left(\frac{s-s'}{M}+\frac{p-p'}{L_p}+\frac{\alpha-\alpha'}{L_c}\right)m}.
\end{aligned}
$$

Appendix B. Derivation of Equation (15)

By using Equations (11), (13) and (14), we obtain Equation (15) as

$$
\begin{aligned}
\tilde{\sigma}^2_{MSE} &= \frac{1}{KR}\sum_{k=0}^{K-1}\sum_{r=0}^{R-1}\frac{E\|\hat{\mathbf{h}}_{r(k),r}-\mathbf{h}_{r(k),r}\|^2}{\beta^2_{r(k),r}N_T} \\
&= \frac{1}{KRN_T}\sum_{k=0}^{K-1}\sum_{r=0}^{R-1}\frac{1}{\beta^2_{r(k),r}}E\left\|\sum_{j=0,j\neq k}^{K-1}\mathbf{h}_{r(j),r}C_{r(k),r(j)}+\sum_{r'=0,r'\neq r}^{R-1}\sum_{j=0}^{K-1}\mathbf{h}_{r'(j),r}C_{r(k),r'(j)}+\frac{1}{M}\sum_{i=0}^{M-1}s^{*}_{r(k)}[i]\mathbf{n}_r[i]\right\|^2 \\
&= \frac{1}{KRN_T}\sum_{k=0}^{K-1}\sum_{r=0}^{R-1}\frac{1}{\beta^2_{r(k),r}}\left\{\sum_{j=0,j\neq k}^{K-1}E\|\mathbf{h}_{r(j),r}\|^2|C_{r(k),r(j)}|^2+\sum_{r'=0,r'\neq r}^{R-1}\sum_{j=0}^{K-1}E\|\mathbf{h}_{r'(j),r}\|^2|C_{r(k),r'(j)}|^2\right. \\
&\qquad\left.+\frac{1}{M^2}\sum_{i=0}^{M-1}E\|\mathbf{n}_r[i]\|^2\right\} \\
&= \frac{1}{KRN_T}\sum_{k=0}^{K-1}\sum_{r=0}^{R-1}\frac{1}{\beta^2_{r(k),r}}\left\{\sum_{j=0,j\neq k}^{K-1}N_T\beta^2_{r(j),r}|C_{r(k),r(j)}|^2+\sum_{r'=0,r'\neq r}^{R-1}\sum_{j=0}^{K-1}N_T\beta^2_{r'(j),r}|C_{r(k),r'(j)}|^2+\frac{1}{M^2}\sum_{i=0}^{M-1}N_T\sigma_n^2\right\} \\
&= \frac{1}{KR}\sum_{k=0}^{K-1}\sum_{r=0}^{R-1}\frac{1}{\beta^2_{r(k),r}}\left\{\sum_{j=0,j\neq k}^{K-1}\beta^2_{r(j),r}|C_{r(k),r(j)}|^2+\sum_{r'=0,r'\neq r}^{R-1}\sum_{j=0}^{K-1}\beta^2_{r'(j),r}|C_{r(k),r'(j)}|^2+\frac{1}{M}\sigma_n^2\right\}.
\end{aligned}
$$

References

1. Ge, X.; Tu, S.; Mao, G.; Wang, C.-X.; Han, T. 5G ultra-dense cellular networks. *IEEE Wirel. Commun.* **2016**, *23*, 72–79. [CrossRef]
2. Ge, X.; Cheng, H.; Guizani, M.; Han, T. 5G wireless backhaul networks: Challenges and research advances. *IEEE Netw.* **2014**, *28*, 6–11. [CrossRef]
3. Nakamura, T.; Nagata, S.; Benjebbour, A.; Kishiyama, Y.; Hai, T.; Xiaodong, S.; Ning, Y.; Nan, L. Trends in small cell enhancements in LTE Advanced. *IEEE Commun. Mag.* **2013**, *51*, 98–105. [CrossRef]
4. Chin, W.H.; Fan, Z.; Haines, R. Emerging technologies and research challenges for 5G wireless networks. *IEEE Commun. Mag.* **2014**, *21*, 106–112. [CrossRef]
5. Angelis, C.T.; Chronopoulos, S.K. System performance of an LTE MIMO downlink in various fading environments. In Proceedings of the Ambient Media and Systems, Berlin, Germany, 27–28 January 2011; pp. 36–43._5. [CrossRef]
6. Votis, C.; Christofilakis, V.; Raptis, V.; Tatsis, G.; Chronopoulos, S.K.; Kostarakis, P. Design and analysis of a multiple-output transmitter based on DDS architecture for modern wireless communications. *AIP Conf. Proc.* **2010**, *1203*, 421–426. [CrossRef]
7. Marzetta, T. Noncooperative cellular wireless with unlimited numbers of base station antennas. *IEEE Trans. Wirel. Comm.* **2010**, *9*, 3590–3600. [CrossRef]
8. Rusek, F.; Persson, D.; Lau, B.; Larsson, E.; Marzetta, T.; Edfors, O.; Tufvesson, F. Scaling up MIMO: Opportunities and challenges with very large arrays. *IEEE Signal Process. Mag.* **2013**, *30*, 40–60. [CrossRef]
9. Larsson, E.G.; Tufvesson, F.; Edfors, O.; Marzetta, T.L. Massive MIMO for next generation wireless systems. *IEEE Commun. Mag.* **2014**, *52*, 186–195. [CrossRef]
10. Pitarokoilis, A.; Mohammed, S.K.; Larsson, E.G. On the optimality of single-carrier transmission in large-scale antenna systems. *IEEE Wirel. Commun. Lett.* **2012**. *1*, 276–279. [CrossRef]
11. Lu, L.; Li, G.Y.; Swindlehurst, A.L.; Ashikhmin, A.; Zhang, R. An overview of massive MIMO: Benefits and challenges. *IEEE J. Sel. Top. Signal Process.* **2013**, *8*, 742–758. [CrossRef]
12. Bjornson, E.; Larsson, E.G.; Marzetta, T.L. Massive MIMO: Ten myths and one critical question. *IEEE Commun. Mag.* **2016**, *54*, 114–123. [CrossRef]
13. Zuo, J.; Zhang, J.; Yuen, C.; Jiang, W.; Luo, W. Multicell multiuser massive MIMO transmission with downlink training and pilot contamination precoding. *IEEE Trans. Veh. Technol.* **2016**, *65*, 6301–6314. [CrossRef]
14. Ge, X.; Zi, R.; Wang, H.; Zhang, J.; Jo, M. Multi-user massive MIMO communication systems based on irregular antenna arrays. *IEEE Trans. Wirel. Commun.* **2016**, *15*, 5287–5301. [CrossRef]
15. Gopalakrishnan, B.; Jindal, N. An analysis of pilot contamination on multi-user MIMO cellular systems with many antennas. In Proceedings of the Signal Processing Advances in Wireless Communications (SPAWC), San Francisco, CA, USA, 26–29 June 2011, doi:10.1109/SPAWC.2011.5990435.
16. Heath, R.; Peters, S.; Wang, Y.; Zhang, J. A current perspective on distributed antenna systems for the downlink of cellular systems. *IEEE Commun. Mag.* **2013**, *51*, 161-167. [CrossRef]
17. Chen, X.; Zhang, Z.; Chen, H.-H. On distributed antenna systems with limited feedback precoding: opportunities and challenges. *IEEE Wirel. Commun.* **2010**, *17*, 80-88. [CrossRef]
18. Yang, C.; Han, S.; Hou, X.; Molisch, A. How do we design CoMP to achieve its promised potential? *IEEE Wirel. Commun.* **2013**, *20*, 67–74. [CrossRef]
19. Kobayashi, M.; Jindal, N.; Caire, G. Training and feedback optimization for multiuser MIMO downlink. *IEEE Trans. Commun.* **2011**, *59*, 2228–2240. [CrossRef]
20. Elijah, O.; Leow, C.Y.; Abdul Rahman, T.; Nunoo, S.; Zakwoi Iliya, S. A comprehensive survey of pilot contamination in massive MIMO-5G system. *IEEE Commun. Surv. Tuts.* **2016**, *18*, 905–923. [CrossRef]
21. Yin, H.; Gesbert, D.; Filippou, M.; Liu, Y. A coordinated approach to channel estimation in large-scale multiple-antenna systems. *IEEE J. Sel. Areas Commun.* **2013**, *31*, 264–273. [CrossRef]
22. Ngo, H.Q.; Larsson, E.G. EVD-based channel estimations for multicell multiuser MIMO with very large antenna arrays. In Proceedings of the 2012 IEEE International Conference on Acoustics, Speech and Signal Processing (ICASSP), Kyoto, Japan, 25–30 March 2012; doi:10.1109/ICASSP.2012.6288608.
23. Jose, J.; Ashikhmin, A.; Marzetta, T.L.; Vishwanath, S. Pilot contamination and precoding in multi-cell TDD systems. *IEEE Trans. Wirel. Commun.* **2011**, *10*, 2640–2651. [CrossRef]

24. 3GPP Technical Specification 36.211. Evolved Universal Terrestrial Radio Access (E-UTRA) Physical Channels and Modulation. Available online: www.3gpp.org (accessed on 20 November 2018).
25. 3GPP Technical Specification 36.213. Evolved Universal Terrestrial Radio Access (E-UTRA) Physical Layer Procedures. Available online: www.3gpp.org (accessed on 20 November 2018).
26. Coleri, S.; Ergen, M.; Puri, A.; Bahai, A. Channel estimation techniques based on pilot arrangement in OFDM systems. *IEEE Trans. Broadcast.* **2002**, *48*, 223–229. [CrossRef]
27. Jafarkhani, H. *Space-Time Coding: Theory and Practice*; Cambridge University Press: Cambridge, UK, 2005; pp. 229–230, ISBN 0-521-84291-3.
28. Recommendation ITR-R M.1225. *Guidelines for Evaluation of Radio Transmission Technologies for IMT-2000*; Int. Telecommun. Union (ITU): Geneva, Switzerland, 1997.
29. Chronopoulos, S.K.; Christofilakis, V.; Tatsis, G.; Kostarakis, P. Performance of turbo coded OFDM under the presence of various noise types. *Wirel. Pers. Commun.* **2016**, *87*, 1319–1336. [CrossRef]

electronics

MDPI

Article

Efficient Pilot Decontamination Schemes in 5G Massive MIMO Systems

Omar A. Saraereh [1], Imran Khan [2], Byung Moo Lee [3,*] and Ashraf Tahat [1]

1 Communication Engineering Department, King Abdullah II School of Engineering, Princess Sumaya University for Technology PSUT, Amman 11941, Jordan; o.saraereh@psut.edu.jo (O.A.S.); tahat@psut.edu.jo (A.T.)

2 Department of Electrical Engineering, University of Engineering & Technology, Peshawar 814, Pakistan; imran_khan@uetpeshawar.edu.pk

3 School of Intelligent Mechatronics Engineering, Sejong University, Seoul 05006, Korea

* Correspondence: blee@sejong.ac.kr

Received: 26 November 2018; Accepted: 27 December 2018; Published: 3 January 2019

Abstract: Massive Multiple-input Multiple-output (MIMO) is an emerging technology for the 5G wireless communication systems which has the potential to provide high spectral efficient and improved link reliability and accommodate large number of users. Aiming at the problem of pilot contamination in massive MIMO systems, this paper proposes two algorithms to mitigate it. The first algorithm is depending on the idea of Path Loss to perform User Grouping (PLUG) which divide the users into the center and edge user groups depending on different levels of pilot contamination. It assigns the same pilot sequences to the center users which slightly suffer from pilot contamination and assign orthogonal pilot sequences to the edge users which severely suffer from pilot contamination. It is assumed that the number of users at the edge of each cell is the same. Therefore, to overcome such limitations of PLUG algorithm, we propose an improved PLUG (IPLUG) algorithm which provides the decision parameters for user grouping and selects the number of central and edge users in each cell in a dynamic manner. Thus, the algorithm prevents the wrong division of users in good channel conditions being considered as an edge user which causes large pilot overhead, and also identifies the users with worst channel conditions and prevents the wrong division of such users from the center user group. The second algorithm for pilot decontamination utilizes the idea of pseudo-random codes in which orthogonal pilot are assigned to different cells. Such codes are deployed to get a transmission pilot by scrambling the user pilot in the cell. Since the pilot contamination is generated because different cells multiplex the same set of orthogonal pilots and the pseudo-random sequences have good cross-correlation characteristics, this paper uses this feature to improve the orthogonality of pilots between different cells. Simulation results show that the proposed algorithms can effectively improve channel estimation performance and achievable rate as compared with other schemes.

Keywords: Massive MIMO; pilot decontamination; MSE; dynamic user scheduling; dynamic pilot allocation

1. Introduction

With the advent of the era of big data and increasing demand by the explosion of growing numbers of subscribers, the demand for communication networks has exploded, and the existing mobile communication networks (4G) are increasingly unable to meet the needs of users for the network [1]. Massive MIMO is a key technology for 5G wireless communications to increase the spectral efficiency [2–4]. It has the ability to be deployed in various communications paradigms such as multicarrier communication, Orthogonal Frequency Division Multiplexing (OFDM) and cooperative communications [5–12]. As massive MIMO utilizes different frequencies in a Frequency Division Duplex (FDD) system. Therefore, the current research on massive MIMO is generally depending on a Time Division Duplex (TDD) system [13], that is, using channel reciprocity to obtain the required channel state information (CSI) [14], but the limited coherence interval limits the number of orthogonal pilots allocated to the user, which inevitably exists. The use of the same pilot by different cell users results in the inability of the Base Station (BS) to distinguish between pilot contamination [15].

In view of the fact that pilot contamination is caused by different cells multiplexing the same pilot, this paper proposes two algorithms for pilot decontamination in massive MIMO systems. The first algorithm is depending on the path loss for performing user grouping (PLUG) while the second algorithm is depending on pseudo-random codes [16]. In PLUG algorithm, the users are divided into a central user group and an edger user group depending on the distance between the users and the BS, and then classified according to the corresponding principle. In the case of a slight loss in the performance of the central user, the communication outage probability of the cell edge user is significantly reduced, and the quality of service (QoS) is significantly improved. On the basis of the PLUG strategy, the improved PLUG (IPLUG) is further proposed, and the decision parameters are dynamically selected to select the number of edge users to realize the dynamic division of the central users and edge users and improve the flexibility of the strategy. The second algorithm is depending on pseudo-random code that assigns different delays to each cell as a code sequence that distinguishes different cells and uses these pseudo-random codes to correspond to the cells. The user pilot performs synchronous scrambling to obtain new user pilots to enhance the orthogonality of user pilots between different cells. At the same time, the mean square error (MSE) of the expected channel estimation under the pilot design scheme is derived and analyzed. The strategy of this paper does not require the large-scale cooperation of BSs. It only needs to know the distance and decision parameters between users and BSs in the current cell and compete for a dynamic grouping by using the division principle. The proposed model has low complexity and requires fewer parameters, which makes it suitable for deploying in massive MIMO systems practical scenarios. It is proved that the proposed pilot design scheme can effectively improve the performance of channel estimation. The numerical simulation results also verify that the proposed pilot design algorithms depending on PLUG and pseudo-random code can greatly improve the performance of channel estimation and effectively reduce the pilot contamination of the system.

2. State-of-the-Art Pilot Decontamination Algorithms

The authors in [8] pointed out that in multi-cell massive MIMO systems, when the number of BS antennas tends to infinity, the performance of the system is mainly limited by pilot contamination, and thereafter the research on pilot contamination in massive MIMO systems has never stopped. The authors in [17] analyzed the influence of pilots on the performance of massive multi-cell MIMO systems and proposed a minimum mean square error (MMSE) multi-cell precoding method. As compared with the traditional zero-forcing (ZF) precoding, the MMSE scheme can significantly improve the gain of the system without pilot contamination, but the pilot contamination is relatively serious. The increase in system gain is limited. In [18], a pilot contamination precoding (PCP) scheme is proposed in which a precoding matrix is designed with ZF to obtain an infinite signal-to-interference plus noise ratio (SINR), but the effect is not ideal when the number of antennas is limited. In [19], the optimal algorithm for finding the optimal pilot contamination precoding matrix and the simple suboptimal algorithm is proposed depending on [18], and the two algorithms are proved to be limited when the number of BS antennas is limited. Compared with traditional MIMO, the gain of the system can be greatly improved. In [20], an intelligent pilot allocation scheme is proposed, which can maximize the uplink SINR of all users in the target cell under the channel of large-scale fading characteristics. The authors in [21,22] proposed a pilot power control method, which was successfully used in the classified cell, which effectively reduced the pilot contamination and improved the downlink reachability and rate of the entire system. Pilot contamination is a problem inherent in a massive MIMO system. The method of solving pilot contamination in different situations is not unique.

Ideally, massive MIMO should use Fully Orthogonal Pilot Scheduling (FOPS) to assign orthogonal pilots to each user, but the length of the pilot sequence and the pilot set size are limited by the channel coherence time. In a typical scenario, the maximum number of orthogonal pilot sequences in a 1ms coherence time is about 200 [23]. Therefore, a massive MIMO system generally adopts Fully Reused Pilot Scheduling (FRPS). Since the pilot between users is non-orthogonal or identical, pilot contamination is unavoidable [16]. When the number of BS antennas tends to be infinite and there is no cooperation, the main factor affecting the system performance is inter-cell interference (ICI) caused by pilot contamination [23], so pilot contamination is critical to the performance improvement of massive MIMO systems.

At present, massive MIMO pilot decontamination is mainly carried out from three aspects: channel estimation method [24–28], matrix precoding method case [29,30], and pilot allocation strategy [21]. The authors in [25] utilize a diagonal Jacket matrix for pilot reduction which has low complexity, excellent eigenvector and a constant diagonal treatment and an energy harvest. The drawback of such a method is that it assumes perfect multipath fast channel estimation. In [26], the authors propose a pilot mitigation algorithm using a low-rate coordination between cells during the channel estimation phase itself. The coordination makes use of the additional second-order statistical information about the user channels, which are shown to offer a power way of discriminating across interfering user with even strongly correlated pilot sequences. The authors in [27] propose a highly efficient Discrete Fourier Transform (DFT) based approximation of the Linear Minimum Mean Square Error (LMMSE) estimator for reducing the pilot contamination problem. The authors in [28] propose an eigenvalue-decomposition-based approach to channel estimation, that estimates the channel blindly from the received data. The approach exploits the asymptotic orthogonality of the channel vectors in massive MIMO systems. It also deploys a short training sequence to resolve the multiplicative ambiguity of the received signals covariance. In [29], the authors propose a zero forcing (ZF) time-shifted pilot scheme, which was known to mitigate the pilot contamination effectively using conjugate beamforming. The authors in [30] proposed interreference-cancellation (IC) precoding scheme for pilot contamination mitigation in massive MIMO systems. They investigated the quality-of-service (QoS) guaranteed user-scheduling which is improved by deploying their proposed scheme. The authors in [31] pilot contamination reduction scheme which is dependent on complex exponential basis expansions. The Linear Time-Varying (LTV) channel is estimated and the optimal pilot symbols are derived following the minimum mean square error (MMSE) criterion and it is shown that the optimal pilot strategy is to group consecutive pilot tones together as a pilot cluster and to distribute uniformly all pilot clusters in frequency-domain. Depending on the pilot allocation strategy, the research found that: In [30], under the principle of maximizing Signal Leakage Noise Ratio (SLNR) precoding, the same guidance is adopted for users with small ICI. Users with high frequency and mutual interference user orthogonal pilots to improve the overall performance of the system in the case of pilot contamination. However, the pilot scheduling scheme requires large-scale cooperation between BSs and needs to know the large-scale fading factor of each user. Such a factor is very difficult in a massive MIMO system with its own complex structure. The authors in [32] proposed a pilot allocation strategy depending on power control, which makes the pilot transmit time slots between cells with relatively large crossover gains staggered, but it is not easy to ensure pilot dynamic synchronization of several cells. The choice of mechanism will directly affect the performance of the strategy. The authors in [33] proposed a pilot coordinated allocation scheme, which allocates pilot sequences by identifying pilot usage conditions and selects user multiplexed pilots that are least affected by pilot contaminations, thereby reducing pilot contamination, but inter-cell cooperation to system brings additional burdens and expenses. The authors in [34] proposed an improved strategy depending on soft pilot multiplexing. On the basis of soft pilot multiplexing technology, packet parameters are introduced for secondary grouping, but the path loss factor, shadow fading, and the size of each user need to be known. The parameters such as the scale fading factor are more complicated, and the computational complexity of implementing the secondary grouping is higher.

3. System Model

3.1. The System Model

Figure 1 shows the model for massive MIMO system. The channel propagation matrix of all users in the cell to the cell site is:

$$\boldsymbol{H}_{ij} = \boldsymbol{G}_{ij}\sqrt{\boldsymbol{D}_{ij}} \tag{1}$$

where: $\boldsymbol{G}_{ij} = \left[\mathbf{g}_{ij1}\ \mathbf{g}_{ij2\ldots}\ \mathbf{g}_{ijK}\right]$, $\mathbf{g}_{ijk} \in \mathbb{C}^{M\times 1}$ is a small-scale fading vector, each vector is independent of each other and obeys a zero mean complex Gaussian distribution with a variance of $\boldsymbol{I}_M$, that is $\mathbf{g}_{ijk} \sim CN(0, \boldsymbol{I}_M)$; $\boldsymbol{D}_{ij} = \mathrm{diag}(\beta_{ij1}\ \beta_{ij2\ldots}\ \beta_{ijK})$ is a $K \times K$ order diagonal matrix, used to describe the jth cell large-scale fading of each user to the ith cell BS, $\beta_{ijk} = z_{ijk}/\left(r_{ijk}/R\right)^{\alpha}$ represents the large-scale fading factor of user k to ith BS in the jth cell. Wherein, the shadow fading z_{ijk} obeys a lognormal distribution, that is, $10\lg\left(z_{ijk}\right)$ obeys a zero mean, the standard deviation is a Gaussian distribution of σ_{shadow}, R is defined as the cell radius, and r_{ijk} represents the distance from the user k to the ith BS in the jth cell, α represents the path loss factor, and z_{ijk}, r_{ijk}, α are independent of each other. At the same time, it is assumed that the antenna arrays in the same BS are sufficiently compact in arrangement, the large-scale fading of specific users is equal in all propagation paths, but the large-scale fading of different users is independent of each other, and the channel is reciprocity, that is, uplink propagation is assumed to be the conjugate transpose of the matrix in the downlink propagation matrix [22]. For a multi-cell multi-user massive MIMO system, the number of antenna arrays installed by the BS is large, and the channel satisfies progressive orthogonality, namely:

$$\left(\frac{\boldsymbol{H}_{ij}^H\boldsymbol{H}_{ij}}{M}\right) = \sqrt{\boldsymbol{D}_{ij}}\left(\frac{\boldsymbol{G}_{ij}^H\boldsymbol{G}_{ij}}{M}\right)_{M\gg K}\sqrt{\boldsymbol{D}_{ij}} \approx \boldsymbol{D}_{ij} \tag{2}$$

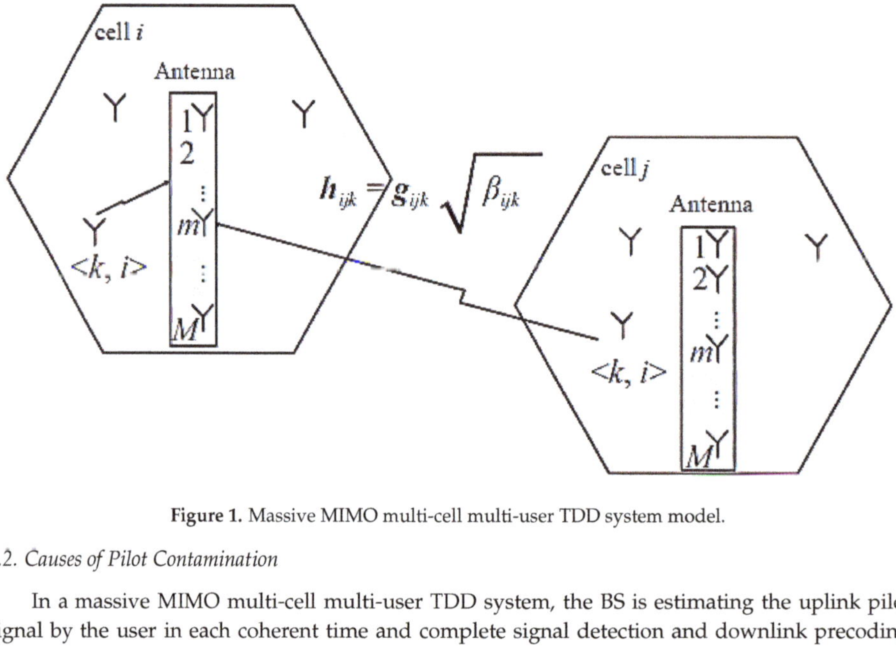

Figure 1. Massive MIMO multi-cell multi-user TDD system model.

3.2. Causes of Pilot Contamination

In a massive MIMO multi-cell multi-user TDD system, the BS is estimating the uplink pilot signal by the user in each coherent time and complete signal detection and downlink precoding.

At the beginning of each coherence time, all users in all cells simultaneously transmit pilot sequences. At the beginning of each coherence time, all users in all cells simultaneously transmit pilot sequences. Suppose $\boldsymbol{\psi}_i = (\boldsymbol{\psi}_{1i}, \boldsymbol{\psi}_{2i}, \ldots, \boldsymbol{\psi}_{Ki})^T$ is the $K \times \tau$ dimension pilot sequence matrix of all users in ith cell (τ is the sequence length), which satisfies $\boldsymbol{\psi}_i \boldsymbol{\psi}_i^H = \boldsymbol{I}_K$, where $\boldsymbol{I}_K$ is the unit matrix of order $K \times K$. Under the FRPS policy, the pilot matrix received by the uplink cell BS is:

$$\boldsymbol{y}_i^p = \sqrt{p_p}\left(\sum_{j=1}^{L} \boldsymbol{H}_{ij}\boldsymbol{\psi}_i\right) + \boldsymbol{n}_i^p \tag{3}$$

where, p_p is the pilot signal transmission power and $\boldsymbol{n}_i^p$ is the additive white Gaussian noise matrix of order $M \times \tau$. After receiving the pilot signal, the BS starts uplink channel estimation. The channel estimation value $\hat{\boldsymbol{H}}_{ii}$ of the target cell is obtained by using the LS channel estimation introduced earlier which is expressed as:

$$\hat{\boldsymbol{H}}_{ii} = \frac{1}{\sqrt{p_p}}\boldsymbol{y}_i^p \boldsymbol{\psi}_i^H = \boldsymbol{H}_{ii} + \sum_{j \neq i} \boldsymbol{H}_{ij} + \frac{1}{\sqrt{p_p}}\boldsymbol{n}_i^p \boldsymbol{\psi}_i^H \tag{4}$$

It can be seen from Equation (4) that the ith cell channel estimation value $\hat{\boldsymbol{H}}_{ii}$ includes the superposition of the channel propagation matrix of other cells to the target cell in addition to the influence of the target channel and noise, which is pilot contamination.

After the user sends the pilot, all users send data signals to the BS and use the same time-frequency resource. The signal received by the ith cell BS is defined as:

$$\boldsymbol{y}_i^u = \sqrt{p_u}\sum_{j=1}^{L}\sum_{k=1}^{K} \boldsymbol{h}_{ijk} x_{jk}^u + \boldsymbol{n}_i^u \tag{5}$$

where x_{jk}^u is the transmitted data symbol of user k in jth cell; p_u is the uplink user data symbol average transmit power; $\boldsymbol{h}_{ijk}$ represents the channel transmission vector of user k to ith cellBS in jth cell, which is the kth column of $\boldsymbol{H}_{ij}$; $\boldsymbol{n}_i^u$ is the additive white Gaussian noise vector of order $M \times 1$. The BS uses the channel estimation value $\hat{\boldsymbol{H}}_{ii}$ of the Equation (4) and the received signal vector $\boldsymbol{y}_i^u$, and the MF detects the original data symbol $\hat{x}_{jk}^u$ transmitted by the user k in the ith cell is expressed as:

$$\hat{x}_{jk}^u = \hat{\boldsymbol{h}}_{iik}^H \boldsymbol{y}_i^u = \left(\sum_{j=1}^{L} \boldsymbol{h}_{ijk}^H + \boldsymbol{v}_{ik}^H\right)\left(\sqrt{p_u}\sum_{j=1}^{L}\sum_{k=1}^{K} \boldsymbol{h}_{ijk} x_{jk}^u + \boldsymbol{n}_i^u\right) \tag{6}$$

where $\boldsymbol{v}_{ik}$ is the column vector of the matrix $\boldsymbol{n}_i^p \boldsymbol{\psi}_i^H / \sqrt{p_p}$. When the number of BS antennas M approaches positive infinity, it is easy to know from Equation (2) that the channel of the massive MIMO system exhibits progressive orthogonality, and the progressive expression of Equation (6) is:

$$\hat{x}_{ik}^u \approx M\sqrt{p_u}\left(\beta_{iik} x_{jk}^u + \sum_{j \neq i} \beta_{ijk} x_{jk}^u\right) \tag{7}$$

It can be known from Equation (7) that when the BS antenna $M \to \infty$, the data symbol $\hat{x}_{ik}^u$ will not be affected by the small-scale fading factor and noise. Therefore, when $M \to \infty$, the signal-to-interference and noise ratio (SINR) of the user k uplink received signal in the ith cell can be defined as:

$$SINR_{ik}^u = \frac{\beta_{iik}^2}{\sum_{j \neq i} \beta_{ijk}^2} \tag{8}$$

It can be seen from Equation (8) that due to the existence of pilot contamination, the uplink SINR is limited by the large-scale fading factor of the same pilot user in the interfering cell. According to Equation (8), the uplink achievable rate of user k in the target ith cell is given by:

$$C_{ik}^{u} = (1 - \mu_0)E\{\text{lb}(1 + SINR_{ik}^{u})\} \tag{9}$$

where μ_0 is the pilot overhead coefficient for full multiplexing, indicating the spectral efficiency loss caused by the pilot sequence used for channel estimation, when other pilot allocation algorithms are used, the adjustment in Equation (9) is to be appropriately made.

3.3. IPLUG Algorithm

The PLUG algorithm significantly improves the QoS of the edge user but lacks certain flexibility. This is because the number of edge users selected by each cell in the PLUG policy is the same, and only compared with the distance between other users in the cell and the BS. The user's own specific environment is not considered. Therefore, we proposed an improved pilot scheduling algorithm depending on path loss to perform user grouping (IPLUG) and dynamically select the number of edge users per cell by introducing decision parameters. Figure 2 is a schematic diagram of the IPLUG algorithm decision parameters. The improved IPLUG algorithm is depending on the PLUG algorithm and is designed to improve the accuracy and legitimacy of performing user groupings. In the IPLUG algorithm, when the decision parameter λ is selected, the BS can complete the dynamic grouping according to the user distance d and $\lambda \times R$. The specific division principle is:

$$d \overset{?}{\geq} \lambda \times R \rightarrow \begin{cases} Yes \rightarrow \text{Edge User} \\ No \rightarrow \text{Center User} \end{cases} \tag{10}$$

Let the central user set be U_c, the edge user set be U_e, the central user pilot set be $\boldsymbol{\psi}_c$, and the edge user pilot set be $\boldsymbol{\psi}_e$, the pilot set of jth cell is $\boldsymbol{\psi}_j = \left(\boldsymbol{\psi}_{1j}, \boldsymbol{\psi}_{2j}, \ldots, \boldsymbol{\psi}_{Kj}\right)^T, j = 1, 2, \ldots L$, then:

$$\boldsymbol{\psi}_{ki}\boldsymbol{\psi}_{kj}^{H} = \begin{cases} 1 \rightarrow \boldsymbol{\psi}_{ki} \subseteq \boldsymbol{\psi}_c \text{ and } \boldsymbol{\psi}_{kj} \subseteq \boldsymbol{\psi}_c \\ 0 \rightarrow \text{otherwise} \end{cases} \tag{11}$$

Then, the pilot vector received by the target ith cell BS is:

$$\overline{\boldsymbol{y}}_i^{p} = \sqrt{p_p}\sum_{j=1}^{L}\left(\sum_{k=1,\, k\in U_c}^{K} \boldsymbol{h}_{ijk}\boldsymbol{\psi}_{kj} + \sum_{k=1,\, k\in U_e}^{L} \boldsymbol{h}_{ijk}\boldsymbol{\psi}_{kj}\right) + \overline{\boldsymbol{n}}_i^{p} \tag{12}$$

For the target ith cell user k, the Least Square (LS) channel estimation is used to obtain the target channel estimation value:

$$\hat{\boldsymbol{h}}_{iik} = \frac{1}{\sqrt{p_p}}\boldsymbol{y}_i^{p}\boldsymbol{\psi}_{ki}^{H} \tag{13}$$

Combining Equations (12) and (13), when the number of antennas $M \rightarrow \infty$ the SINR of the central user under the IPLUG algorithm is:

$$SINR_{ik}^{uc} = \frac{\left|\boldsymbol{h}_{iik}^{H}\boldsymbol{h}_{iik}\right|^2}{\sum_{j\neq i,\, \boldsymbol{\psi}_{ki}\subseteq\boldsymbol{\psi}_c,\, \boldsymbol{\psi}_{kj}\subseteq\boldsymbol{\psi}_c}\left|\boldsymbol{h}_{jik}^{H}\boldsymbol{h}_{jik}\right|^2} \tag{14}$$

The edge user is not affected by the pilot contamination, so when $M \rightarrow \infty$, the edger user $SINR_{ik}^{ue} \rightarrow \infty$.

Figure 3 illustrates the flowchart for the IPLUG algorithm.

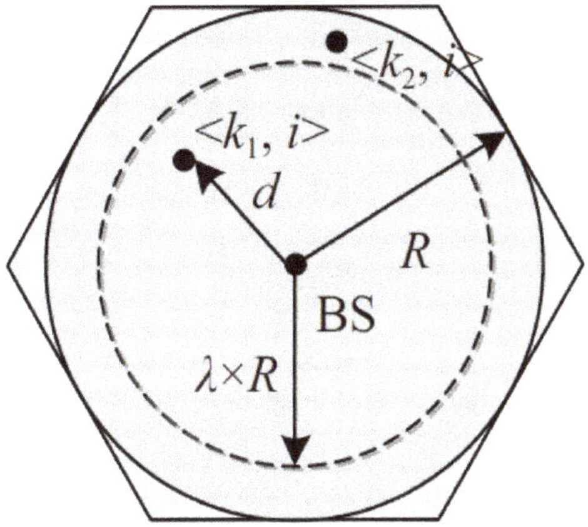

Figure 2. IPLUG policy decision parameters for user grouping.

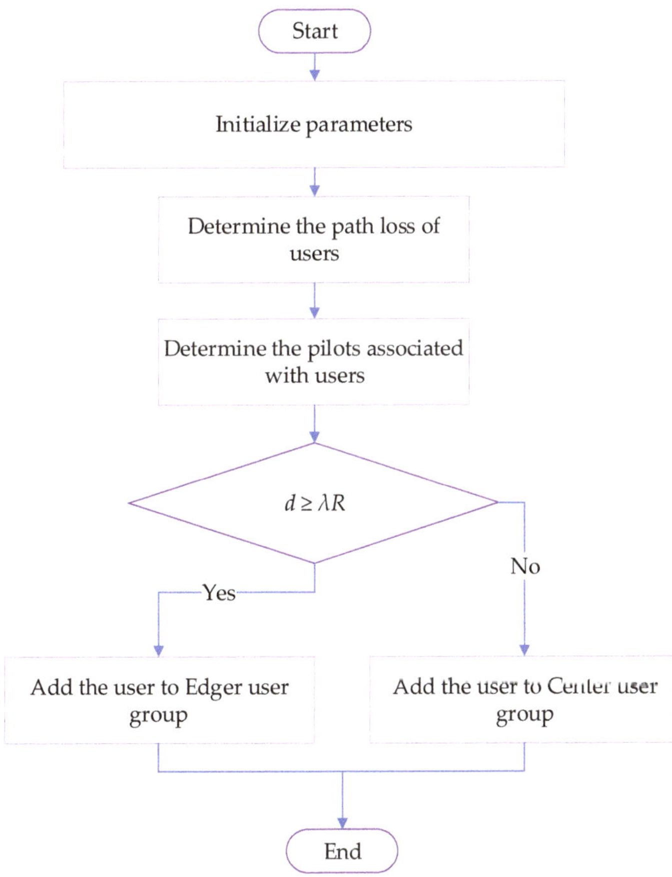

Figure 3. Proposed IPLUG algorithm flowchart.

3.4. Pseudo-Random Code-Based Pilot Scheduling

3.4.1. Uplink

Assume that all users in any cell send a pilot sequence of length τ, if the pilot transmitted by K users in the jth cell is $\psi_j = [\phi_{j1}\phi_{j2} \ldots \phi_{jK}]$, where $\phi_{jk} \in \mathbb{C}^{\tau\times 1}$ is the pilot transmitted by the kth user in the cell ($\phi_{jk}^H\phi_{jk} = 1$), when the average transmit power of the user is ρ_r, the pilot signal received by the lth cell BS is:

$$Y_l = \sqrt{\rho_r\tau}\sum_{j=1}^{L} G_{jl}\psi_j^T + n_l \tag{15}$$

where $G_{jl} = H_{jl}\sqrt{D_{jl}}$ is the channel between all users of the jth cell and the lth cell BS. Let β_{jlk} denote the large-scale fading coefficient of the user k of the jth cell to the lth cell BS, then D_{jl} can be expressed as a diagonal matrix, and the diagonal element is $\beta_{jl} = \left[\beta_{jl1}, \beta_{jl2}, \ldots, \beta_{jlK}\right]$, n_l denotes the additive white Gaussian noise of the lth cell. The term H_{jl} can be determined from the following matrix expression:

$$H_{jl} = \begin{bmatrix} h_{j_1}l_1 & \cdots & h_{j_1}l_M \\ \vdots & \ldots & \vdots \\ h_{j_K}l_1 & \cdots & h_{j_K}l_M \end{bmatrix} \tag{16}$$

The elements of H_{jl} are independently and identically distributed (i.i.d), and $h_{j_k}l_m$ represents a small-scale fading coefficient between the kth user in the jth cell and the mth antenna of the BS in the lth cell.

The BS estimates the channel using the received pilot signal, which is known from the standard results in the estimation theory [24]. When the number of BS antennas tends to infinity, the MMSE estimate of the channel is:

$$\hat{G}_{jl} = \sqrt{\rho_r\tau}Y_l\left(I + \rho_r\tau\sum_{i=1}^{L}\psi_i^* D_{il}\psi_i^T\right)^{-1}\psi_j^* D_{jl} \tag{17}$$

3.4.2. Downlink

Considering the BS of the lth cell, it is assumed that the information symbol transmitted by the lth cell to the user is $S_l = [S_{l1}S_{l2} \ldots S_{lK}]^T$, and $A_l = f(\hat{G}_{ll})$ is a linear precoding matrix of order $M \times K$, where $f(\cdot)$ represents a specific linear precoding technique at the BS. After precoding, the signal matrix sent by the BS can be represented as A_lS_l, and the BS satisfies the average power constraint, that is, $tr\{E[A_lS_lS_l^HA_l^H]\} \leq P$, then the data information received by the user of the jth cell can be expressed as:

$$X_j = \sqrt{\rho_f}\sum_{l=1}^{L} G_{jl}^T A_l S_l + n_j \tag{18}$$

where ρ_f is the downlink transmit power and n_j is the AWGN noise of the corresponding cell. Assuming a simple MF precoding technique at the BS, i.e., $A_l = \hat{G}_{ll}^*$, Equation (25) can be expressed as:

$$X_j = \sqrt{\rho_f}\sum_{l=1}^{L} G_{jl}^T\hat{G}_{ll}^*S_l + n_j = \sqrt{\rho_f\rho_r\tau}\sum_{l=1}^{L} G_{jl}^T\left(\rho_r\tau\sum_{j'=1}^{L} G_{j'l}\psi_{j'}^T + n_l\right)^* Z^*S_l + n_j \tag{19}$$

where $Z = \left(I + \rho_r\tau\sum_{i=1}^{L}\psi_i^*D_{il}\psi_i^T\right)^{-1}\psi_j^*D_{jl}$. When the number of BS antennas increases and tends to infinity then:

$$\frac{1}{M\sqrt{\rho_f\rho_r\tau}}X_j \to \sqrt{\rho_r\tau}\sum_{l=1}^{L} D_{jl}\left(\psi_l^H\psi_j + \rho_r\tau\sum_{i=1}^{L}\psi_l^H\psi_i D_{il}^*\psi_i^H\psi_j\right)^{-1} D_{ll}^*S_l \tag{20}$$

It can be obtained from Equation (20) that when all cells multiplex the same set of orthogonal pilots, the right side of Equation (20) can be equated to:

$$\sum_{l=1}^{L} D_{jl}\left(I+\rho_r\tau\sum_{i=1}^{L} D_{il}^{*}\right)^{-1} D_{ll}^{*}S_l \tag{21}$$

At this time, the signal received by the jth cell will be interfered by the signals transmitted by other cells, especially when the large-scale fading coefficient is large; this phenomenon is called pilot contamination. When all cells use orthogonal pilots, the right side of Equation (20) can be equated to:

$$D_{jj}\left(I+\rho_r\tau D_{jj}^{*}\right)^{-1} D_{jj}^{*}S_l \tag{22}$$

In this case, the signal received by the user is a scaling of the signal transmitted by the BS of the cell, and there is no pilot contamination. However, in actual communication, because the coherence interval is limited, it cannot guarantee that all cells can adopt an orthogonal pilot. Therefore, pilot contamination has become an important factor limiting the performance of massive MIMO systems.

3.4.3. Pseudo-Random Code Scheme Description

The pseudo-random code has the similarity of noise sequence. It is a seemingly random but actually regular periodic binary sequence, including m sequence, a Gold sequence, M sequence and combined sequence [35]. Pseudo-random codes have good autocorrelation and cross-correlation properties and can be used as address codes in CDMA communication technology. In addition, thanks to its pseudo-random characteristics, pseudo-random codes are also widely used in information encryption [36]. The pseudo-random sequence is generated by the linear feedback shift register shown in Figure 4, where $c_i = 0$, indicates that the line is off; $c_i = 1$ indicates that the line is on. The characteristic polynomial of generating a pseudo-random sequence is:

$$f(x)=\sum_{i=0}^{n} c_i x^i = c_0 + c_1 x + \ldots + c_n x^n \tag{23}$$

where x^i indicates the ith shift register; $c_0 = c_1 = 1$, for any $i \neq 0$ and $i \neq n$, $c_i \in \{0,1\}$.

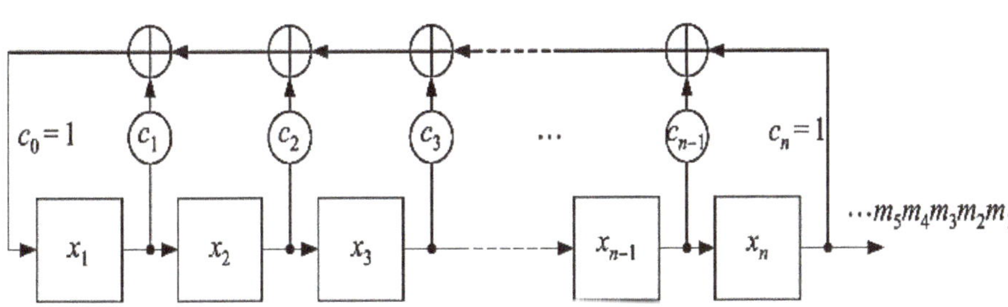

Figure 4. Pseudo-random code linear feedback shifter register.

Since the pilot contamination is generated because different cells multiplex the same set of orthogonal pilots and the pseudo-random sequences have good cross-correlation characteristics, this paper uses this feature to improve the orthogonality of pilots between different cells. Depending on the above considerations, this paper proposes a pilot design scheme depending on pseudo-random code: in the case that all cells multiplex the same set of orthogonal pilots, the pseudo-random code is used as the code sequence for distinguishing different cells, for each cell allocates a pseudo-random code with different delays, and synchronously scrambles the pseudo-random code in each cell to the user pilot of the corresponding cell to obtain a new pilot.

The pilot design scheme depending on the pseudo-random code is shown in Figure 5. Since the flow of the pilot design of different cells is the same, in order to facilitate the representation without loss of generality, the pilot design scheme is specifically described here. Assume that the pilot of the kth user in a cell is $\phi_k = [\phi_{k1}, \phi_{k2}, \ldots, \phi_{k\tau}]^T$, and the pseudo-random sequence generated by the linear feedback shifter register in the corresponding cell is $m = [m_1, m_2, \ldots, m_N]$. The scrambler in Figure 5 indicates that the pseudo-random matrix is multiplied by the matrix of the corresponding user's pilot matrix. In general, a pseudo-random sequence is a 0–1 bit stream whose length is much larger than the length of the pilot sequence.

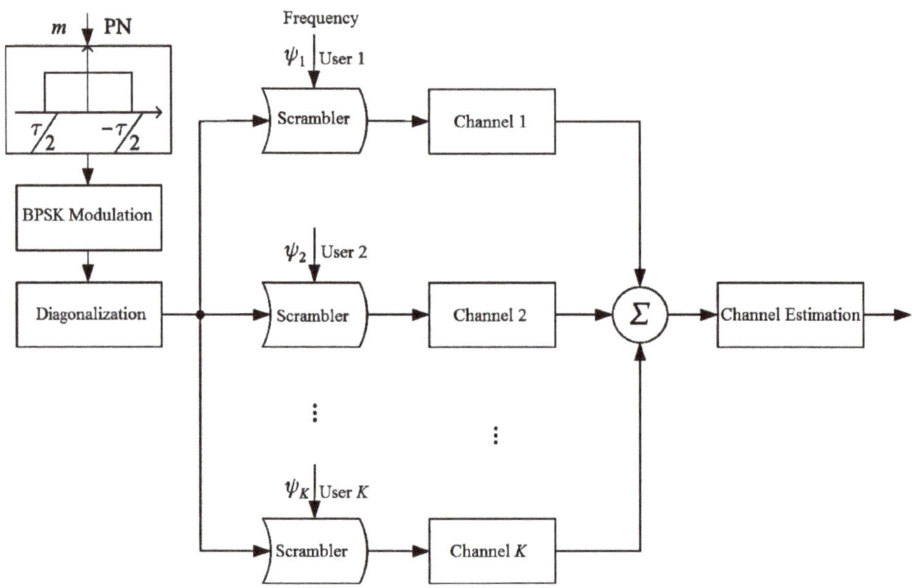

Figure 5. Pseudo-random code pilot design scheme.

In order to ensure that the generated pseudo-random sequence can be used for scrambling, it needs to process as follows:

1. The pseudo-random sequence generated by the linear feedback shift register is truncated, and the method is to generate a rectangle depending on the pilot length τ and using a rectangular window for pseudo-random sequences to make truncation.
2. Performing BPSK modulation on the truncated pseudo-random sequence.
3. The pseudo-random sequence is diagonalized after modulation, that is, τ pseudo-random numbers are generated as pseudo-random matrices on the diagonals of the diagonal matrix.

The pseudo-random matrix used for scrambling after the above processing can be expressed as:

$$P = \begin{bmatrix} m_1 & 0 & \cdots & 0 \\ 0 & m_2 & \cdots & 0 \\ \vdots & \vdots & \ddots & \vdots \\ 0 & 0 & \cdots & m_\tau \end{bmatrix} \tag{24}$$

After the pseudo-random matrix is scrambled by the scrambler to the user pilot, the pilot of the kth user outputted by the scrambler is obtained as:

$$\Psi_k = P\phi_k = [m_1\ m_2\phi_{k2} \ldots\ m_\tau\phi_{k\tau}\ ]^T \tag{25}$$

The scrambling method used in this paper is to use the same pseudo-random sequence to synchronously scramble the pilots of all users in the cell, and the pilot of K users in the cell can be obtained after the pilot design as:

$$\Psi = \begin{bmatrix} m_1\phi_{11} & \cdots & m_1\phi_{K1} \\ \vdots & \cdots & \vdots \\ m_\tau\phi_{1\tau} & \cdots & m_\tau\phi_{K\tau} \end{bmatrix} \tag{26}$$

The designed pilot is transmitted to the BS through the channel, and the BS can obtain the required Channel State Information (CSI) by using channel estimation.

When the pilot design is performed on users of other cells, considering the pilot contamination is related to the distance between users, when the distance between two users is large enough, the influence of pilot contamination is neglected. Therefore, the proposed pseudo-random code pilot scheduling scheme only considers the pilot contamination of the target cell and ignores the pilot contamination of the farther cell. Under the above assumptions, only the pseudo-random sequence needs to be allocated to the target cell and the neighboring cell, and the pseudo-random sequence can be reused for the farther cell, which greatly reduces the usage and implementation complexity of the pseudo-random sequence and makes it easy to design a better pilot sequence. Depending on the above considerations, the pilot design of all cells can be designed by using the proposed pilot design scheme. The only difference is that the pseudo-random sequence used by the target cell and the neighboring cell cannot be the same, which involves a pseudo-random sequence selection problem. Since the cross-correlation between different pseudo-random sequences is different, if the cross-correlation value between the selected pseudo-random sequences is large, even if the user pilot is designed, the pilot contamination between the cells is still serious, so that the pilot design seems to be meaningless, so it is necessary to follow the certain criteria when selecting the pseudo-random sequence as the code sequence to distinguish difference cells. Theoretically, the cross-correlation values between pseudo-random sequences are connected. Nearly 0, the effect of the pilot design is more obvious, but the complexity of acquiring these pseudo-random sequences will become very high. In order to obtain the desired effect within a certain complexity range, this paper adopts the following criteria to obtain the required pseudo-random sequence. Assuming that the number of cells requiring different pseudo-random sequences is M, the corresponding M pseudo-random sequences need to be satisfied:

$$\left|\rho\left(m_i, m_j\right)\right|_{i \neq j} < \gamma_i,\ j = 1,\ 2,\ \ldots,\ M \tag{27}$$

where $\rho\left(m_i, m_j\right) = \frac{1}{n}\sum_{k=1}^{n} m_{ik}m_{jk}$, is a function that represents a normalized cross-correlation value between two pseudo-random sequences. $\gamma \in [0,1]$ is a constant values indicating the upper limit of the normalized cross-correlation value between any two selected pseudo-random sequences.

It can be seen from Equation (27) that in the case where the pseudo-random sequence used is determined, the setting of the value is an important factor affecting the impact of the pilot design scheme. The smaller the value, the smaller the cross-correlation value between pseudo-random sequences. After the pilot design, the orthogonality of user pilots between different cells is stronger, and the effect of reducing pilot contamination is more obvious. Otherwise, it cannot effectively reduce the pilot contamination. The BS and the user in each cell share a pseudo-random sequence, and the BS can distinguish pilots from different cells by using detection techniques. At the same time, since each cell has the same pseudo-random sequence to scramble a set of orthogonal pilots of all users in the cell, after pilot design, the user pilots in each cell remain orthogonal, and in addition, no additional intra-cell interference will be introduced. Moreover, increasing the pilot length is beneficial to improve the efficiency of the pilot design. This is because the length of the pseudo-random sequence is the same as the length of the pilot, and the longer the pseudo-random sequence, the better the orthogonality between them, so increasing the pilot length is equivalent to improving the orthogonality between the pseudo-random sequences. Thereby, the orthogonality of user pilots between different cells after the pilot design is improved.

3.4.4. Mean Square Error (MSE) Performance Analysis of Expected Channel Estimation

This section will use the new pilot to derive and analyze the MSE of the channel estimate and explore the pilot design method described in Section 3.3. It can be seen from [15] that when the pilot contamination is relatively serious, the channel estimation performance of the system drops sharply. In this case, even if a complex multi-cell MMSE precoding scheme is applied, the performance improvement of the massive MIMO system is limited. In order to facilitate derivation and analysis, this section only studies simple single-cell ZF precoding and its precoding matrix is defined as:

$$A_l = \frac{\hat{G}_{ll}\left(\hat{G}_{ll}^H \hat{G}_{ll}\right)^{-1}}{\sqrt{tr\left[\left(\hat{G}_{ll}^H \hat{G}_{ll}\right)^{-1}\right]}} \tag{28}$$

It can be found from (28) that a single-cell ZF precoding matrix can be designed as long as the information of the desired channel is obtained. Therefore, the subsequent derivation and analysis in this section are depending on the estimated expected channel.

As can be seen from Section 3.4, the user pilot of the jth cell is $\Psi_j = P_j\Psi$, then Equation (15) can be expressed as:

$$Y_l = \sqrt{\rho_r\tau}\sum_{j=1}^{L} G_{jl}\Psi_j^T + n_l = \sqrt{\rho_r\tau}\sum_{j=1}^{L} G_{jl}\psi^T P_j^T + n_l \tag{29}$$

In the case where the number of BS antennas is limited, the MMSE estimation of the desired channel is:

$$\hat{G}_{ll} = \sqrt{\rho_r\tau}\, Y_l \left(C_n + \rho_r\tau\sum_{i=1}^{L} P_i^*\psi^* C_{il}\psi^T P_i^T\right)^{-1} P_i^*\psi^* C_{il} \tag{30}$$

where $C_n = E\{n_l^H n_l\}$, is the autocorrelation matrix, representing the received noise n_l; $C_{jl} = E\left\{G_{jl}^H G_{jl}\right\}$ is the autocorrelation matrix representing the channel transfer matrix G_{jl}.

The MMSE defining the expected channel estimation of the lth cell BS is:

$$M^{mse} \triangleq E\left\{||\hat{G}_{ll} - G_{ll}||_F^2\right\} \tag{31}$$

Let $R = \sqrt{\rho_r\tau}\left(C_n + \rho_r\tau\sum_{i=1}^{L} P_i^*\psi^* C_{il}\psi^T P_i^T\right)^{-1} P_i^*\psi^* C_{il}$, then $\hat{G}_{ll} = Y_l R$, substituting into (38) we get:

$$\begin{aligned} M^{mse} &= E\left\{tr\left\{(Y_l R - G_{ll})^H (Y_l R - G_{ll})\right\}\right\} \\ &= tr\{R^H E\{Y_l^H Y_l\} R - R^H E\{Y_l^H G_{ll}\} - E\{G_{ll}^H Y_l\} R + E\{G_{ll}^H G_{ll}\}\} \end{aligned} \tag{32}$$

Using the model shown in Equation (29), there is:

$$E\left\{Y_l^H Y_l\right\} = C_n + \rho_r\,\tau\sum_{i=1}^{L} P_i^*\psi^* C_{il}\psi^T P_i^T \tag{33}$$

$$E\left\{Y_l^H G_{ll}\right\} = \sqrt{\rho_r\tau} P_i^*\psi^* C_{ll} \tag{34}$$

$$E\left\{G_{ll}^H Y_l\right\} = \sqrt{\rho_r\tau} C_{ll}\psi^T P_l^T \tag{35}$$

Substituting Equations (32)–(34) into Equation (35), we can rewrite M^{mse} as:

$$\begin{aligned} M^{mse} &= tr\left\{R^H\left(C_n+\rho_r\,\tau\sum_{i=1}^{L}P_i^*\psi^*C_{il}\psi^TP_i^T\right)R-\sqrt{\rho_r\tau}R^HP_i^*\psi^*C_{ll}-\sqrt{\rho_r\tau}C_{ll}\psi^TP_l^TR+C_{ll}\right\} \\ &= tr\{C_{ll}-\sqrt{\rho_r\tau}C_{ll}\psi^TP_l^TR\} \\ &= tr\left\{C_{ll}-\rho_r\tau\,C_{ll}\psi^TP_l^T\left(C_n+\rho_r\tau\sum_{i=1}^{L}P_i^*\psi^*C_{il}\psi^TP_i^T\right)^{-1}P_l^*\psi^*C_{ll}\right\} \end{aligned} \tag{36}$$

Since the matrix P_i is a diagonal matrix composed of diagonal elements 1 and -1, and $\psi^H\psi=I$, so $(P_i\psi)^H(P_i\psi)=I$ can be obtained according to the properties of the matrix conjugate transpose. Using the matrix inversion principle, we can express Equation (36) as:

$$M^{mse}=tr\left\{C_{ll}-\rho_r\tau\,C_{ll}\left(\Psi_l^TC_n\Psi_l^*+\rho_r\tau\sum_{i=1}^{L}\Psi_l^T\Psi_i^*C_{il}\Psi_i^T\Psi_i^*\right)^{-1}C_{ll}\right\} \tag{37}$$

where $\Psi_i=P_i\psi$, $i=1,\,2,\,\ldots,\,L$. When $M\to\infty$ then:

$$C_{ll}=E\left\{G_{ll}^HG_{ll}\right\}=\sqrt{D_{ll}}E\left\{H_{ll}^HH_{ll}\right\}\sqrt{D_{ll}}=MD_{ll} \tag{38}$$

$$C_n=E\left\{n_l^Hn_l\right\}=MI_\tau \tag{39}$$

Substituting Equations (38) and (39) into Equation (37), we get:

$$\frac{1}{M}M^{mse}=tr\left\{D_{ll}-\rho_r\tau\,D_{ll}\left(I+\rho_r\tau\sum_{i=1}^{L}\Psi_l^T\Psi_i^*D_{il}\Psi_i^T\Psi_i^*\right)^{-1}D_{ll}\right\} \tag{40}$$

When the number of users in each cell is $K=1$, $\Psi_l^T\Psi_i^*=(P_l\psi)^T(P_i\psi)^*=\gamma_{li}$, where γ_{li} is a normalized cross-correlation value between the pseudo-random code in the lth cell and the pseudo-random code in the ith cell, which can be obtained from equation (27). Assuming that the large-scale fading coefficient $D_{il}=[\beta_{il1}]$ at this time, then Equation (40) can be simplified as:

$$\frac{1}{M}M^{mse}=\beta_{ll1}-\frac{\rho_r\tau\beta_{ll1}^2}{1+\rho_r\tau\beta_{ll1}+\rho_r\tau\sum_{i\neq l}\gamma_{li}^2\beta_{li1}} \tag{41}$$

It can be seen from the results of Equation (41) that when the large-scale fading coefficient and the uplink transmit power are constant and the pilot length is constant, the MSE performance of the channel estimation that can be obtained by applying the pilot scheme depending on the pseudo-random code and it is mainly limited by the cross-correlation performance between the pseudo-random sequences used: when the normalized cross-correlation value between the pseudo-random sequences used is large ($|\gamma_{li}|$ is close to 1), the MSE performance of the channel estimation is relatively poor. When the normalized cross-correlation value between the pseudo random sequences used is small ($|\gamma_{li}|$ is close to 0), the MSE performance of the channel estimation is better. It can be concluded that if a suitable pseudo-random sequence is selected, the above method is adopted. The proposed pilot design scheme can effectively improve the performance of channel estimation, thus achieving the purpose of reducing pilot contamination. Figure 6 shows the flowchart of the proposed pseudo-random pilot code algorithm.

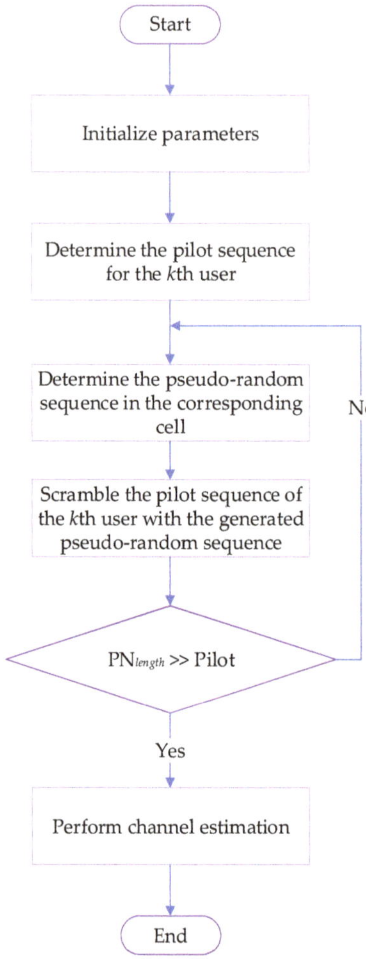

Figure 6. Proposed pseudo-random algorithm flowchart.

4. Simulation Results

4.1. Simulation Scenario and Parameters

Taking the MATLAB software (R2017b, The MathWorks, Natick, MA, USA) as the simulation platform, it is assumed that the massive MIMO multi-cell multi-user TDD system includes 7 regular hexagonal cells, as shown in Figure 7.

Table 1 shows the details of simulation parameters which are used for simulation analysis of the proposed pilot decontamination schemes. It is assumed that the antenna spacing is $\frac{1}{2}$ times the carrier wavelength, that is, there is a correlation between the antennas, but the correlation is not considered in this paper, and it is assumed that all the antennas are omnidirectional antennas. The number of users in each cell is $K = 10$, the user is uniformly distributed in the cell, the pilot length is 128, the path loss factor is 3, the pilot overhead coefficient is $\mu_0 = 0.05$, the cell radius $R = 500$ m, and the concentrated antennas is $16 \leq M \leq 1024$, and the lognormal shadow fading is 4 dB (only the last normal logarithmic normal shadow fading is set to 2 dB), and the number of simulations is 2000.

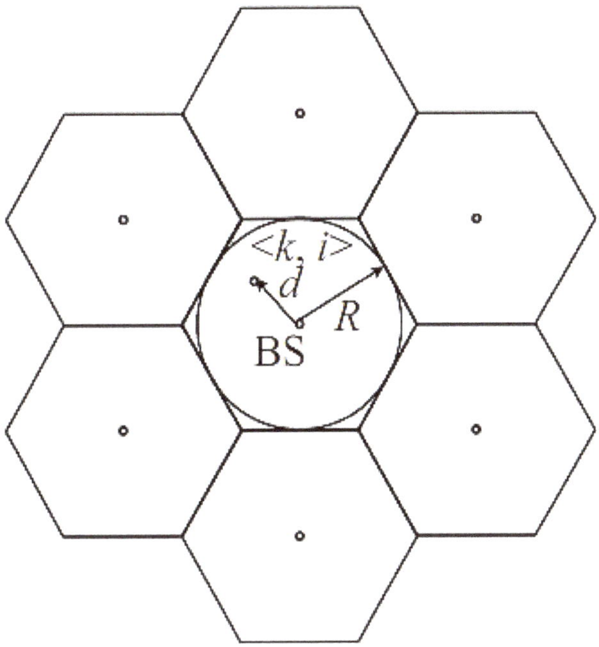

Figure 7. Simulation scenario of massive MIMO system of seven cells.

Table 1. Simulation Parameters.

S. No	Parameter Name	Value
1	Number of BS antennas (M)	16~1024
2	Number of users per cell (K)	10
3	Antenna Spacing	$\frac{1}{2}\lambda_c$
4	Pilot Length (τ)	128
5	Path Loss Factor (α)	3
6	Pilot overhead coefficient (μ_0)	0.05
7	Cell Radius (R)	1000 m
8	Lognormal Shadow Fading	4 dB
9	Number of simulations	5000
10	Average transmit power at BS (ρ_f)	20 dB
11	Average transmit power at User (ρ_r)	10 dB
12	The upper limit of normalized cross-correlation (γ)	0.3
13	Cross-gain (β_{jlk})	0.7

According to the above simulation parameters, the Monte Carlo method is used to simulate the uplink SINR, and the normalized channel estimation MSE and the target cell achievable rate, thereby comparing the performance between the PLUG algorithm, the IPLUG, and the FRPS algorithm respectively.

4.2. Analysis of PLUG Algorithm

Figures 8 and 9 respectively show the performance of the channel estimation normalized MSE and uplink SINR of the central user and the edge user. The performance of the uplink SINR varies with the number of antennas.

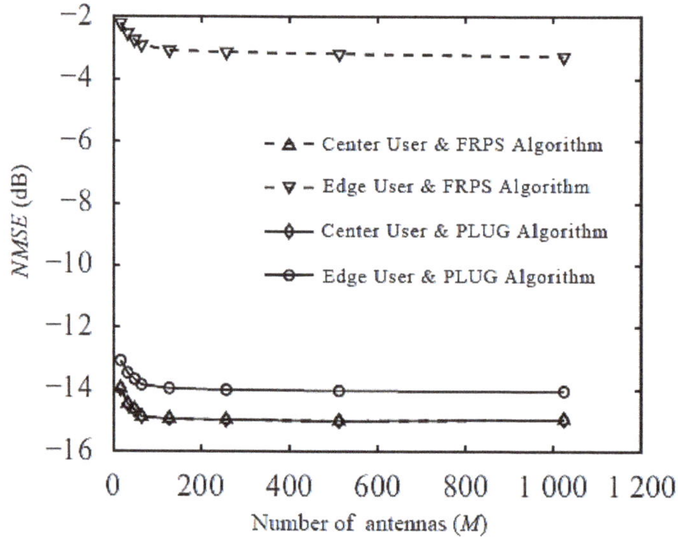

Figure 8. NMSE comparison of FRPS and PLUG Algorithms.

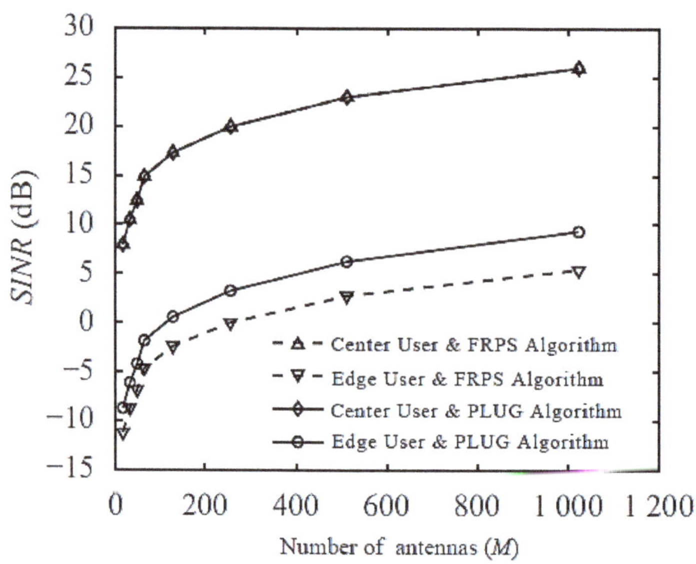

Figure 9. NMSE comparison of FRPS and PLUG Algorithms.

It can be seen from Figures 8 and 9 that the channel estimation normalized MSE of the edge user and the central user under both algorithms has a decreasing trend with the increase in the number of antennas. However, the MSE and SINR curves of the central users in both algorithms overlap, indicating that the center user multiplexes the pilots and causes the same pilot contamination. The edge user allocates orthogonal pilots under the PLUG algorithm, so there is no pilot contamination. When the number of antennas is 256, the NMSE performance of the edge users is increased by 10.88 dB, and the SINR performance is improved by 3.23 dB. Figure 10 is a graph showing the average achievable sum-rate performance of the center user and the edge user as the number of antennas increases when the number of edge users is one. Figure 12 shows the performance variation of the uplink achievable rate of the target cell of the central user and the edge user when the number of edge users is 1. It can be seen from Figure 10 that the average achievable rate of the central user in the PLUG algorithm is slightly lost. This is because the orthogonal pilot set is added under such algorithm, and the pilot overhead is increased, resulting in a decrease in spectral efficiency (SE), but the performance of the edge user is very good. With a large boost, the average achievable is increased by 0.42 bps/Hs when the number of antennas is 256. It can be seen from Figure 12 that the uplink achievable rate curves of the target cell are basically coincident under the two algorithms, which indicates that the PLUG algorithm does not bring loss to the overall performance of the target cell, but only increases the fairness of the central user and the edge use. It is also obvious from the results that the proposed PLUG algorithm reduces the probability of edge user communication interruption. Figure 13 compares the uplink target cell achievable rate in the case where the number of edge users of the FRPS and PLUG algorithms is different and for the number of antennas of 256. As can be seen from Figure 13, when the number of edge users increases, the achievable rate of the PLUG algorithm gradually decreases. This is because when the number of edge users is greater than one, the cost of pilot overhead exceeds the performance gains of the PLUG algorithm.

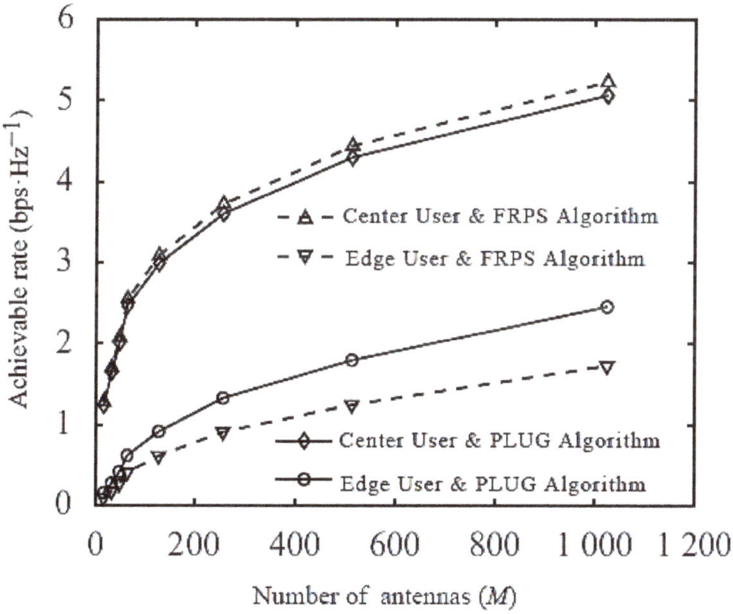

Figure 10. Comparison achievable rate of target cell of FRPS and PLUG algorithms.

4.3. Analysis of IPLUG Algorithm

Figure 11 compares the NMSE of the central user and the edge user against the number of antennas for the FRPS, PLUG, and IPLUG algorithms, wherein the IPLUG algorithm corresponds to a decision parameter 0.9, and the PLUG algorithm corresponds to an edge user of 1. As can be seen from Figure 11, when the number of antennas is 256, the NMSE performance of the edge users under the IPLUG algorithm is 2.95 dB higher than that of PLUG algorithm, and the central user NMSE is improved by 4.68 dB. The improvement of channel estimation performance greatly improves the performance of uplink signal detection and downlink coding. This is due to the flexibility brought by the IPLUG algorithm.

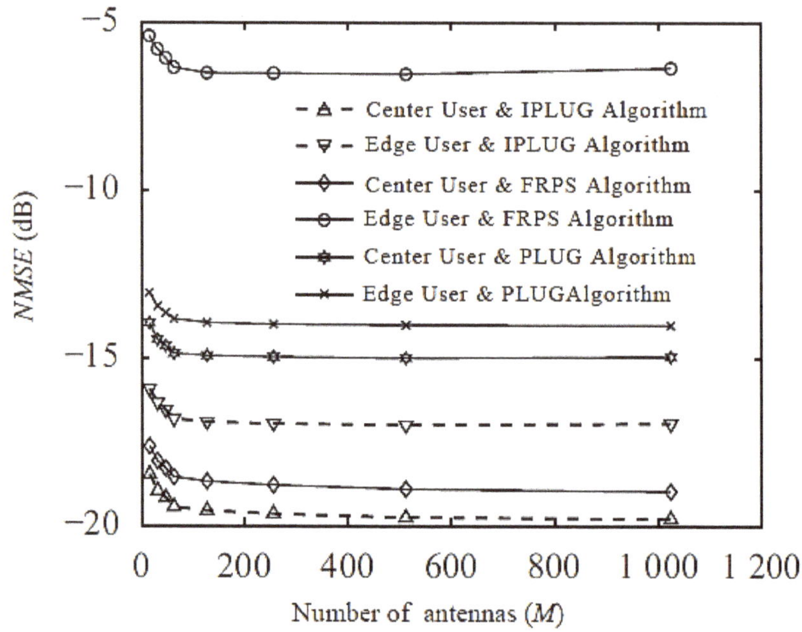

Figure 11. NMSE comparison of IPLUG, FRPS, and PLUG algorithms.

Figure 14 compares the uplink target cell achievable rate for FRPS and IPLUG with different decision parameters when the number of antennas is 256. As can be seen from Figure 14, when the decision parameter increases to $\lambda \geq 0.68$, the achievable rate of the IPLUG algorithm gradually increases. Therefore, the achievable rate of the proposed IPLUG algorithm have significantly exceeded the FRPS algorithm, which is a clear advantage of the IPLUG algorithm, avoiding the waste of pilot overhead caused by some users with good channel conditions being misclassified as edge users, or some users with poor channel conditions are being misclassified as the central user which results in communication interruption.

4.4. Analysis of Pseudo-Random Pilot Scheme

Figure 15 shows the probability density function (PDF) curves of the channel estimation MSE for the proposed pseudo-random scheme and compares it with no pilot contamination and with pilot contamination cases. The parameters set used for these results are: the number of users per cell is $K = 4$, the number of BS antennas $M = 100$ and pilot length $\tau = 8$. It can be seen from Figure 15 that the MSE of the channel estimation that can be obtained by all cells multiplexing the same set of orthogonal pilots is around 1.63, and the MSE distribution of the channel estimation obtained by the pilot after the proposed pseudo-random pilot design is around 0.82. It can be seen that the proposed pseudo-random code scheme is opposite to the channel when all the cells are multiplexed with the same set of orthogonal pilots. The estimated MSE performance has increased nearly two times. Although all cells use orthogonal pilots, there is no pilot contamination, but there are certain uncorrelated noise and fast fading effects when the number of antennas is limited. The MSE of channel estimation is not 0. As the number of BS antennas increases, the uncorrelated noise and fast fading effects are gradually averaged, and the MSE of channel estimation gradually approaches 0.

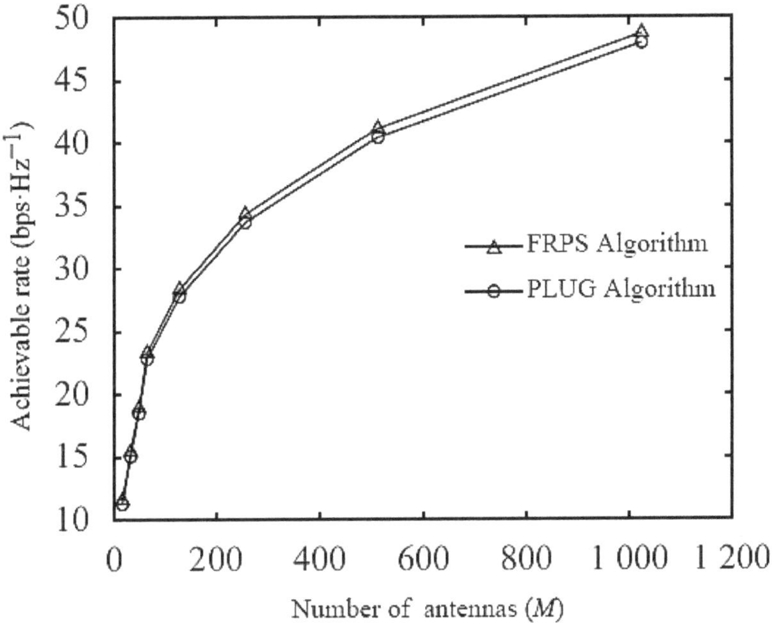

Figure 12. Achievable rate comparison of FRPS and PLUG algorithms for user fairness analysis.

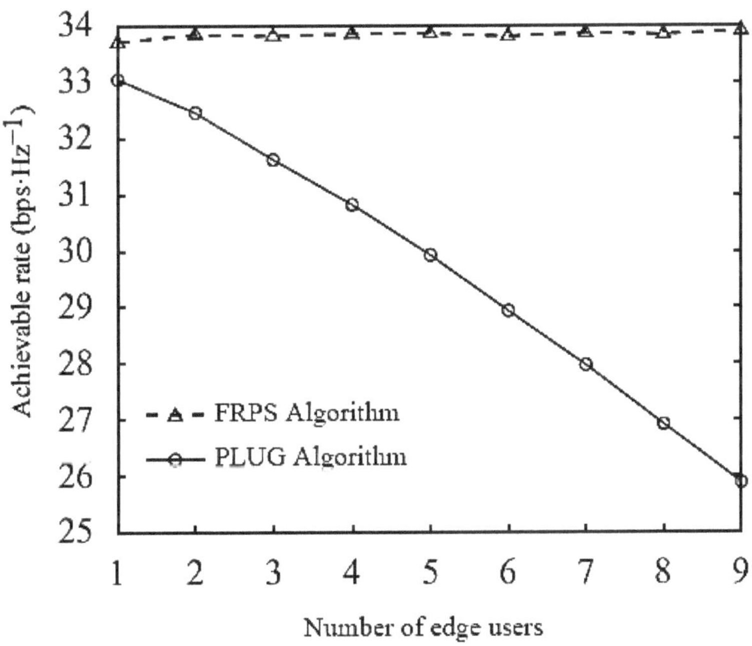

Figure 13. Comparison of achievable sum rate against a different number of edge users for FRPS and PLUG algorithms.

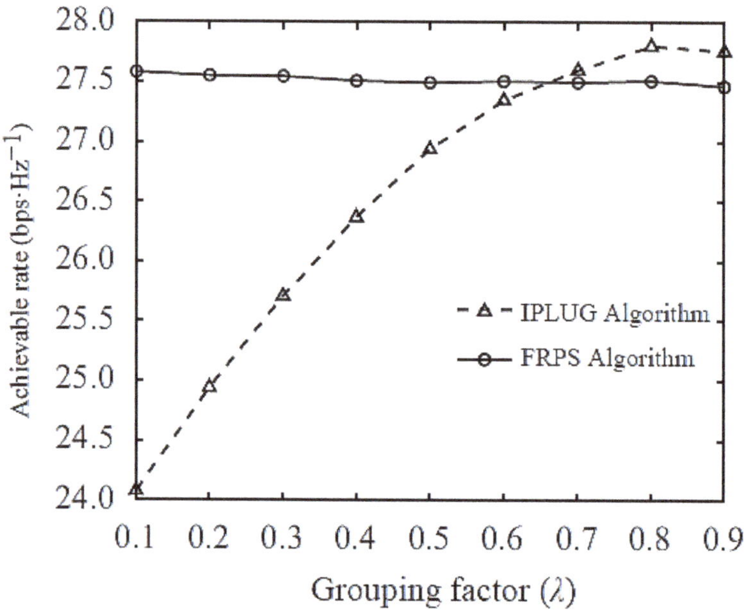

Figure 14. Comparison of achievable sum rate against a different number of user grouping factor λ for IPLUG and FRPS algorithms.

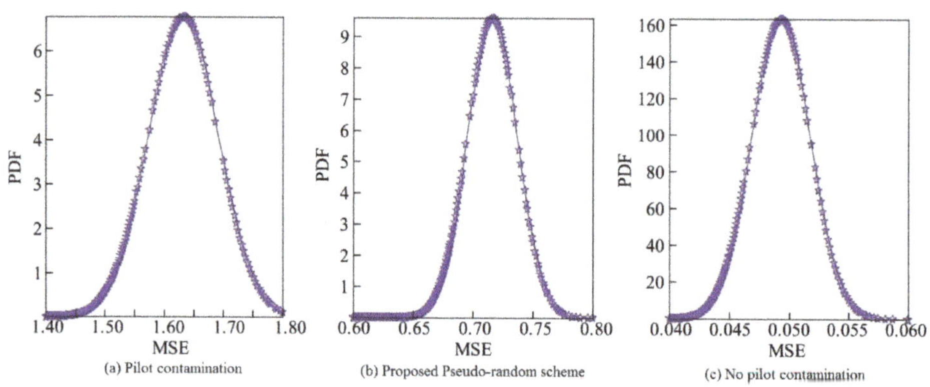

Figure 15. PDF comparison of the proposed pseudo-random code scheme with other cases.

Figure 16 depicts the trend of the system's downlink transmission BER as the number of antennas increases. The number of users per cell is $K = 8$, the number of BS antennas $M = 100$ and pilot length $\tau = 32$. It can be seen from Figure 16, as the number of BS antennas increases, the performance of the system is improved. In the case of full pilot contamination, the pilot performance is too serious, and the improvement of system performance is not obvious with the increase of the number of antennas. Therefore, it can be foreseen that when the number of antennas reaches a certain value, the system performance will not be improved.

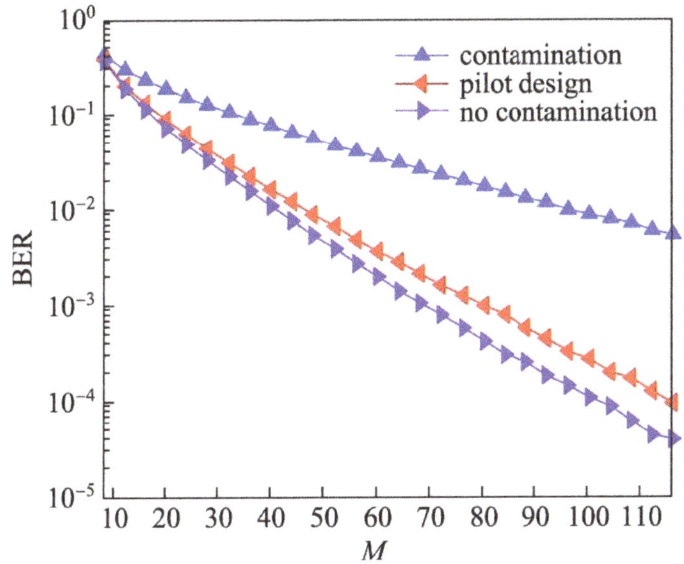

Figure 16. Comparison of the BER of the proposed pseudo-random pilot design scheme with other schemes against the number of BS antennas.

When the proposed pilot design scheme or no contamination is used, the performance of the system is significantly improved with the increase of the number of antennas. Figure 17 depicts the trend of the downlink BER of the system as the average transmit power ρ_f of the BS increases. The results are analyzed for different pilot lengths. The number of users per cell $K = 8$, the number of antennas configured by the BS $M = 100$. As can be seen from Figure 17 that when the number of BS antennas is large, the increase of the BS transmit power is effective for improving the system performance, but when it exceeds a certain value, the system performance will not be improved, and this value will be affected by the severity of pilot contamination. It can be seen from Figure 17 that in the case of complete pilot contamination, the system performance will not be improved after the BS transmission power reaches 25 dB. In the case of the proposed pilot design scheme (pseudo-random code) or no contamination, this value is greater than 25 dB. At the same time, it can be found that in the case of complete pilot contamination, the system performance will not be improved with the increase of the pilot length τ; in the absence of pilot contamination, the increase of the pilot length can improve the system performance to a small extent, but there is no need to sacrifice pilot overhead to boost such tiny performance. After using the proposed pseudo-random pilot design scheme, the performance of the system will increase significantly with the increase of the pilot length. Furthermore, when the pilot length $\tau = 64$, the performance achieved by the proposed pilot design scheme is very close to the case of no pilot contamination. Figure 18 compares the BER of the proposed pseudo-random pilot design scheme with other cases versus the number of cell users K. The number of antennas configured by the BS is $M = 100$, and pilot length $\tau = 128$. It can be seen from Figure 18 that in the case where the number of BS antennas is large, the increase in the number of cell users may deteriorate the performance of the system, but the performance of the system is still far better when the proposed pseudo-random pilot design or no pilot contamination scheme is used. At the same time, when the number of users is small and the pilot length $\tau = 128$, the same performance of the system can be almost achieved in the case of no contamination when the proposed pilot design scheme is adopted. This is because the increase in the length of the pseudo-random code can only make them infinitely close to the orthogonal but cannot completely remove the pilot contamination. In the case of a small number of users, the pilot contamination is small enough to affect the performance of the entire system.

In the case of a large number of users, the impact on the overall system performance is considerable due to the superposition of pilot contamination. Overall, the proposed pilot design scheme is very obvious for the improvement of system performance.

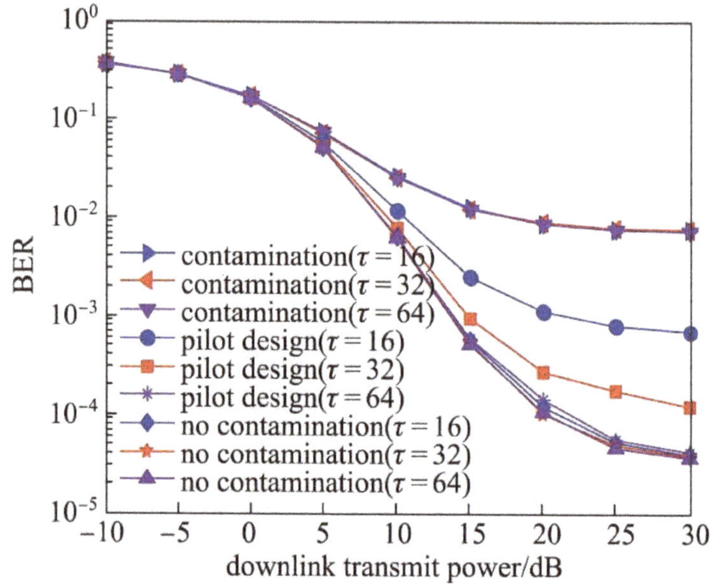

Figure 17. Comparison of the BER of the proposed pseudo-random code and another scheme versus the average BS power ρ_f.

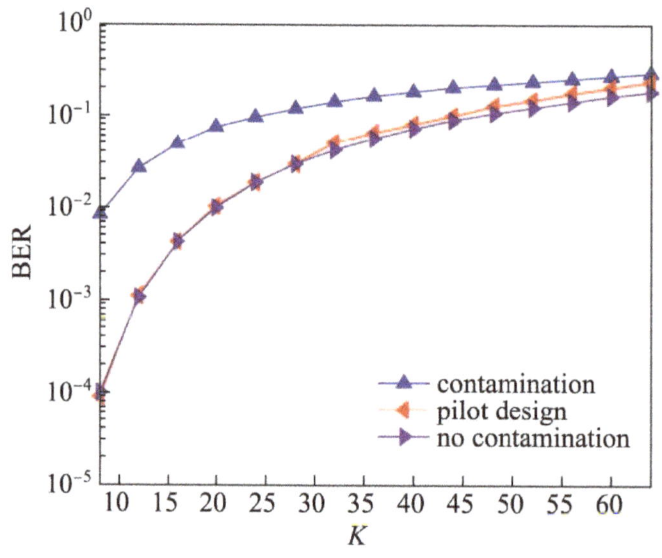

Figure 18. BER comparison against the number of cell users for the proposed pseudo-random pilot design and other cases.

Figure 19 compares the NMSE of the proposed IPLUG algorithm with the conventional state-of-the-art algorithms [29,30,32] wherein the IPLUG algorithm corresponds to a decision parameter

0.9. As can be seen from the figure that the proposed IPLUG algorithm shows better NMSE performance than the conventional algorithms for increasing number of base station antennas. Figure 20 compares.

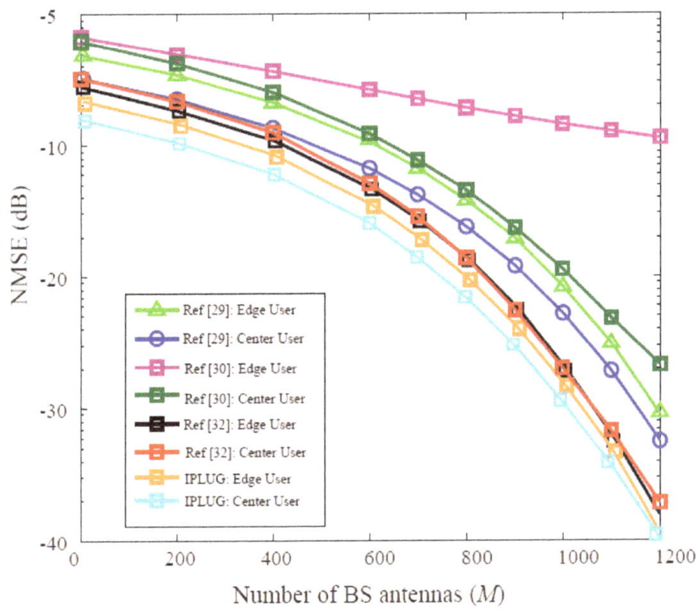

Figure 19. NMSE comparison of IPLUG and conventional algorithms.

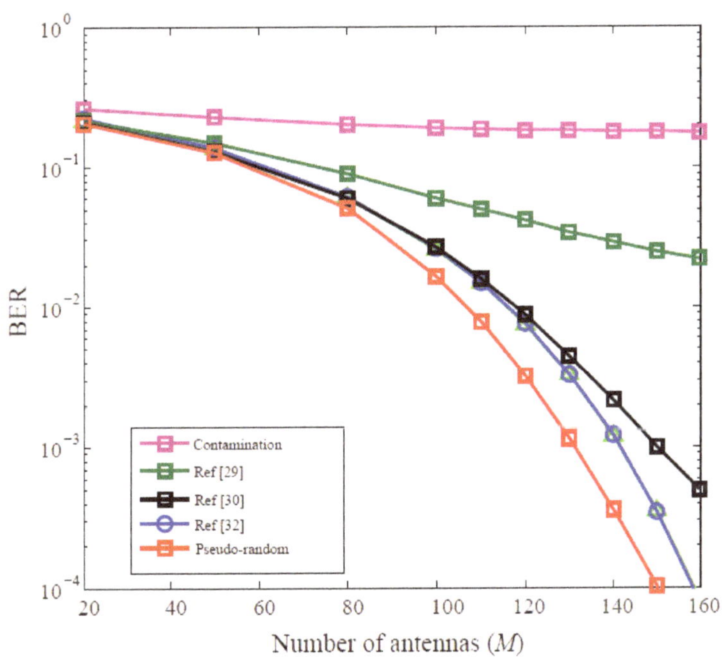

Figure 20. Comparison of the BER of the proposed pseudo-random pilot design scheme with other schemes against the number of BS antennas.

Figure 20 illustrates the BER comparison of the proposed pseudo-random code scheme and the conventional schemes as the number of antennas increases. The number of users per cell is $K = 8$, the number of BS antennas $M = 160$ and pilot length $\tau = 32$. It can be seen from Figure 20, as the number of BS antennas increases, the performance of the system is improved. The proposed pseudo-random code scheme shows better BER performance as compared with the conventional algorithms with increasing number of antennas.

5. Conclusions

This paper proposes a robust approach for effective pilot decontamination in massive MIMO systems. Two efficient pilot decontamination schemes are proposed. The first scheme is depending on Path Loss to perform User Grouping (PLUG) method while the second scheme is depending on the pseudo-random code. The PLUG scheme divides users into central and edge users. Edger users are allocated orthogonal pilots, and central users are assigned multiplex pilots, which improves edge user performance. The Improved PLUG scheme (IPLUG) is further proposed to overcome the deficiency of the PLUG scheme as it dynamically selects and correctly classifies the edge users and central users so that there is no wrong misclassification and therefore, the communication quality of service is improved. The analytical and simulation results show that the proposed IPLUG scheme can avoid the waste of pilot overhead caused by users with good channel conditions being misclassified as edge users, or the users with poor channel conditions being misclassified as central users, resulting in communication interruption, therefore, the proposed IPLUG scheme increases the fairness of communication for each user. The MSE of the expected channel estimation after the proposed pseudo-random code pilot design scheme is deduced and analyzed. It is found that this scheme can not only effectively improve the performance of channel estimation, but a more accurate channel estimation can be obtained by selecting an appropriate pseudo-random code and pilot length. Thereby improving the performance of the entire downlink system. The above conclusions are verified by numerical simulation. The numerical results also show that the proposed pilot design scheme depending on pseudo-random code can greatly improve the performance of the entire system limited by pilot contamination. The proposed PLUG and IPLUG schemes are focused only on the cells of the omnidirectional antenna, and other sector cells are not considered. At the same time, the pilot allocation is performed in the target cell, and the pilot allocation between the cells is independent of each other and has certain limitations. Therefore, the future research direction will be PLUG and IPLUG schemes depending on pilot decontamination in sectoral cells. The problem is to consider the mutual connection of each cell to reasonably allocate pilots throughout the system.

Author Contributions: Conceptualization, O.A.S., I.K. and B.M.L.; Data curation, O.A.S. and I.K.; Formal analysis, O.A.S., I.K. and A.T.; Funding acquisition, I.K. and B.M.L.; Investigation, O.A.S., I.K. and A.T.; Methodology, O.A.S., I.K., B.M.L. and A.T.; Project administration, O.A.S. and I.K.; Software, O.A.S., I.K. and A.T.; Writing—original draft, O.A.S., I.K. and B.M.L.; Writing—review & editing, O.A.S., I.K., B.M.L. and A.T.

Funding: This work was supported by the Basic Science Research Program through the National Research Foundation of Korea (NRF) funded by the Ministry of Education (grant number: NRF-2017R1D1A1B03028350).

Conflicts of Interest: The authors declare no conflict of interest.

References

1. Khan, I. A Robust Signal Detection Scheme for 5G Massive Multiuser MIMO Systems. *IEEE Trans. Veh. Technol.* **2018**, *67*, 9597–9604. [CrossRef]
2. Khan, I.; Singh, D. Efficient Compressive Sensing Based Sparse Channel Estimation for 5G Massive MIMO Systems. *AEU Int. J. Electron. Commun.* **2018**, *89*, 181–190. [CrossRef]
3. Khan, I.; Zafar, M.H.; Jan, M.T.; Lloret, J.; Basheri, M.; Singh, D. Spectral and Energy Efficient Low-Overhead Uplink and Downlink Channel Estimations for 5G Massive MIMO Systems. *Entropy* **2018**, *20*, 92. [CrossRef]
4. Khan, I.; Singh, M.; Singh, D. Compressive Sensing-Based Sparsity Adaptive Channel Estimation for 5G Massive MIMO Systems. *Appl. Sci.* **2018**, *8*, 754. [CrossRef]

5. Molisch, A.F.; Win, M.Z.; Winters, J.H. Space-time-frequency (STF) coding for MIMO-OFDM systems. *IEEE Commun. Lett.* **2002**, *6*, 370–372. [CrossRef]
6. Bolcskei, H.; Paulraj, A.J. Space-frequency coded broadband OFDM systems. In Proceedings of the IEEE Wireless Communications and Networking Conference (WCNC), Chicago, IL, USA, 23–28 September 2000; pp. 1–6.
7. Darsena, D.; Gelli, G.; Paura, L.; Verde, F. Blind Channel Shortening for Space-Time-Frequency Block Coded MIMO-OFDM Systems. *IEEE Trans. Wirel. Commun.* **2012**, *11*, 1022–1033. [CrossRef]
8. Redferm, A.J. Receiver window design for multicarrier communication systems. *IEEE J. Sel. Areas Commun.* **2002**, *20*, 1029–1036. [CrossRef]
9. Darsena, D.; Gelli, G.; Paura, L.; Verde, F. NBI-resistant zero-forcing equalizers for OFDM systems. *IEEE Commun. Lett.* **2005**, *9*, 744–746. [CrossRef]
10. Mergen, B.S.; Scaglione, A. Randomized Space-Time Coding for Distributed Cooperative Communication. *IEEE Trans. Signal Process.* **2007**, *55*, 5003–5017. [CrossRef]
11. Knopp, R.; Humblet, P.A. Information capacity and power control in single-cell multiuser communications. In Proceedings of the IEEE International Conference on Communications (ICC), Seattle, WA, USA, 18–22 June 1995; pp. 331–335.
12. Mastronarde, N.; Verder, F.; Darsena, D.; Scaglione, A.; van der Schaar, M. Transmitting Important Bits and Sailing High Radio Waves: A Decentralized Cross-Layer Approach to Cooperative Video Transmission. *IEEE J. Sel. Areas Commun.* **2012**, *30*, 1597–1604. [CrossRef]
13. Rusek, F.; Persson, D.; Buon, K.L.; Larsson, E.G.; Marzetta, T.L.; Edfors, O.; Tufvesso, F. Scaling up MIMO: Opportunities and challenges with very large arrays. *IEEE Signal Process. Mag.* **2013**, *30*, 40–60. [CrossRef]
14. Biswas, S.; Xue, J.; Khan, F.A.; Ratnarajah, T. Performance Analysis of Correlated Massive MIMO Systems with Spatially Distributed Users. *IEEE Syst. J.* **2016**, *12*, 1850–1861. [CrossRef]
15. Khan, I. Channel modeling and analysis of OWC-massive MIMO Systems. *Opt. Commun.* **2018**, *434*, 209–217. [CrossRef]
16. Marzetta, T.L. Noncooperative cellular wireless with unlimited numbers of base station antennas. *IEEE Trans. Wirel. Commun.* **2010**, *9*, 3590–3600. [CrossRef]
17. Jose, J.; Ashikhmin, A.; Marzetta, T.L.; Vishwanath, S. Pilot contamination and precoding in multi-cell TDD systems. *IEEE Trans. Wirel. Commun.* **2011**, *10*, 2640–2651. [CrossRef]
18. Ashikhmin, A.; Marzetta, T.L. Pilot contamination precoding in multi-cell large-scale antenna systems. In Proceedings of the IEEE International Symposium on Information Theory Proceedings (ISIT), Cambridge, MA, USA, 1–6 July 2012; Volume 1, pp. 1137–1141.
19. Liangbin, L.; Ashikhmin, A.; Marzetta, T.L. Pilot contamination precoding for interference reduction in large-scale antenna systems. In Proceedings of the 51th Annual Allerton Conference on Communication, Control, and Computing, Monticello, IL, USA, 2–4 October 2013; Volume 2, pp. 226–232.
20. Nguyen, T.M.; Le, L.B. Joint pilot assignment and resource allocation in multicell massive MIMO network: Throughput and energy efficiency maximization. In Proceedings of the IEEE Wireless Communications and Networking Conference (WCNC), New Orleans, LA, USA, 9–12 March 2015; pp. 393–398.
21. Cheng, H.V.; Björnson, E.; Larsson, E.G. Optimal Pilot and Payload Power Control in Single-Cell Massive MIMO Systems. *IEEE Trans. Signal Process.* **2017**, *65*, 2363–2378. [CrossRef]
22. Dao, H.T.; Kim, S. Pilot power allocation for maximizing the sum rate in massive MIMO systems. *IET Commun.* **2018**, *12*, 1367–1372. [CrossRef]
23. Elijah, O.; Leow, C.Y.; Rahman, T.A.; Nunoo, S.; Iliya, S.Z. A Comprehensive Survey of Pilot Contamination in Massive MIMO—5G System. *IEEE Commun. Surv. Tutor.* **2015**, *18*, 905–923. [CrossRef]
24. Hoydis, J.; Brink, S.T.; Debbah, M. Massive MIMO in the UL/DL of cellular networks: How many antennas do we need? *IEEE J. Sel. Areas Commun.* **2013**, *31*, 160–171. [CrossRef]
25. Khan, I.; Zafar, M.H.; Ashraf, M.; Bayati, A.K.S. Computationally Efficient Channel Estimation for 5G Massive Multiple-Input Multiple-Output Systems. *Electronics* **2018**, *7*, 382. [CrossRef]
26. Yin, H.; Gesbert, D.; Filippou, M.; Liu, Y. A coordinated approach to channel estimation in large-scale multiple-antenna systems. *IEEE J. Sel. Areas Commun.* **2012**, *31*, 264–273. [CrossRef]
27. Neumann, D.; Joham, M.; Weiland, L.; Utschick, W. Low-complexity computation of LMMSE channel estimates in massive MIMO. In Proceedings of the 19th International ITG Workshop on Smart Antennas, Ilmenau, Germany, 3–5 March 2015; pp. 1–6.

28. Ngo, H.; Larsson, E.G. EVD-based channel estimation in multicell multiuser MIMO systems with very large antenna arrays. In Proceedings of the IEEE International Conference on Acoustics, Speech and Signal Processing, Kyoto, Japan, 25–30 March 2012; pp. 3249–3252.
29. Jin, S.; Wang, Z.; Zheng, L.; Wong, K.; Huang, Y.; Tang, X. On massive MIMO zero-forcing transceiver using time-shifted pilots. *IEEE Trans. Veh. Technol.* **2016**, *65*, 59–74. [CrossRef]
30. Xiong, X.; Jiang, B.; Gao, X.; You, X. QoS-Guaranteed User Scheduling and Pilot Assignment for Large-Scale MIMO-OFDM Systems. *IEEE Trans. Veh. Technol.* **2015**, *65*, 6275–6289. [CrossRef]
31. Dai, X. Optimal training design for linearly time-varying MIMO/OFDM channels modeled by a complex exponential basis expansion. *IET Commun.* **2007**, *1*, 945–953. [CrossRef]
32. Saxena, V.; Fodor, G.; Karipidis, E. Mitigating pilot contamination by pilot reuse and power control schemes for massive MIMO systems. In Proceedings of the IEEE 81st Vehicular Technology Conference, Glasgow, UK, 11–14 May 2015; pp. 1–6.
33. Zhu, X.; Wang, Z.C.; Dai, L.L.; Qian, C. Smart pilot assignment for massive MIMO. *IEEE Commun. Lett.* **2015**, *19*, 1–4. [CrossRef]
34. Jin, S.; Li, M.; Huang, Y.; Du, Y.; Gao, X. Pilot scheduling schemes for multi-cell massive multiple-input-multiple-output transmission. *IET Commun.* **2015**, *9*, 689–700. [CrossRef]
35. Litvinenko, L. OFDM signal PAPR reduction by pre-scrambling and clipping for frequency domain comb-type channel estimation case. In Proceedings of the 23rd International Conference Radioelectronika, Pardubice, Czech Republic, 16–17 April 2013; pp. 273–277.
36. Yang, B.; Deng, C.J.; Wu, P.H.; Xi, J.; Shi, L. Image encryption algorithm depending on two-one-dimensional logistic chaotic inter-scrambling systems and m-sequence. In Proceedings of the IEEE International Conference on Software Engineering and Service Science (ICSESS), Beijing, China, 27–29 June 2014; pp. 521–524.

electronics

MDPI

Article

An Efficient Pilot Assignment Scheme for Addressing Pilot Contamination in Multicell Massive MIMO Systems, Shamala Subramaniam and Ali Mohammed Mansoor

Ahmed S. Al-hubaishi [1,*], Nor Kamariah Noordin [1,2], Aduwati Sali [1,2], Shamala Subramaniam [3,4] and Ali Mohammed Mansoor [5]

1 Department of Computer and Communication Engineering, Universiti Putra Malaysia, Serdang 43400, Malaysia; nknordin@upm.my (N.K.N.); aduwati@upm.edu.my (A.S.)
2 Wireless and Photonics Networks Laboratory, Universiti Putra Malaysia, Serdang 43400, Malaysia
3 Department of Communication Technology and Network, Universiti Putra Malaysia, Serdang 43400, Malaysia; drshamala@gmail.com
4 Sports Academy, Universiti Putra Malaysia, Serdang 43400, Malaysia
5 Faculty of Computer Science & Information Technology, University of Malaya, Kuala Lumpur 50603, Malaysia; ali.mansoor@um.edu.my
* Correspondence: alhubaishiahmed033@gmail.com; Tel.: +60-967-777-888-129

Received: 25 February 2019; Accepted: 21 March 2019; Published: 27 March 2019

Abstract: The reuse of the same pilot group across cells to address bandwidth limitations in a network has resulted in pilot contamination. This causes severe inter-cell interference at the targeted cell. Pilot contamination is associated with multicell massive multiple-input multiple-output (MIMO) systems which degrades the system performance even when extra arrays of antennas are added to the network. In this paper, we propose an efficient pilot assignment (EPA) scheme to address this issue by maximizing the minimum uplink rate of the target cell's users. To achieve this, we exploit the large-scale characteristics of the fading channel to minimize the amount of outgoing inter-cell interference at the target cell. Results from the simulation show that the EPA scheme outperforms both the conventional and the smart pilot assignment (SPA) schemes by reducing the effect of inter-cell interference. These results, show that the EPA scheme has significantly improved the system performance in terms of achievable uplink rate and cumulative distribution function (CDF) for both signal-to-interference-plus-noise ratio (SINR), and uplink rate.

Keywords: pilot contamination; massive MIMO; pilot assignment; large-scale fading coefficients

1. Introduction

Equipping the base station (BS) with a large number of antennas (also known as massive multiple-input multiple-output (MIMO)) has been considered one of the fundamental technologies that leads to 5G [1]. The introduction of this technology is to meet the increasing demand for mobile data in 5G [2]. Although the use of massive MIMO systems increases spectral efficiency, enhances energy efficiency, and reduces the effect of small scale fading [3–7], but invariably promotes pilot contamination. In massive MIMO, time-division duplex (TDD) protocol is preferred over the frequency-division duplex (FDD) [8,9], as the former allows channel estimation in one direction (i.e., uplink) and avoids the estimation of the other side (i.e., downlink) due to channel reciprocity property. In other words, the use of TDD based channel reciprocal minimizes the overhead signals used for channel estimation, which largely saves network bandwidth. Although, the channel estimation ensures high utilization of TDD massive MIMO via uplink transmission, but its channel coherence blocks are restricted in size (limited size). Therefore, the orthogonal pilot sequences cannot be allocated for all users among the cells. To overcome this problem, the orthogonal pilot sequences have to be reused across the cells. Although, pilot reuse approach is a remarkable way forward in addressing the associated problem, however, the channel estimate obtained in a given cell will be contaminated by pilots transmitted by users in other cells. Specifically, the inter-cell interference exacerbates the estimation error and also makes sure the channel estimation of two or more users sharing the same pilot sequence is correlated at a given cell [10]. Thus, with multicell massive MIMO systems, its performance deteriorates during uplink and downlink transmission. This issue is referred to as pilot contamination, and depicted in Figure 1.

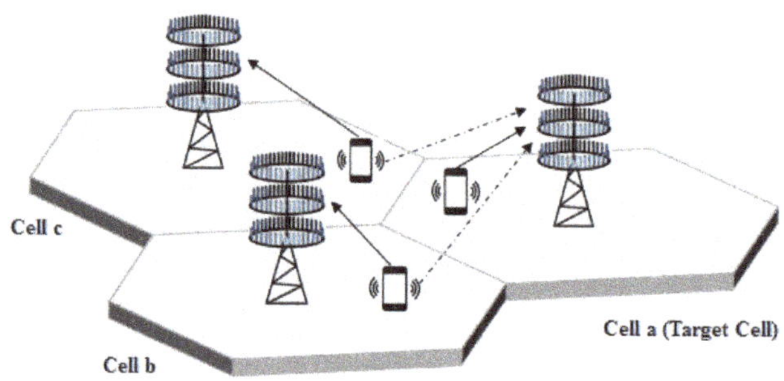

Figure 1. The effect of pilot contamination in multicell massive MIMO systems at a cell a, where the solid line represents the direct gain and the dotted line represents the inter-cell interference.

To address the issue associated with pilot contamination, several research methods have been proposed to eliminate/relieve pilot contamination. Among these methods, the pilot assignment technique is identified to be a potential technique for solving this problem. Smart pilot assignment (SPA) method proposed by [11], focused on adjusting the combination between the users and pilot sequences, but did not consider inter-cell interference which causes the pilot contamination. In this paper, we propose an efficient pilot assignment mechanism to improve the performance of users with respect to intense pilot contamination in multicell massive MIMO systems. We summarize our contributions below:

- We formulate the pilot assignment as an optimization problem and develop a heuristic algorithm, in order to maximize the minimum throughput considering the reduction in the inter-cell interference pilot contamination.
- We evaluate the performance of the proposed mechanism in terms of signal-to-interference-plus-noise ratio (SINR), and uplink rate with an extensive MATLAB simulation.
- We compare our work with SPA and other conventional schemes.

The rest of this paper is organized as follows. The related work is summarized in Section 2, the system model is described in Section 3, the pilot contamination phenomenon and the achievable uplink rate are illustrated in Section 4, the EPA scheme is explained in Section 5, the simulation results are depicted in Section 6, and finally, this paper is concluded in Section 7.

Notation: Throughout this paper, the bold lower case letters represent vectors and matrices are represented by bold upper case letters. $\mathbf{I}_M$ denotes the identity matrix of dimensions $M \times M$. The operators

$(.)^{-1}$, $(.)^{T}$, and $(.)^{H}$ are defined for inverse, transpose and conjugate transpose operations, respectively. The expectation operator is represented by $E[.]$.

2. Related Work

Different traditional algorithms based on pilot assignment have been proposed for pilot decontamination [12,13]. A vertex graph-coloring-based pilot assignment has been proposed in [12], the pilot sequences are allocated to the users according to the inter-cell interference (ICI) graph. The evaluation of ICI graph depends on both angle of arrival (AoA) correlation and distances between users. However, this scheme requires a second order channel information to construct ICI graph. A deep learning-based pilot allocation scheme (DL-PAS) is proposed in [13] to address the pilot contamination problem in massive MIMO systems. This algorithm aims at learning from the relationship between pilot assignment and users' location. However, the DL algorithm requires high data and subsequently takes a longer time to process the data.

The authors in [14,15] developed the location-based pilot assignment approaches for pilot decontamination. A new expression for line of sight (LOS) interference is derived in [14] which is considered as the criteria for pilot allocation. Although, there was an improvement in the sum spectral efficiency (SE), but the pilot assignment process takes a longer time to be implemented, especially in large networks. The work in [15] characterizes the angular region of the targeted user, and the pilot assignment process was implemented with the aim of making this region interference-free. This angular region is characterized by both the number of BS antennas and the location of the targeted user. However, the pilot assignment problem is formulated by the joint optimization problems which subsequently introduce high computational complexity.

In [16,17], the pilot allocation based pilot reuse (reuse factor more than 1) is also considered for pilot contamination's elimination technique. A systematically-constructed pilot reuse method is proposed in [16]. In this approach, the neighbor cells are allowed to use different sets of pilot sequences according to the tree division. To improve performance, it ensures larger distance between cells that share similar pilot sets, the depth of the tree is increased as the pilot contamination severity increases. This approach offers an effective performance when the ratio of the channel coherence time to the number of users in each cell is relatively large. For the purpose of improving the quality of service (QoS) of the edge users, a soft pilot reuse (SPR) scheme was proposed by [17]. The channel quality for each user is initially compared with a determined threshold before the pilot allocation procedure, but an increase in complexity was recorded due to additional computational cost incurred by finding the optimal threshold value.

By considering a fairness among users in order to mitigate the pilot contamination, pilot allocation schemes were proposed in [18,19]. Specifically, to maximize the sum rate of the system and guarantee fairness among users, a pilot allocation scheme was proposed by [18]. An optimization problem is formulated based on a max-product criterion, then both min-leakage algorithm and user- exchange algorithm based on greedy (UEBG) pilot allocation were suggested to solve the optimization problem. Although this scheme almost achieves the same performance as the optimal exhaustive search algorithm (ESA), it still suffers a setback due to high complexity. For the purpose of pilot contamination mitigation in [19], the pilot assignment scheme based on the harmonic SINR utility function was introduced to regulate the fairness among users. However, the system complexity increases as the number of users and network size grows (more than two cells).

Based on performance degradation of users, a pilot assignment scheme has been proposed in [20], the degradation performance is initially evaluated for all users according to the value of the uplink achievable rate. Therefore, the optimal pilot sequences were assigned to users who suffered from the highest degradation in a greedy way. Obviously, this scheme is not effective in bad channel conditions.

In [11,21], the pilot allocation approaches aim at enhancing the performance of users who suffer from bad SINR. The pilot allocation in [21] focused on maximizing the sum capacity of the whole system for pilot decontamination. In this work, the pilot sequences were assigned initially to the users who have bad channel condition. However, the complexity of the pilot assignment procedure increases as the network size is increased. A SPA scheme is proposed in [11] to improve the performance of users with poor SINR. Users with low channel quality were assigned to pilot sequences which resulted in a low interference. However, the achievement of this scheme is limited as it did not consider inter-cell interference which causes the pilot contamination.

Some authors have tried to make a combination of two schemes to get an improved performance as shown in [22,23]. As such, a joint pilot assignment scheme has been proposed by [22], in which time-shifted [24] and the SPA [11] schemes were combined in order to mitigate the effect of pilot contamination. Inter-group interference is suppressed according to [24] strategy, whereas SPA is used to reduce intra-group interference. Although an improved overall performance was recorded, the mutual interference between downlink data and uplink pilot signals cannot be eliminated despite the use of SPA scheme. New pilot assignment schemes such as greedy-based and swapping-based were implemented together with pilot contamination precoding design (PCP) for massive MIMO downlinks [23]. This combination offers a considerable improvement over the random pilot assignment, but the PCP matrix is changed according to the update in pilot assignment information.

By exploiting the channel sparsity for wideband massive MIMO system, the pilot contamination can be removed with the help of pilot assignment policy in [25]. The pilot assignment policy is designed to help identify the subspace of the desired channel. The difficulty in this approach, lies on how to deal with the subspace estimation, which can be realized through multiple frames after randomizing the pilot contamination.

Differing from the aforementioned works [11,20], we consider the source of inter-cell interference throughout pilot assignment, which is essentially the cause of the pilot contamination. In some other works [12–15], the availability of some factors (e.g., user location, AoA, or LOS interference) are needed for pilot assignment which are not always easy to estimate, while our approach requires only large-scale fading coefficients, which can be tracked easily as they do not frequently change during coherence interval. Besides, comparing to previous works [17–19,21], our algorithm is not computationally intensive, and therefore it can be applied for large-scale networks.

3. System Model

In this section, we describe the system model under which the TDD-massive MIMO systems are implemented. In this model, the uplink comprises L cells, in which each cell contains a BS equipped with M antennas. Furthermore, in each cell coverage area K single-antenna users communicate simultaneously to their designated BS, assuming that $M \gg K$ [2,5]. The propagation channels connecting the *k-th* user located in the *j-th* cell to the BS in *i-th* cell is modeled as Rayleigh block fading [26] and the channel vector $\mathbf{h}_{jk}^{i} \in \mathbb{C}^{M\times 1}$ is denoted as:

$$\mathbf{h}_{jk}^{i} = \mathbf{g}_{jk}^{i}\sqrt{\beta_{jk}^{i}}. \tag{1}$$

where $\mathbf{g}_{jk}^{i}$ and β_{jk}^{i} denote the small scale-fading vector and large-scale fading coefficient, respectively. The small scale-fading vector has a complex Gaussian distribution with zero mean and unity variance, $\mathcal{CN}$ $(0, \boldsymbol{I}_M)$, while the large-scale fading coefficient is referred to the effect of both path-loss and shadowing and it can be tracked easily as it changes slowly during coherence interval $\tau_c = B_c T_c$ [27–29]. We use B_c and T_c to denote the coherence bandwidth and the coherence time, respectively. Figure 2 illustrates the coherence block for TDD protocol. We also consider that large-scale fading coefficient is equal for all antenna elements, assuming that the distance between user k and BS is significantly larger than the distances between antenna elements.

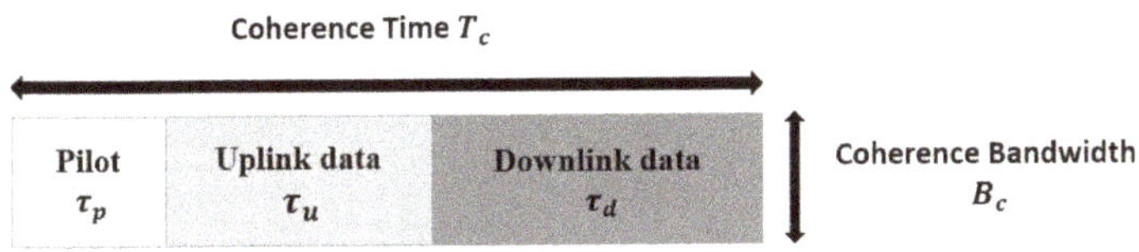

Figure 2. Time-division duplex (TDD) Protocol.

4. Pilot Contamination and Achievable Uplink Rate

Since the size of channel coherence blocks is limited, it is difficult to assign orthogonal pilot sequences to all users in order to prevent pilot contamination. Thus, it is necessary to reuse the pilot sequences in all cells to overcome this limitation [2]. The pilot sequences $\Phi = [\phi_1, \phi_2, ..., \phi_K]^T \in \mathbb{C}^{M\times\tau_p}$ are assumed mutually orthogonal $\Phi^T\Phi = \tau_p \mathbf{I}_K$ with length of τ_p. During the pilot phase, the pilot sequences are distributed randomly to all users. Thus, the received signal $\mathbf{U}_i^{\phi} \in \mathbb{C}^{M\times\tau_p}$ at the BS in the *i-th* cell can be written as:

$$\mathbf{U}_i^{\phi} = \sqrt{\rho_\phi}\sum_{j=1}^{L}\sum_{k=1}^{K}\mathbf{h}_{jk}^{i}\phi_k^T + \mathbf{N}_i^{\phi} \tag{2}$$

$$\mathbf{U}_i^{\phi} = \sqrt{\rho_\phi}\sum_{k=1}^{K_i}\mathbf{h}_{i_k}^{i}\phi_k^T + \sqrt{\rho_\phi}\sum_{\substack{j=1\\ j\neq i}}^{L}\sum_{k=1}^{K_j}\mathbf{h}_{jk}^{i}\phi_k^T + \mathbf{N}_i^{\phi} \tag{3}$$

where ρ_ϕ denotes the pilot transmission power, and $\mathbf{N}_i^{\phi} \in \mathbb{C}^{M\times\tau_p}$ denotes the additive white Gaussian noise (AWGN) matrix which is assumed independent and identically distributed (i.i.d) random variables whose elements have zero mean and variance σ_N^2. The received signal $\mathbf{U}_i^{\phi}$ is called the observation, in which the BS in the cell i can use it to estimate the channel responses. The first term in (3) represents the received pilot signals from users in the serving cell, whereas, the middle term represents the inter-cell interference signal from the neighbor cells, which causes the pilot contamination. Correspondingly, the received uplink data $\mathbf{u}_i^d \in \mathbb{C}^M$ at the BS in the *i-th* cell can be represented by:

$$\mathbf{u}_i^d = \sqrt{\rho_u}\sum_{j=1}^{L}\sum_{k=1}^{K}\mathbf{h}_{jk}^{i}x_{jk}^u + \mathbf{n}_i^u \tag{4}$$

where x_{jk}^u denotes the uplink transmitted symbol from user k located in the *j-th* cell, ρ_u denotes the power of the uplink transmitted symbol with $E[\mid x_{jk}^u \mid^2] = 1$, and $\mathbf{n}_i^u \in \mathbb{C}^{M\times\tau_u}$ denotes the AWGN vector with variance σ_n^2 and zero mean value. The minimum mean square error (MMSE) is exploited for the purpose of the channel estimation $\hat{\mathbf{h}}_{jk}^i \in \mathbb{C}^{M\times 1}$ [9]. Therefore, the MMSE estimated channel vector $\hat{\mathbf{h}}_{jk}^i$ based on the observation $\mathbf{U}_i^{\phi}$ in (2) can be given as [10]:

$$\hat{\mathbf{h}}_{jk}^i = \sqrt{\rho_u}\mathbf{R}_{jk}^i\mathbf{\Psi}_{jk}^i\mathbf{u}_{ijk}^p \tag{5}$$

and

$$\mathbf{\Psi}_{jk}^i = \left(\sum_{j,k}\rho_u\tau_p\,\mathbf{R}_{jk}^i + \sigma_u^2\mathbf{I}_K\right)^{-1} \tag{6}$$

where $\mathbf{u}_{ijk}^{p} = \mathbf{U}_i^{\phi}\ \phi_k^*$ which is called the received proceed signal, $\boldsymbol{\Psi}_{jk}^{i}$ denotes the inverse of the normalized correlation matrix, and $\mathbf{R}_{jk}^{i}$ denotes the spatial correlation matrix of the channel to be estimated, $\mathbf{R}_{jk}^{i} = E[\mathbf{h}_{jk}^{i}\mathbf{h}_{jk}^{iH}]$.

The estimated channel is then used to detect the uplink data symbol and precode the downlink data. Herein, we consider both maximum ratio combining (MRC) and zero forcing (ZF) as a linear detectors at the BS which are given by [30]:

$$\mathbf{A}_i = \begin{cases} \hat{\mathbf{H}}_i^i & MRC \\ \hat{\mathbf{H}}_i^i\ (\hat{\mathbf{H}}_i^{i\,H}\ \hat{\mathbf{H}}_i^i)^{-1} & ZF \end{cases} \tag{7}$$

The received detected signal is evaluated by multiplying the received uplink data signal $\mathbf{u}_i^d$ by the decoding vector $\mathbf{a}_{ik}^{iH}$, which represents the *k-th* column of the matrix $\mathbf{A}_i$ and $\mathbf{h}_{ik}^i$ is the *k-th* column of the matrix $\mathbf{H}_i^i$. Therefore, the detected symbol of user k at a given BS located in a cell i can be expressed as:

$$z_{iik}^u = \mathbf{a}_{ik}^{iH}\mathbf{u}_i^d = \mathbf{a}_{ik}^{iH}\left(\sqrt{\rho_u}\sum_{j=1}^{L}\sum_{k=1}^{K}\mathbf{h}_{jk}^i x_{jk}^u + \mathbf{n}_i^u\right) \tag{8}$$

$$z_{iik}^u = \sqrt{\rho_u}\mathbf{a}_{ik}^{iH}\mathbf{h}_{ik}^i x_{ik}^u + \sqrt{\rho_u}\sum_{\substack{n=1\\n\neq k}}^{K_i}\mathbf{a}_{ik}^{iH}\mathbf{h}_{in}^i x_{in}^u + \sqrt{\rho_u}\sum_{\substack{j=1\\j\neq i}}^{L}\sum_{k=1}^{K_j}\mathbf{a}_{ik}^{iH}\mathbf{h}_{jk}^i x_{jk}^u + \mathbf{a}_{ik}^{iH}\mathbf{n}_i^u \tag{9}$$

The first term in (9) represents the desired signal, the second one represents the intra-cell interference, the third term is the effect of pilot contamination (inter-cell interference), and the last one represents the uncorrelated noise.

Consequently, the average SINR of the *k-th* user in the target cell i can be evaluated as:

$$SINR_{ik}^u = \frac{\rho_u\ |\mathbf{a}_{ik}^{iH}\mathbf{h}_{ik}^i|^2}{\rho_u\ \sum_{\substack{j=1\\j\neq i}}^{L}\sum_{k=1}^{K_j}|\mathbf{a}_{ik}^{iH}\ \mathbf{h}_{jk}^i|^2 + \frac{v_{ik}^i}{\rho_u}} \tag{10}$$

and

$$v_{ik}^i = {\rho_u}^2\sum_{\substack{n=1\\n\neq k}}^{K_i}|\mathbf{a}_{ik}^{iH}\ \mathbf{h}_{in}^i|^2 + \rho_u\ ||\mathbf{a}_{ik}^i||^2$$

where v_{ik}^i denotes the intra-cell interference and uncorrelated noise, in which their effect is almost neglected as the number of antennas increases ($M \to \infty$) [5]. Then, the uplink SINR can be described by large-scale fading coefficients β_{jk}^i as follows:

$$SINR_{ik}^u = \frac{{\beta_{ik}^i}^2}{\sum_{j\neq i}{\beta_{jk}^i}^2},\quad when\ (M \to \infty) \tag{11}$$

It is clear from the above expression that the effect of small-scale fading and thermal noise are averaged out as the number of antennas is increased [5]. Therefore, the ergodic achievable uplink rate of the user k according to [26] is:

$$R_{ik}^u = \frac{1}{T_c}\sum_{\tau_u}(1 + SINR_{i\lambda}^u) \tag{12}$$

where R^u_{ik} is calculated in bit/channel use and τ_u refers to the uplink duration. From (12), it is obvious that the average uplink rate of multicell massive MIMO systems is limited due to pilot contamination and it cannot be boosted by increasing either the number of serving antennas or both ρ_u and ρ_p.

5. Proposed Scheme

In this section, an efficient heuristic algorithm is developed for addressing the multicell massive MIMO associated problem. To do this, the assignment and the reuse of pilot group across cells in the network is formulated as an optimization problem.

5.1. Problem Formulation

Formally, we formulate an optimization problem as depicted in:

$$\mathcal{P} \xrightarrow{M\to\infty} \hat{\mathcal{P}} : \max_{\forall k \in i} \left(min \frac{{\beta^i_{ik}}^2}{\min_{\forall \phi_k} \sum_{j\neq i} {\beta^i_{jk}}^2} \right) \tag{13}$$

The above optimization problem is based on the method proposed by [11]. In this method, it is assumed the number of antennas is very large and as such make use of the large-scale fading coefficients β^i_{jk}. To address problems related to pilot contamination, this study concentrates on assigning the pilot sequences for a specific cell in multicell massive MIMO systems. In the target cell, the number of possible iterations is defined by the number of K users which is usually very high. In contrast, the conventional scheme assigns the pilot sequences $\Phi = [\phi_1, \phi_2, ..., \phi_K]^T$ randomly to K users.

The performance of multicell massive MIMO systems is much degraded by the effect of the strong inter-cell interference from the neighbor cells and is exacerbated when the channel quality of the users in target cell is poor. Specifically, in the SPA scheme, the set of users with the worst channel quality are assigned pilot sequences with the lowest inter-cell interference. Although these pilot sequences have the lowest interference, they are still considered high interference pilot sequences when used by users which have bad channel quality. Therefore, the interference that is associated with such pilot sequences must be minimized.

5.2. Proposed Solution

To achieve minimal outgoing inter-cell interference among neighbor cells for the target cell, we ensure a weak channel cross gain of the interfering users against desired users. The large-scale fading coefficients are used to measure the effect of inter-cell interference at the target BS. Thus, the effect of inter-cell interference can be measured by using these fading coefficients as it changes progressively during the coherence interval τ_c, as every user measured result is sent to its corresponding BS. The required conditions for finding the large-scale fading coefficients can be met in long term evolution-advanced (LTE-A) systems. These corresponding BSs contain the channel's information for the available BSs. The user keeps tracking these BSs until a reliable BS is identified for suitable handover. To enhance the cooperation among the BSs, we assume acquisition of the coordinated multi points (CoMP). Furthermore, a mobility management entity (MME) is connected to BSs by S1 interface and has a huge ability for computing. As a result, this unit can collect the large-scale fading coefficients from the connected BSs [31,32].

To abate the effect of setback suffered by users due to poor channel quality or high interference, the SINR is optimized. This is done by assigning the pilot sequence, which is associated with low interference, to users having poor channel quality.

In order to achieve this, we propose a heuristic algorithm based on SPA to solve the optimization problem in (13). Before illustrating the algorithm, we need to define a set of parameters η_{jk} which characterizes the squared cross gain of the interfering users from neighboring cells:

$$\eta_{jk} = {\beta^{i}_{jk}}^{2}, \quad k = 1,2,...,K, \;\; j = 1,2,...,L \; and \; j \neq i$$

The interference that is produced by users who shared the same pilot sequence ϕ_k can be evaluated at the target cell as:

$$\xi_k = \sum_{j \neq i} \eta_{jk} \tag{14}$$

In addition, the set of parameters ω_k is used to characterize the square channel quality of the target cell's users which can be expressed by:

$$\omega_k = {\beta^{i}_{ik}}^{2}, \qquad k=1,2,...,K$$

So, the optimization problem in (13) can be re-written as the following:

$$\mathcal{P} \xrightarrow{M \to \infty} \hat{\mathcal{P}} : \max_{\forall k \in i} \left(\min_{\min \forall \phi_k} \frac{\omega_k}{\sum_{j \neq i} \eta_{jk}} \right) \tag{15}$$

The proposed algorithm EPA is summarized in Algorithm 1 to solve the above optimization problem.

Algorithm 1 Efficient Pilot Assignment (EPA).

1: **Input:**
2: $\beta^{i}_{jk} \; \forall i, \; j \; and \; k$
3: **Output:**
4: Assigning pilot sequences Φ for all users in all j cells $\forall j = 1,2,...,L$
5: **Procedure:**
  for *each neighbor cells* $j \neq i$ **do**
  for *all users K in cell j* **do**
6: Evaluate: $\eta_{jk} = {\beta^{i}_{jk}}^{2}, k = 1,2,...,K$
  end for
7: Classifying the users into different levels: $V_1, V_2, ..., V_k, ..., V_K$.
  $V_k = [\; \eta_{1k}, \eta_{2k}, ..., \eta_{jk}, ..., \eta_{(L-1)k} \;]$
8: Assign the pilot sequence ϕ_k to the users in V_k.
  end for
9: Find the sum: $\xi_k = \sum_{j \neq i} \eta_{jk} \;, \xi_k \in [\xi_1, \xi_2, ..., \xi_K] \; \forall \Phi$
  for *each user in the target cell i* **do**
10: Evaluate $\omega_k = {\beta^{i}_{ik}}^{2}, \qquad k = 1,2,...,K$
  end for
11: Sort ω_k in descending order: $\omega_1 \geq \omega_2, ..., \geq \omega_k, ..., \geq \omega_K$
12: Sort ξ_k in descending order: $\xi_1 \geq \xi_2 \geq, ..., \geq \xi_k, ..., \geq \xi_K$
13: Assigning the pilot sequence ϕ_k ,which associates with ξ_k, to the user who has ω_k.

The available large scale fading coefficients are exploited to measure the interference from the neighbor cells. From the above algorithm, the users in the neighbor cells are classified into different levels according to the value of squared cross gain (η_{jk}), which gives an indication of the strength of the interference at the target cell i. The users that cause the highest interference (which have the largest η_{jk}) are classified as the level V_1 users. This level involves the worst interfering users from each neighbor cell. The second level V_2 contains the users which cause less interference than that in V_1. This classification process will continue until the last level V_K, which contains the users that produce the smallest interference. The *k-th* interference level can be represented by:

$$V_k = [\eta_{1k}, \eta_{2k}, \ldots, \eta_{(L-1)k}] \tag{16}$$

The amount of interference that is produced by the users in each level is described by (14). After that, the interfering users in each level are assigned the same pilot sequence. For instance, the users in V_1 and V_K are assigned the pilot sequences ϕ_1 and ϕ_K, respectively. As a result, the pilot sequence ϕ_1 is suffering from the highest interference, whereas ϕ_K is the one with the lowest interference. The remaining pilot sequences have different levels of interference between ϕ_1 and ϕ_K. After minimizing the inter-cell interference at the serving BS, the second step is to assign pilot sequences to its users and this can be achieved by solving the following formula:

$$\mathcal{P} \xrightarrow{M\to\infty} \hat{\mathcal{P}} : \underset{\forall k \in i}{max} \left(min \frac{\varpi_k}{\xi_k} \right) \tag{17}$$

Obviously, from the EPA algorithm, the pilot assignment process for the users of the target cell depends on, both the squared channel quality ϖ_k and the minimized outgoing interference ξ_k, which is caused by users sharing the same pilot sequence in the level V_k . The optimization problem in (17) can be solved with the help of the SPA algorithm. In this algorithm, users that suffer setbacks due to bad channel quality are exempted from the pilot sequence as it will cause severe interference. Thus, the sets of users with the worst channel quality are assigned pilot sequence with the lowest inter-cell interference. For the remaining cells, the process will continue in a sequential way, excluding the cells that are already included with the target cell.

Furthermore, our algorithm is not computationally intensive in the sense that it ultimately relies on cell sorting, thus the time complexity it incurs is $O\,(L\,K \log K)$, and therefore it works faster if compared to recent schemes. For example, EPA shows less computational complexity than the work in [19,21], which incur $O\,(L\,K^3)$ and $O\,(L^2\,K \log K)$, respectively. In addition, the scheme in [17] incurs $O\,(M({K^2}_e + {K^2}_{CS})$, where K_e denotes the number of edge users in the network, and K_{CS} represents the number of users in the largest cell. So apparently [17] is much more intense than EPA. The SPA scheme [11], as it is fundamentally limited to only a target cell optimization, unsurprisingly it incurs only $O\,(\,K \log K)$.

6. Simulation Results

The base code implemented is obtained from [26], while Monte Carlo simulation is used to evaluate the performance of the EPA scheme. A typical hexagonal cellular network made up of L cells is considered in the EPA scheme. Each of these cells comprises of a BS which is equipped with M number of antennas and K users with single antennas under its coverage area [2,5]. A center cell surrounded by all other cells is considered as a target cell. The system parameters are summarized in Table 1. The parameter β^i_{ik} is modeled in decibel as [10]:

$$\beta^i_{ik} = \Upsilon + 10\,\alpha\,\log_{10}\left(\frac{d^i_{jk}}{1\text{ km}}\right) + F^i_{jk} \tag{18}$$

where d^i_{jk} (km) is the distance between the *k-th* user in the *j-th* cell and the BS in the *i-th* cell, α is the path-loss exponent, Y determines the median channel gain at 1 km as a reference distance which can be calculated according to many propagation models [33], and $F^i_{jk} \backsim \mathcal{N}(0, \sigma^2_{sf})$ is the shadow fading which creates log-normal random variations around the nominal value $\Upsilon + 10\ \alpha\ \log_{10}(d^i_{jk}/1\ \text{km})$.

Table 1. Simulation parameters.

Parameter	Value (for Figures 3–12)	Value (for Figure 13)
Number of Cells L	7	7
Number of BS Antennas M	$8 \leq M \leq 512$	$8 \leq M \leq 512$
Number of users in each cell K	8	20
Cell Radius$R$	500 m	300 m
Cell edge SNR	15 dB	20 dB
Path-loss Exponentα	3	3
Shadow Fading Standard Deviation σ^2_{sf}	8	8
Thermal Noise Variance	−174 dBm/H	−174 dBm/H

We evaluate the SPA [11] and the conventional schemes [2,5] against the EPA scheme. Figure 3 depicts the average uplink rate per user of the EPA, SPA and conventional schemes against the number of BS's antennas using the ZF as a linear detector. Obviously, the average uplink rate of the EPA scheme outperforms the other schemes. This improvement can be attributed to the policy implemented for pilot assignment in the neighbor cells. This implemented policy ensures a significant reduction of the inter-cell interference at the serving BS, which invariably leads to a better throughput. Due to the pilot assignment in the target cell which was executed according to the users' channel quality, the SPA scheme achieves better performance than the other conventional scheme. However, the performance of both SPA and conventional schemes changes slightly when the number of antennas exceeds certain points (e.g., greater than 150).

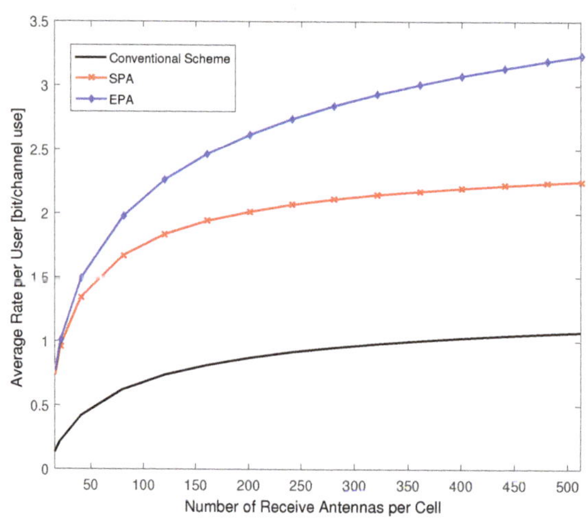

Figure 3. The average uplink rate per user with zero forcing (ZF) for different numbers of antennas.

Figure 4 shows the impact of the EPA scheme when using the MRC as a linear detector. It can be clearly observed that the average uplink rate per user (bits/channel use) is substantially enhanced by the EPA scheme when the number of antennas is increased. The superiority of the EPA scheme over other schemes, arose as a result of the minimization of the inter-cell interference that comes from the neighbor cells. This is achieved by allowing the users in each interference level V_k to share the same pilot sequence. Consequently, EPA scheme has shown a low interference from the neighbor cells compared to the other schemes.

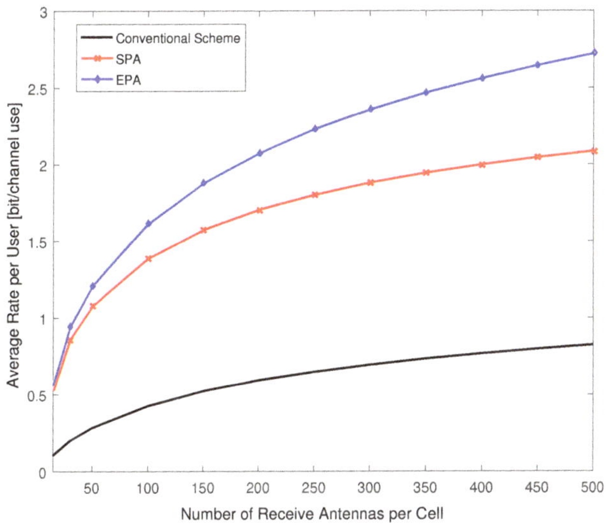

Figure 4. The average uplink rate per user with maximum ratio combining (MRC) for different numbers of antennas.

Figure 5 depicts the performance of the EPA scheme when compared with both conventional and SPA schemes in terms of cumulative distribution function (CDF) of the average SINR. When the number of BS's antennas is 64 with ZF detector, the probabilities of the average uplink SINR being less than −10 dB for the conventional, the SPA, and the proposed EPA schemes are almost 80%, 26.25%, and 10%, respectively. The improvement is achieved because the effect of the interference, which is associated with the pilot sequences, on channel quality of the users in the target cell became slight, which effectively increased the SINR of the system.

Figure 6 depicts the CDF of the minimum SINR when M is 64. It is evident that the minimum SINR of the EPA scheme is significantly improved when compared with SPA and conventional schemes. For example, the probability of the minimum SINR to be less than −20 dB for the EPA scheme is approximately 16.25%, while this probability is about 34.6% and 79.6% for the SPA and the conventional schemes, respectively. The reason behind this improvement is due to assigning the pilots of the users with the lowest interference, in the neighbor cells, to the users who have bad channel quality in the target cell. In consequence, the performance of these users was improved due to the reduction of their interference.

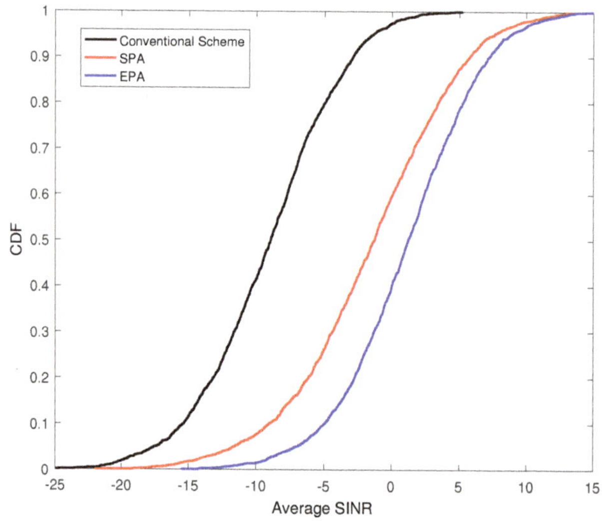

Figure 5. The cumulative distribution function (CDF) of the average signal-to-interference-plus-noise ratio (SINR) when $M = 64$ using ZF.

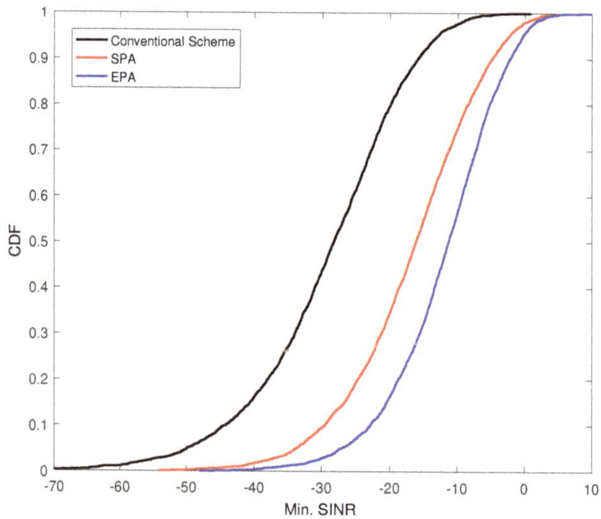

Figure 6. The CDF of the min. SINR when $M = 64$ using ZF.

Figures 7 and 8 depict the CDF of average and minimum SINR, respectively, using MRC detector when M is 64. As observed from Figures 7 and 8, the EPA scheme outperforms the SPA and the conventional schemes. As shown in Figure 7, the EPA scheme increases the average SINR by 1.8 dB over the SPA scheme, whereas it increases up to 4.69 dB for minimum SINR, as illustrated in Figure 8.

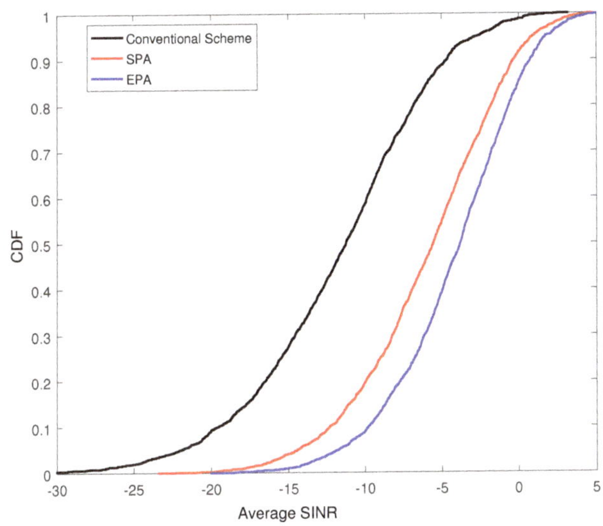

Figure 7. The CDF of the average SINR when $M = 64$ using MRC.

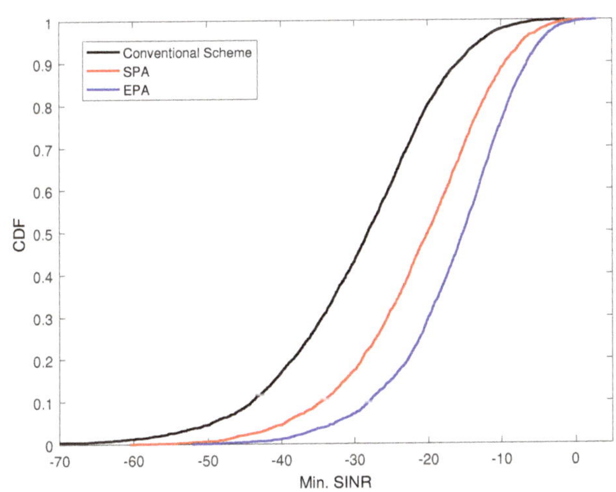

Figure 8. The CDF of the min. SINR when $M = 64$ using MRC.

From Figures 5–8, the minimum SINR always achieves a better performance. In other words, the performance of edge users is significantly enhanced. This is due to the fact that the inter-cell interference has been greatly reduced at the target cell while the users with poor channel quality are assigned to the suitable pilot sequences in order to maximize its SINR. Moreover, the results obtained as a result of using ZF and MRC linear detections, are approximately comparable when run on the same parameters setting. This is because the inter-cell interference is greatly reduced by the EPA scheme that runs before the process of signal detection.

By using ZF detector, the performance of the EPA scheme has been examined in terms of the CDF of the average uplink rate when M is 64, as shown in Figure 9. It can be seen that the performance of the CDF in the conventional scheme is highly influenced by the pilot contamination. The assignment of the pilot randomly, has led to the worst performance compared to the SPA and the EPA schemes. On the other hand, the EPA scheme outperforms the SPA and the conventional schemes, since the effect of users who cause the highest interference is considered weak compared to users having good channel quality when they are assigned the same pilot sequence. As a result, these interfering users are excluded from sharing the same pilots of users with bad channel quality.

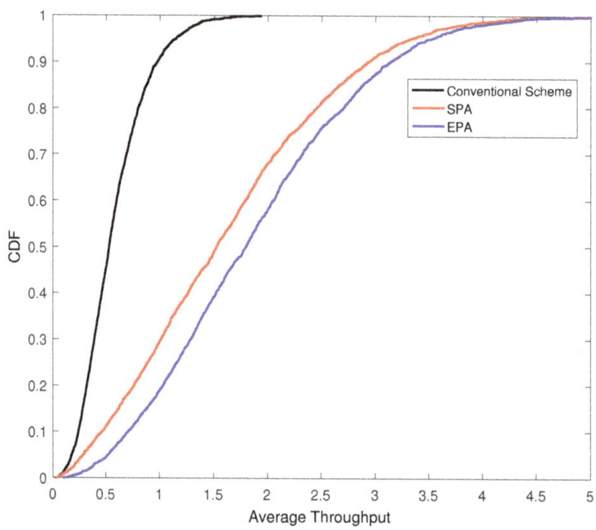

Figure 9. The CDF of the average uplink rate when $M = 64$ using ZF.

Result of evaluation for the CDF of the minimum uplink rate is depicted in Figure 10. It is clear that the EPA scheme performs better than the other schemes. For example, the minimum uplink rate of the EPA scheme is doubled when compared to the SPA scheme. This improvement has been achieved because the interference associated with pilot sequences, which is allocated to users with bad channel quality, was reduced effectively by the EPA scheme.

Figures 11 and 12 represent the CDF of the average and the minimum uplink rate, respectively, when the MRC is utilized and M is 64. The EPA scheme achieves the highest performance when compared with other schemes, especially in the minimum uplink rate. Specifically, the achieved gain in minimum uplink rate is doubled while it is 1.2 times in average uplink rate in comparison with SPA. The reason for this improvement in the minimum uplink rate is due to the priority given to the users having the worst channel quality during the pilot assignment process.

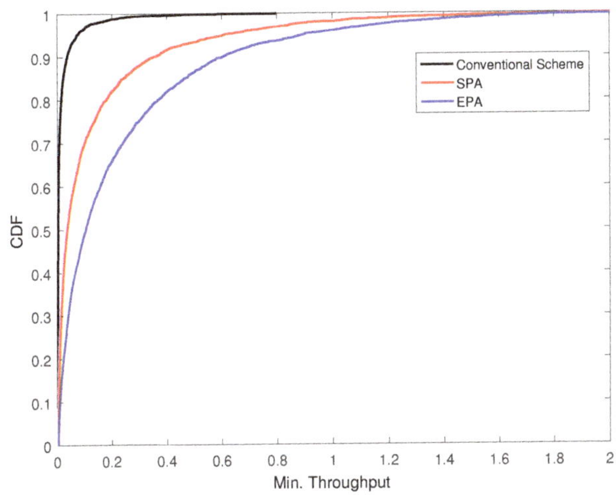

Figure 10. The CDF of min. uplink rate when $M = 64$ using ZF.

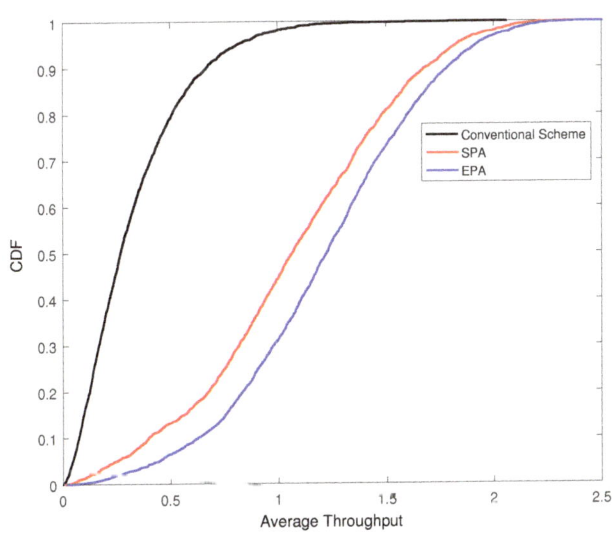

Figure 11. The CDF of average uplink rate when $M = 64$ using MRC.

In order to verify the effectiveness of the EPA scheme, the average uplink rate against the number of antennas has been evaluated in Figure 13 with different parameters, considering ZF detector. These parameters, which are shown in Table 1, increase the interference severity at the target cell. Obviously, the average uplink rate of EPA schema is higher than other schemes, despite the intensity of interference.

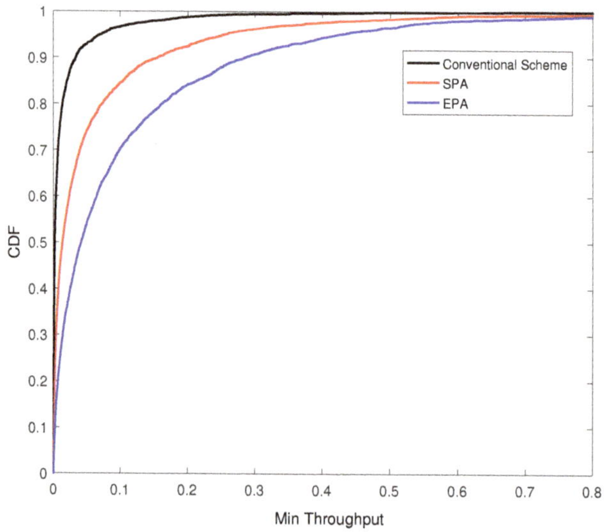

Figure 12. The CDF of min. uplink rate when $M = 64$ using MRC.

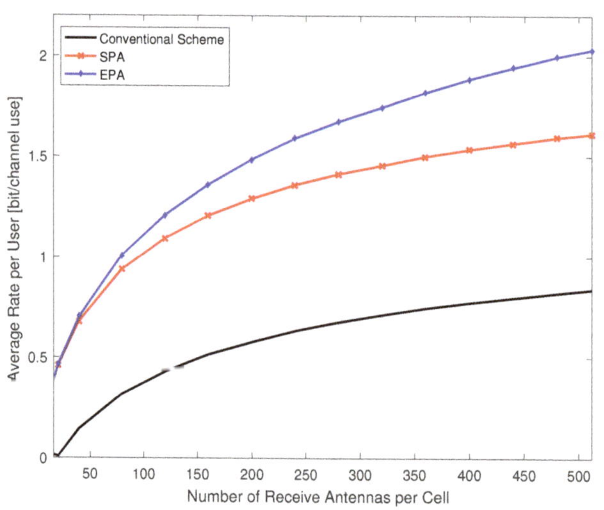

Figure 13. The average uplink rate per user with ZF for different numbers of antennas, K – 20 , and R = 300 m.

7. Conclusions

In this paper, we propose a new pilot assignment approach to address the pilot contamination in multicell massive MIMO systems. An optimization problem is formulated in order to improve the minimum uplink rate for users in the target cell. Our approach to solving this optimization problem ensures an overall reduction of the outgoing inter-cell interference of neighbor cells. This reduction is achieved by assigning the pilot sequences to the neighbor cell's users and maximizing the minimum uplink rate of the target cell's users based on SPA algorithm. The numerical results have clearly shown that the EPA scheme is more effective than the other schemes in both MRC and ZF linear detections. Additionally, using such an efficient assigning approach entitles the new EPA scheme to achieve significant performance when the typical parameter M is 64 compared to the SPA and the conventional schemes. Likewise, the minimum uplink rate is greatly enhanced by the new EPA scheme than the SPA scheme. Furthermore, the proposed scheme has also proved high effectiveness and performance even in severe interference environments.

Author Contributions: Conceptualization, A.S.A.-h. and N.K.N.; Formal analysis, A.S.A.-h. and N.K.N.; Methodology, A.S.A.-h., N.K.N., A.S., S.S. and A.M.M.; Writing—review & editing, A.S.A.-h. and N.K.N.

Funding: This work was funded by [Advancing the State of the Art of MIMO: The Key to Successful Evolution of Wireless Networks (ATOM)] grant number [690750-ATOM-H2020-MSCA- RISE-2015, UPM: 6388800-10801] and The APC was funded by [Research Management Centre (RMC)].

Conflicts of Interest: The authors declare no conflicts of interest.

References

1. Boccardi, F.; Heath, R.W.; Lozano, A.; Marzetta, T.L.; Popovski, P. Five disruptive technology directions for 5G. *IEEE Commun. Mag.* **2014**, *52*, 74–80. [CrossRef]
2. Marzetta, T.L. Noncooperative cellular wireless with unlimited numbers of base station antennas. *IEEE Trans. Wirel. Commun.* **2010**, *9*, 3590–3600. [CrossRef]
3. Lim, Y.G.; Chae, C.B.; Caire, G. Performance analysis of massive MIMO for cell-boundary users. *IEEE Trans. Wirel. Commun.* **2015**, *14*, 6827–6842. [CrossRef]
4. Larsson, E.G.; Edfors, O.; Tufvesson, F.; Marzetta, T.L. Massive MIMO for next generation wireless systems. *IEEE Commun. Mag.* **2014**, *52*, 186–195. [CrossRef]
5. Rusek, F.; Persson, D.; Lau, B.K.; Larsson, E.G.; Marzetta, T.L.; Edfors, O.; Tufvesson, F. Scaling up MIMO: Opportunities and challenges with very large arrays. *IIEEE Signal Process. Mag.* **2013**, *30*, 40–60. [CrossRef]
6. Björnson, E.; Larsson, E.G.; Debbah, M. Optimizing multi-cell massive MIMO for spectral efficiency: How many users should be scheduled? In Proceedings of the 2014 IEEE Global Conference on Signal and Information Processing (GlobalSIP), Atlanta, GA, USA, 3–5 December 2014. [CrossRef]
7. Rajatheva, N.; Osseiran, A.; Monserrat, J.F.; Marsch, P. *5G Mobile and Wireless Communications Technology*; Osseiran, A., Monserrat, J.F., Marsch, P., Eds.; Cambridge University Press: Cambridge, UK, 2016; ISBN 978-1-107-13009-8.
8. Björnson, E.; Larsson, E.G.; Marzetta, T.L. Massive MIMO: Ten myths and one critical question. *IEEE Commun. Mag.* **2016**, *54*, 114–123. [CrossRef]
9. Jose, J.; Ashikhmin, A.; Marzetta, T.; Vishwanath, S. Pilot Contamination and Precoding in Multi-Cell TDD Systems. *IEEE Trans. Wirel. Commun.* **2011**, *10*, 2640–2651. [CrossRef]
10. Björnson, E.; Hoydis, J.; Sanguinetti, L. Massive MIMO Networks: Spectral, Energy, and Hardware Efficiency. *IFound. Trends Signal Process.* **2017**, *11*, 154–655. [CrossRef]
11. Zhu, X.; Wang, Z.; Dai, L.; Qian, C. Smart Pilot Assignment for Massive MIMO. *IIEEE Commun. Lett.* **2015**, *19*, 1644–1647. [CrossRef]

12. Dao, H.T.; Kim, S. Vertex Graph-Coloring-Based Pilot Assignment With Location-Based Channel Estimation for Massive MIMO Systems. *IEEE Access* **2018**, *6*, 4599–4607. [CrossRef]
13. Kim, K.; Lee, J.; Choi, J. Deep Learning Based Pilot Allocation Scheme (DL-PAS) for 5G Massive MIMO System. *IIEEE Commun. Lett.* **2018**, *22*, 828–831. [CrossRef]
14. Akbar, N.; Yan, S.; Yang, N.; Yuan, J. Location-Aware Pilot Allocation in Multicell Multiuser Massive MIMO Networks. *IEEE Trans. Veh. Technol.* **2018**, *67*, 7774–7778. [CrossRef]
15. Muppirisetty, L.S.; Charalambous, T.; Karout, J.; Fodor, G.; Wymeersch, H. Location-Aided Pilot Contamination Avoidance for Massive MIMO Systems. *IEEE Trans. Wirel. Commun.* **2018**, *17*, 2662–2674. [CrossRef]
16. Sohn, J.; Yoon, S.W.; Moon, J. On Reusing Pilots Among Interfering Cells in Massive MIMO. *IEEE Trans. Wirel. Commun.* **2017**, *16*, 8092–8104. [CrossRef]
17. Zhu, X.; Wang, Z.; Qian, C.; Dai, L.; Chen, J.; Chen, S.; Hanzo, L. Soft Pilot Reuse and Multi-Cell Block Diagonalization Precoding for Massive MIMO Systems. *IEEE Trans. Veh. Technol.* **2016**, *65*, 3285–3298. [CrossRef]
18. ZLiu, J.; Li, Y.; Zhang, H.; Guo, S. Efficient and Fair Pilot Allocation for Multi-cell Massive MIMO Systems. In Proceedings of the 2017 9th International Conference on Wireless Communications and Signal Processing (WCSP), Nanjing, China, 11–13 October 2017. [CrossRef]
19. Ma, S.; Xu, E.L.; Salimi, A.; Cui, S. A Novel Pilot Assignment Scheme in Massive MIMO Networks. *IEEE Commun. Lett.* **2018**, *7*, 262–265. [CrossRef]
20. Wang, H.; Huang, Y.; Jin, S.; Du, Y.; Yang, L. Pilot Contamination Reduction in Multi-cell TDD Systems with Very Large MIMO Arrays. *Wirel. Pers. Commun.* **2017**, *96*, 5785–5808. [CrossRef]
21. Du, T.; Wang, Y.; Wang, J.; Du, Y.; Yang, L. Multi-cell Joint Optimization to Mitigate Pilot Contamination for Multi-Cell Massive MIMO Systems. In Proceedings of the 2017 IEEE 85th Vehicular Technology Conference (VTC Spring), Sydney, NSW, Australia, 4–7 June 2017. [CrossRef]
22. Wang, P.; Zhao, C.; Zhang, Y.; Stuber, G.L. A Novel Pilot Assignment Approach for Pilot Decontaminating in Massive MIMO Systems. In Proceedings of the 2017 IEEE Wireless Communications and Networking Conference (WCNC), San Francisco, CA, USA, 19–22 March 2017. [CrossRef]
23. Zhao, M.; Zhang, H.; Guo, S.; Yuan, D. Joint Pilot Assignment and Pilot Contamination Precoding Design for Massive MIMO Systems. In Proceedings of the 2017 26th Wireless and Optical Communication Conference (WOCC), Newark, NJ, USA, 7–8 April 2017. [CrossRef]
24. Fernandes, F.; Ashikhmin, A.; Marzetta, T.L. Inter-Cell Interference in Noncooperative TDD Large Scale Antenna Systems. *IEEE J. Sel. Areas Commun.* **2013**, *31*, 192–201. [CrossRef]
25. Chen Z.; Yang, C. Pilot decontamination in wideband massive MIMO systems by exploiting channel sparsity. *IEEE Trans. Wirel. Commun.* **2016**, *15*, 5087–5100. [CrossRef]
26. Björnson, E.; Matthaiou, M.; Debbah, M. Massive MIMO with Non-Ideal Arbitrary Arrays: Hardware Scaling Laws and Circuit-Aware Design. *IEEE Trans. Wirel. Commun.* **2015**, *14*, 4353–4368.. [CrossRef]
27. Li, M.; Jin, S.; Gao, X. Spatial Orthogonality-Based Pilot Reuse for Multi-cell Massive MIMO Transmission. In Proceedings of the 2013 International Conference on Wireless Communications and Signal Processing, Hangzhou, China, 24–26 October 2013. [CrossRef]
28. Ashikhmin, A.; Marzetta, T. Pilot Contamination Precoding in Multi-Cell Large Scale Antenna Systems. In Proceedings of the 2012 IEEE International Symposium on Information Theory Proceedings, Cambridge, MA, USA, 1–6 July 2012. [CrossRef]
29. Muller, R.; Cottatellucci, L.; Vehkapera, M. Blind pilot decontamination. *IEEE J. Sel. Top. Signal Process.* **2014**, *8*, 773–786. [CrossRef]
30. NGO, H.Q.; Larsson, E.G.; Marzetta, T.L. Energy and Spectral Efficiency of Very Large Multiuser. *IEEE Trans. Wirel. Commun.* **2013**, *61*, 1436–1449. [CrossRef]
31. Sesia, S.; Toufik, I.; Baker, M. Network Architecture and Protocols. In *LTE-The UMTS Long Term Evolution: From Theory to Practice*; John Wiley & Sons Ltd.: Chichester, UK, 2011; ISBN 978-0-470-66025-6.

32. Zhu, X.; Dai, L.; Wang, Z. Graph coloring based pilot allocation to mitigate pilot contamination for multi-cell massive mimo systems. *IEEE Commun. Lett.* **2015**, *19*, 1842–1845. [CrossRef]
33. Sarkar, T.K.; Ji, Z.; Kim, K.; Medouri, K.; Salazar-Palma, M. A survey of various propagation models for mobile communication. *IEEE Antennas Propag. Mag.* **2003**, *45*, 51–82. [CrossRef]

electronics

MDPI

Article

Downlink Channel Estimation in Massive Multiple-Input Multiple-Output with Correlated Sparsity by Overcomplete Dictionary and Bayesian Inference

Wei Lu [1,*], Yongliang Wang [1], Xiaoqiao Wen [1], Shixin Peng [2] and Liang Zhong [3]

1. Air Force Early Warning Academy, Wuhan 430019, China; ylwangkjld@163.com (Y.W.); evacassidy@126.com (X.W.)
2. National Engineering Research Center for E-Learning, Central China Normal University, Wuhan 430079, China; pengshixin@mail.ccnu.edu.cn
3. Department of communication system, China University of Geoscience, Wuhan 430074, China; hustzhongliang@gmail.com

* Correspondence: lvweiwhut@aliyun.com; Tel.: +86-027-85695800

Received: 24 March 2019; Accepted: 24 April 2019; Published: 28 April 2019

Abstract: We exploited the temporal correlation of channels in the angular domain for the downlink channel estimation in a massive multiple-input multiple-output (MIMO) system. Based on the slow time-varying channel supports in the angular domain, we combined the channel support information of the downlink angular channel in the previous timeslot into the channel estimation in the current timeslot. A downlink channel estimation method based on variational Bayesian inference (VBI) and overcomplete dictionary was proposed, in which the support prior information of the previous timeslot was merged into the VBI for the channel estimation in the current timeslot. Meanwhile the VBI was discussed for a complex value in our system model, and the structural sparsity was utilized in the Bayesian inference. The Bayesian Cramér–Rao bound for the channel estimation mean square error (MSE) was also given out. Compared with other algorithms, the proposed algorithm with overcomplete dictionary achieved a better performance in terms of channel estimation MSE in simulations.

Keywords: massive MIMO; channel estimation; Bayesian inference; overcomplete dictionary

1. Introduction

Massive multiple-input multiple-output (MIMO) is the key technology for next generation wireless communication. The large number of antennas enable high spectrum efficiency and lower power consumption [1]. To get these benefits, the base station (BS) needs to acquire the channel stated information (CSI) for uplink and downlink. Pilot-based channel estimation is widely used in wireless communication systems. In the time division duplex (TDD) system, the channel reciprocity is used to get the CSI by only estimating the uplink channel at BS. In the frequency division duplex (FDD) system, the channel reciprocity cannot be used directly. In FDD massive MIMO system it is challenging to get the downlink CSI with the conventional feedback scheme. In the conventional feedback scheme each user estimates its channel and then feeds back the estimated CSI to the BS. The pilot and feedback overheads are high for massive MIMO, since they are scaling linearly with the number of antennas. Hence, it is important to design an efficient downlink channel estimation and feedback scheme for a FDD massive MIMO system.

By exploiting the sparsity in massive MIMO channel, compressed sensing (CS) was applied in the channel estimation and feedback. The users could feed the compressed training measurements back to the BS, and an orthogonal matching pursuit (OMP) was used for downlink CSI recovery in [2]. In [3] the modified basis pursuit (MBP) was proposed by utilizing the partial priori signal support information to improve the recovery performance. In [4] the support information of a signal in the discrete fourier transform domain was incorporated into the weighted l_1 minimization approach for CS recovery, which could reduce the number of measurements by the size of the known part of support. In [5] a three-level weighting scheme based on the support information was used for the weighted l_1 minimization and the simulation results showed superiority. In [6] we exploited the reciprocity between uplink and downlink channels in the angular domain, and diagnosed the supports of the downlink channel from the estimated uplink channel, and proposed a weighted subspace pursuit (SP) channel estimation algorithm for FDD massive MIMO. It can be seen that CS was effective in the channel estimation for massive MIMO.

However, most of these algorithms need the sparsity level in the estimation algorithm, which is not practical in engineering scenarios. The Bayesian framework can be applied to the compressive channel estimation. In [7], Bayesian estimation of sparse massive MIMO channel was developed in which neighboring antennas shared among each other their information about the channel support. In [8] a variational expectation maximization strategy was used for massive MIMO channel estimation, and a Gaussian mixture prior model was designed to capture the individual sparsity for each channel and the joint sparsity among users. In [9] a sparse Bayesian learning algorithm was proposed for FDD massive MIMO channel estimation with arbitrary 2D-array. By the Bayesian framework in compressive channel estimation the sparsity level is unnecessary, and it has relatively better recovery performance. Additionally, there exists angular reciprocity in massive MIMO. For example, the channel covariance matrices for uplink and downlink are reconstructed by making use of the angle reciprocity between uplink and downlink channels in [10]. Hence it is promising to apply the angular reciprocity and Bayesian framework in the compressive massive MIMO channel estimation.

Additionally, there exists angular reciprocity in the FDD massive MIMO. There is also time correlation of channels. In [11] a differential compressive feedback in FDD massive MIMO was proposed based on the channel impulses response (CIR) between timeslots, which were slow time-varying and sparse, and the differential CIR between two CIRs in adjacent timeslots was sparse. Inspired by the sparsity in the angular domain and time correlation of channels, the correlated angular sparsity can also be exploited for massive MIMO channel estimation.

In this paper we proposed a downlink channel estimation in a TDD/FDD massive MIMO system. The timeslots were divided into groups. In each group the estimated channel support information of the previous timeslot was utilized by the following timeslot. The correlated angular sparsity between timeslots in the downlink channel was utilized in the Bayesian inference for channel recovery. We transformed the complex sparse vector to the real sparse vector recovery by Bayesian inference, and the structural sparsity of the transformed real sparse vector was utilized. Meanwhile, the prior support information from the estimated channel in the previous timeslot was made use of in modeling the hidden hyperparameters in the Bayesian model. A Bayesian Cramér–Rao bound analysis is presented, and simulations are given out to verify the performance of the proposed algorithm. The main contributions were as follows: (1) a group-based channel estimation scheme was proposed, in which previous estimated channel support information was used as the priori information in the following timeslot due to the sparsity correlation; (2) priori information was merged into the Bayesian inference algorithm for channel recovery; (3) the Bayesian Cramér–Rao bound for the channel estimation mean square error (MSE) was analyzed.

The system model is illustrated in Section 2, while the proposed channel estimation algorithm based on Bayesian inference is presented in Section 3. The Bayesian Cramér–Rao bound (BCRB) for the channel estimation of mean square error (MSE) is given out in Section 4. Simulations and conclusions are presented in Sections 5 and 6.

In the paper, we used the following notations. Scalars, vectors and matrices were denoted by lower-case, boldface lower-case and boldface upper-case symbols. The probability density function of a given random variable was denoted by $p(\cdot)$. Gamma($x|a$, b) was the Gamma probability density function (PDF) with shape parameters a and b for x, while Normal($x|c$, d) was the Gaussian PDF with parameters mean c and variance d for x. $\Gamma(\cdot)$ was the Gamma function, and $\ln(\cdot)$ was the logarithm function. $\mathrm{Tr}(\cdot)$ stood for the trace operator. $\mathbb{E}_a(\cdot)$ denoted the expectation operation with the PDF of variable a.

2. System Model

We considered a massive MIMO TDD/FDD system with a single user, and assumed that the BS was equipped with N antennas and the user terminal (UT) had a single antenna. For the downlink channel estimation in the massive MIMO system, the BS transmitted the pilots to UT. The UT received the pilots and fed back the received signal to the BS directly. The received signal $\mathbf{y}^d(t)$ at the UT in the t-th timeslot was written as

$$\mathbf{y}^d(t) = \sqrt{\rho^d}\mathbf{A}\mathbf{h}^d(t) + \mathbf{n}^d(t) \tag{1}$$

where $\mathbf{h}^d(t) \in \mathbb{C}^{N\times 1}$ is the downlink channel, $\mathbf{A} \in \mathbb{C}^{T_d\times N}$ is the downlink pilots, T_d is the pilot length, ρ^d is the downlink received power, $\mathbf{n}^d \in \mathbb{C}^{T_d\times 1}$ is the received noise with each element to be i.i.d Gaussian with mean 0 and variance σ^2, $\mathbf{y}^d(t) \in \mathbb{C}^{T_d\times 1}$ is the received signal at UT.

Since in the massive MIMO there existed sparsity, when $\mathbf{D}^d \in \mathbb{C}^{N\times M}$ was the channel dictionary for downlink channel which could be unitary dictionary or overcomplete dictionary ($M > N$, their column vector had the form of steering vector with a different sampling angle), $\mathbf{h}_a^d(t)$ was the sparse representation with $\mathbf{h}^d(t) = \mathbf{D}^d\mathbf{h}_a^d(t)$. In this paper we applied the overcomplete dictionary to present the sparse angular channel to get a better recovery performance. In the downlink channel estimation, we needed to obtain $\hat{\mathbf{h}}_a^d(t)$ the estimated downlink channel in the angular domain in the t-th timeslot.

By utilizing the sparse channel representation we then had

$$\mathbf{y}^d(t) = \sqrt{\rho^d}\mathbf{A}\mathbf{D}^d\mathbf{h}_a^d(t) + \mathbf{n}^d(t) \tag{2}$$

For simplicity, the timeslot mark is omitted in the following equations. Since $\mathbf{y}^d(t)$, $\mathbf{h}_a^d(t)$, and $\mathbf{n}^d(t)$ are complex number vectors, we could rewrite Equation (2) into real number vectors as

$$\begin{bmatrix} \mathrm{Re}(\mathbf{y}^d) \\ \mathrm{Im}(\mathbf{y}^d) \end{bmatrix} = \begin{bmatrix} \mathrm{Re}(\sqrt{\rho^d}\mathbf{A}\mathbf{D}^d) & -\mathrm{Im}(\sqrt{\rho^d}\mathbf{A}\mathbf{D}^d) \\ \mathrm{Im}(\sqrt{\rho^d}\mathbf{A}\mathbf{D}^d) & \mathrm{Re}(\sqrt{\rho^d}\mathbf{A}\mathbf{D}^d) \end{bmatrix} \begin{bmatrix} \mathrm{Re}(\mathbf{h}_a^d(t)) \\ \mathrm{Im}(\mathbf{h}_a^d(t)) \end{bmatrix} + \begin{bmatrix} \mathrm{Re}(\mathbf{n}^d(t)) \\ \mathrm{Im}(\mathbf{n}^d(t)) \end{bmatrix} \tag{3}$$

where $\mathrm{Re}(\cdot)$ and $\mathrm{Im}(\cdot)$ denote the real and imaginary parts respectively. For simplicity, we rewrote Equation (3) as

$$\overline{\mathbf{y}} = \overline{\mathbf{A}}\overline{\mathbf{h}} + \overline{\mathbf{n}} \tag{4}$$

where $\overline{\mathbf{y}} = \begin{bmatrix} \mathrm{Re}(\mathbf{y}^d) \\ \mathrm{Im}(\mathbf{y}^d) \end{bmatrix}$, $\overline{\mathbf{A}} = \begin{bmatrix} \mathrm{Re}(\sqrt{\rho^d}\mathbf{A}\mathbf{D}^d) & -\mathrm{Im}(\sqrt{\rho^d}\mathbf{A}\mathbf{D}^d) \\ \mathrm{Im}(\sqrt{\rho^d}\mathbf{A}\mathbf{D}^d) & \mathrm{Re}(\sqrt{\rho^d}\mathbf{A}\mathbf{D}^d) \end{bmatrix}$, $\overline{\mathbf{h}} = \begin{bmatrix} \mathrm{Re}(\mathbf{h}_a^d(t)) \\ \mathrm{Im}(\mathbf{h}_a^d(t)) \end{bmatrix}$ and $\overline{\mathbf{n}} = \begin{bmatrix} \mathrm{Re}(\mathbf{n}^d(t)) \\ \mathrm{Im}(\mathbf{n}^d(t)) \end{bmatrix}$.

On the other hand, we considered the meaning of sparse angular channel representation $\mathbf{h}_a^d(t)$. If the transmission angles were allocated exactly at the sampling points in the channel dictionary $\boldsymbol{D}^d$, then the corresponding coefficient in the $\mathbf{h}_a^d(t)$ was nonzero. If the path number was smaller than the antenna number, then $\mathbf{h}_a^d(t)$ was sparse. However, there was leakage effect induced by dictionary mismatch which will have deteriorated the sparsity of the angular channel representation [12]. When the movement velocity of UT was not very high, e.g., v = 12 km/h, and the typical timeslot duration τ = 0.5 ms, the movement distance of UT in one timeslot was 0.017 m. When the distance of UT and BS was 200 m, the angle change for the line of sight (LoS) transmission in one timeslot was 0.0049° which was much smaller than the sampling interval in the dictionary. For the non-LoS (NLoS) transmission, the angle change was also small which is discussed in Section 4.1. Hence the transmission angle change between two timeslots is very small if the transmission environment doesn't change dramatically, and there is correlation in the angular channel sparsity between adjacent timeslots. In other words, the information regarding the estimated angular channel in the previous timeslot could be utilized in the current channel estimation.

It was proven that the prior support information could improve the channel recovery performance [3–6]. Hence in this paper we made use of the prior support information from the previous timeslot to improve the Bayesian channel estimation. In the following section we have discussed how to merge the prior information into the Bayesian inference algorithm for channel estimation.

3. Proposed Algorithm

We designed a three-layer hierarchical graphical model as shown in Figure 1. In the first layer, $\overline{\mathbf{h}}$ was assigned a Gaussian prior distribution

$$p(\overline{\mathbf{h}}|\boldsymbol{\alpha}) = \prod_{i=1}^{2N} p(\overline{h}_i|\alpha_i) \tag{5}$$

where $\overline{h}_i$ and α_i are the i-th entry in $\overline{\mathbf{h}}$ and $\boldsymbol{\alpha}$ respectively, $p(\overline{h}_i|\alpha_i) = Normal(\overline{h}_i|0, \alpha_i)$ and α_i is the inverse variance of the Gaussian distribution. When $\overline{h}_i$ is close to 0, then α_i is very large, and vice versa.

In the second layer, we assumed a Gamma distribution as hyperpriors over the hyperparameters α_i, and it can be presented as

$$p(\boldsymbol{\alpha}) = \prod_{i=1}^{2N} Gamma(\alpha_i|a_i, b_i) \tag{6}$$

where Gamma(·) is the Gamma PDF, and the parameters a_i and b_i characterize the shape of Gamma PDF. For fixed a_i, the larger b_i is, the smaller α_i is; then $\overline{h}_i$ tends to be nonzero. In the sparse Bayesian learning a_i and b_i were set to be very small for non-informative hyperprior over α_i [13].

In our model, we set a_i to be constant with a predefined value, and we modeled b_i as random parameters. In Figure 1 it could be found that the entries of $\overline{y}$ were divided into two sets by their indices, i.e., $\{S, S_{+N}\}$ and $\{S, S_{+N}\}^c$, where S was the set with channel support indices from the previous timeslot, and S_{+N} was the set with each index in S added by N, since we converted the complex system model to the real system model as Equation (3).$\{S, S_{+N}\}^c$ was the complementary set of $\{S, S_{+N}\}$. For example, in the $(t-1)$-th timeslot, the positions of nonzero entries or called supports in $\mathbf{h}_a^d(t-1)$ were S = {4, 5, 6}, then $S_{+N} = \{4+N, 5+N, 6+N\}$. The probable supports for $\mathbf{h}_a^d(t)$ in the current t-th timeslot can be assumed to be the same as those for previous $(t-1)$-th timeslot for simplicity. On the other hand, we could have also diagnosed the probable channel supports further by taking the angle deviation and leakage effects into consideration. In this paper we adopted the support diagnosis algorithm, and the details can be found in [6].

For $\overline{y}_j$, $j \in \{S, S_{+N}\}$, we employed a Gamma distribution over the hyperparameters b_i in the third layer as

$$\text{Gamma}(b_i|c, d) = \Gamma(c)^{-1} d^c b_i^{c-1} e^{-db_i} \tag{7}$$

where c and d characterize the shape of Gamma PDF. By the system model and assumptions for massive MIMO, we could use a Bayesian inference to perform the sparse channel recovery.

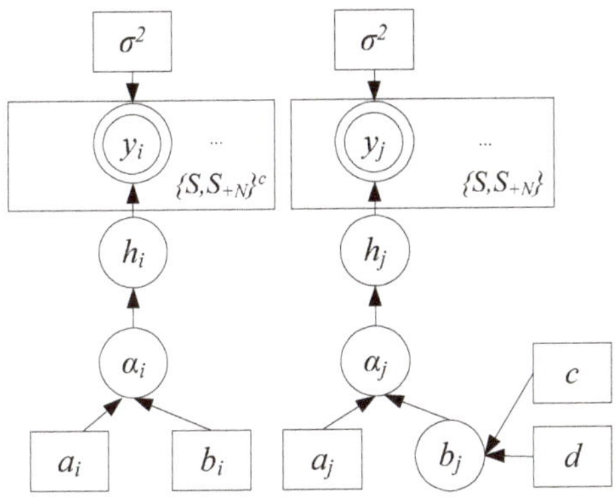

Figure 1. Graphical model for the channel estimation with Bayesian inference. The nodes with double circle, single circle and square correspond to the observed data, hidden variables and parameters, respectively.

According to the standard Bayesian inference [14], let $\mathbf{z} \triangleq \left\{\overline{\boldsymbol{h}}, \boldsymbol{\alpha}, \boldsymbol{b}\right\}$, we have

$$\begin{aligned} \ln p(\mathbf{z}_i) &= \mathbb{E}_{\mathbf{z}_i, i \neq j}[\ln p(\overline{\mathbf{y}}, \mathbf{z})] + \text{constant} \\ &\propto \mathbb{E}_{\mathbf{z}_i, i \neq j}[\ln p(\overline{\mathbf{y}}, \mathbf{z})] \end{aligned} \tag{8}$$

where constant is a constant used for $p(\mathbf{z}_i)$ normalization, $p(\overline{y}, z)$ is the joint pdf for $\overline{\boldsymbol{h}}$ and $\mathbf{z}$, and $\mathbf{z}_i$ can be $\overline{\boldsymbol{h}}, \boldsymbol{\alpha}, \text{and} \boldsymbol{b}$. We have $p(\overline{\mathbf{y}}, \mathbf{z}) = p(\mathbf{z}|\overline{\mathbf{y}})p(\overline{y})$. We assume $p(\mathbf{z}|\overline{\mathbf{y}})$ posterior independence among the hidden variables $\mathbf{z}$, then $p(\mathbf{z}|\overline{\mathbf{y}}) \approx p(\mathbf{z})$, and $p(\mathbf{z})$ is the product of PDF of $\overline{\boldsymbol{h}}, \boldsymbol{\alpha}$, and $\boldsymbol{b}$.

In order to make use of the prior support information from the previous timeslot and the structure sparsity in Equation (4), we needed to make some modifications to the standard Bayesian inference. The main considerations for the modifications were as follows:

(I) Since we rewrote Equation (2) as Equation (4), if $h^d_{a,i}$ was nonzero, then $\overline{h}_i$ and $\overline{h}_{i+N}$ were nonzero simultaneously. Hence it was wise to assume that b_i and b_{i+N} were the same;

(II) In the standard Bayesian learning a_i and b_i were set to be very small for non-informative hyperprior over α_i. This assumption was valid if no prior information was provided. If the prior support information was available, such as that the support information of the previous timeslot could be used for channel estimation in the coming timeslot by sparsity correlation, it was wise to assume that the supports between adjacent timeslots were partially common. If the *i-th* element in the angular channel vector was nonzero, then the hyperparameter b_i and b_{i+N} tended to be variables rather than to be fixed small numbers, which meant only for the indices from the prior support set S the third layer prior model was adopted.

It can be seen that the consideration (II) was similar to [15]. However, our proposed algorithm was extended for a complex number system and the structure sparsity was considered. However, on the other hand, the overcomplete dictionary was adopted in our algorithm.

The proposed uplink-aided downlink channel estimation based on Bayesian inference was as follows:

(i) Update of p($\overline{h}$)

According to Equation (8), by ignoring the terms which are independent of $\overline{\boldsymbol{h}}$, we have

$$\begin{aligned}\ln p(\overline{\mathbf{h}}) &\propto \mathbb{E}_{\alpha,\mathbf{b}}\left[\ln p(\overline{\mathbf{y}}|\overline{\mathbf{h}}) + \ln p(\overline{\mathbf{h}}|\alpha) + \ln p(\mathbf{b})\right] \\ &\propto \mathbb{E}_{\alpha,\mathbf{b}}\left[\ln p(\overline{\mathbf{y}}|\overline{\mathbf{h}}) + \ln p(\overline{\mathbf{h}}|\alpha)\right] \\ &= \frac{-1}{2\sigma^2}(\overline{\mathbf{y}} - \overline{\mathbf{A}}\overline{\mathbf{h}})^T(\overline{\mathbf{y}} - \overline{\mathbf{A}}\overline{\mathbf{h}}) - \frac{1}{2}\overline{\mathbf{h}}^T\Lambda\overline{\mathbf{h}}\end{aligned} \tag{9}$$

where $\Lambda = \text{diag}\{\mathbb{E}_\alpha[\alpha_i]\}$, σ^2 is the noise variance in the system model, the vectors $\mathbf{b}$ and α are comprised by b_i and α_i respectively. Since $p(\overline{\mathbf{y}}|\overline{\mathbf{h}})$ and $p(\overline{h}|\alpha)$ are a Gaussian distribution, then $p(\overline{h})$ follows a Gaussian distribution with the mean μ and covariance ϕ given by

$$\mu = \frac{1}{\sigma^2}\Phi\overline{\mathbf{A}}^T\overline{\mathbf{y}} \tag{10}$$

$$\Phi = \left(\frac{1}{\sigma^2}\overline{\mathbf{A}}^T\overline{\mathbf{A}} + \Lambda\right) \tag{11}$$

(ii) Update of $p(\alpha)$

According to Equation (8), by ignoring the terms which are independent of α, we have

$$\begin{aligned}\ln p(\alpha) &\propto \mathbb{E}_{\overline{\mathbf{h}},\mathbf{b}}\left[\ln p(\overline{\mathbf{y}}|\overline{\mathbf{h}}) + \ln p(\overline{\mathbf{h}}|\alpha) + \ln p(\alpha|\mathbf{a},\mathbf{b}) + \ln p(\mathbf{b})\right] \\ &\propto \mathbb{E}_{\overline{\mathbf{h}},\mathbf{b}}\left[\ln p(\overline{\mathbf{h}}|\alpha) + \ln p(\alpha|\mathbf{a},\mathbf{b})\right] \\ &= \sum_{i=1}^{2N}\mathbb{E}_{\overline{h},b}\left\{(a_i - 0.5)\ln\alpha_i - \left(0.5\overline{h}_i^2 + b_i\right)\alpha_i\right\} \\ &= \sum_{i\in\{S,S_{+N}\}}\mathbb{E}_{\overline{h},b}\left\{(a_i + 0.5 - 1)\ln\alpha_i - \left(0.5\overline{h}_i^2 + b_i\right)\alpha_i\right\} + \\ &\sum_{i\in\{S,S_{+N}\}^c}\mathbb{E}_{\overline{h},b}\left\{(a - 0.5)\ln\alpha_i - \left(0.5\overline{h}_i^2 + b_i\right)\alpha_i\right\} \\ &= \sum_{i\in\{S,S_{+N}\}}\left\{(a_i + 0.5 - 1)\ln\alpha_i - \left(\frac{\mathbb{E}_{\overline{h},b}(b_i + b_{i+N})}{2} + \frac{\mathbb{E}_{\overline{h},b}(\overline{h}_i^2 + \overline{h}_{i+N}^2)}{4}\right)\alpha_i\right\} + \\ &\sum_{i\in\{S,S_{+N}\}^c}\left\{(a_i + 0.5 - 1)\ln\alpha_i - \left(b_i + \frac{\mathbb{E}_{\overline{h},b}(\overline{h}_i^2 + \overline{h}_{i+N}^2)}{4}\right)\alpha_i\right\}\end{aligned} \tag{12}$$

where S is the estimated support set from the previous timeslot. Since the complex system model was converted in Equation (4). By (II), $S_{+N} \triangleq \{s_i + N\}$ was also the support set in the converted system model in Equation (4). For $i \in \{S, S_{+N}\}$, b_i is variable number, b_i and b_{i+N} were assumed to be the same, we used $0.5\mathbb{E}_{\overline{h},b}(b_i + b_{i+N})$ to present $\mathbb{E}_{\overline{\mathbf{h}},\mathbf{b}}(b_i)$. The same assumption was applied to $\overline{h}_i$ and $\overline{h}_{i+N}$ with $\mathbb{E}_{\overline{\mathbf{h}},\mathbf{b}}(\overline{h}_i^2) = 0.5\mathbb{E}_{\overline{\mathbf{h}},\mathbf{b}}(\overline{h}_i^2 + \overline{h}_{i+N}^2)$. In this way the structural sparsity was utilized.

Since $p(\alpha|a,b)$ is the Gamma distribution and $p(\overline{h}|\alpha)$ is the Gaussian distribution, $p(\alpha)$ is the Gamma distribution. Then $p(\alpha_i)$ is also the Gamma distribution with the updated parameters $\widetilde{a}_i$ and $\widetilde{b}_i$ given by

$$\widetilde{a}_i = a_i + 0.5 \tag{13}$$

$$\widetilde{b}_i = \begin{cases}\frac{\mathbb{E}_{\overline{\mathbf{h}},\mathbf{b}}(b_i + b_{i+N})}{2} + \frac{\mathbb{E}_{\overline{\mathbf{h}},\mathbf{b}}(\overline{h}_i^2 + \overline{h}_{i+N}^2)}{4}, i \in \{S, S_{+N}\} \\ b_i + \frac{\mathbb{E}_{\overline{\mathbf{h}},\mathbf{b}}(\overline{h}_i^2 + \overline{h}_{i+N}^2)}{4}, i \in \{S, S_{+N}\}^c\end{cases} \tag{14}$$

(iii) Update of $p(\mathbf{b}_{\{S,S_{+N}\}})$

According to Equation (8), by ignoring the terms which are independent of $\mathbf{b}$, we have

$$\begin{aligned} \ln p(\mathbf{b}_{\{S,S_{+N}\}}) &\propto \mathbb{E}_{\alpha,\overline{\mathbf{h}}}\left[\ln p(\overline{\mathbf{y}}|\overline{\mathbf{h}}) + \ln p(\overline{\mathbf{h}}|\alpha) + \ln p(\alpha|\mathbf{a},\mathbf{b}) + \ln p(\mathbf{b}|\mathbf{c},\mathbf{d})\right] \\ &\propto \mathbb{E}_{\alpha,\overline{\mathbf{h}}}[\ln p(\alpha|\mathbf{a},\mathbf{b}) + \ln p(\mathbf{b}|\mathbf{c},\mathbf{d})] \\ &= \sum_{i\in\{S,S_{+N}\}} \{-b_i\mathbb{E}_{\alpha}(\alpha_i) + (c_i-1)\ln b_i - d_i b_i\} \end{aligned} \tag{15}$$

where $\mathbf{b}_{\{S,S_{+N}\}}$ is comprised by the entries indicated by $\{S, S_{+N}\}$ in $\mathbf{b}$. In (15) the α, $\boldsymbol{a}$, $\boldsymbol{b}$, $\boldsymbol{c}$, $\boldsymbol{d}$ are also comprised by their indicated $\{S, S_{+N}\}$, the subscript $\{S, S_{+N}\}$ is omitted for simplicity. As shown in Figure 1, $\mathbf{b}_{\{S,S_{+N}\}}$ was modelled as a Gamma distribution. Since $p(\alpha_i|a_i, b_i)$ and $p(b_i|c_i, d_i)$ were a Gamma distribution, $p(\widetilde{\mathbf{b}}_{\{S,S_{+N}\}})$ was Gamma$(\widetilde{b}_{i\in\{S,S_{+M}\}}|\widetilde{c}_i, \widetilde{d}_i)$, and the updated $\widetilde{c}_i$ and $\widetilde{d}_i$ were given by

$$\widetilde{c}_i = a_i + c_i \tag{16}$$

$$\widetilde{d}_i = d_i + \mathbb{E}_{\alpha}(\alpha_i) \tag{17}$$

Then the Bayesian inference for the channel estimation was executed iteratively among (i), (ii), and (iii). The details of the algorithm are summarized in step 3 of Algorithm 1. When the estimated channel vector $\overline{\mathbf{h}\prime}$ was recovered, we needed to convert it to the complex vector $\boldsymbol{h}_a^d$ according to Equation (3).

Algorithm 1 Downlink channel estimation with variational inference algorithm and overcomplete dictionary.

Input: $\overline{\mathbf{A}}, \overline{\mathbf{y}}, \sigma^2$
Output: $\overline{\mathbf{h}\prime}$

1. Divide the timeslots into groups, and with each group comprised by t_g timeslots.
2. For the first timeslot in the group, use variational Bayesian inference (VBI) for channel estimation, and obtain the angular channel supports.
3. For the rest of the timeslots in the group, utilize the support information from the previous timeslot for channel estimation one by one. The recovery algorithm in each timeslot is as follows:
   3.1. Initialize $\boldsymbol{\alpha}$, $\boldsymbol{a}$, $\boldsymbol{b}$, $\boldsymbol{c}$, $\boldsymbol{d}$.
   3.2. $\mu = \frac{1}{\sigma^2}\phi\overline{\mathbf{A}}^{\mathrm{T}}\overline{\mathbf{y}}$, $\phi = (\frac{1}{\sigma^2}\overline{\mathbf{A}}^{\mathrm{T}}\overline{\mathbf{A}} + \Lambda)$, $\mathbb{E}_{\overline{\mathbf{h}},\mathbf{b}}(\overline{\mathbf{h}}_{\mathrm{i}}^{2}) = \mu_{\mathrm{i}}^2 + \phi_{\mathrm{i,i}}$, where $\Lambda = \mathrm{diag}\{\mathbb{E}_{\alpha}[\alpha_i]\}$, μ_i is the i-th entry in μ, and $\phi_{\mathrm{i,i}}$ is the i-th diagonal entry in ϕ.
   3.3. Update $\widetilde{a}_i$ and $\widetilde{b}_i$ according to Equations (13) and (14) in (ii) ($\widetilde{a}_i$ and $\widetilde{b}_i$ are the updated a_i and b_i, and a_i and b_i are the results from last iteration); then according to the property of the Gamma distribution variable, $\mathbb{E}_{\alpha}(\alpha_i) = \widetilde{a}_i/\widetilde{b}_i$.
   3.4. Update $\widetilde{c}$ and $\widetilde{d}$ according to Equations (16) and (17) in (iii) ($\widetilde{c}$ and $\widetilde{d}$ are the updated c and d, and c and d are the results from last iteration); then according to the property of the Gamma distribution variable, $\mathbb{E}_{\alpha}(\widetilde{b}_i) = \widetilde{c}/\widetilde{d}$.
   3.5. Go to step 3.2 until stop criteria meets.
   3.6. Then $\overline{h\prime} = \mu$.
4. Go back to step 2 for a new group of timeslots.

In a practical massive MIMO system, the transmission environment may change suddenly, in this way the correlation of sparsity between adjacent timeslots will deteriorate, and the previous channel support information cannot be utilized. On the other hand, the error will accumulate if the previous channel support information is utilized timeslot by timeslot. Hence, the initialization is important for the robustness and efficiency of the algorithm. As shown in Figure 2 divided the timeslots into groups, and each group was comprised of several timeslots. During the channel estimation for each group, the VBI was used for the channel estimation in the first timeslot, and then the proposed algorithm was executed for the remaining timeslots in which the channel support information of the previous timeslot was made use of by the current timeslot. This procedure is detailed in steps 1, 2 and 4 in Algorithm 1.

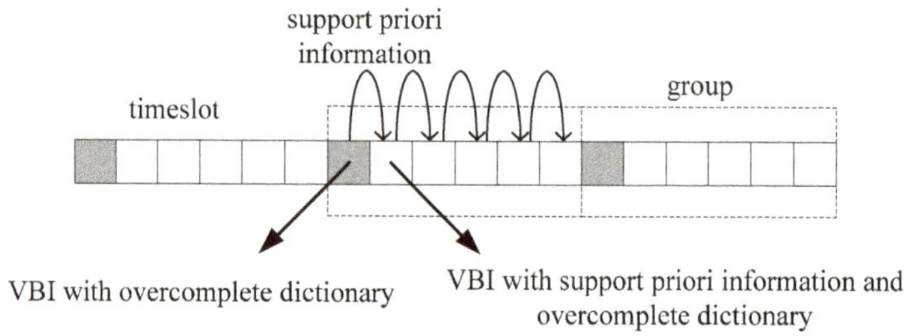

Figure 2. Channel estimations by group. Each block represents one timeslot, and the block filled with grey is the timeslot with variational Bayesian inference (VBI) for the channel estimation, while the blank blocks are the timeslots with the proposed algorithm for channel estimation.

4. Discussion

4.1. Sparsity Correlation Analysis

The UT movement distance was very small when the velocity of UT was small and the timeslot was 0.5 ms. The reflector for the transmission was static during the UT moving between timeslots. The ellipse geometry channel model is shown in Figure 3. The line of sight (LoS) distance between BS and UT was d_{Los}, the non-LoS (NLoS) distance by reflector between BS and UT was d_{NLoS}, and the UT movement distance in one timeslot was d_Δ. If the transmission path was still reflected by the same reflector as shown in Figure 3, the maximum and minimum NLoS distances from BS to UT between timeslots were $d_{NLoS}+d_\Delta$ and $d_{NLoS}-d_\Delta$. The transmission angle change was Δ_θ. The distance between the reflector and BS was d_1. By some mathematical manipulations shown in Appendix A, we got

$$\Delta_\theta \approx \frac{2d_\Delta(d_{NLoS}-d_1)}{2d_1 d_{LoS}}\frac{1}{\sqrt{1-\cos^2\theta}} \tag{18}$$

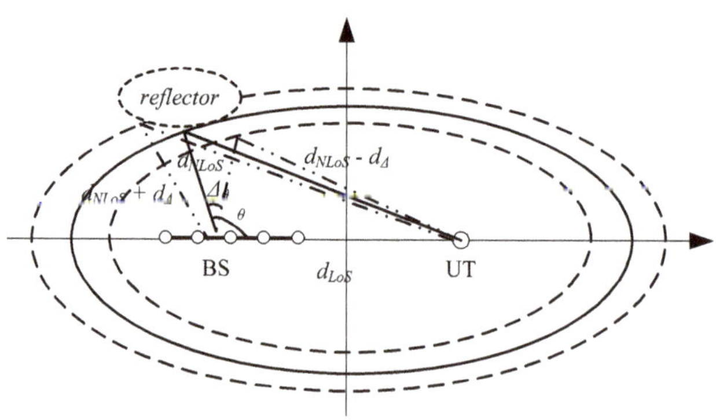

Figure 3. Ellipse geometry channel model for line of sight (LoS) and non-LoS (NLoS) transmission.

In order to illustrate the angle change Δ_θ during one timeslot, we assumed that d_{NLoS} was 800 m, the velocity of UT was 14.4 km/h, and the typical timeslot duration τ = 0.5 ms, then the movement distance of UT in one timeslot was 0.02 m. By changing the distance between BS and reflector, as shown in Figure 4, the angle change was not more than 0.025°. It should be noted that when the LoS distance and the d_{NLoS} were fixed, BS and reflector distance could not be arbitrary vales due to triangle inequality. Hence, the angle of arrival or departure changed slowly and then there was sparsity correlation among the angular channels for adjacent timeslots.

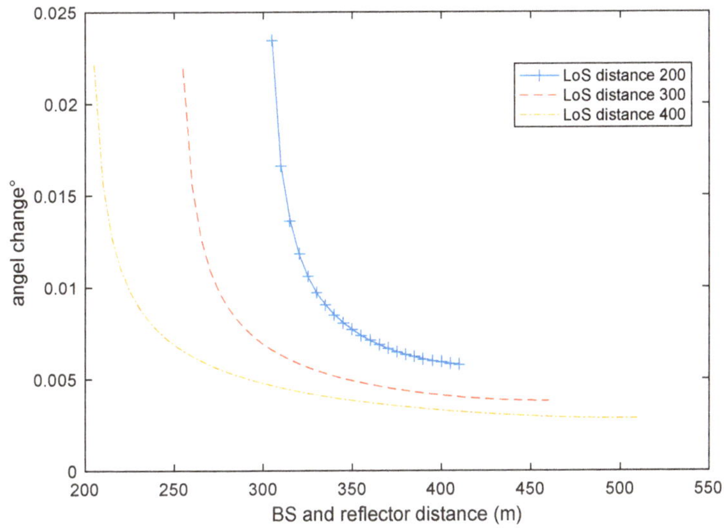

Figure 4. Transmission angle change during one timeslot with a different LoS distance and different distances between the base station (BS) and reflector.

4.2. Bayesian Cramér-Rao Bound Analysis

In this section we have discussed the Bayesian Cramér–Rao bound (BCRB) for the channel estimation with the proposed algorithm. Let $\mathbf{z} \triangleq \{\overline{h}, \sigma\}$, then the BCRB for the channel vector $\overline{\mathbf{h}}$ is given by the inverse of the Fisher information matrix $\mathbf{J}$ as:

$$\mathbf{J} = \mathbb{E}_{\mathbf{z}}\left\{-\frac{\partial^2 \log p(\overline{\mathbf{y}}, \mathbf{z})}{\partial z_i \partial z_j}\right\} \tag{19}$$

According to the system model in Section 2, $\overline{h}, \sigma$ are independent, the Fisher information matrix $\mathbf{J}$ is block diagonal. We can rewrite $p(\overline{y}, z)$ as

$$p(\overline{\mathbf{y}}, \mathbf{z}) = p(\overline{\mathbf{y}}|\mathbf{z})p(\overline{\mathbf{h}}|\boldsymbol{\alpha})p(\boldsymbol{\alpha}|\mathbf{b})p(\mathbf{b}) \tag{20}$$

Then the BCRB on the MSE of the estimated channel vector $\overline{\mathbf{h}'}$ is given by

$$\mathbb{E}\left\{||\overline{\mathbf{h}'} - \overline{\mathbf{h}}||^2\right\} \geq tr\left(\mathbf{J}^{-1}_{\overline{h}_i\overline{h}_j}\right) \tag{21}$$

where $\mathbf{J}_{\overline{h}_i\overline{h}_j} = \mathbb{E}_{\mathbf{z}}\left\{-\frac{\partial^2 \log p(\overline{\mathbf{y}}, \mathbf{z})}{\partial \overline{h}_i \partial \overline{h}_j}\right\}$ is the fisher information sub-matrix. Thus, we can obtain the Bayesian Cramér–Rao bound of the minimum mean square error for the estimated channel $\overline{\mathbf{h}'}$ as shown in Proposition 1.

Proposition 1. *The BCRB of MSE for the channel estimation $\overline{\mathbf{h}'}$ is represented as*

$$\mathbb{E}\left\{\|\overline{\mathbf{h}'}-\overline{\mathbf{h}}\|^2\right\} \geq tr\left(\left(diag(\mathbb{E}\left(\frac{1}{\alpha_i}\right)) + \frac{1}{\sigma^2}\overline{\mathbf{A}}^T\overline{\mathbf{A}}\right)^{-1}\right) = \sum_{i\in S}\frac{1}{\frac{1+c}{ad_i}+\frac{\lambda_i}{\sigma}} + \sum_{i\notin S}\frac{1}{\frac{b_i}{a}+\frac{\lambda_i}{\sigma}} \tag{22}$$

where S is the diagnosed support set, λ_i is the eigenvalues of $\overline{\mathbf{A}}^T\overline{\mathbf{A}}$, and $\overline{\mathbf{A}}^T\overline{\mathbf{A}} \in \mathbb{R}^{2N\times 2N}$, and a, b_i, c, and d_i are the parameters in the Bayesian model in Figure 1. *When $T_d, M \to \infty$ and $\frac{T_d}{M} = \beta$,* according to the random matrix theory, we have

$$\begin{aligned} \mathbb{E}\left\{\left(\overline{\mathbf{h}}-\hat{\overline{\mathbf{h}}}\right)^H\left(\overline{\mathbf{h}}-\hat{\overline{\mathbf{h}}}\right)\right\} & \geq |S|\cdot\frac{1}{|S|}\sum_{i\in S}\frac{1}{\frac{1+c}{a\min(d)}++\frac{\lambda_i}{\sigma}} + (N-|S|)\cdot\frac{1}{(N-|S|)}\sum_{i\notin S}\frac{1}{\frac{\max(b)}{a}+\frac{\lambda_i}{\sigma}} \\ & \to |S|\frac{a\min(\mathbf{d})}{1+c}\left(1-\frac{F(snr_1,\beta)}{4\beta snr_1}\right) + (N-|S|)\frac{a}{\max(\mathbf{b})}\left(1-\frac{F(snr_2,\beta)}{4\beta snr_2}\right) \end{aligned} \tag{23}$$

*where $snr_1 = \frac{amin(\mathbf{d})}{(1+c)\sigma}$, $snr_2 = \frac{a}{\sigma max(\mathbf{b})}$, $F(x,z) = \left(\sqrt{x(1+\sqrt{z})^2+1} - \sqrt{x(1-\sqrt{z})^2+1}\right)^2$, $min(\mathbf{d})$ and $max(\mathbf{b})$ are the minimum and maximum entries in **d** and **b**.*

The proof of proposition 1 is presented in Appendix B. From proposition 1, we can see that the MSE lower bound is related to the priori support size $|S|$, $(1+c)/\min(\mathbf{d})$ and $\max(\mathbf{b})$ for the massive MIMO channel estimation.

5. Simulations

In the simulation, the support diagnosis algorithm in [6] was adopted, and we assumed that the transmission angle change between timeslots was within 1 degree. The pilot length was 50, and antenna number at the BS was 100. The channel was generated according to the spatial model as defined in 3GPP TR25.996. We compared our proposed algorithm with a unitary dictionary with a size of 100 and the overcomplete dictionary with a size of 150, 200, and 250, and compared this with a Bayesian sparse learning (SL) [16], weighted subspace pursuit (WSP) [6], weighted l_1 minimization (W-l_1 min) [5], weighted iteratively reweighted least square(W-IRLS), IRLS [17], compressive sampling matched pursuit (COSAMP) in [11], and l_1 minimization (l_1 min) [18].

In order to evaluate the channel estimation performance, we used a normalized mean-square error (MSE) between true and estimated channel vectors as follows:

$$MSE = \frac{1}{T}\sum_{T}\frac{\|\hat{h}^d - h^d\|^2}{\|h^d\|^2} \tag{24}$$

where T is the number of trials, $\hat{\mathbf{h}}^d$ and $\mathbf{h}^d$ are the estimated and original channel vector, respectively for each trial. In the simulations the trial number T was 250.

In Figure 5 the overcomplete dictionary size was 150 in the proposed algorithm. It could be seen that when the unitary dictionary was used, our proposed algorithm outperformed WSP, COSAMP and IRLS, but was a little worse than W-l_1 with a small gap. However, when the overcomplete dictionary was used, our proposed algorithm outperformed other algorithms, but almost had the same performance as SL with a little performance improvement which could be seen in the zoomed-in subfigure. The overcomplete dictionary in the proposed algorithm can dramatically improve the MSE performance due to the fact that there are more atoms in the overcomplete dictionary than in the unitary dictionary which can improve the sparsity in the angular channel; however, it doesn't mean that the larger the overcomplete dictionary size is, the better performance it has, which is shown in Figure 6.

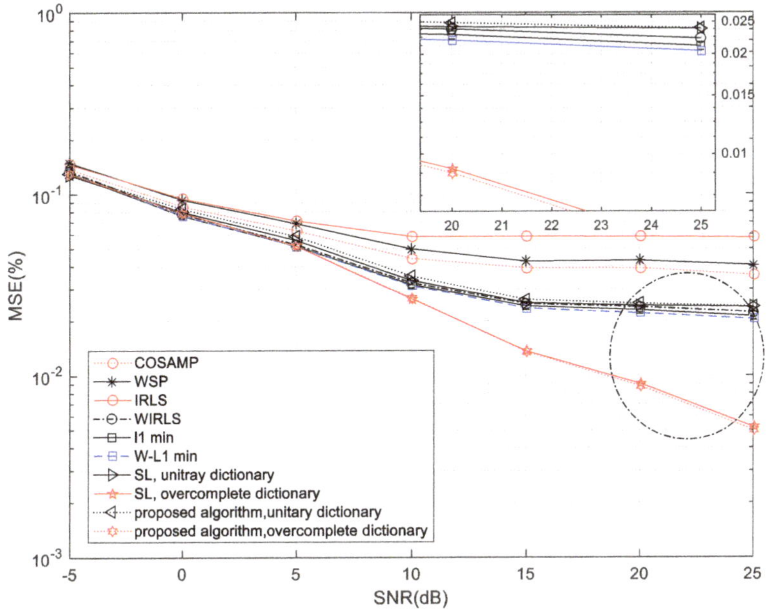

Figure 5. Comparisons of channel estimation mean square error (MSE) for different algorithms.

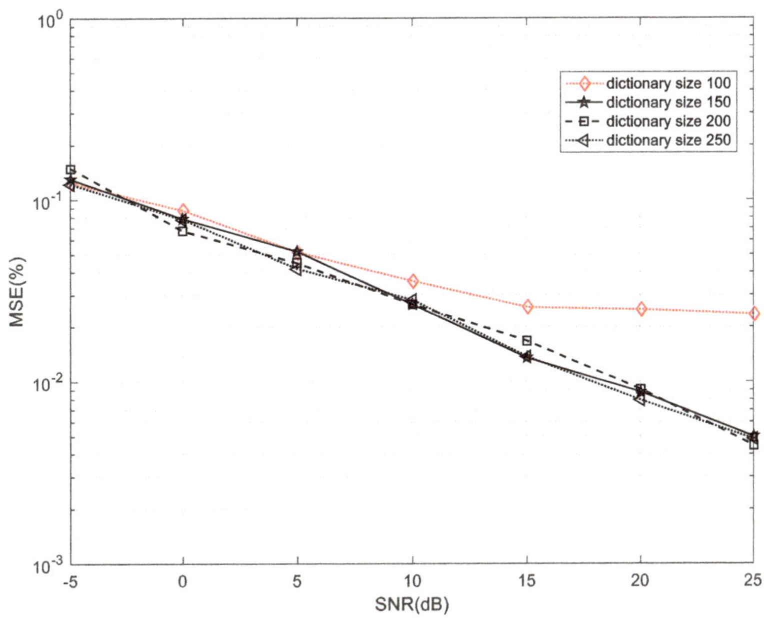

Figure 6. Comparisons of channel estimation of MSE for the proposed algorithm with different dictionary sizes.

We compared the performance of the proposed algorithm with different dictionary sizes in Figure 6. It can be seen that in the high SNR region the performance improved when an overcomplete dictionary was used, but the MSE performance gain did not improve when increasing the dictionary size. For example, the algorithm with a dictionary size of 150 had a relatively better performance than with a dictionary size of 100. However, the performances with a dictionary size of 200 and 250 almost gave the same trends as that with a dictionary size of 150. This was because the larger dictionary would induce angel ambiguity because the correlation of atoms increased. Hence, in the practical engineering, the dictionary size is not recommended to be very large. A large dictionary size is computationally expensive and the benefit is limited. It also should be noted that in the low SNR region the MSE performance with a larger dictionary size did not always do better than those with a smaller dictionary size. For example, when the SNR was 0 dB, they hadsimilar performance. The reason was that in the low SNR region the estimated channel support of the previous timeslot was not accurate enough, and on the other hand larger dictionary size would have deteriorated the dictionary incoherence.

We compared the runtime and convergence performance of the proposed algorithm with a different dictionary size in Figure 7. The relative error was defined as the ratio of the difference of adjacent iteration results to the previous iteration result. It can be seen that the proposed algorithm with dictionary size 150 converged fast than with a dictionary size of 100. However, the improvement had its price, and the runtime for the proposed algorithm with dictionary size 150 was longer which meant that the computational complexity was higher with a larger dictionary size. Based on the simulation results shown in Figures 6 and 7, when the antenna at BS is 100, the dictionary size is recommended to be set at 150 or so to balance the performance improvement and computation complexity.

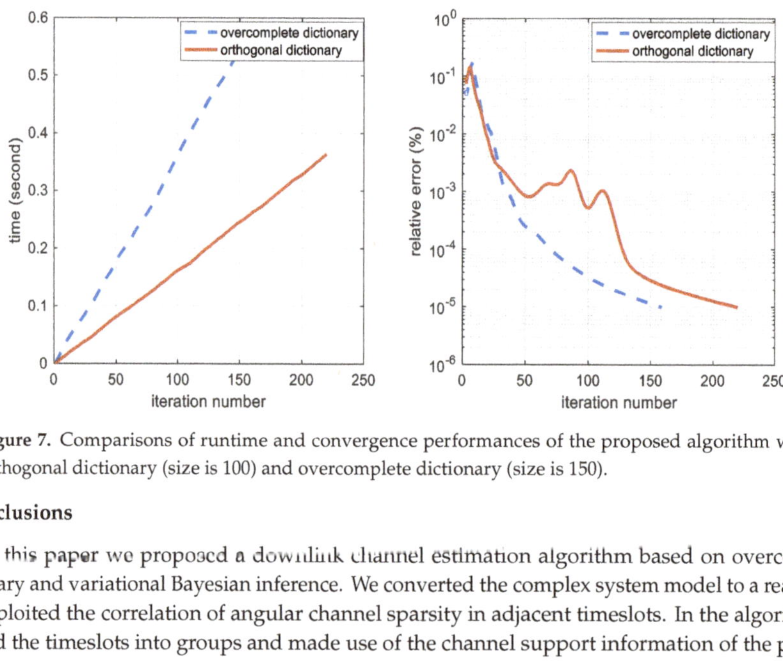

Figure 7. Comparisons of runtime and convergence performances of the proposed algorithm with orthogonal dictionary (size is 100) and overcomplete dictionary (size is 150).

6. Conclusions

In this paper we proposed a downlink channel estimation algorithm based on overcomplete dictionary and variational Bayesian inference. We converted the complex system model to a real model and exploited the correlation of angular channel sparsity in adjacent timeslots. In the algorithm we divided the timeslots into groups and made use of the channel support information of the previous timeslot to the channel estimation in the current timeslot within each group. The sparsity correlation and Bayesian Cramér–Rao bound for the MSE of channel estimation was analyzed. Compared with other recovery algorithms, such as WSP, IRLS, WIRLS, l_1 min, W-l_1 min and COSAMP, our proposed algorithm with overcomplete dictionary had a relatively better performance. Moderate overcomplete dictionary can improve the MSE performance of channel estimation to balance the computational complexity and performance gain.

Author Contributions: Conceptualization and methodology, W.L.; validation, X.W., S.P. and L.Z.; formal analysis, W.L.; writing—original draft preparation, W.L.; supervision, Y.W.

Funding: This work was supported in part by the National Science Foundation of China (No.61601509 and 61601334), the China Postdoctoral Science Foundation Grant (No.2016M603045 and 2018M632889) and the self-determined research funds of CCNU(CCNU18QN007) from the colleges basic research and operation of MOE.

Acknowledgments: Thank you to the reviewers for their valuable comments.

Conflicts of Interest: The authors declare no conflicts of interest.

Appendix A

Proof of angle change with UT movement.

According to the cosine law, we have

$$\cos\theta = \frac{d_1^2 + d_{LoS}^2 - (d_{NLoS} - d_1)^2}{2d_1 d_{LoS}}, \tag{A1}$$

$$\cos(\theta \pm \Delta_\theta) = \frac{d_1^2 + d_{LoS}^2 - (d_{NLoS} \pm d_\Delta - d_1)^2}{2d_1 d_{LoS}}. \tag{A2}$$

Then we can get

$$\theta \pm \Delta_\theta = \arccos\left(\frac{d_1^2 + d_{LoS}^2 - (d_{NLoS} - d_1)^2 - d_\Delta^2 \mp 2d_\Delta(d_{NLoS} - d_1)}{2d_1 d_{LoS}}\right). \tag{A3}$$

Since d_Δ^2 is very small compared with d_1 and d_{LoS}, by the first-order approximation we have

$$\begin{aligned} \theta \pm \Delta_\theta &\approx \arccos\left(\frac{d_1^2+d_{LoS}^2-(d_{NLoS}-d_1)^2 \mp 2d_\Delta(d_{NLoS}-d_1)}{2d_1 d_{LoS}}\right) \\ &\approx arc\cos\left(\frac{d_1^2+d_{LoS}^2-(d_{NLoS}-d_1)^2}{2d_1 d_{LoS}}\right) \pm \frac{2d_\Delta(d_{NLoS}-d_1)}{2d_1 d_{LoS}} \frac{1}{\sqrt{1-\cos^2\theta}} \\ &= \theta \pm \frac{2d_\Delta(d_{NLoS}-d_1)}{2d_1 d_{LoS}} \frac{1}{\sqrt{1-\cos^2\theta}} \end{aligned}. \tag{A4}$$

Then we have

$$\Delta_\theta \approx \frac{2d_\Delta(d_{NLoS} - d_1)}{2d_1 d_{LoS}} \frac{1}{\sqrt{1 - \cos^2\theta}}. \tag{A5}$$

□

Appendix B

Proof of Proposition 1.

Let $\mathbf{z} \triangleq \{\overline{h}, \sigma\}$, the we have

$$\mathbb{E}_{\mathbf{z}}\left\{(\mathbf{z} - \hat{\mathbf{z}})(\mathbf{z} - \hat{\mathbf{z}})^T\right\} \geq \mathbf{J}^{-1}. \tag{A6}$$

Since $\overline{h}, \sigma$ are independent, the Fisher information matrix $\mathbf{J}$ is block diagonal, and can be presented as

$$\mathbf{J} = \begin{bmatrix} \mathbf{J}_{\overline{\mathbf{h}},\overline{\mathbf{h}}} & 0 \\ 0 & \mathbf{J}_{\sigma,\sigma} \end{bmatrix}. \tag{A7}$$

Then the inverse of matrix $\mathbf{J}$ is

$$\mathbf{J}^{-1} = \begin{bmatrix} \mathbf{J}_{\overline{\mathbf{h}},\overline{\mathbf{h}}}^{-1} & 0 \\ 0 & \mathbf{J}_{\sigma,\sigma}^{-1} \end{bmatrix}. \tag{A8}$$

Because $p(\overline{y}, z) = p\left(\overline{y}\middle|z\right)p\left(\overline{h}\middle|\alpha\right)p\left(\alpha\middle|b\right)p(b)p(\sigma)$, we have

$$\begin{aligned}\mathbf{J} &= \mathbb{E}_{\mathbf{z}}\left\{-\frac{\partial^2 \log p(\overline{\mathbf{y}},\mathbf{z})}{\partial z_i \partial z_j}\right\} \\ &= \mathbb{E}_{\mathbf{z}}\left\{-\frac{\partial^2 \log p(\overline{\mathbf{y}}|\mathbf{z})}{\partial z_i \partial z_j}\right\} + \mathbb{E}_{\mathbf{z}}\left\{-\frac{\partial^2 \log p(\overline{\mathbf{h}}|\alpha)}{\partial z_i \partial z_j}\right\} + \mathbb{E}_{\mathbf{z}}\left\{-\frac{\partial^2 \log p(\alpha|\mathbf{b})}{\partial z_i \partial z_j}\right\} + \\ &\mathbb{E}_{\mathbf{z}}\left\{-\frac{\partial^2 \log p(\mathbf{b})}{\partial z_i \partial z_j}\right\} + \mathbb{E}_{\mathbf{z}}\left\{-\frac{\partial^2 \log p(\sigma)}{\partial z_i \partial z_j}\right\}\end{aligned} \tag{A9}$$

Since we mainly focus on the MSE of $\overline{h}$, we only need to analyze $\mathbf{J}_{\overline{\mathbf{h}},\overline{\mathbf{h}}}$. We discuss the above formula part by part as follows:

1) Let $\mathbf{J}_{\overline{\mathbf{h}},\overline{\mathbf{h}}}(\overline{\mathbf{y}}) = \mathbb{E}_{\mathbf{z}}\left\{-\frac{\partial^2 \log p(\overline{\mathbf{y}}|\mathbf{z})}{\partial z_i \partial z_j}\right\}$, according to the Bayesian model in Figure 1, we have

$$p\left(\overline{y}\middle|z\right) \sim Normal(\overline{\mathbf{y}}|\overline{\mathbf{A}}\overline{\mathbf{h}}, \sigma\mathbf{I})\text{then}\mathbf{J}_{\overline{\mathbf{h}},\overline{\mathbf{h}}}(\overline{\mathbf{y}}) = \mathbb{E}_{\mathbf{z}}\{\frac{\overline{\mathbf{A}}^T\overline{\mathbf{A}}}{\sigma}\} = \frac{\overline{\mathbf{A}}^T\overline{\mathbf{A}}}{\sigma}.$$

2) Let $\mathbf{J}_{\overline{\mathbf{h}},\overline{\mathbf{h}}}(\overline{\mathbf{h}}) = \mathbb{E}_{\mathbf{z}}\left\{-\frac{\partial^2 \log p(\overline{\mathbf{h}}|\alpha)}{\partial z_i \partial z_j}\right\}$, and we have $P\left(\overline{\mathbf{h}}\middle|\alpha\right) = \prod_{i=1}^{2N} Normal\left(\overline{h}_i\middle|0, \alpha_i\right)$, then we get $\mathbf{J}_{\overline{\mathbf{h}},\overline{\mathbf{h}}}(\overline{\mathbf{h}}) = \mathbb{E}_{\mathbf{z}}\left\{\frac{1}{\alpha_i}\right\}$.

3) Because $\mathbb{E}_{\mathbf{z}}\left\{-\frac{\partial^2 \log p(\alpha|\mathbf{b})}{\partial z_i \partial z_j}\right\}$, $\mathbb{E}_{\mathbf{z}}\left\{-\frac{\partial^2 \log p(\mathbf{b})}{\partial z_i \partial z_j}\right\}$ and $\mathbb{E}_{\mathbf{z}}\left\{-\frac{\partial^2 \log p(\sigma)}{\partial z_i \partial z_j}\right\}$ are independent with $\overline{h}$, they are all 0. Then in summary, we get $\mathbf{J}_{\overline{\mathbf{h}},\overline{\mathbf{h}}} = diag\left(\mathbb{E}\left(\frac{1}{\alpha_i}\right)\right) + \frac{1}{\sigma^2}\overline{\mathbf{A}}^T\overline{\mathbf{A}}$.

Since the priori support set information is used in our proposed algorithm, a three-layer model is constructed for the elements belonging to the priori support set, and a two-layer model is used for the elements not belonging to the priori support set, so $\mathbb{E}_Z\left\{\frac{1}{\alpha_i}\right\}$ has different expressions for the two cases. $\mathbb{E}_Z\left\{\frac{1}{\alpha_i}\right\}$ in the two cases are discussed as follows:

1) When i belongs to the priori support set, according to the three-layer graph model we have

$$p(\alpha) = \prod_{i=1}^{2N} Gamma\left(\alpha_i\middle|a, b_i\right), \tag{A10}$$

$$p(b_i) = \mathrm{Gamma}\left(b_i\middle|c, d_i\right) = \Gamma(c)^{-1} d_i^c b_i^{c-1} e^{-d_i b_i}. \tag{A11}$$

Then we get

$$\begin{aligned}p(\alpha_i) &= \int_0^\infty p\left(\alpha_i\middle|b_i\right)p(b_i)db_i \\ &= \int_0^\infty \Gamma(a)^{-1} b_i^a \alpha_i^{a-1} e^{-b_i\alpha_i}\Gamma(c)^{-1} d_i^c b_i^{c-1} e^{-d_i b_i} db_i \\ &= \Gamma(a)^{-1}\Gamma(c)^{-1}\alpha_i^{a-1} d_i^c \frac{\Gamma(a+c)}{(a_i+d_i)^{a+c}}\end{aligned}. \tag{A12}$$

Accordingly, we have

$$\begin{aligned}\mathbb{E}\left\{\frac{1}{\alpha_i}\right\} &= \int_0^\infty \frac{1}{\alpha_i}\Gamma(a)^{-1}\Gamma(c)^{-1}\alpha_i^{a-1} d_i^c \frac{\Gamma(a+c)}{(a_i+d_i)^{a+c}} d\alpha_i \\ &= \int_0^\infty \frac{1}{\alpha_i}\frac{\Gamma(a+c)}{\Gamma(a)\Gamma(c)}\left(\frac{\alpha_i}{d_1}\right)^{a-1}\left(\frac{\alpha_i}{d_1}+1\right)^{-a-c} d\frac{\alpha_i}{d_1}\end{aligned}, \tag{A13}$$

where $\frac{\Gamma(a+c)}{\Gamma(a)\Gamma(c)}\left(\frac{\alpha_i}{d_i}\right)^{a-1}\left(\frac{\alpha_i}{d_i}+1\right)^{-a-c}$ satisfies the probability density function of Beta prime distribution. According to the properties of the Beta prime distribution, when $-a < -1 < c$, we have

$$\begin{aligned}\mathbb{E}\left\{\left(\overline{\mathbf{h}}-\hat{\overline{\mathbf{h}}}\right)^H\left(\overline{\mathbf{h}}-\hat{\overline{\mathbf{h}}}\right)\right\} &\geq |S|\cdot\frac{1}{|S|}\sum_{i\in S}\frac{1}{\frac{1+c}{a\min(d)}++\frac{\lambda_i}{\sigma}}+(N-|S|)\cdot\frac{1}{(N-|S|)}\sum_{i\notin S}\frac{1}{\frac{\max(b)}{a}+\frac{\lambda_i}{\sigma}} \\ &\rightarrow |S|\frac{a\min(\mathbf{d})}{1+c}\left(1-\frac{F(snr_1,\beta)}{4\beta snr_1}\right)+(N-|S|)\frac{a}{\max(\mathbf{b})}\left(1-\frac{F(snr_2,\beta)}{4\beta snr_2}\right)\end{aligned}. \tag{A14}$$

2) When i does not belong to the priori support set, according to the high-order moment properties for the general gamma distribution, we have

$$\mathbb{E}\left\{\frac{1}{\alpha_i}\right\}=\frac{b_i}{a}. \tag{A15}$$

Then in summary, we have

$$\mathbb{E}\left\{\|\overline{\mathbf{h}}'-\overline{\mathbf{h}}\|^2\right\}\geq tr\left(\left(diag(\mathbb{E}\left(\frac{1}{\alpha_i}\right))+\frac{1}{\sigma^2}\overline{\mathbf{A}}^T\overline{\mathbf{A}}\right)^{-1}\right)=\sum_{i\in S}\frac{1}{\frac{1+c}{ad_i}+\frac{\lambda_i}{\sigma}}+\sum_{i\notin S}\frac{1}{\frac{b_i}{a}+\frac{\lambda_i}{\sigma}}, \tag{A16}$$

where S is the diagnosed support set, λ_i is the eigenvalues of $\overline{\mathbf{A}}^{\mathrm{T}}\overline{\mathbf{A}}$,and $\overline{\mathbf{A}}^{\mathrm{T}}\overline{\mathbf{A}}\in\mathbb{R}^{2\mathrm{M}\times 2\mathrm{M}}$.

When overcomplete dictionary is as $\mathbf{D}^d=\left\{\frac{1}{\sqrt{N}}e^{-j\frac{2\pi}{M}kn}\right\}_{n,k}, k\in\{1,\cdots,M\}, n\in\{1,\cdots,N\}$, and $\mathbf{A}$ is Gaussian random matrix with each element is mean 0 and variance $\frac{1}{T_d}$, then $\mathbf{AD}^{\mathrm{d}}$ is complex Gaussian random matrix. Then $\overline{\mathbf{A}}$ is Gaussian random matrix with mean 0 and variance $\frac{\rho^d}{2T_d}$.

According to the random matrix theory, for $N\times K$ dimensional random matrix $\mathbf{H}$ with each element is independent and is variable with mean 0 and variance $1/N$, when $K,N\rightarrow\infty$ and $\frac{K}{N}\rightarrow\beta$,then the empirical distribution of eigenvalues of $\mathbf{H}^{\mathrm{T}}\mathbf{H}$ converges almost surely as $f_\beta(x)=\left(1-\frac{1}{\beta}\right)^+\delta(x)+\frac{\sqrt{(x-a)^+(b-x)^+}}{2\pi\beta x}$, where $(x)^+=\max(0,x), a=\left(1-\sqrt{\beta}\right)^2, b=\left(1+\sqrt{\beta}\right)^2$.

Since $\overline{\mathbf{A}}\in\mathbb{R}^{2T_d\times 2M}$, and its element is Gaussian random variable with mean 0 and variance $\frac{\rho^d}{2T_d}$ By applying the above results for the empirical distribution of eigenvalues of $\mathbf{H}^{\mathrm{T}}\mathbf{H}$, when $T_d,M\rightarrow\infty$ and $\frac{T_d}{M}=\beta$, the empirical distribution of eigenvalues λ of $\overline{\mathbf{A}}^{\mathrm{T}}\overline{\mathbf{A}}$ converges almost surely as

$$f_\beta(\lambda)=\left(1-\frac{1}{\beta}\right)^+\delta(\lambda)+\frac{\sqrt{(\lambda-a\prime)^+(b\prime-\lambda)^+}}{2\pi\beta\lambda\sqrt{\rho^d}} \tag{A17}$$

where $a\prime=\sqrt{\rho^d}\left(1-\sqrt{\beta}\right)^2, b\prime=\sqrt{\rho^d}\left(1+\sqrt{\beta}\right)^2$. When $s,M\rightarrow\infty$ and $\frac{s}{M}=\mu$, we have

$$\begin{aligned}\mathbb{E}\left\{\left(\overline{\mathbf{h}}-\hat{\overline{\mathbf{h}}}\right)^H\left(\overline{\mathbf{h}}-\hat{\overline{\mathbf{h}}}\right)\right\} &\geq |S|\cdot\frac{1}{|S|}\sum_{i\in S}\frac{1}{\frac{1+c}{a\min(d)}++\frac{\lambda_i}{\sigma}}+(N-|S|)\cdot\frac{1}{(N-|S|)}\sum_{i\notin S}\frac{1}{\frac{\max(b)}{a}+\frac{\lambda_i}{\sigma}} \\ &\rightarrow |S|\frac{a\min(\mathbf{d})}{1+c}\left(1-\frac{F(snr_1,\beta)}{4\beta snr_1}\right)+(N-|S|)\frac{a}{\max(\mathbf{b})}\left(1-\frac{F(snr_2,\beta)}{4\beta snr_2}\right)\end{aligned}, \tag{A18}$$

where $snr_1=\frac{a\min(\mathbf{d})}{(1+c)\sigma}$, $snr_2=\frac{a}{\sigma\max(\mathbf{b})}$ and $F(x,z)=\left(\sqrt{x\left(1+\sqrt{z}\right)^2+1}-\sqrt{x\left(1-\sqrt{z}\right)^2+1}\right)^2$.

Then the proofs are complete. □

References

1. Lu, L.; Li, G.Y.; Swindlehurst, A.L.; Ashikhmin, A.; Zhang, R. An overview of Massive MIMO: Benefits and challenges. *IEEE J. Sel. Top. Signal Process.* **2014**, *8*, 742–758. [CrossRef]
2. Rao, X.; Lau, V.K.N. Distributed compressive CSIT estimation and feedback for FDD multi-user massive MIMO systems. *IEEE Trans. Signal Process.* **2014**, *12*, 3261–3271.
3. Vaswani, N.; Lu, W. Modified-CS: Modifying compressive sensing for problems with partially known support. *IEEE Trans. Signal Process.* **2010**, *9*, 4595–4607. [CrossRef]
4. Borries, R.V.; Miosso, C.; Potes, C. Compressed sensing using priori information. In Proceedings of the 2nd IEEE International Workshop on Computational Advances in Multi-Sensor Adaptive, St. Thomas, VI, USA, 12–14 December 2007; pp. 121–124.
5. Tseng, C.C.; Wu, J.Y.; Lee, T.S. Enhanced compressive downlink CSI Recovery for FDD Massive MIMO systems using weighted Block l_1 minimization. *IEEE Trans. Commun.* **2016**, *3*, 1055–1066. [CrossRef]
6. Lu, W.; Wang, Y.; Fang, Q.; Peng, S. Downlink compressive channel estimation with support diagnosis in FDD massive MIMO. *J. Wirel. Commun. Netw.* **2018**, *115*, 1–12. [CrossRef]
7. Masood, M.; Afify, L.H.; Al-Naffouri, T.Y. Efficient coordinated recovery of sparse channels in massive MIMO. *IEEE Trans. Signal Process.* **2015**, *1*, 104–118. [CrossRef]
8. Cheng, X.; Sun, J.; Li, S. Channel estimation for FDD multi-user massive MIMO: A variational Bayesian inference-based approach. *IEEE Trans. Wirel. Commun.* **2017**, *11*, 7590–7602. [CrossRef]
9. Dai, J.; Liu, A.; Lau, V.K.N. FDD massive MIMO channel estimation with arbitrary 2D-Array Geometry. *IEEE Trans. Signal Process.* **2018**, *10*, 2584–2599. [CrossRef]
10. Xie, H.; Gao, F.; Jin, S.; Fang, J.; Liang, Y. Channel Estimation for TDD/FDD Massive MIMO Systems with Channel Covariance computing. *IEEE Trans. Wirel. Commun.* **2018**, *6*, 4206–4218. [CrossRef]
11. Shen, W.; Dai, L.; Shi, Y.; Shim, B.; Wang, Z. Joint Channel Training and Feedback for FDD Massive MIMO Systems. *IEEE Trans. Veh. Technol.* **2016**, *10*, 8762–8767. [CrossRef]
12. Tauböck, G.; Hlawatsch, F.; Eiwen, D.; Rauhut, H. Compressive Estimation of Doubly Selective Channels in Multicarrier Systems: Leakage Effects and Sparsity-Enhancing Processing. *IEEE J. Sel. Top. Signal Process.* **2010**, *2*, 255–271. [CrossRef]
13. Tipping, M.E. Sparse Bayesian learning and the relevance vector machine. *J. Mach. Learn. Res.* **2001**, *1*, 211–244.
14. Christopher, M. Bishop. Bayesian Linear Regression. In *Pattern Recognition and Machine Learning*; Springer-Verlag: Heidelberg, Germany, 2006; pp. 152–160.
15. Fang, J.; Shen, Y.; Li, F.; Li, H.; Chen, Z. Support knowledge-aided sparse Bayesian learning for compressed sensing. In Proceedings of the 2015 IEEE International Conference on Acoustics, Speech and Signal Processing, Brisbane, Australia, 19–24 April 2015; pp. 3786–3790.
16. Ji, S.; Xue, Y.; Carin, L. Bayesian compressive sensing. *IEEE Trans. Signal Process.* **2008**, *6*, 2346–2356. [CrossRef]
17. Lu, W.; Wang, Y.; Fang, Q.; Peng, S. Compressive Channel Estimation Based on Weighted IRLS in FDD Massive MIMO. *Wirel. Pers. Commun.* **2018**, *2*, 1–10. [CrossRef]
18. Candes, E.J.; Romberg, J.K.; Tao, T. Stable signal recovery from incomplete and inaccurate measurements. *Commun. Pure Appl. Math.* **2006**, *8*, 1207–1223. [CrossRef]

electronics

MDPI

Article

Multiple-Symbol Non-Coherent Detection for Differential QAM Modulation in Uplink Massive MIMO Systems

Hieu Trong Dao and Sunghwan Kim *

School of Electrical Engineering, University of Ulsan, 93 Daehak-ro, Nam-gu, Ulsan 44610, Korea; hieu.dtvt10@gmail.com

* Correspondence: sungkim@ulsan.ac.kr; Tel.: +82-52-259-1401

Received: 21 May 2019; Accepted: 17 June 2019; Published: 20 June 2019

Abstract: In this paper, we propose a novel multiple-symbol detector based on maximum likelihood metric for differential quadrature amplitude modulation in massive multiple-input multiple-output (MIMO) systems. While current research on differential modulation in massive MIMO has focused on two consecutive symbols, our proposed detector is based on multiple-symbol, which is larger than or equal to two. Moreover, we derive new distance based on the proposed detector. To encode and decode data, we apply existing look-up table algorithm using the proposed distance, which is known as optimum encoding algorithm for differential modulation. Simulation results show the improvement based on the bit-error-rate performance since the proposed detector and distance vary according to the channel statistic information.

Keywords: 5G wireless networks; massive MIMO; non-coherent detection; QAM

1. Introduction

Massive multiple-input multiple-output (MIMO) transmission technique has gained a lot of attention in recent decades [1–13], since it can achieve significant improvement in terms of the energy and spectral efficiency while using simple signal processing [1–4]. Massive MIMO systems prefer operating in the time division duplex (TDD) mode in which users must synchronously send mutually orthogonal pilot signals to the corresponding base station (BS) so that the BS can estimate the channels. This method uses the estimated channels to perform signal processing [5–13], so that pilot signals account for a significant part of the total coherent interval, which decreases the spectral efficiency. In addition, when the number of users is large, the orthogonal pilot set has to be reused in every cell, which leads to pilot contamination problems; this is considered as a performance bottleneck in massive MIMO systems.

The authors of [8] investigate the power allocation to improve the spectral efficiency, this require a large information exchange in backhaul of system between BSs or between BSs and users. Besides, the algorithm to optimize the power is quite complex. Some semi-blind and blind channel estimation methods in uplink massive MIMO have been proposed. In [9], the authors proposed an eigenvalue decomposition-based method to blindly estimate the uplink channel from the data signal. However, they assumed that the number of antennas was very large such that the channel vectors become mutually orthogonal. The authors of [10] derived a new channel estimator based on subspace projection. However, this channel estimation algorithm relies heavily on the eigenvalues of the channel matrix. Interestingly, the authors of [11] proposed an energy detection scheme in which data symbols could be detected without relying on estimated channels. The scheme used in [11] requires designing unique modulated signal constellation for each user in the system. Another promising technique that does not require estimated channels is differential modulation, but it has not received much attention in the massive MIMO research field until now. In [12], differential quadrature amplitude modulation (DQAM) was proposed for massive MIMO systems. The modulation scheme used in [12] was based on the asymptotic behavior of the channel when the number of BS antennas goes to infinity; however, the authors of [12] did not show the detector when the number of BS antennas is finite. The authors of [13] generalized the QAM detector in [12] and proposed a new detector and non-coherent distance with better performance when the number of BS antennas is not very large. The differential encoding part of [13] was done via the look-up table algorithm used in [14,15]; this is known as the optimum encoding algorithm for differential modulation. Besides, both [12,13] can only detect two consecutive symbols at a time. Recently, the authors of [16] developed a new differential detector based on multiple-symbol differential detection (MSDD) and the generalized likelihood ratio test (GLRT) criterion. However, the authors of [16] only consider the case of M-ary Phase Shift Keying modulation (M-ary PSK).

In this paper, we propose a novel multiple-symbol detector for DQAM based on the maximum likelihood metric, which can detect more than two symbols at a time and varies following channel condition to adapt better with the change in environment. In addition, we propose a novel distance which can be used to encode and decode data by using the look-up table algorithm in [14,15] for DQAM encoding. Since the proposed scheme varies following the change in channel statistic information while the schemes in [12,13] are unchanged, they adapt better to the change of environment and show significantly better performance when compared to previous works [12,13].

2. System Model and Previous Works

Similar to [12,13], we also consider an uplink massive MIMO system consisting of a single-antenna user and a base station equipped with a large number of antennas $M(M \gg 1)$. We consider the TDD mode and the block fading model which are popular used in research on massive MIMO system [1,3,5,7,8], in which the channel is unchanged in one coherent interval T. The received signal vector at the mth BS antenna is modeled as [1]

$$\mathbf{y}_m = \sqrt{\rho} h_m \mathbf{x} + \mathbf{n}_m, \tag{1}$$

where ρ presents the average signal-to-noise ratio (SNR); $\mathbf{x} = [x_t, x_{t+1}, ..., x_{t+L-1}]^T$ is the transmit signal vector with length $L(L \leq T)$ and $E[||\mathbf{x}||^2] = L$, where elements of $\mathbf{x}$ are taken from the conventional QAM constellation as shown in Figure 1a, $\mathbf{n}_m$ is the additive white Gaussian noise vector at the mth BS antenna whose entries follow $CN(0,1)$; and h_m is the channel coefficient with $CN(\mu_h, \sigma_h^2)$. An example of the block fading model and signal vector is illustrated as in Figure 1b.

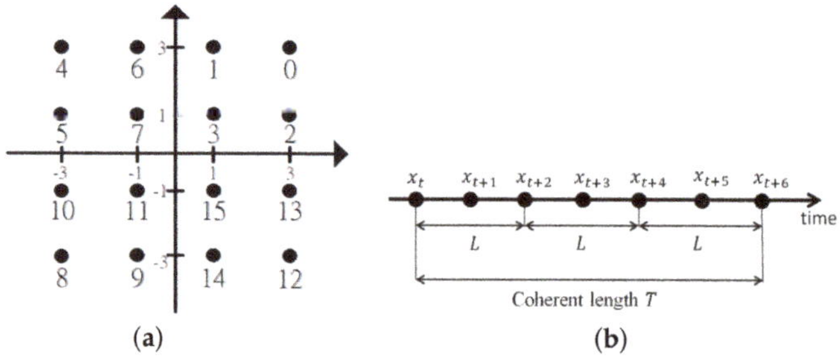

Figure 1. Conventional 16-QAM constellation and illustration of block fading. (**a**) 16-QAM constellation; (**b**) an illustration of block fading length $T = 7$ and signal vector length $L = 3$.

For simplicity, we normalize the channel so that $\mu_h^2 + \sigma_h^2 = 1$. Since Rayleigh and Rician fading models are very popular in evaluating system performance on both massive and regular MIMO system [17,18], we focus on these two models with $\mu_h = \sqrt{\frac{K_r}{1+K_r}}$ and $\sigma_h = \sqrt{\frac{1}{1+K_r}}$ [12]. Thus, the channel vector from a user to a BS now can be modeled as

$$\mathbf{h} = \mu_h \mathbf{h}_{\mathbf{LOS}} + \sigma_h \mathbf{h}_{\mathbf{NLOS}}, \tag{2}$$

where K_r represents the Rician factor; In a special case, when $K_r = 0$, the channel becomes a Rayleigh fading channel. Additionally, $\mathbf{h}_{\mathbf{LOS}} = [1, \exp^{-j\pi \sin(\theta)}, ..., \exp^{-j\pi(M-1)\sin(\theta)}]^T \in C^{M\times 1}$ with the arrival angle θ; is the light-of-sight (LOS) component when the antenna spacing is a half of wavelength. $\mathbf{h}_{\mathbf{NLOS}} \in C^{M\times 1}$ denotes the non-light-of-sight (NLOS) component whose elements follow i.i.d Gaussian variables with zero mean and unit variance.

In [12,13], the authors considered two consecutive tth and $(t-1)$th instants, with channel vectors $\mathbf{h}_t$; and $\mathbf{h}_{t-1}$, and assumed $\mathbf{h}_t \approx \mathbf{h}_{t-1}$. The received signal vector at the tth instant is given as

$$\mathbf{y}_t = \sqrt{\rho}\mathbf{h}_t x_t + \mathbf{n}_t. \tag{3}$$

where x_t is taken from a 16-DQAM constellation based on [19]. With a very large number of BS antennas M, they have

$$\begin{aligned} &\lim_{M\to\infty} \frac{1}{M}\mathbf{h}_t^H \mathbf{h}_{t-1} = 1, \lim_{M\to\infty} \frac{1}{M}\mathbf{n}_t^H \mathbf{h}_{t-1} = 0, \\ &\lim_{M\to\infty} \frac{1}{M}\mathbf{h}_t^H \mathbf{n}_{t-1} = 0, \lim_{M\to\infty} \frac{1}{M}\mathbf{n}_t^H \mathbf{n}_{t-1} = 0, \end{aligned} \tag{4}$$

Eventually, the signal symbol at the tth instant can be detected as

$$r_t = \frac{1}{M}\mathbf{y}_t^T \mathbf{y}_{t-1}^* = \rho x_t x_{t-1}^* \text{ for very large } M, \tag{5}$$

in which $x_t x_{t-1}^*$ can be mapped back to the information symbol by the encoding rule of [19]. However, the authors of [12] did not propose a detector for when the number of BS antennas M is finite. The authors of [13] generalized the detector in [12] as ([13], Equation (6)), which can be applied for any value of M. After that, they proposed a new two consecutive-symbol detector based on the conditional probability and a new non-coherent distance as ([13], Equations (9) and (10)). The new distance in [13] is used for look-up table algorithm in [14,15] for differential encoding.

Particularly, the authors of [14] had already proved that any differential encoding techniques can be transform equivalently to a differential encoding via a look-up table. Using the algorithm in [14,15] to create look-up table for encoding and decoding 16-DQAM signal, a brief explanation of a look-up table is as follows. The readers should refer to [14,15] for the details of the algorithm.

- Sort all possible codeword pair c_x and c_y in descending order based on their proposed distance.
- Arrange all groups: Take codeword pair one by one from the sorted list L. Put two codewords of a pair into the same group if possible; Otherwise, put them into two different groups and the two groups are written into L, which is called the list of near group pairs (near groups means the distance between members in two groups are small, which means there is a big chance that these two groups may be wrongly estimated as each other). Two codewords cannot be in the same group if their first symbols are identical.
- Assign modulated symbol to each group $G_1, G_2 \ldots, G16$ based on list L: Take group pairs one by one in order from list L. The number of different bits assigned for the considered two groups taken from L should be as small as possible (since two near groups have big change to be wrongly estimated as each other, this step's goal is to minimize the bit error).

Some properties of the look-up table are listed up below.

- The number of rows in the table are numbers of groups of codewords.
- Codewords belong to the same group will encode for the same information.
- Number of codewords in each group are the same.
- The important metric to generalize the look-up table is the non-coherent distance between any codewords, which is one of the new propose in this paper.
- General rules of generating the look-up table is that two codewords with small non-coherent distance should be placed in the same group. Otherwise, they should be putted into different groups.

Moreover, this look-up table can be optimized by using algorithms proposed in [14,15]. In this paper, we propose new non-coherent detector, new non-coherent distance and apply the algorithm in [14,15] to generate the look-up table. We compare the performance of the proposed detector and distance to the existing detectors and distances proposed in [12,13]. Due to the limitation of length, we would like to skip the detail of the look-up table algorithm and refer interested readers to [14,15].

3. New Differential Detector and Non-Coherent Distance

Consider the received signal at mth BS antenna as in Equation (1), the conditional probability of the received signal vector $\mathbf{y}_m$, given transmitted signal vector $\mathbf{x}$, is calculated as

$$p(\mathbf{y}_m|\mathbf{x}) = \frac{1}{2\pi \det(R_y)} \exp\left\{ -\frac{1}{2}(\mathbf{y}_m - \bar{\mathbf{y}}_m)^H R_y^{-1} (\mathbf{y}_m - \bar{\mathbf{y}}_m) \right\} \tag{6}$$

where $\bar{\mathbf{y}}_m$ is the mean of $\mathbf{y}_m$ and is given as

$$\bar{\mathbf{y}}_m = E\{\mathbf{y}_m\} = \sqrt{\rho}\mu_h \mathbf{x}, \tag{7}$$

where $\det(R_y)$ is the determinant of R_y, and R_y is the covariance matrix of $\mathbf{y}_m$, which can be calculated as

$$\begin{aligned} R_y &= E\left\{ (\mathbf{y}_m - \bar{\mathbf{y}}_m)(\mathbf{y}_m - \bar{\mathbf{y}}_m)^H \right\} \\ &= E\left\{ \rho(h_m - \mu_h)\mathbf{x}\mathbf{x}^H (h_m - \mu_h)^H + \sqrt{\rho}(h_m - \mu_h)\mathbf{x}\mathbf{n}_m^H \right. \\ &\quad \left. + \sqrt{\rho}(h_m - \mu_h)^H \mathbf{n}_m \mathbf{x}^H + \mathbf{n}_m \mathbf{n}_m^H \right\} = (\rho\sigma_h^2 \mathbf{x}\mathbf{x}^H + I_N). \end{aligned} \tag{8}$$

Since the proposed detector aims to maximize the summation of the conditional probability of received signal vector $\mathbf{y}_m$ at all BS antennas $1 \leq m \leq M$, given transmitted signal vector $\mathbf{x}$, the estimated signal vector $\hat{\mathbf{x}}$ can be calculated as

$$\hat{\mathbf{x}} = \arg\max_{\mathbf{x}\in\chi} \sum_{m=1}^{M} p(\mathbf{y}_m|\mathbf{x}), \tag{9}$$

where χ presents the vector space of all possible transmitted signal vectors $\mathbf{x}$. Since the natural logarithm function is monotonically increasing, maximizing $p(\mathbf{y}_m|\mathbf{x})$ is equivalent to maximize $\ln p(\mathbf{y}_m|\mathbf{x})$. Finally, the proposed detector is given as

$$\hat{\mathbf{x}} = \arg\max_{\mathbf{x}\in\chi} \left\{ \sum_{m=1}^{M} -(\mathbf{y}_m - \bar{\mathbf{y}}_m)^H R_y^{-1} (\mathbf{y}_m - \bar{\mathbf{y}}_m) - \ln(\det(R_y)) \right\}, \tag{10}$$

Specially, when the channel is Rayleigh fading, we have $\mathbf{y}_m = 0$, and the proposed detector becomes

$$\hat{\mathbf{x}} = \arg\max_{\mathbf{x}\in\chi, x_t\in Q_1} \left\{ \sum_{m=1}^{M} -\mathbf{y}_m^H R_y^{-1} \mathbf{y}_m - \ln(\det(R_y)) \right\}, \tag{11}$$

where Q_1 is the first quadrant. This means the number of decision values that need to be calculated is reduced by a factor of four, from N^L to $\frac{N^L}{4}$, with N-QAM. The reduction occurs since when $\bar{\mathbf{y}}_m = 0$, for any codeword $\mathbf{x} = [x_t, x_{t+1}, ..., x_{t+L-1}]^T$ with the first symbol x_t belongs to the first quadrant of the constellation, there are also three other codewords which have the first symbol x_t' belong to three other quadrants, that have the same estimated vector $\hat{\mathbf{x}}$ as $\mathbf{x}$. In other words, we only need to calculate the decision values of codewords which have the first symbol x_t belongs to the first quadrant of the constellation in case of Rayleigh fading. To calculate the proposed non-coherent distance, we propose the distance from $\mathbf{x}_1$ to $\mathbf{x}_2$ and the distance from $\mathbf{x}_2$ to $\mathbf{x}_1$ as in Equations (12) and (13) based on the proposed detector, Equation (10), as below:

$$d(\mathbf{x}_1 \rightarrow \mathbf{x}_2) = ||[-\rho(c_h\mathbf{x}_1)^H R_{x_1}^{-1}(c_h\mathbf{x}_1) - \ln(\det(R_{x_1}))] - [-\rho(c_h\mathbf{x}_1)^H R_{x_2}^{-1}(c_h\mathbf{x}_1) - \ln(\det(R_{x_2}))]||; \quad (12)$$

$$d(\mathbf{x}_2 \rightarrow \mathbf{x}_1) = ||[-\rho(c_h\mathbf{x}_2)^H R_{x_2}^{-1}(c_h\mathbf{x}_2) - \ln(\det(R_{x_2}))] - [-\rho(c_h\mathbf{x}_2)^H R_{x_1}^{-1}(c_h\mathbf{x}_2) - \ln(\det(R_{x_1}))]||. \quad (13)$$

R_{x_1}, R_{x_2}andc_h in Equations (12) and (13) are calculated as

$$R_{x_1} = \rho\sigma_h^2\mathbf{x}_1\mathbf{x}_1^H + I_L; R_{x_2} = \rho\sigma_h^2\mathbf{x}_2\mathbf{x}_2^H + I_L; c_h = \sigma_h + \mu_h; \quad (14)$$

The Equation (12) is the non-coherent distance from $\mathbf{x}_1$ to $\mathbf{x}_2$, which is based on the assumption that we did send the codeword $\mathbf{x}_1$ but the detector wrongly estimated that $\mathbf{x}_2$ was sent. Inversely, the Equation (13) is the non-coherent distance from $\mathbf{x}_2$ to $\mathbf{x}_1$ in which, the $\mathbf{x}_2$ was actually sent but the detector wrongly estimated that $\mathbf{x}_1$ was sent. In other words, Equations (12) and (13) can be used by likelihood estimator as a distance between two codewords $\mathbf{x}_1$ and $\mathbf{x}_2$. The larger the values of Equations (12) and (13) are, the less chance the detector wrongly estimates between $\mathbf{x}_1$ and $\mathbf{x}_2$. Eventually, the proposed non-coherent distance is calculated as

$$d(\mathbf{x}_1, \mathbf{x}_2) = \min(d(\mathbf{x}_1 \rightarrow \mathbf{x}_2), d(\mathbf{x}_2 \rightarrow \mathbf{x}_1)). \quad (15)$$

In the differential encoding part, we apply the look-up table algorithm for DQAM as in [14,15] by using the proposed non-coherent distance in Equation (15).

The main contribution of these above steps and equations are summed up as follow.

- Equation (10) is our proposed detector which is derived based on the conditional probability of received signal and contains channel statistical information, so that it adapts better with the change in environment.
- Equation (11) is a simplified detector when the channel is Rayleigh fading.
- Equations (12) and (13) are derived based on Equation (10), and they are used to evaluate the distance between two signal vectors (a small distance means that it is easy to wrongly decode between two signal vectors).
- $d(\mathbf{x}_1 \rightarrow \mathbf{x}_2)$ is distance from $\mathbf{x}_1$ to $\mathbf{x}_2$ which is used to evaluate how likely it is to wrongly estimate $\mathbf{x}_1$ as $\mathbf{x}_2$. Similar explanation is applied to $d(\mathbf{x}_2 \rightarrow \mathbf{x}_1)$.
- Eventually, we use Equation (15) to take the minimum value between $d(\mathbf{x}_1 \rightarrow \mathbf{x}_2)$ and $d(\mathbf{x}_2 \rightarrow \mathbf{x}_1)$ as the distance between $\mathbf{x}_1$ and $\mathbf{x}_2$.

4. Numerical Results

In simulations, we use the conventional 16-QAM constellation and apply the look-up table algorithm in [14,15] to differentially encode the information. Particularly, one 4-bit information symbol is encoded into two consecutive 16-QAM points; thus, the non-coherent distance as in Equation (15) calculated with the length of transmitted signal vectors $\mathbf{x}_1, \mathbf{x}_2$ is 2. Finally, the look-up table for 16-DQAM has 16 rows presenting 16 different groups; each group contains 16 different vectors $\mathbf{x}$, and all transmitted signal vectors in the same group correspond to the same information symbol. The look-up table for the proposed 16-DQAM scheme with Rician fading channel $K_r = 1$ and an average SNR = −4 dB is given in Table 1 as an example. Notice that, in [14,15] after generating the look-up table, there is one more step that maps information bit to each group. This step adds a little more improvement in bit error rate performance since groups with small non-coherent distance are mapped to information bit symbols with a small difference in the number of bits. However, in this simulation, we focus on comparing the performance of detector and distance between our proposed ones and previous ones in [12,13] so that we skip this step and add the information bit symbols sequentially from first group to last group in the look-up table.

Table 1. Look-up table for 16-DQAM using proposed distance, $L = 2$.

G_1	(0;0)	(1;1)	(2;2)	(3;0)
G_2	(0;1)	(1;6)	(2;0)	(3;1)
G_3	(0;2)	(1;0)	(2;13)	(3;2)
G_4	(0;3)	(1;3)	(2;3)	(3;3)
G_5	(0;4)	(1;5)	(2;6)	(3;4)
G_6	(0;5)	(1;10)	(2;4)	(3;5)
G_7	(0;6)	(1;4)	(2;1)	(3;6)
G_8	(0;7)	(1;7)	(2;7)	(3;7)
G_9	(0;8)	(1;9)	(2;10)	(3;8)
G_{10}	(0;9)	(1;14)	(2;8)	(3;9)
G_{11}	(0;10)	(1;8)	(2;5)	(3;10)
G_{12}	(0;11)	(1;11)	(2;11)	(3;11)
G_{13}	(0;12)	(1;13)	(2;14)	(3;12)
G_{14}	(0;13)	(1;2)	(2;12)	(3;13)
G_{15}	(0;14)	(1;12)	(2;9)	(3;14)
G_{16}	(0;15)	(1;15)	(2;15)	(3;15)

Figure 2 illustrates the simulation results of the proposed 16-DQAM scheme with different lengths of estimated signal vector $L = 2$ and 3, as well as the 16-DQAM schemes in [12,13] where the number of BS antennas are $M = 128$ and 500, and the coherent length $T = 7$. Since the authors of [12] did not show the detector when M is finite, we suppose that the 16-DQAM scheme in [12] uses the generalized detector as in [13], Equation (6), and the corresponding non-coherent distance in [13]. As previously shown in [12,13], the 16-DQAM scheme of [12] shows an error floor when M is not very large. The proposed scheme significantly outperforms the schemes in [12,13] for both $M = 128$ and 500.

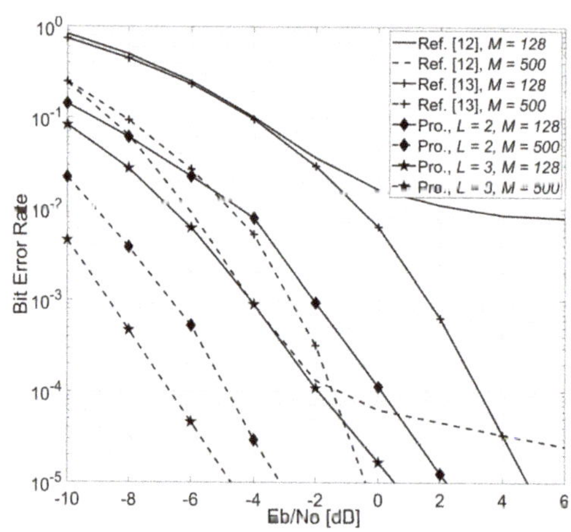

Figure 2. Performance comparison between the proposed 16-DQAM scheme and the previous works of References [12,13] under Rayleigh fading.

Since the schemes in [12,13] can only detect two consecutive symbols at a time, we simulate our proposed scheme when the length of signal vector is $L = 2$. We can clearly see that, with the same channel condition and signal vector's length, the proposed scheme outperforms the other schemes for $M = 128$ with nearly 3 dB when BER = 10^{-4}. With not so large number of BS antenna $M = 128$, the significant improvement of the proposed scheme shows a huge potential that it can be deployed in real system. When M increases to 500, the BER performance increases much more further with nearly 5 dB at BER = 10^{-4}, which also shows the advantages of massive MIMO with very large number of BS antennas.

Noticeably, the performance of the proposed scheme is improved significantly when the length of the estimated signal vector L increases, regardless of the value of M. The gain is nearly 1.5 dB at BER $= 10^{-5}$ when the length L increases from 2 to 3. When $M = 500$, the scheme of [13] only performs better than [12] at low BER ($\leq 10^{-4}$) while the performance of the proposed scheme is remarkably better than both [12,13]; the gain is approximately 3 dB at BER $= 10^{-5}$ in comparison with [13].

Figure 3 shows the simulation results for the aforementioned schemes under Rician fading with $T = 7, L = 2, M = 128$ and different values of the Rician factor $K_r = 0, 1,$ and 10. Notice that when $K_r = 0$, the Rician channel becomes a Rayleigh channel. The error floor still happens when $K_r = 1$ in the case of [12]. However, when the LOS component of the Rician channel becomes stronger with $K_r = 10$, the error floor seems to disappear and the performance of [12] is improved much more than [13]; the performance gap is nearly 1 dB at BER $= 10^{-5}$. The performances of scheme of [13] are nearly the same with different values of K_r, this is because the detector in [13] cancels out the channel coefficient between two consecutively received symbols. The performance of the proposed scheme is the best among three schemes. Even with $K_r = 0$ (i.e., Rayleigh fading), the performance of the proposed scheme remains better than the other schemes with $K_r = 1$ or 10. When K_r is increased from 0 to 1 and 10, the performance of the proposed scheme is improved significantly with gains 5.5 dB and 9.5 dB, respectively. In summary, we conclude that the proposed scheme shows much better performance in Rician channels than in Rayleigh channels.

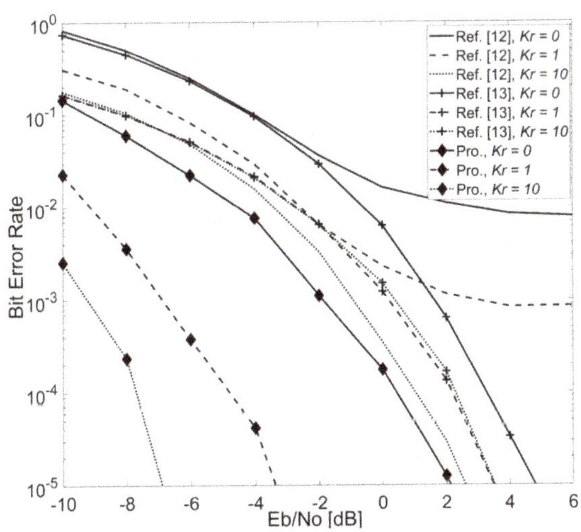

Figure 3. Performance comparison between the proposed 16-DQAM scheme and the previous works of References [12,13] under Rician fading with different Rician factors.

5. Conclusions

In this paper, we propose a new detector and non-coherent distance for differential QAM modulation in massive MIMO systems. We also apply the well-known look-up table algorithm for DQAM encoding using the proposed non-coherent distance. The proposed detector can detect multiple symbols ($\geq$ 2) at a time. The proposed scheme varies following the change in channel information statistics, allowing them to adapt better to the change in environment. Additionally, they can be applied in a wide class of channels with a not too large number of base station antennas. This paper focuses on massive MIMO system with single cell. Therefore, as a future work, it will be very attractive to investigate the performance and how to improve the proposed scheme in the multiple-cell environment where there is the presence of interference between users in nearby cells.

Author Contributions: All authors discussed the contents of the manuscript and contributed to its presentation. H.T.D. designed and implemented the proposed scheme, analyzed the simulation results and wrote the paper under the supervision of S.K.

Funding: This research was funded by the Research Program through the National Research Foundation of Korea (NRF-2016R1D1A1B03934653, NRF-2019R1A2C1005920).

Conflicts of Interest: The authors declare no conflict of interest.

Abbreviations

The following abbreviations are used in this manuscript:

MIMO	Multiple-input multiple-output
BS	Base station
TDD	Time Division Duplex
DQAM	Differential Quadrature Amplitude Modulation
SNR	Signal-to-noise
LOS	Light-of-Sight
NLOS	non Light-of-Sight
AWGN	Additive white Gaussian noise

References

1. Marzetta, T.L. Noncooperative cellular wireless with unlimited numbers of base station antennas. *IEEE Trans. Wirel. Commun.* **2010**, *9*, 3590–3600. [CrossRef]
2. Kim, K.J.; Choi, K.J.; Lee, S.R.; Kim, K.S. Multi-user massive MIMO for next-generation WLAN systems. *IET Electron. Lett.* **2015**, *51*, 792–794. [CrossRef]
3. Lu, L.; Li, G.Y.; Swindlehurst, A.L.; Ashikhmin, A.; Zhang, R. An overivew of massive MIMO: Benefits and challenges. *IEEE J. Sel. Top. Signal Process.* **2014**, *8*, 742–758. [CrossRef]
4. Hassan, N.; Fernando, X. Massive MIMO Wireless Networks: An Overview. *Electronics* **2017**, *6*, 63. [CrossRef]
5. Trong Dao, H.; Kim, S. Worst Cell Based Pilot Allocation in Massive MIMO Systems. *Electronics* **2018**, *7*, 197. [CrossRef]
6. Nyarko, J.K.N.; Mbom, C.A. A Performance Study of Massive MIMO Heterogeneous Networks with Ricean/Rayleigh Fading. *Electronics* **2018**, *7*, 79. [CrossRef]
7. Zhu, X.; Dai, L.; Wang, Z.; Wang, X. Weighted-graph-coloring-based pilot decontamination for multicell massive MIMO systems. *IEEE Trans. Veh. Technol.* **2017**, *66*, 2829–2834. [CrossRef]
8. Dao, H.T.; Kim, S. Pilot power allocation for maximising the sum rate in massive MIMO systems. *IET Commun.* **2018**, *12*, 1367–1372. [CrossRef]
9. Ngo, H.Q.; Larsson, E.G. EVD-based channel estimation in multicell multiuser MIMO systems with very large antenna arrays. In Proceedings of the 2012 IEEE International Conference on Acoustics, Speech and Signal Processing (ICASSP 2012) , Kyoto, Japan, 25–30 March 2012; pp. 3249–325.
10. Muller, R.R.; Cottatellucci, L.; Vehkapera, M. Blind pilot decontamination. *IEEE J. Sel. Top. Signal Process* **2014**, *8*, 773–786. [CrossRef]
11. Manolakos, A.; Chowdhury, M.; Goldsmith, A. Energy-Based Modulation for Noncoherent Massive SIMO Systems. *IEEE Trans. Wirel. Commun.* **2016**, *15*, 773–786. [CrossRef]
12. Kong, D.; Xia, X.; Jiang, T. A Differential QAM Detection in Uplink Massive MIMO Systems. *IEEE Trans. Wirel. Commun.* **2016**, *15*, 6371–6383. [CrossRef]

13. Wei, R.; Wang, X. Differential 16-QAM and 16-APSK for Uplink Massive MIMO Systems. *IEEE Wirel. Commun. Lett.* **2017**, *7*, 170–173. [CrossRef]
14. Wei, R.-Y. Differential encoding by a look-up table for quadrature amplitude modulation. *IEEE Trans. Commun.* **2011**, *59*, 84–94. [CrossRef]
15. Wei, R.-Y.; Chen, L.-T. Further results on differential encoding by a table. *IEEE Trans. Commun.* **2012**, *60*, 2580–2590. [CrossRef]
16. Wang, Y.; Tian, Z. Multiple Symbol Differential Detection for Noncoherent Communications With Large-Scale Antenna Arrays. *IEEE Wirel. Commun. Lett.* **2018**, *7*, 190–193. [CrossRef]
17. Angelis, C.T.; Chronopoulos, S.K. System Performance of an LTE MIMO Downlink in Various Fading Environments. In *International Conference on Ambient Media and Systems*; Springer: Berlin, Germany, 2011; pp. 36–43.
18. Chronopoulos, S.K.; Christofilakis, V.; Tatsis, G.; Kostarakis, P. Preliminary BER Study of a TC-OFDM system operating under noisy conditions. *J. Eng. Sci. Technol. Rev.* **2016**, *9*, 13–16. [CrossRef]
19. Egri, R.G.; Horrigan, F.A. A finite group of complex integers and its application to differentially coherent detection of QAM signals. *IEEE Trans. Inf. Theory* **1994**, *40*, 216–219. [CrossRef]

electronics

MDPI

Article

Throughput Enhancement in Downlink MU-MIMO Using Multiple Dimensions

Jong-Gyu Ha, Jae-Hyun Ro and Hyoung-Kyu Song *

Department of Information and Communication Engineering, uT Communication Research Institute, Sejong University, Gunja-dong 98, Gwangjin-gu, Seoul 05006, Korea

* Correspondence: songhk@sejong.ac.kr; Tel.: +82-2-3408-3890

Received: 13 May 2019; Accepted: 4 July 2019; Published: 5 July 2019

Abstract: This paper focuses on the throughput performance enhancement in the single cell multi-user MIMO (MU-MIMO) downlink system model. For better quality of service, this paper proposes the scheme that increases system throughput and improves the spectral efficiency. Specifically, the signal transmission and detection schemes are proposed by using multiple dimensions. At the transmitter side, two dimensions (power and space) are adopted at the same time. To achieve multiple access (MA), the space domain is exploited by using a block diagonalization (BD) precoding technique, and the power domain is exploited to transmit more data symbols. At the receiver, the signal detection structure corresponding to a transmitter is also proposed. In the simulation results, comparisons of throughput performance are presented in various aspects. As a result, the proposed scheme outperforms the conventional schemes using only one dimension in terms of throughput. This paper shows strong performance in MU-MIMO senarios by adopting multiple dimensions.

Keywords: multi-user MIMO; space division multiple access (SDMA); block diagonalization (BD); non-orthogonal multiple access (NOMA); broadcast channel

1. Introduction

In the future, in order to handle explosive data traffic, studies will probably aim to increase channel capacity and data rate in the overall wireless communication system [1]. Multiple-input multiple-output (MIMO) has been studied in wireless systems since it has dramatic gains in channel capacity [2]. Also, multiuser MIMO (MU-MIMO) has been studied widely as a potential for improving the overall throughput [3–5]. In downlink broadcasting (BC) channel, MU-MIMO is accomplished by multiuser beamforming that eliminates the multiuser interference (MUI) completely [6–9]. A number of users can be served by one base station (BS) simultaneously and the spectral efficiency can be increased. Therefore, the use of space-division multiple access (SMDA) in the downlink channel provides a considerable gain in system capacity. The sum rate of the MU-MIMO broadcast channel is achieved by dirty paper coding (DPC). However, the critical drawback of DPC is extreme high complexity to implement in practice [10]. Another promising technique in an MU-MIMO system is block diagonalization (BD). In this paper, the proposed scheme considers BD as a generalization of the channel inversion [11,12]. BD supports the multiple data stream with low complexity and approaches the sum capacity of DPC using user selection algorithms [13].

On the other hand, non-orthogonal multiple access (NOMA) is one of the most promising techniques for improving the overall spectral efficiency [14,15]. NOMA shares the same resources with multiple users by exploiting the domains. The well-known NOMA schemes can be divided into power-domain and code-domain NOMA. Power-domain NOMA multiplexing is achieved by different allocated power for users according to the channel conditions. Symbols are superposed to each user and receivers perform successive interference cancellation (SIC) [16]. Furthermore,

MIMO-NOMA is another popular technique to increase sum capacity. In MIMO-NOMA work, the signal processing techniques are investigated and two main routes exist: single-cluster and multi-cluster MIMO-NOMA [17–19]. Using MIMO-NOMA, a full benefit of system capacity and user fairness is achieved by well-chosen power allocation, user clustering, and beamforming [20,21]. Most of these papers consider the user grouping and clustering. However, the proposed scheme fundamentally does not consider user grouping and clustering. Due to the fundamental difference, the proposed scheme has additional advantages. In this paper, the proposed signal transmission and detection scheme use superposition coding (SC) and SIC in conjunction with MU-MIMO to improve the system spectral efficiency. With the SDMA scheme, a number of data symbols can be served to each user by SC and SIC. The power and space domain are fully exploited at the same time and the overall system throughput can be improved in single cell MU-MIMO scenario. The main contributions of this paper are summarized as follows:

1. First, the proposed scheme is written in a different view from the existing MU-MIMO papers. Recently, many researches in MU-MIMO and NOMA have been mainly focused on increasing spectral efficiency by using user clustering, power allocation, and beamforming. User clustering means that close and far users are clustered within each beam and result in intra-beam interference. Intra-beam interference is a major cause of performance degradation. In existing papers, the resource reuse approach is highly affected by intra-beam interference and is critical to overall system performance. However, the proposed scheme does not need to consider intra-beam interference. The proposed scheme makes a single user in each beam and power domain is exploited in a single user. In conclusion, power and spatial domains are exploited within a single user.
2. Second, the proposed scheme improves performance through joint design that uses dimensions appropriately according to wireless communication systems. In a wireless communication system, it is important to decide which technique to use according to the system requirements and characteristics along with cost and complexity constraints. While there are various combinations of techniques that make up the wireless communication system, it is difficult to get a clear answer as to the best scheme for a complex multiuser system under a range of typical operating conditions. This paper proposes a technique that can appropriately bring out the merits of various combinations of techniques.
3. Finally, the proposed idea can be described from a different perspective and provides insights into other system implementations. The key idea here is to use multiple dimensions appropriately. The beauty of this approach is that it can even be applied to other multiple access systems. The various system models can be implemented by using additional dimensions. This paper demonstrates that the system throughput can be increased by using multiple dimensions. This concept can be extended further. If the additional dimensions are used without degradation or with a little tradeoff, the system throughput can be increased. The idea presented in this paper provides a multitude of interesting avenues for future research.

The remainder of this paper is organized as follows. Section 2 describes the problems of the existing scheme and explains the solution and the reasonability of the proposed scheme. Section 3 describes the overall system model and the proposed scheme. Also, this section presents the performance of the proposed scheme. Section 4 presents simulation results and compares with the conventional scheme under various conditions. Section 5 considers some extensions about the implementation of the proposed scheme. Finally, conclusion is provided in Section 6.

2. Motivation

The novelty of the proposed scheme is to use multiple dimensions as much as possible. Figure 1a,b represent examples of resource usage in NOMA and BD schemes using only one dimension, respectively. Figure 1c represents example of resource usage for the proposed scheme. In Figure 1a,b which represent existing techniques, the two techniques cannot be simply combined since all existing schemes utilize each technique for multiple access. However, in the proposed scheme, the SC is used to separate each symbol by utilizing the power domain. In conclusion, it is possible to utilize both power and space domain at the same time and improve performance by changing usage in SC application. In addition, through proper combination, the proposed scheme compensates the shortcomings of existing schemes and has advantages. First, overhead of the BS is reduced. A dynamic user scheduling and grouping strategy needs the feedback information at the BS. The proposed scheme can reduce the feedback overhead significantly by not considering intra-beam interference. Second, proper fairness among users is assured. The interference is treated as noise to users and does not guarantee fairness among users. In the proposed scheme, a certain error probability is ensured for all users. Third, the complexity is reduced for each user. In the case of the existing NOMA scheme, it is necessary to decode even if it is not the user's own signal. However, in the proposed scheme, all the decoded symbols become the user's own data. Therefore, it is possible to reduce the complexity of the user in demodulating unnecessary data. As a result, the proposed scheme allows more flexibility in spreading user signals over the multiusers.

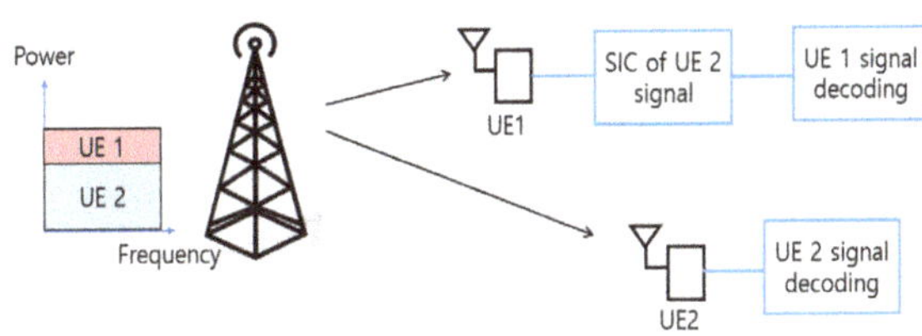

(a)

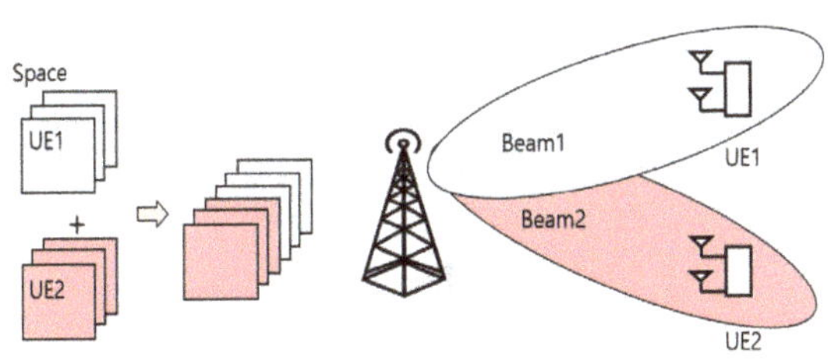

(b)

Figure 1. *Cont.*

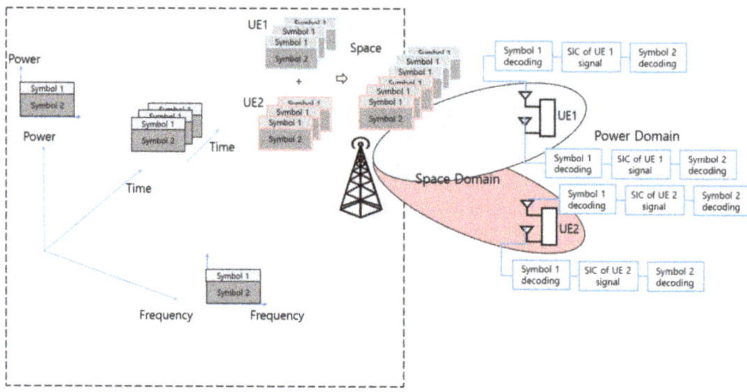

(c)

Figure 1. Examples of resource usage in various dimensions (**a**) Example of resource usage for conventional NOMA; (**b**) Example of resource usage for conventional BD; (**c**) Example of resource usage for proposed scheme.

By adopting the multiple dimensions, the same resources can be shared at each domain. In this paper, both the power and space domain are exploited. Figure 2 shows an example of additional dimension usage. Four dimensions (frequency, time, power, space) are used and the same resources can be appropriately shared. Therefore, the more data symbols can be transmitted to more users.

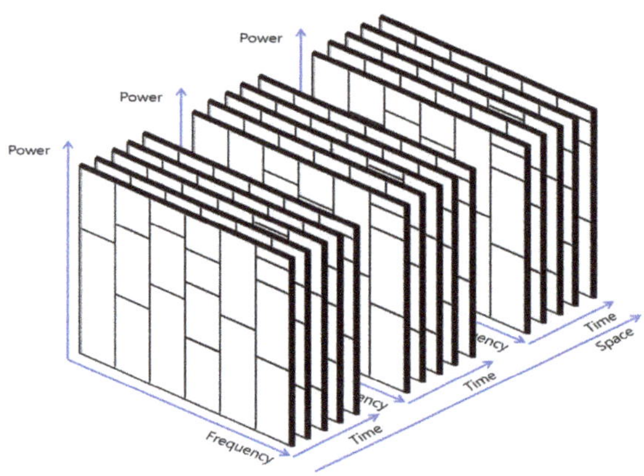

Figure 2. The spectral efficiency using multiple dimensions.

This paper considers a downlink single-cell MU-MIMO system as shown in Figure 3. In the proposed scheme, multiple access is accomplished by SDMA and the power domain is exploited to transmit more data symbols to each user. As a result, the proposed scheme has significant potential to improve spectral efficiency and provide better wireless services to many users. Also, this paper offers advantages in various design issues to meet the requirements and characteristics of the system by using multiple dimensions.

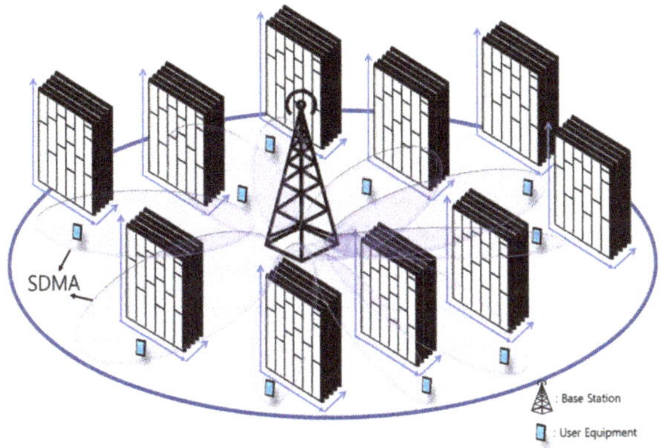

Figure 3. The downlink single-cell MU-MIMO system.

3. Proposed Scheme

This section describes the proposed scheme from data transmission to signal detection. Using the proposed system, more data symbols can be transmitted and more users can be served simultaneously. First, the transmission model explains how two additional domains can be used to improve spectral efficiency. Second, the signal detection model explains how each signal is detected and reliability can be satisfied. Additionally, in the received SINR, some considerations for the proposed scheme are presented. Finally, the sum throughput between the proposed scheme and conventional schemes is compared. As a result, the throughput performance of the proposed scheme is superior to that of the conventional scheme.

3.1. System Model

This paper considers a downlink MU-MIMO broadcast system which consists of one BS and K users as shown in Figure 4. BS is equipped with N_t transmitting antennas and each user has N_r receiving antennas. The MIMO channel of each user is assumed to be flat fading, since frequency selective fading channel can be easily overcome by using orthogonal frequency division multiplexing (OFDM) modulation. The system model can be further extended to frequency selective fading MIMO channel considering all subcarriers. In this system model, the transmit signal for the k-th user can be denoted as follows,

$$\mathbf{x}_k = \mathbf{W}_k\mathbf{s}_k. \tag{1}$$

The received signal at the k-th user is given by

$$\mathbf{y}_k = \underbrace{\mathbf{H}_k\mathbf{W}_k\mathbf{s}_k}_{\text{desired signal}} + \underbrace{\sum_{j=1, j\neq k}^{K} \mathbf{H}_k\mathbf{W}_j\mathbf{s}_j}_{\text{undesired signals}} + \mathbf{n}_k, k = 1, \cdots, K, \tag{2}$$

where k and j are user indices, $\mathbf{W}_k$ is $N_t \times N_r$ precoding matrix for user k, $\mathbf{s}_k$ is a $N_r \times 1$ data symbol vector, $\mathbf{x}_k$ is a $N_t \times 1$ precoded signal vector for the k-th user. $\mathbf{y}_k$ is a received signal vector for the k-th user and $\mathbf{n}_k$ is $N_r \times 1$ zero-mean additive white Gaussian noise (AWGN) vector with variance σ^2. In the Equation (2), the first term denotes signal in the intended direction (desired user $\mathbf{s}_k$) and the second term denotes multi-user interference caused due to undesired signals (undesired users $\mathbf{s}_j$)

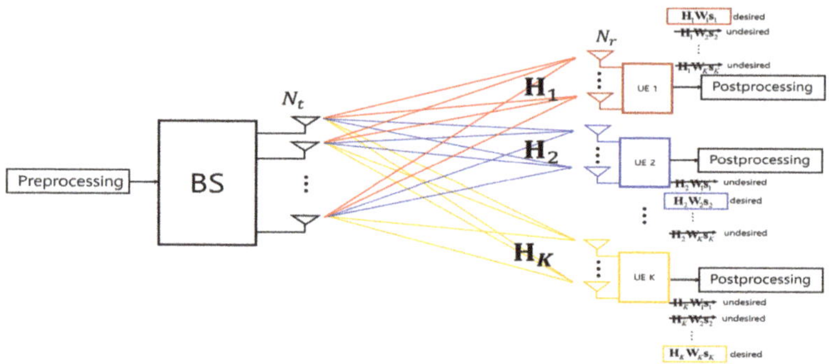

Figure 4. The downlink MU-MIMO broadcasting model.

3.2. Data Transmission Model

To exploit both the power and space domain, BD and SC scheme are used. The overall transmission model is shown in Figure 5. First, SC is used for exploiting power domain. The transmit symbols are superposed with different powers in one signal. Therefore, the transmitter can transmit more data stream at the same time. In the existing NOMA scheme, SC and SIC are used to suppress the MUI by allocating the different powers to different users. However, the proposed scheme uses SC for separating the symbols in one superposed signal on the same user which BS transmits. In the proposed scheme, the transmit signals for the i-th receiving antenna can be written as follows,

$$s_i = \sqrt{P_1}\tilde{s}_{i,1} + \sqrt{P_2}\tilde{s}_{i,2} + \cdots + \sqrt{P_N}\tilde{s}_{i,N}, \tag{3}$$

where i is a receiving antenna index, N is the number of symbols in one superposed signal. $\tilde{s}$ are the symbols in one superposed signal. The proposed scheme allocates optional power to symbols to detect each symbol, leading $P_1 < P_2 < \cdots < P_N$. Then, transmit signals for the k-th user are defined as follows,

$$\mathbf{s}_k = \begin{bmatrix} s_1(k) \\ s_2(k) \\ \vdots \\ s_{N_r}(k) \end{bmatrix} = \begin{bmatrix} \sqrt{P_1}\tilde{s}_{1,1}(k) + \sqrt{P_2}\tilde{s}_{1,2}(k) + \cdots + \sqrt{P_N}\tilde{s}_{1,N}(k) \\ \sqrt{P_1}\tilde{s}_{2,1}(k) + \sqrt{P_2}\tilde{s}_{2,2}(k) + \cdots + \sqrt{P_N}\tilde{s}_{2,N}(k) \\ \vdots \\ \sqrt{P_1}\tilde{s}_{N_r,1}(k) + \sqrt{P_2}\tilde{s}_{N_r,2}(k) + \cdots + \sqrt{P_N}\tilde{s}_{N_r,N}(k) \end{bmatrix}. \tag{4}$$

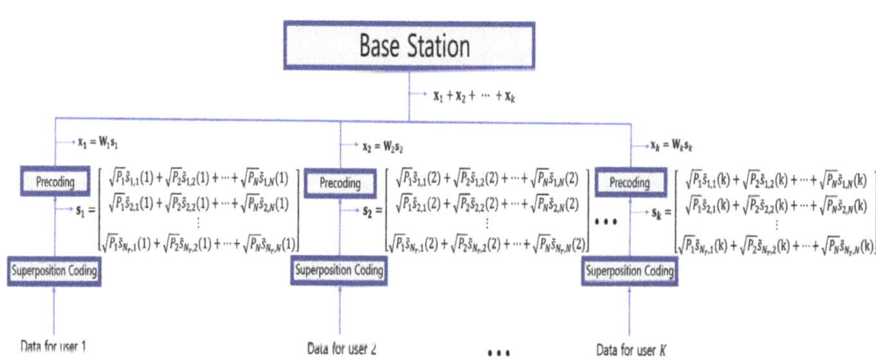

Figure 5. The proposed signal transmission model.

For exploiting space domain, the precoding matrix $\mathbf{W}$ should be designed in SDMA. In the scheme, BD beamforming is adopted to suppress the MUI since BD method shows a strong performance in terms of capacity and has a good flexibility. The objective of the BD method is to completely elminate the MUI by employing the precoding matrix $\mathbf{W}$. Then, the precoding matrix design for MUI elimination can be defined as follows,

$$\tilde{\mathbf{H}}_k \mathbf{W}_k = 0, k = 1, \cdots, K. \tag{5}$$

where $\tilde{\mathbf{H}}_k$ is represented as the channel matrix for all users except for user k,

$$\tilde{\mathbf{H}}_k = [\mathbf{H}_1^T \cdots \mathbf{H}_{k-1}^T \ \mathbf{H}_{k+1}^T \cdots \mathbf{H}_K^T]^T. \tag{6}$$

With the help of singular value decomposition (SVD), the precoding matrix for eliminating the MUI is designed. SVD is used to decompose a matrix into matrices representing rotation and scaling. By applying the SVD, the $\tilde{\mathbf{H}}_k$ is defined as follows

$$\tilde{\mathbf{H}}_k = \mathbf{U}_k \mathbf{\Lambda}_k [\mathbf{V}_k^{(1)} \ \mathbf{V}_k^{(0)}]^H, \tag{7}$$

where $\mathbf{\Lambda}_k$ is the diagonal matrix of which the diagonal elements are singular value of $\tilde{\mathbf{H}}_k$. $\mathbf{V}_k^{(0)}$ contains vectors of the zero singular values, $\mathbf{V}_k^{(1)}$ and contains vectors of the non-zero singular values. $\mathbf{V}_k^{(0)}$ is an orthogonal basis for the null space of $\tilde{\mathbf{H}}_k$ and the required precoding matrix. As a result, intended user's channel is projected on the null space in order to have the transmission under the constraint of zero-interference (i.e.,$\tilde{\mathbf{H}}_k \tilde{\mathbf{V}}_k^{(0)} = 0\,, k = 1, \cdots, K.$)

The received signal of the k-th user after eliminating MUI is defined as follows,

$$\begin{aligned}
\mathbf{y}_k &= \mathbf{H}_k \mathbf{W}_k \mathbf{s}_k + \mathbf{n}_k = \mathbf{H}_{eff,k} \mathbf{s}_k + \mathbf{n}_k \\
&= \mathbf{H}_{eff,k} \begin{bmatrix} s_1(k) \\ s_2(k) \\ \vdots \\ s_{N_r}(k) \end{bmatrix} + \begin{bmatrix} n_1(k) \\ n_2(k) \\ \vdots \\ n_{N_r}(k) \end{bmatrix} \\
= \mathbf{H}_{eff,k} & \begin{bmatrix} \sqrt{P_1}\tilde{s}_{1,1}(k) + \sqrt{P_2}\tilde{s}_{1,2}(k) + \cdots + \sqrt{P_N}\tilde{s}_{1,N}(k) \\ \sqrt{P_1}\tilde{s}_{2,1}(k) + \sqrt{P_2}\tilde{s}_{2,2}(k) + \cdots + \sqrt{P_N}\tilde{s}_{2,N}(k) \\ \vdots \\ \sqrt{P_1}\tilde{s}_{N_r,1}(k) + \sqrt{P_2}\tilde{s}_{N_r,2}(k) + \cdots + \sqrt{P_N}\tilde{s}_{N_r,N}(k) \end{bmatrix} + \begin{bmatrix} n_1(k) \\ n_2(k) \\ \vdots \\ n_{N_r}(k) \end{bmatrix}.
\end{aligned} \tag{8}$$

where $\mathbf{H}_{eff,k}$ denotes the effective channel of the k-th user. According to the Equation (8), the MUI is perfectly eliminated and the k-th user receives its own data. Finally, the users can be considered as point-to-point MIMO.

As a result, by exploiting both the spatial and power domains additionally, more data symbols can be transmitted in the same resource (frequency/time). In the proposed scheme, as the number of symbols at each superposed signal is increased, sum throughput at each user is linearly improved. Although the allocated power to each symbol is reduced, the total throughput is improved since the number of transmit symbols is increased. Unlike other techniques that use only one dimension, such as space or power domains, the proposed scheme can achieve significant gains in overall system throughput by using additional domains.

3.3. Signal Detection Model

The MUI is perfectly eliminated by using the precoding matrix. Since the k-th user receives its own data without MUI, the appropriate receiver structure for each user is similar to the point-to-point MIMO. In Equation (8), MIMO detection is performed. In MIMO detection algorithm, there exist linear and non-linear algorithms. In the proposed scheme, linear detection algorithms such as zero-forcing

(ZF) and minimum mean squared error (MMSE) can be applied simply. If the ZF detection scheme is used, the filter matrix for the k-th user is as follows,

$$\mathbf{G}_k = \mathbf{H}_{eff,k}^H (\mathbf{H}_{eff,k} \mathbf{H}_{eff,k}^H)^{-1}. \tag{9}$$

The detected superposed signals of the k-th user using ZF MIMO detection can be represented as follows,

$$\begin{aligned} \hat{\mathbf{s}}_k &= \mathbf{G}_k \mathbf{y}_k = \mathbf{H}_{eff,k}^H (\mathbf{H}_{eff,k} \mathbf{H}_{eff,k}^H)^{-1} (\mathbf{H}_{eff,k} \mathbf{s}_k + \mathbf{n}_k) \\ &= \mathbf{H}_{eff,k}^H \left(\mathbf{H}_{eff,k}^H\right)^{-1} \left(\mathbf{H}_{eff,k}\right)^{-1} \mathbf{H}_{eff,k} \mathbf{s}_k + \mathbf{H}_{eff,k}^H (\mathbf{H}_{eff,k} \mathbf{H}_{eff,k}^H)^{-1} \mathbf{n}_k = \mathbf{s}_k + \mathbf{G}_k \mathbf{n}_k. \end{aligned} \tag{10}$$

In the Equation (10), $\mathbf{s}_k$ can be detected since it satisfies $\mathbf{G}_k \mathbf{H}_{eff,k} = \mathbf{I}$. However, if the linear detection algorithms are used, the bit error rate (BER) performance is too poor since the power of symbols in one superposed signal is low. Low performance detection techniques in terms of BER can cause a negative effect on performing SIC. Therefore, the non-linear detection algorithms can be applied such as maximum likelihood (ML), ordered successive interference cancellation (OSIC), decision feedback equalizer (DFE), QRD-M.

From Equation (10), the estimated superposed signals of the k-th user after performing MIMO detection can be reconstructed as follows,

$$\hat{\mathbf{s}}_k = \begin{bmatrix} \hat{s}_1(k) \\ \hat{s}_2(k) \\ \vdots \\ \hat{s}_{N_r}(k) \end{bmatrix} = \begin{bmatrix} \sqrt{P_1}\hat{s}_{1,1}(k) + \sqrt{P_2}\hat{s}_{1,2}(k) + \cdots + \sqrt{P_N}\hat{s}_{1,N}(k) \\ \sqrt{P_1}\hat{s}_{2,1}(k) + \sqrt{P_2}\hat{s}_{2,2}(k) + \cdots + \sqrt{P_N}\hat{s}_{2,N}(k) \\ \vdots \\ \sqrt{P_1}\hat{s}_{N_r,1}(k) + \sqrt{P_2}\hat{s}_{N_r,2}(k) + \cdots + \sqrt{P_N}\hat{s}_{N_r,N}(k) \end{bmatrix}, \tag{11}$$

where $\hat{s}$ are the symbols at each estimated superposed signal. In Equation (11), the superposed signal can be decoded by conducting SIC. Therefore, SIC is performed at each receiving antenna. In estimated superposed signal, the strong symbol, i.e., $\sqrt{p_N}\hat{s}_{Nr,N}$ is first decoded. The first decoded symbol can be represented as follows,

$$\hat{\mathbf{s}}_k^d = \begin{bmatrix} \hat{s}_{1,N}^d(k) \\ \hat{s}_{2,N}^d(k) \\ \vdots \\ \hat{s}_{N_r,N}^d(k) \end{bmatrix}. \tag{12}$$

The decoded symbol is then subtracted from the superposed signal.

$$\hat{\mathbf{s}}_k - \hat{\mathbf{s}}_k^d = \begin{bmatrix} \sqrt{P_1}\hat{s}_{1,1}(k) + \sqrt{P_2}\hat{s}_{1,2}(k) + \cdots + \sqrt{P_N}\hat{s}_{1,N}(k) \\ \sqrt{P_2}\hat{s}_{2,1}(k) + \sqrt{P_2}\hat{s}_{2,2}(k) + \cdots + \sqrt{P_N}\hat{s}_{2,N}(k) \\ \vdots \\ \sqrt{P_1}\hat{s}_{N_r,1}(k) + \sqrt{P_2}\hat{s}_{N_r,2}(k) + \cdots + \sqrt{P_N}\hat{s}_{N_r,N}(k) \end{bmatrix} - \begin{bmatrix} \hat{s}_{1,N}^d(k) \\ \hat{s}_{2,N}^d(k) \\ \vdots \\ \hat{s}_{N_r,N}^d(k) \end{bmatrix} - \begin{bmatrix} \hat{s}_{1,N-1}^d(k) \\ \hat{s}_{2,N-1}^d(k) \\ \vdots \\ \hat{s}_{N_r,N-1}^d(k) \end{bmatrix}. \tag{13}$$

Finally, all the symbols in the superposed signal can be decoded by performing SIC. The receiver model is described in Figure 6.

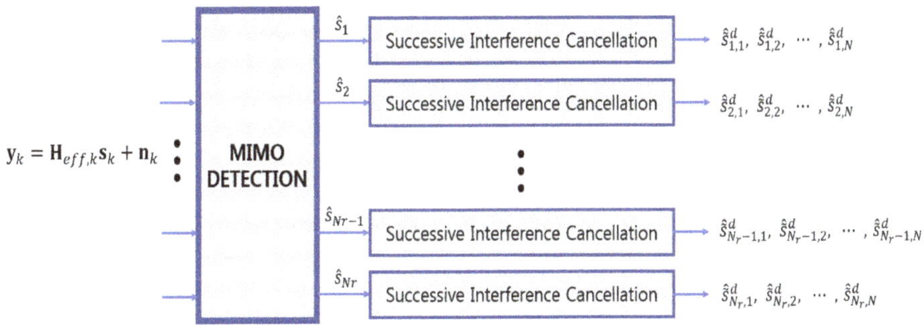

Figure 6. The proposed signal detection model.

3.4. Received SINR

In this subsection, signal to interference plus noise ratio (SINR) for the symbols in a superposed signal is represented to consider the performance of proposed scheme. If it is assumed that the two symbols are superposed in one superposed signal, i.e., $s_i = \sqrt{P_1}\tilde{s}_{i,1} + \sqrt{P_2}\tilde{s}_{i,1}$, the received signal at the i-th receiving antenna for each user after eliminating MUI can be defined as follows,

$$y_i = \lambda_i(\sqrt{P_1}\tilde{s}_{i,1} + \sqrt{P_2}\tilde{s}_{i,2}), \tag{14}$$

where i is received antenna index, λ_i is a channel gain in one MIMO parallel channel. In this case, $P_1 < P_2$ subject to total power constraint. Therefore, $\tilde{s}_{i,1}$ is a weak symbol and $\tilde{s}_{i,2}$ is a strong symbol. Then the received SINR for the strong symbol is defined as follows,

$$\mathrm{SINR}_{\tilde{s}_2} = \frac{\lambda_i P_2}{\sigma^2 + \lambda_i P_1}. \tag{15}$$

The weak symbol can be decoded by employing SIC. If it assumed that perfect SIC decoding is conducted, the received SINR for weak symbol is defined as follows,

$$\mathrm{SINR}_{\tilde{s}_1} = \frac{\lambda_i P_1}{\sigma^2}. \tag{16}$$

In the proposed scheme, SINR shows two negative effects in terms of system throughput. First, the weak symbol acts as an interference for received SINR of the strong symbol. As shown in Equation (15), $\lambda_i P_2$ term is considered as noise. Second, strong symbols is likely be decoded incorrectly.If a strong symbol is not properly decoded, it can adversely affect the decoding of weak symbol. The negative effects for the system throughput are restated in the simulation result. Although there are some degradations in terms of throughput on exploiting the power domain, the system throughput is increased since the number of transmit symbols is linearly increased at high SNR.

4. Simulation Results

This section shows simulation results to demonstrate the throughput gain of the proposed scheme and compares the results with other conventional schemes. Simulation results also provide some considerations for the proposed scheme. The simulations are performed in seven multi-path Rayleigh fading and time-invariant channel model. OFDM overcomes the frequency selectivity of the wideband channel and multiple carriers enable the high rate data transmission [22,23]. The OFDM symbol is composed of 128 FFT size, four pilots and the 108 data subcarriers based on specification for IEEE 802.11n. The remaining 16 subcarriers are zero padding and OFDM symbol duration is 4 microseconds.

All the simulation are simulated with quadrature phase-shift keying (QPSK) modulation (data bits per subcarrier is 2).

Figure 7a,b compare the sum throughput between the proposed schemes and the conventional schemes for the case of two users as a function of SNR. The throughput T is calculated as follows,

$$\begin{aligned} T &= N_b \times (1-E)^L \times K \div T_s \\ N_b &= N_s \times O \times N_r \times s, \\ L &= N_s \times O, \end{aligned} \tag{17}$$

where N_b is the number of transmit data bits and L is the number of data bits in one OFDM symbol. E is the BER for each user and N_s is the number of data subcarriers. O is the number of data bits per subcarrier and s is the number of symbols in one superposed signal. T_s is the OFDM symbol duration. In the simulation, proposed scheme and conventional BD are the MIMO system that BS and each user have multiple antennas. On the other hand, the conventional NOMA has one antenna at the BS and each user since multiple access is accomplished by the power domain. The proposed scheme and BD scheme are simulated in the case of $N_t = 4$, $N_r = 2$, $K = 2$ $(4,2,2)$ and NOMA scheme is $N_t = 1$, $N_r = 1$ $K = 2$ $(1,1,2)$. And in the proposed scheme, the number of symbols at each superposed signal is two $(N = 2)$. Therefore, the total number of transmit symbols to each user is 8. The power of each symbol is allocated at a ratio of 8:2 from the total power $P_t = 1$. In MIMO system, the ML MIMO detection is applied before conducting SIC. The summary of the simulation parameters is shown in the Table 1.

Table 1. Simulation Parameters.

Scheme	N_s	O	N_r	s	K
Conventional BD	108	2	2	1	2
Conventional NOMA	108	2	1	1	2
Proposed Scheme	108	2	2	2	2

In the simulations, there are two types of simulation results: with and without SIC error. Figure 7a,b show the simulation result with SIC error and Figure 7c,d show the simulation result without SIC error. Both cases of the proposed scheme outperform the conventional schemes in terms of maximum throughput since the proposed scheme exploits both the spatial and power dimensions. The conventional BD shows better performance than the proposed scheme in terms of BER. However, the proposed scheme has higher throughput since the proposed scheme transmits more symbols. In Figure 7b, the BER performance of the proposed scheme is better than that of NOMA since the effect of MIMO detection and SIC decoding is improved at high SNR. Figure 7c,d show the impact of SIC error on the proposed scheme. Without SIC error, throughput and BER performance are better than when there is SIC error. If the strong symbol is wrongly decoded at low SNR, the weak signal is also wrongly decoded. As a result, the error propagation occurs. Therefore, the system needs to be designed to avoid error propagation and reduce SIC error. As the proposed scheme minimizes the impact of SIC error, the performance of the proposed scheme approaches the case where there is no SIC error.

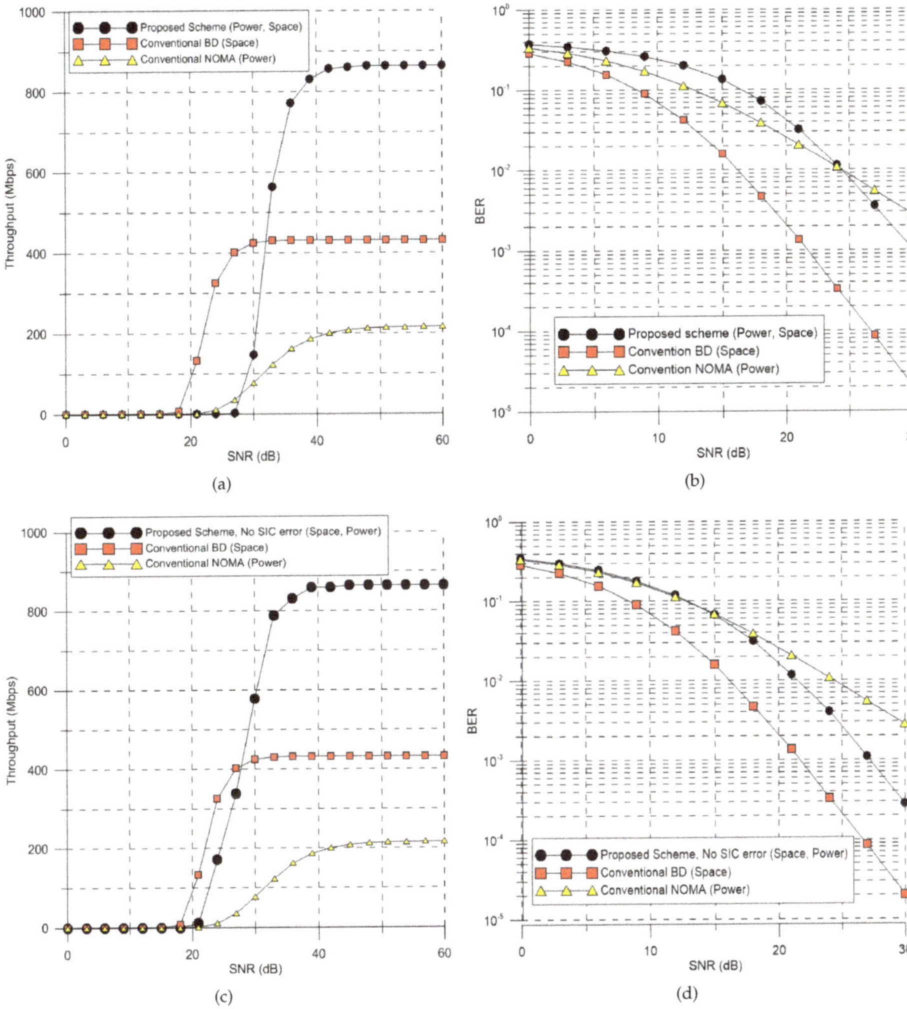

Figure 7. Throughput and BER performance of conventional and proposed schemes: (**a**) Throughput performance with SIC error; (**b**) BER performance with SIC error; (**c**) Throughput performance without SIC error; (**d**) BER performance without SIC error.

Figure 8a, b show the performance difference of the proposed scheme according to the number of users and antennas (N_t, N_r, K). The cases of $(6,3,2)$ and $(6,2,3)$ have the higher throughput performance than $(4,2,2)$ case since $(6,3,2)$ and $(6,2,3)$ transmit more symbols by exploiting both space and power dimension. $(6,2,3)$ case has lower throughput performance in low SNR than $(6,3,2)$ case since $(6,3,2)$ case allocates more symbols to each user. On the other hand, in case of $(6,2,3)$, one more user can be serviced. In terms of BER, $(6,3,2)$ has better BER performance than $(6,2,3)$ since $(6,3,2)$ transmits more symbols to each user than $(6,2,3)$. $(4,2,2)$ has better BER performance than $(6,2,3)$ since more power is allocated to each symbol. However, $(6,2,3)$ has higher throughput than $(4,2,2)$ since $(6,2,3)$ transmits more symbols.

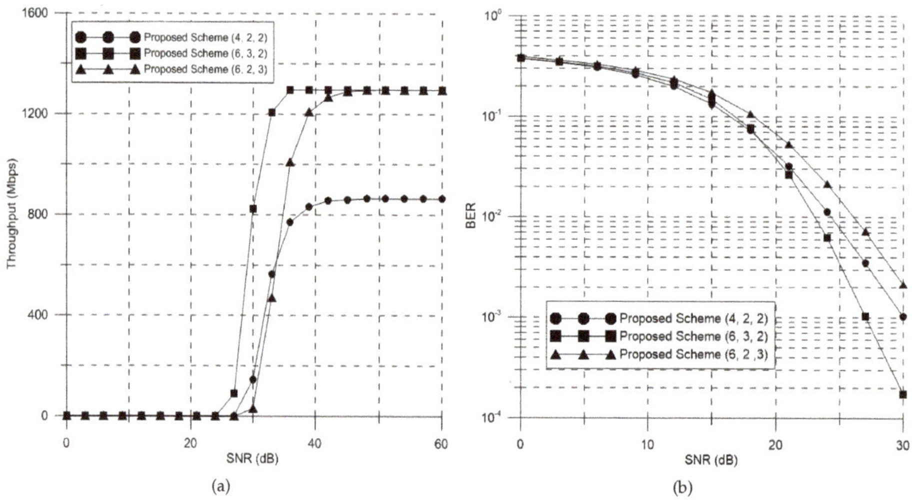

Figure 8. Throughput and BER performance of proposed scheme according to the number of users and antennas: (**a**) Throughput performance; (**b**) BER performance.

Figure 9a, b show the difference of the performance between the cases of linear MIMO detection and non-linear MIMO detection. The method with ML detection outperforms the method with ZF detection. The ML technique has better detection performance than the ZF technique before performing SIC on each antenna. Additionally, the ML detection technique mitigates the error propagation compared to the ZF detection scheme.

As a result, the overall simulation results show better performance compared to conventional schemes by using two dimensions simultaneously. In the proposed scheme, by exploiting both power and space domains at the same time, the transmitted symbol is increased. The results show that the superiority of the proposed scheme, and the proposed scheme uses the dimensions appropriately.

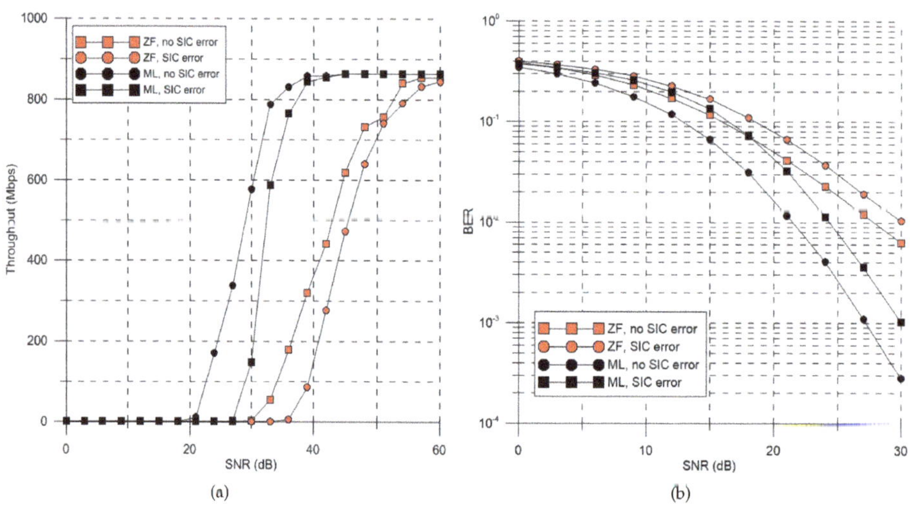

Figure 9. Throughput and BER performance of ML detection and ZF detection: (**a**) Throughput performance; (**b**) BER performance.

5. Implementation Issue

This section presents some additional considerations for system application and limitation under practical constraints. In addition, this section gives some ideas for additional performance gain and presents some methods to reduce some negative effects in the proposed scheme.

5.1. Complexity

In the proposed scheme, the additional implementation complexity in typical NOMA is not needed since SIC is performed for the own user's data. However, even though the superposed signals are the user's own data, the symbols with small allocation power in superposed signal have poor BER performance. To solve this problem, a detection scheme with better performance should be used. However, a non-linear algorithm has high complexity. For this problem, a complexity-reduced detection algorithm can be considered. The main consideration is achieving higher throughput with lower complexity.

5.2. SIC-Error Propagation

SIC is often assumed to be successful with perfect decoding. However, for systems with actual modulation and coding, decoding error inevitably occurs, causing error propagation and remarkable performance degradation. As shown in the simulation results, there are performance differences with and without SIC errors. Ultimately, well-designed system for the proposed scheme which avoids error propagation and decoding error should be considered for optimum performance.

5.3. Power Allocation

The achievable throughput is affected by the transmit power allocation. If it is assumed that two symbols are superposed in one signal, we should consider how much power should be allocated for each symbol. Basically, allocating more power to strong symbols can reduce the error propagation. However, allocating more power to strong symbols increases the error probability of weak symbols since the power of a weak symbol is too low. As a result, power allocation to each symbol should be considered according to the number of the symbols at the superposed signal subject to total power constraint.

5.4. Optimal Parameters

In the proposed scheme, if the number of the symbols in a superposed signal is increased, the sum throughput can be increased linearly. However, the sum throughput cannot be increased without limit since there is a limitation that a receiver can detect the symbols. As more symbols are transmitted, the power allocated to the symbols is reduced. Therefore, if too many symbols are transmitted, the receiver can not detect each symbol. Furthermore, there is also the degradation of sum throughput because of the interference in performing SIC. Therefore, the important issue is to find a near-optimal parameter between the number of users and the number of symbols in a superposed signal in the overall system.

6. Conclusions

This paper suggests multi-dimensionality and a methodology to improve the throughput in an MU-MIMO system. This paper also presents the transceiver structure of the proposed scheme. As a result, the multiple dimensions (space and power) are exploited at the same time, and the overasll system spectral efficiency is improved. If more dimensions are used without degradation or with a little tradeoff in performance, the system throughput can be increased. Also, various system models can be implemented by using additional dimensions.

Author Contributions: J.-G.H. proposed throughput enchancement scheme for MU-MIMO downlink channel and processed the simulation; J.-H.R. analyzed the simulation results and made the figure; H.-K.S. reviewed the algorithm and provided the experimental materials for better computational simulations and revised critical errors of the manuscript

Funding: This research was supported by Institute for Information & communications Technology Promotion(IITP) grant funded by the Korea government(MSIT) (No.2017-0-00217, Development of Immersive Signage Based on Variable Transparency and Multiple Layers) and was supported by the MSIT(Ministry of Science and ICT), Korea, under the ITRC(Information Technology Research Center) support program(IITP-2018-2018-0-01423) supervised by the IITP(Institute for Information & communications Technology Promotion).

Conflicts of Interest: The authors declare no conflict of interest.

References

1. Gupta, A.; Jha, R.K. A survey of 5G network: Architecture and emerging technologies. *IEEE Access* **2015**, *3*, 1206–1232. [CrossRef]
2. Goldsmith, A. Capacity Limits of MIMO Channels. *IEEE JSAC* **2003**, *5*, 684–702. [CrossRef]
3. Weingarten, H.; Steinberg, Y.; Shamai, S. The Capacity Region of the Gaussian MIMO Broadcast Channel. *IEEE Trans. Inf. Theory* **2006**, *52*, 3936–3964. [CrossRef]
4. Caire, G.; Shamai, S. On the achievable throughput of a multi-antenna Gaussian broadcast channel. *IEEE Trans. Inf. Theory* **2003**, *43*, 1691–1706. [CrossRef]
5. Angelis, C.T.; Chronopoulos, S.K. System Performance of an LTE MIMO Downlink in Various Fading Environments. In Proceedings of the Ambient Media and Systems, Berlin, Germany, 27–28 January 2011; pp. 36–43.
6. Choi, L.U.; Murch, R.D. A transmit preprocessing technique for multiuser MIMO systems using a decomposition approach. *IEEE Trans. Wirel. Commun.* **2004**, *3*, 20–24. [CrossRef]
7. Spencer, Q.H.; Swindlehurst, A.L.; Haardt, M. Zero-forcing methods for downlink spatial multiplexing in multiuser MIMO channels. *IEEE Trans. Signal Process.* **2004**, *2*, 461–471. [CrossRef]
8. Viswanathan, H.; Venkatesan, S.; Huang, H. Downlink capacity evaluation of cellular networks with known-interference cancellation. *IEEE J. Sel. Areas Commun.* **2003**, *21*, 802–811. [CrossRef]
9. Votis, C.; Christofilakis, V.; Raptis, V.; Tatsis, G.; Chronopoulos, S.K.; Kostarakis, P. Design and analysis of a multiple-output transmitter based on DDS architecture for modern wireless communications. *AIP Conf. Proc.* **2010** *1203*, 421–426.
10. Jindal, N.; Goldsmith, A. Dirty-paper coding versus TDMA for MIMO broadcast channels. *IEEE Trans. Inf. Theory* **2005**, *51*, 1783–1794. [CrossRef]
11. Rim, M. Multi-user downlink beamforming with multiple transmit and receive antennas. *Electron. Lett.* **2002**, *38*, 1725–1726. [CrossRef]
12. Chen, R.; Andrews, J.G.; Heath, R.W. Multiuser Space-Time Block Coded MIMO System with Downlink Precoding. In Proceedings of the 2004 IEEE International Conference on Communications, Paris, France, 20–24 June 2004
13. Shen, Z.; Chen, R.; Andrews, J.G.; Heath, R.W.; Evans, B.L. Low complexity user selection algorithms for multiuser MIMO systems with block diagonalization. *IEEE Trans. Signal Process.* **2006**, *54*, 3658–3663. [CrossRef]
14. Saito, Y.; Kishiyama, Y.; Benjebbour, A.; Nakamura, T.; Li, A.; Higuchi, K. Non-orthogonal multiple access (NOMA) for cellular future radio access. In Proceedings of the 2013 IEEE 77th Vehicular Technology Conference, Dresden, Germany, 2–5 June 2013
15. Ali, M.S.; Hossain, E.; Kim, D. Non-orthogonal multiple access (NOMA) for downlink multiuser MIMO systems: User clustering, beamforming, and power allocation. *IEEE Access* **2017**, *5*, 565–577. [CrossRef]
16. Islam, S.M.R.; Avazov, N.; Dobre, O.A.; Kwak, K.-S. Power-domain non-orthogonal multiple access (NOMA) in 5G systems: Potentials and challenges. *IEEE Commun. Surv. Tutor.* **2016**, *19*, 721–742. [CrossRef]
17. Zeng, M.; Yadav, A.; Dobre, O.A.; Tsiropoulos, G.I.; Poor, H.V. On the sum rate of MIMO-NOMA and MIMO-OMA systems. *IEEE Wirel. Commun. Lett* **2017**, *6*, 534–537. [CrossRef]
18. Ding, Z.; Adachi, F.; Poor, H.V. The application of MIMO to nonorthogonal multiple access. *IEEE Trans. Wirel. Commun.* **2016**, *15*, 537–552. [CrossRef]
19. Ding, Z.; Schober, R.; Poor, H.V. A general MIMO framework for NOMA downlink and uplink transmission based on signal alignment. *IEEE Trans. Wirel. Commun.* **2016**, *15*, 4438–4454. [CrossRef]

20. Choi, Y.I.; Lee, J.; Rim, M.; Kang, C.G. On the performance of beam division non orthogonal multiple access for FDD-based large-scale multi-user MIMO systems. *IEEE Trans. Wirel. Commun.* **2017**, *15*, 5077–5089. [CrossRef]
21. Yang, Q.; Wang, H.-M.; Ng, D.W.K.; Lee, M.H. NOMA in downlink SDMA with limited feddback: Performance analysis and optimization. *IEEE J. Sel. Areas Commun.* **2017**, *35*, 2281–2294. [CrossRef]
22. Van Nee, R.; Prasad, R. *OFDM for Wireless Multimedia Communications*; Artech House, Inc.: Norwood, MA, USA, 2000
23. Chronopoulos, S.K.; Votis, C.; Raptis, V.; Tatsis, G.; Kostarakis, P. In depth analysis of noise effects in orthogonal frequency division multiplexing systems, utilising a large number of subcarriers. *AIP Conf. Proc.* **2010**, *1203*, 967–972.

electronics

MDPI

Article

LOS-Based Equal Gain Transmission and Combining in General Frequency-Selective Ricean Massive MIMO Channels

Qiuna Yan [1,2], Yu Sun [1] and Dian-Wu Yue [1,*]

1 School of Information Science and Technology, Dalian Maritime University, Dalian 116026, China; qiu7589@dlmu.edu.cn (Q.Y.); suny@dlmu.edu.cn (Y.S.)
2 State Key Laboratory of Integrated Services Networks, Xidian University, Xi'an 710071, China
* Correspondence: dwyue@dlmu.edu.cn; Tel.: +86-411-8472-4261

Received: 13 December 2018; Accepted: 7 January 2019; Published: 10 January 2019

Abstract: In general frequency-selective Ricean fading environments with doubly-ended spatial correlation, this paper investigates the spectral efficiency of a broadband massive multiple-input multiple-output (MIMO) system. In particular, in order to reduce overhead of channel estimation effectively, it proposes a scheme of equal gain transmission and combining, which is only based on line- of-sight (LOS) component and has low hardware complexity. With the scheme, several interesting transmit power scaling properties without and with spatial correlation are derived when the number of antennas at the transmitter or the number of antennas at the receiver grows in an unlimited way. Furthermore, the asymptotical rate analysis is extended to the cooperative relaying scenarios with decode-and-forward and amplify-and-forward protocols, respectively, and then two novel power scaling laws are given.

Keywords: massive MIMO; beamforming; line-of-sight; Ricean fading; frequency-selective; power scaling

1. Introduction

Recently, massive multiple-input multiple-output (MIMO) has attracted great interest in both academia and industry, and has been a promising solution to meet the demanding spectral efficiency requirement of 5G systems [1]. Its promising benefits includes significant increase of both spectral and energy efficiencies [2–4]. Interestingly, the two benefits of massive MIMO can be achieved by maximum-ration transmission/maximum-ration combining (MRT/MRC) or zero-forcing (ZF) precoding/detection [3,5].

With MRT/MRC and ZF linear processing, many scholars have given various asymptotic performance analyses. In particular, the power scaling law in the limit of the large number of antennas has been widely studied in order to quantify the power savings. For Rayleigh and Ricean fading environments, with MRC and ZF detectors, authors in [3,6] analyzed uplink massive MIMO system performance. If perfect channel state information (CSI) is available, they showed that, when the number of base station (BS) antennas grows large and the transmit power of each user is scaled down proportionally to it, the ergodic achievable rate can asymptotically be equal to a positive constant.

In order to obtain the needed CSI, channel estimation must be obviously carried on [7]. However, the channel estimation will result in not only heavy overhead but also pilot contamination in multi-cells, which will become a serious problem [2,8]. For a point to point massive MIMO system in Ricean fading, to reduce the heavy overhead to estimate the CSI, we investigated a scheme with equal gain transmission /equal gain combining (EGT/EGC), which is only based on the line-of-sight (LOS) component (or say specular component) and has low hardware complexity [9]. It was showed that, with this scheme, the ergodic achievable rate can converge to that of the corresponding MRT/MRC based on the perfect CSI as the two numbers of antennas at the transmitter and receiver go to infinity. After that, we further considered the novel linear processing scheme for a downlink or uplink multiuser massive MIMO system [10,11] and showed that each user in the downlink or uplink system can have asymptotically the same rate as in the single-user case when the number of BS antennas goes without bound.

It should be pointed out that the above-mentioned results with the novel scheme have only considered uncorrelated Ricean frequency-flat fading channels without a relay [12]. Recently, we tried to develop our analysis to frequency-selective Ricean fading channels [13], but only for a very simple and special scenario [14]. The EGT/EGC linear transmission scheme is very attractive for massive MIMO systems since it enables low-complexity and inexpensive hardware [15–18]. Motivated by these facts, in this paper, we make use of a comparatively complicated and general frequency-selective Ricean fading channel model [14,19,20] to investigate further the LOS-based EGT/EGC scheme. We firstly derive several interesting power scaling properties for broadband massive MIMO systems with and without spatial correlation. In particular, it is shown that the ergodic achievable rate of LOS-based EGT/EGC scheme can have the same asymptotic value as the ergodic achievable rate of the whole CSI-based MRT/MRC scheme if the two numbers of transmit and receive antennas go without bound and with a fixed ratio. Then, we extend our asymptotical performance analysis to the cooperative relaying scenarios with decode-and-forward (DF) protocol and with amplify-and-forward (AF) protocol, respectively, and obtain two novel power scaling laws for the two scenarios. In particular, it is also shown that the ergodic achievable rate of LOS-based scheme can have the same asymptotic value as the ergodic achievable rate of the CSI-based scheme if the number of source antennas and the two numbers of transmit and receive relay antennas go without bound and with two fixed ratios, respectively.

The manuscript is organized as follows: in Section 2, the system model is introduced. In Section 3, the proposed LOS-based transmission scheme is presented and its power scaling law without correlation and with correlation is derived, respectively. Extension of our analysis to a cooperative relaying system is given in Section 4. In Section 5, the analysis results are verified by simulation. Finally, in Section 6, some concluding remarks are given.

Notation: boldface lower and upper case letters denote column vectors and matrices, respectively. The superscripts $(\cdot)^{\dagger}$ and $(\cdot)^{\mathrm{T}}$ stand for conjugate-transpose and transpose operations, respectively. The expectation operator is denoted by $\mathbb{E}\{\cdot\}$. $\alpha \sim \mathtt{CN}(0, \delta^2)$ stands for a circularly symmetric complex Gaussian variable α which has zero mean and variance δ^2.

2. System Model

Since a set of parallel independent frequency flat MIMO channels can be used to describe a frequency selective MIMO channel, we start with introducing the frequency flat channels [14,19].

For a point-to-point MIMO system over frequency-flat channels, we assume that it has N transmit antennas and M receive antennas. Then, we can represent a $M \times 1$ received signal vector as

$$\mathbf{y} = \sqrt{p}\mathbf{H}_0\mathbf{x} + \mathbf{z}, \tag{1}$$

where $\mathbf{z}$ denotes the additive white Gaussian noise (AWGN) vector that has zero-mean and covariance matrix $\sigma^2\mathbf{I}_M$ with $\mathbf{I}_M$ being the $M \times M$ identity matrix, $\mathbf{x}$ denotes the transmitted signal vector, $\mathbf{H}_0 = [h_{mn}]_{m,n=1}^{M,N}$ stands for the $M \times N$ channel matrix whose element h_{mn} denotes the channel gain between the m-th antenna at the receiver and the n-th antenna at the transmitter, and p is the average transmitted power. The channel matrix $\mathbf{H}_0$ under Ricean fading consists of a LOS matrix and a scattered matrix, i.e.,

$$\mathbf{H}_0 = \sqrt{\bar{\kappa}_0}\overline{\mathbf{H}}_0 + \sqrt{\tilde{\kappa}_0}\widetilde{\mathbf{H}}_0, \tag{2}$$

where $\bar{\kappa}_0 = \frac{\kappa_0}{1+\kappa_0}$, $\tilde{\kappa}_0 = \frac{1}{1+\kappa_0}$. Note that $\kappa_0 > 0$ represents the Ricean K-factor. The LOS matrix $\overline{\mathbf{H}}_0$ can be written as

$$\overline{\mathbf{H}}_0 = \mathbf{r}_0\mathbf{t}_0^{\mathrm{T}}. \tag{3}$$

Here, $\mathbf{r}_0$ denotes the specular array response at the receiver and can be expressed as

$$\mathbf{r}_0 = [1, e^{j2\pi d_r \sin(\theta)}, \ldots, e^{j2\pi(M-1)d_r,\sin(\theta)}]^{\mathrm{T}}, \tag{4}$$

where θ is the angle of arrival of the LOS component and d_r is the antenna spacing normalized by wavelength at the receiver. Similarly, $\mathbf{t}_0$ denotes the specular array response at the transmitter and can be given by

$$\mathbf{t}_0 = [1, e^{j2\pi d_t \sin(\phi)}, \ldots, e^{j2\pi(N-1)d_t \sin(\phi)}]^{\mathrm{T}} \tag{5}$$

where ϕ is the angle of departure of the LOS component and d_t is the antenna spacing normalized by wavelength at the transmitter. The entries in the scattering matrix $\widetilde{\mathbf{H}}_0$, $[\widetilde{\mathbf{H}}_0]_{mn} \sim \mathtt{CN}(0,1)$, i.e., they are circular complex Gaussian random variables with zero mean and unit variance. Furthermore, we assume that they are independent and identically distributed (i.i.d).

Now, we are concerned with a broadband orthogonal frequency-division multiplexing (OFDM)-MIMO system with K subcarriers, where ideal OFDM transmission with proper cyclic prefix extension is assumed. For the k-th subcarrier, the input–output relationship is expressed as

$$\mathbf{y} = \sqrt{p}\mathbf{H}\mathbf{x} + \mathbf{z}, \tag{6}$$

where $\mathbf{x}$ is just the normalized signal vector, $\mathbf{z}$ is the AWGN vector, and $\mathbf{H}$ is the channel matrix. The channel matrix can be given by as

$$\mathbf{H} = \sum_{\ell=0}^{L-1} \rho_\ell \mathbf{H}_\ell \exp(-j2\pi\frac{k}{K}\ell), \tag{7}$$

where L represents the channel delay spread, $\{\rho_\ell^2\}$ is the power delay profile satisfying $\sum_{\ell=0}^{L-1}\rho_\ell^2 = 1$, and $\mathbf{H}_\ell$ stands for the channel matrix at time delay ℓ. Furthermore, $\mathbf{H}_\ell$, $\ell = 0, 1, \cdots, L-1$ are mutually uncorrelated, Ricean distributed, and can be expressed as in Label (2)

$$\mathbf{H}_\ell = \sqrt{\bar{\kappa}_\ell}\overline{\mathbf{H}}_\ell + \sqrt{\tilde{\kappa}_\ell}\widetilde{\mathbf{H}}_\ell. \tag{8}$$

In particular, $\overline{\mathbf{H}}_\ell = \mathbf{r}_\ell\mathbf{t}_\ell^{\mathrm{T}}$ is just as in Label (3) and $\widetilde{\mathbf{H}}_\ell$ is also modeled as a random matrix consisting of i.i.d. elements.

3. LOS-Based EGT/EGC and Power Scaling Laws

3.1. The Scenario without Correlation

The scattered component of k-th subcarrier's channel matrix can be described as

$$\widetilde{\mathbf{H}} = \sum_{\ell=0}^{L-1} \rho_\ell \sqrt{\tilde{\kappa}_\ell}\widetilde{\mathbf{H}}_\ell \exp(-j2\pi\frac{k}{K}\ell). \tag{9}$$

Now, it is assumed that $\widetilde{\mathbf{H}}$ is not available, but the LOS component

$$\overline{\mathbf{H}} = \sum_{\ell=0}^{L-1} \rho_\ell \sqrt{\bar{\kappa}_\ell} \mathbf{H}_\ell \exp(-j2\pi \frac{k}{K} \ell) \tag{10}$$

can be available. In what follows, by employing only $\overline{\mathbf{H}}$, we will present a linear processing scheme with EGT/EGC and then compare it with the MRT/MRC scheme based on the perfect CSI.

Since $\widetilde{\mathbf{H}}$ is not available, a couple of the normalized weighting vectors $\mathbf{w}_t$ and $\mathbf{w}_r$ can be chosen in such a way that the effective output signal-to-noise ratio (SNR) can become maximum. The largest eigenvalue of matrix $\overline{\mathbf{H}}^\dagger \overline{\mathbf{H}}$ is now denoted by $\lambda_{\max}(\overline{\mathbf{H}}^\dagger \overline{\mathbf{H}})$. Due to the fact that $\mathbf{w}_r^\dagger \mathbf{z}^k \sim \mathtt{CN}(0, \sigma^2)$, the effective output signal-to-interference-plus-noise ratio (SINR) can be described as

$$\gamma_S^{(k)} = \frac{p|\mathbf{w}_r^\dagger \overline{\mathbf{H}} \mathbf{w}_t^{\mathrm{T}\dagger}|^2}{p|\mathbf{w}_r^\dagger \widetilde{\mathbf{H}} \mathbf{w}_t^{\mathrm{T}\dagger}|^2 + \sigma^2} = \frac{p\lambda_{\max}(\overline{\mathbf{H}}^\dagger \overline{\mathbf{H}})}{p|\mathbf{w}_r^\dagger \widetilde{\mathbf{H}} \mathbf{w}_t^{\mathrm{T}\dagger}|^2 + \sigma^2}. \tag{11}$$

We denote by R_S the ergodic achievable rate of the LOS-based scheme. Then,

$$R_S = \mathbb{E}\{\frac{1}{K} \sum_{k=0}^{K-1} \log_2(1 + \gamma_S^{(k)})\} = \frac{1}{K} \sum_{k=0}^{K-1} R_S^{(k)}, \tag{12}$$

where $R_S^{(k)} = \mathbb{E}\{\log_2(1 + \gamma_S^{(k)})\}$. We have the following results through the derivation.

Lemma 1. *Define* $\lambda_{\max}^{(k)} = \lambda_{\max}(\overline{\mathbf{H}}^\dagger \overline{\mathbf{H}})$ *and* $\tilde{\kappa}_S = \sum_{\ell=0}^{L-1} \rho_\ell^2 \tilde{\kappa}_\ell$. *Then,*

$$\log_2(1 + \frac{p\lambda_{\max}^{(k)}}{p\tilde{\kappa}_S + \sigma^2}) \le R_S^{(k)} \le \log_2(1 + \frac{p\lambda_{\max}^{(k)}}{\sigma^2}). \tag{13}$$

Proof of Lemma 1. Regarding the ergodic achievable rate of the k-th subcarrier, it is easy for us to derive its following lower bound with the help of the well-known Jensen's inequality:

$$R_S^{(k)} \ge \log_2(1 + \frac{1}{\mathbb{E}(1/\gamma_S^{(k)})}) = \log_2(1 + \frac{p\lambda_{\max}^{(k)}}{p\sum_{\ell=0}^{L-1} \rho_\ell^2 \mathbb{E}_\ell + \sigma^2}), \tag{14}$$

where $\mathbb{E}_\ell = \mathbb{E}|\mathbf{w}_r^\dagger \widetilde{\mathbf{H}}_\ell \mathbf{w}_t^{\mathrm{T}\dagger}|^2 = \tilde{\kappa}_\ell$ for $0 \le \ell \le L-1$. Thus,

$$R_C^k \ge \log_2(1 + \frac{p\lambda_{\max}^{(k)}}{p\tilde{\kappa}_S + \sigma^2}). \tag{15}$$

Moreover, we can obtain from Label (11) that

$$R_S^{(k)} = \mathbb{E}\log_2(1 + \frac{p\lambda_{\max}^{(k)}}{p|\mathbf{w}_r^\dagger \widetilde{\mathbf{H}} \mathbf{w}_t^{\mathrm{T}\dagger}|^2 + \sigma^2}) \le \log_2(1 + \frac{p\lambda_{\max}^{(k)}}{\sigma^2}). \tag{16}$$

Thus, Lemma 1 holds. □

Lemma 2. *Let* $\bar{\kappa}_U = \sum_{\ell=0}^{L-1} \rho_\ell^2 \bar{\kappa}_\ell$. *and* $\bar{\kappa}_L = \max\{\rho_\ell^2 \bar{\kappa}_\ell, 0 \le \ell \le L-1\}$. *Then,*

$$\kappa_L \le \lim_{MN \to \infty} \frac{\lambda_{\max}^{(k)}}{MN} \le \kappa_U. \tag{17}$$

Proof of Lemma 2. For $0 \le \ell \le L-1$ and $0 \le b \le L-1$, we define $\varphi_{\ell b} = 2\pi d_r(\sin(\theta_b) - \sin(\theta_\ell))$, and then have that

$$\frac{\bar{\mathbf{H}}_{\ell}^{\dagger}\bar{\mathbf{H}}_{b}}{MN} = \varrho_{\ell b}\mathbf{U}_{\ell b}, \tag{18}$$

where

$$\varrho_{\ell b} = \frac{\mathbf{r}_{\ell}^{\dagger}\mathbf{r}_{b}}{M} = \begin{cases} 1, & \text{for } l = b; \\ \frac{1-e^{jM\varphi_{\ell b}}}{M(1-e^{j\varphi_{\ell b}})}, & \text{for } \ell \neq b \end{cases} \tag{19}$$

and

$$\mathbf{U}_{\ell b} = \frac{\mathbf{t}_{\ell}^{\mathrm{T}\dagger}\mathbf{t}_{b}^{\mathrm{T}}}{N} = [u_{an}]_{a,n=1}^{N,N} \tag{20}$$

with $u_{an} = \frac{1}{N}e^{2\pi d_t((a-1)\sin(\theta_b)-(n-1)\sin(\theta_\ell))}$. Now, suppose that $M \geq N$. Noting that $\lim_{M\to\infty}\varrho_{\ell b} = 0$ for $\ell \neq b$ and $\mathrm{tr}(\mathbf{U}_{\ell\ell}) = 1$, we can obtain that

$$\lim_{MN\to\infty}\frac{\lambda_{\max}(\overline{\mathbf{H}}^{\dagger}\overline{\mathbf{H}})}{MN} \leq \sum_{\ell=0}^{L-1}\rho_{\ell}^{2}\bar{\kappa}_{\ell}\lim_{MN\to\infty}\lambda_{\max}(\frac{\overline{\mathbf{H}}_{\ell}^{\dagger}\overline{\mathbf{H}}_{\ell}}{MN}) = \bar{\kappa}_{U}. \tag{21}$$

Moreover, we also have

$$\lim_{MN\to\infty}\frac{\lambda_{\max}(\overline{\mathbf{H}}^{\dagger}\overline{\mathbf{H}})}{MN} \geq \bar{\kappa}_{L} \tag{22}$$

since, for any ℓ, we can get when $\mathbf{w}_t = \frac{\mathbf{t}_\ell}{\sqrt{N}}$ and $\mathbf{w}_r = \frac{\mathbf{r}_\ell}{\sqrt{M}}$

$$\lim_{MN\to\infty}\frac{\lambda_{\max}(\overline{\mathbf{H}}^{\dagger}\overline{\mathbf{H}})}{MN} \geq \lim_{MN\to\infty}\frac{|\mathbf{w}_{r}^{\dagger}\widetilde{\mathbf{H}}\mathbf{w}_{t}^{\mathrm{T}\dagger}|^{2}}{MN} = \bar{\kappa}_{\ell}. \tag{23}$$

Therefore, Lemma 2 holds when $M \geq N$. When $N \geq M$, we can also similarly prove that Lemma 2 holds, based on the fact that $\lambda_{\max}(\overline{\mathbf{H}}^{\dagger}\overline{\mathbf{H}}) = \lambda_{\max}(\overline{\mathbf{H}\mathbf{H}}^{\dagger})$. □

Proposition 1. *If $E = MNp$ be fixed as $MN \to \infty$, then we have*

$$\log_2(1 + \frac{E\bar{\kappa}_L}{\sigma^2}) \leq \lim_{MN\to\infty} R_S \leq \log_2(1 + \frac{E\bar{\kappa}_U}{\sigma^2}). \tag{24}$$

Proof of Proposition 1. If $E = MNp$ is fixed when $MN \to \infty$, we readily show that Proposition 1 holds by using Lemmas 1 and 2. □

Remark 1. *This proposition gives the lower and upper bounds of the ergodic achievable rate of the LOS-based scheme. In the following special cases, we can obtain further the exact expressions of the ergodic achievable rate.*

Corollary 1. *When $N = 1$, we have, if $E = Mp$ be fixed as $M \to \infty$,*

$$\lim_{M\to\infty} R_S = \log_2(1 + \frac{E\bar{\kappa}_U}{\sigma^2}). \tag{25}$$

Similarly, when $M = 1$, we also have if $E = Np$ be fixed as $N \to \infty$

$$\lim_{N\to\infty} R_S = \log_2(1 + \frac{E\bar{\kappa}_U}{\sigma^2}). \tag{26}$$

Proof of Corollary 1. When $M = 1$ or $N = 1$, due to the fact that $\lim_{MN\to\infty}\frac{\lambda_{\max}(\overline{\mathbf{H}}^{\dagger}\overline{\mathbf{H}})}{MN} = \bar{\kappa}_U$, it easily follows that $\lim_{MN\to\infty} R_S = \log_2(1 + \frac{E\bar{\kappa}_U}{\sigma^2})$. □

Corollary 2. *If $E = MNp$ be fixed as $M \to \infty$ and $N \to \infty$, then we have*

$$\lim_{M,N\to\infty} R_S = \log_2(1 + \frac{E\bar{\kappa}_L}{\sigma^2}). \tag{27}$$

Proof of Corollary 2. Without loss of generality, $\overline{\mathbf{H}}$ can be rewritten as

$$\overline{\mathbf{H}} = \sum_{\ell=0}^{L-1} \rho_\ell \sqrt{\bar{\kappa}_\ell} \mathbf{r}_\ell \mathbf{t}_\ell^{\mathrm{T}} \exp(-j2\pi\frac{k}{K}\ell), \tag{28}$$

where $\rho_0\sqrt{\bar{\kappa}_0} \geq \rho_1\sqrt{\bar{\kappa}_1} \geq \cdots \geq \rho_{L-1}\sqrt{\bar{\kappa}_{L-1}}$. We can rewrite $\overline{\mathbf{H}}$ in a matrix form as

$$\overline{\mathbf{H}} = \mathbf{A}_r \mathbf{D} \mathbf{A}_t^{\mathrm{T}}, \tag{29}$$

where $\mathbf{D}$ is a $L \times L$ diagonal matrix, $[\mathbf{D}]_{ll} = \rho_\ell\sqrt{MN\bar{\kappa}_\ell}$, and $\mathbf{A}_r$ and $\mathbf{A}_t$ are defined as follows:

$$\mathbf{A}_r = \frac{1}{\sqrt{M}}[\mathbf{r}_0, \mathbf{r}_1, \ldots, \mathbf{r}_{L-1}] \tag{30}$$

and

$$\mathbf{A}_t = \frac{1}{\sqrt{N}}[\mathbf{t}_0, \mathbf{t}_1 \exp(-j2\pi\frac{k}{K}), \ldots, \mathbf{t}_{L-1}\exp(-j2\pi\frac{k}{K}(L-1))]. \tag{31}$$

Since both $\{\mathbf{r}_0, \mathbf{r}_1, \ldots, \mathbf{r}_{L-1}\}$ and $\{\mathbf{t}_0, \mathbf{t}_1, \ldots, \mathbf{t}_{L-1}\}$ are orthogonal vector sets when $M \to \infty$ and $N \to \infty$ [21], $\mathbf{A}_r$ and $\mathbf{A}_t$ are asymptotically unitary matrices. For matrix $\overline{\mathbf{H}}$, thus we can form a singular value decomposition (SVD) as follows

$$\overline{\mathbf{H}} = \mathbf{U}\boldsymbol{\Sigma}\mathbf{V}^\dagger = [\mathbf{A}_r|\mathbf{A}_r^\perp]\boldsymbol{\Sigma}[\mathbf{A}_t|\mathbf{A}_t^\perp]^{\mathrm{T}}, \tag{32}$$

where $\boldsymbol{\Sigma}$ is a diagonal matrix including all singular values on its diagonal, i.e.,

$$[\boldsymbol{\Sigma}]_{ll} = \begin{cases} \rho_\ell\sqrt{MN\bar{\kappa}_\ell}, & \text{for } 0 \leq l \leq L-1, \\ 0, & \text{for } l > L-1. \end{cases} \tag{33}$$

Then,

$$\lim_{M,N\to\infty} \frac{\lambda_{\max}^{(k)}}{MN} = \rho_0^2\bar{\kappa}_0 = \bar{\kappa}_L. \tag{34}$$

Thus, we finally obtain the desired result. □

On the other hand, suppose that the perfect CSI is known, i.e., both of the LOS and scattered components are available at the transmitter and the receiver. Then, the weighting vectors $\mathbf{w}_t$ and $\mathbf{w}_r$ should be chosen in such a way that the exact output SNR is maximized. Thus, the resulting output SNR can be written as [11]

$$\gamma_P^{(k)} = \frac{p}{\sigma^2}\lambda_{\max}(\mathbf{H}^\dagger\mathbf{H}), \tag{35}$$

where $\lambda_{\max}(\mathbf{H}^\dagger\mathbf{H})$ stands for the largest eigenvalue of $\mathbf{H}^\dagger\mathbf{H}$. For the MRT/MRC scheme based on the perfect CSI, let R_P represent its ergodic achievable rate, i.e., $R_P = \mathbb{E}\{\frac{1}{K}\sum_{k=0}^{K-1}\log_2(1+\gamma_P^{(k)})\}$. Now, we obtain the following power scaling law.

Proposition 2. *When $M \to \infty$ and $N \to \infty$, suppose that $E = MNp$ is fixed and $N/M \to \mu$. We have that*

$$\lim_{M,N\to\infty} R_P = \lim_{M,N\to\infty} R_S. \tag{36}$$

Proof of Proposition 2. Due to the fact $\mathbf{H} = \overline{\mathbf{H}} + \widetilde{\mathbf{H}}$, we can have

$$\begin{aligned} \frac{1}{M}[\mathbf{H}^{\dagger}\mathbf{H}] &= \frac{1}{M}[(\overline{\mathbf{H}} + \widetilde{\mathbf{H}})^{\dagger}(\overline{\mathbf{H}} + \widetilde{\mathbf{H}})] \\ &= \frac{\overline{\mathbf{H}}^{\dagger}\overline{\mathbf{H}}}{M} + \frac{\overline{\mathbf{H}}^{\dagger}\widetilde{\mathbf{H}}}{M} + \frac{\widetilde{\mathbf{H}}^{\dagger}\overline{\mathbf{H}}}{M} + \frac{\widetilde{\mathbf{H}}^{\dagger}\widetilde{\mathbf{H}}}{M}. \end{aligned} \quad (37)$$

If we let

$$\mathbf{G} = \frac{\overline{\mathbf{H}}^{\dagger}\widetilde{\mathbf{H}}}{M} = [g_{uv}]_{u,v=1}^{N,N}, \quad (38)$$

it follows that

$$g_{uv} = \frac{1}{M}\sum_{k=1}^{M}[\overline{\mathbf{H}}^{\dagger}]_{uk}[\widetilde{\mathbf{H}}]_{kv}. \quad (39)$$

With the aid of (9) and (10), we can have that $|[\overline{\mathbf{H}}^{\dagger}]_{uk}|^2 \leq \bar{\kappa}_S \leq 1$, and $[\widetilde{\mathbf{H}}]_{kv} \sim \mathtt{CN}(0, \delta^2)$ with $\delta^2 = \tilde{\kappa}_S \leq 1$. As $[\widetilde{\mathbf{H}}]_{kv}$, $1 \leq k \leq M$ are independent each other, we know that

$$g_{uv} \sim \mathtt{CN}(0, \sigma_{uv}^2), \;\; \sigma_{uv}^2 \leq \frac{1}{M}. \quad (40)$$

Thus, it can follow that, when $M \to \infty$, $\mathbf{G} \to \mathbf{Q}$, where $\mathbf{Q}$ denotes a matrix with all zero elements. Similarly, we can have that, if $M \to \infty$,

$$\mathbf{G}^{\dagger} = \frac{\widetilde{\mathbf{H}}^{\dagger}\overline{\mathbf{H}}}{M} \to \mathbf{Q}. \quad (41)$$

Now, M is assumed to be large enough. Then, we certainly have that

$$\begin{aligned} \lambda_{\max}(\frac{\mathbf{H}^{\dagger}\mathbf{H}}{M}) &= \lambda_{\max}(\frac{\overline{\mathbf{H}}^{\dagger}\overline{\mathbf{H}}}{M} + \frac{\widetilde{\mathbf{H}}^{\dagger}\widetilde{\mathbf{H}}}{M}) \\ &\leq \lambda_{\max}(\frac{\overline{\mathbf{H}}^{\dagger}\overline{\mathbf{H}}}{M}) + \lambda_{\max}(\frac{\widetilde{\mathbf{H}}^{\dagger}\widetilde{\mathbf{H}}}{M}). \end{aligned} \quad (42)$$

When $M \to \infty$, suppose that $N/M \to \mu$. Then, we easily derive from ([22], Theorem 2.37), only noting that $[\widetilde{H}]_{mn} \sim \mathtt{CN}(0, \tilde{\kappa}_S)$

$$\lambda_{\max}(\frac{1}{M}\widetilde{\mathbf{H}}^{\dagger}\widetilde{\mathbf{H}}) \to \tilde{\kappa}_S(1 + \sqrt{\mu})^2. \quad (43)$$

Thus, we further get

$$\lambda_{\max}(\frac{1}{MN}\mathbf{H}^{\dagger}\mathbf{H}) \leq \lambda_{\max}(\frac{1}{MN}\overline{\mathbf{H}}^{\dagger}\overline{\mathbf{H}}) + \tilde{\kappa}_S(1 + \sqrt{\mu})^2/N. \quad (44)$$

In addition, we can obtain

$$\lambda_{\max}(\frac{1}{MN}\mathbf{H}^{\dagger}\mathbf{H}) \geq \lambda_{\max}(\frac{1}{MN}\overline{\mathbf{H}}^{\dagger}\overline{\mathbf{H}}). \quad (45)$$

When $M \to \infty$ and $N \to \infty$, we can get, by combining (44) with (45),

$$\lambda_{\max}(\frac{1}{MN}\mathbf{H}^{\dagger}\mathbf{H}) \to \lim_{M,N \to \infty} \frac{\lambda_{\max}^{(k)}}{MN}. \quad (46)$$

It should be noticed that

$$R_P = \frac{1}{K}\sum_{k=1}^{K} \mathbb{E}\log_2(1+\frac{p}{\sigma^2}\lambda_{\max}(\mathbf{H}^\dagger(k)\mathbf{H}(k)))$$
$$= \frac{1}{K}\sum_{k=1}^{K} \mathbb{E}\log_2(1+\frac{pMN}{\sigma^2}\lambda_{\max}(\frac{\mathbf{H}^\dagger(k)\mathbf{H}(k)}{MN})). \tag{47}$$

Therefore, when $M \to \infty$ and $N \to \infty$, if $E = pMN$ is fixed, we can have finally

$$\lim_{M,N\to\infty} R_P = \lim_{M,N\to\infty} R_S. \tag{48}$$

Thus, Proposition 2 holds. □

Remark 2. *This proposition implies that, when the two numbers of antennas at the transmitter and the receiver grow large with a fixed ratio, the ergodic achievable rate of the LOS-based scheme has the same asymptotic value as the ergodic achievable rate of the whole CSI-based scheme.*

3.2. The Scenario with Correlation

Now, we consider extending the proposed LOS-based ECT/EGC without spatial correlation to the scenario in which there exists doubly-ended spatial correlation. The MIMO system model presented in Section 2 is necessarily modified. $\widetilde{\mathbf{H}}_\ell, \ell = 0,1,2,\cdots,L-1$ is now modeled as doubly-correlated Rayleigh fading, with transmit and receive correlation matrices Ψ_ℓ and Φ_ℓ, i.e., [12],

$$\widetilde{\mathbf{H}}_\ell = [\Phi_\ell]^{1/2}\widetilde{\mathbf{H}}_\ell^{\omega}[\Psi_\ell]^{1/2}, \tag{49}$$

where $\widetilde{\mathbf{H}}_\ell^{\omega}$ is an i.i.d. matrix with each entry $\sim \mathtt{CN}(0,1)$. Since the scattered component of the channel matrix remains unchanged, the needed weighting vectors $\mathbf{w}_t$ and $\mathbf{w}_r$ should also remain unchanged. The ergodic achievable rate of the scenario with spatial correlation is denoted by

$$R_C = \frac{1}{K}\sum_{k=0}^{K-1} R_C^{(k)},\ \ R_C^{(k)} = \mathbb{E}\{\log_2(1+\gamma_C^{(k)})\}. \tag{50}$$

With respect to R_C, we have the following results by a similar derivation.

Lemma 3.

$$\log_2(1+\frac{p\lambda_{\max}^{(k)}}{p\tilde{\kappa}_C+\sigma^2}) \le R_C^{(k)} \le \log_2(1+\frac{p\lambda_{\max}^{(k)}}{\sigma^2}), \tag{51}$$

where $\tilde{\kappa}_C = \sum_{\ell=0}^{L-1}\rho_\ell^2\tilde{\kappa}_\ell\|\mathbf{w}_r^\dagger[\Phi_\ell]^{\frac{1}{2}}\|^2\|[\Psi_\ell]^{\frac{1}{2}}\mathbf{w}_t^{T\dagger}\|^2$.

Proposition 3. *If* $E = MNp$ *is fixed as* $MN \to \infty$, *then*

$$\lim_{MN\to\infty} R_C = \lim_{MN\to\infty} R_S. \tag{52}$$

Remark 3. *This proposition implies that the two ergodic achievable rates with and without spatial correlation have the same asymptotic value when MN goes without bound.*

4. Cooperative Relaying Systems

4.1. The Scenario with Decode-and-Forward Protocol

Ricean fading often happens in cooperative MIMO systems [23]. Therefore, we can use a Ricean MIMO channel model to describe both the source-relay and relay-destination links. Still in frequency-selective Ricean fading environments, we especially study a classical cooperative relay system with a source node, a destination node and a relay node. The relay node can be equipped with a large-scale antenna array while both the source node and the destination node can be also equipped with a large-scale antenna array. The cooperative system with the three nodes is assumed to operate in a half-duplex mode, and the replay node employs the DF protocol for transmission. Each transmission for the system can be completed through two stages. Obviously, the cooperative system is a composite of two massive MIMO subsystems: one subsystem working at the first stage and the other subsystem at the second stage. Therefore, the rate analysis results mentioned above can be applied to the cooperative relay system.

We suppose that the relay makes use of M antennas to receive and transmit data and also suppose that the source has N_1 antennas and the destination has N_2 antennas. We denote the average transmitted power at the source and the relay by p_1 and p_2, respectively. In addition, we still let R_P and R_S represent the ergodic achievable rates for the whole CSI-based MRT/MRC scheme and the LOS-based EGT/EGC scheme, respectively. Then, we can obtain the following power scaling property for the cooperative system.

Proposition 4. *When $M \to \infty$ and $N_1 \to \infty$, let $E_1 = MN_1 p_1$ be fixed and $N_1/M \to \mu_1$ for the source-relay link. When $M \to \infty$ and $N_2 \to \infty$, let $E_2 = MN_2 p_2$ be fixed and $N_2/M \to \mu_2$ for the relay-destination link. Then,*

$$\lim_{M,N_1,N_2 \to \infty} R_P = \lim_{M,N_1,N_2 \to \infty} R_S. \tag{53}$$

Proof of Proposition 4. From [24], it follows that the ergodic achievable rate with the perfect SCI-based MRT/MRC is written as

$$R_P = \min\{R_P^{(1)}/2, R_P^{(2)}/2\}, \tag{54}$$

where $R_P^{(1)}$ and $R_P^{(2)}$ are the corresponding ergodic achievable rates of the source-relay and relay-destination transmission links, respectively. Similarly, we also have that the ergodic achievable rate with the only LOS-based EGT/EGC can be given by

$$R_S = \min\{R_S^{(1)}/2, R_S^{(2)}/2\}, \tag{55}$$

where $R_S^{(1)}$ and $R_S^{(2)}$ are the corresponding ergodic achievable rates of the source-relay and relay-destination transmission links, respectively. Under the condition of Proposition 4, we get by Proposition 2

$$\lim_{M,N_1 \to \infty} R_P^{(1)} = \lim_{M,N_1 \to \infty} R_S^{(1)} \tag{56}$$

and

$$\lim_{M,N_2 \to \infty} R_P^{(2)} = \lim_{M,N_2 \to \infty} R_S^{(2)}. \tag{57}$$

Thus, it is easy to obtain the desired result (53). □

Remark 4. *This proposition implies that, when the number of source antennas and the two numbers of relay antennas at the transmitter and the receiver grow large with fixed ratios, the ergodic achievable rate of the LOS-based scheme also has the same asymptotic value as the ergodic achievable rate of the whole CSI-based scheme.*

4.2. The Scenario with Amplify-and-Forward Protocol

The DF is a regenerative relaying transmission strategy. Now, we consider employing a nonregenerative strategy involving AF to replace the DF. Then, we can have a power scaling law as follows.

Proposition 5. *Suppose that* $N_1 = N_2 = 1$*. When* $M \to \infty$*, let* $E_1 = Mp_1$ *and* $E_2 = MN_2p_2$ *be fixed. Then,*

$$\lim_{M\to\infty} R_S = \log_2(1 + (\frac{E_1\bar{\kappa}_{U1}}{\sigma_1^2} \cdot \frac{E_2\bar{\kappa}_{U2}}{\sigma_2^2})/(\frac{E_1\bar{\kappa}_{U1}}{\sigma_1^2} + \frac{E_2\bar{\kappa}_{U2}}{\sigma_2^2} + 1)). \tag{58}$$

Proof of Proposition 5. We denote by $\gamma_{S1}^{(k)}$ and $\gamma_{S2}^{(k)}$ the output instantaneous SNR of the source-relay and relay-destination links, respectively. From [25], we obtain that

$$\begin{aligned} R_S &= \frac{1}{K}\sum_{k=0}^{K-1} \mathbb{E}\{\log_2(1+\gamma_S^{(k)})\} \\ &= \frac{1}{K}\sum_{k=0}^{K-1} \mathbb{E}\{\log_2(1+(\gamma_{S1}^{(k)}\cdot\gamma_{S2}^{(k)})/(\gamma_{S1}^{(k)}+\gamma_{S2}^{(k)}+1))\}. \end{aligned} \tag{59}$$

Based on the proof of Lemma 2, we can have the following asymptotical SNR expressions

$$\lim_{M\to\infty} \gamma_{S1}^{(k)} = \frac{E_1\bar{\kappa}_{U1}}{\sigma_1^2} \tag{60}$$

and

$$\lim_{M\to\infty} \gamma_{S2}^{(k)} = \frac{E_1\bar{\kappa}_{U2}}{\sigma_2^2} \tag{61}$$

Thus, the power scaling law (58) holds. □

5. Simulation Results

For OFDM-MIMO systems in frequency-selective Ricean fading channels, we in this section provide our analytical results and simulation results. In all simulations, we assume that all of these spacings between adjacent antennas at the transmitter and the receiver are 0.5. We set the number of subcarriers $K = 256$, the channel delay spread $L = 3$, and the noise variance as $\sigma^2 = 1$. In addition, we let $\rho_0^2 = \rho_1^2 = \rho_2^2 = 1/3$, $\theta_0 = \phi_0 = \pi/6$, $\theta_1 = \phi_1 = \pi/4$, and $\theta_2 = \phi_2 = \pi/3$. In Figures 1 and 2, the Ricean K-factor κ is fixed and is equal to 5 dB.

In order to verify Propositions 1 and 3, we consider firstly the scenario with spatial correlation when $N = 3$ and $E = 20$ dB. The spatial correlation among antennas is assumed to follow the exponential model, i.e., the correlation magnitude between antenna p and q can be determined by $c(p,q) = g^{|p-q|}$, where g denotes the correlation coefficient [12]. Therefore, we represent the correlation matrices (i,j)-th of Φ_ℓ by $[\Phi_\ell]_{ij} = (g_r^\ell)^{|i-j|}$ and (i,j)-th of Ψ_ℓ by $[\Psi_\ell]_{ij} = (g_t^\ell)^{|i-j|}$, respectively, $\ell = 0,1,2$. Moreover, we set $g_r^0 = g_r^1 = g_r^2 = g_r$ and $g_t^0 = g_t^1 = g_t^2 = g_t$. For the correlation coefficients $g_t = g_r = g = 0, 0.3, 0.6, 0.9$, as M increases from 6 to 60, Figure 1 provides a curve of the exact average rate R_C and two curves of the upper and lower bounds of R_S. It can be observed that the exact ergodic rate R_C increases as the number of receive antenna M grows large, and is always between the two bounds of R_S. As both of the correlation coefficients (g_t, g_r) increase, R_C is closer to the upper bound, and becomes higher and higher than R_S. This indicates that, compared to the uncorrelated scenario, the presence of spatial correlation results in improving the rate performance under the LOS-based EGT/EGC scheme. Therefore, if the LOS-based scheme can be employed, we can achieve performance benefits from the spatial correlation, which is obviously different from the traditional point of view. This implies that it would be practical if a large-scale antenna array is compactly arranged.

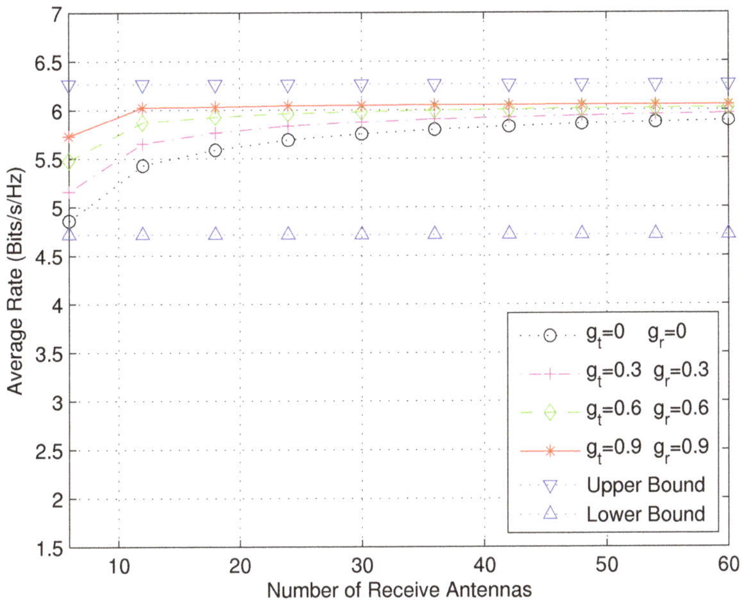

Figure 1. The ergodic achievable rate versus the number of receive antennas for comparing the case with correlation and the case without correlation.

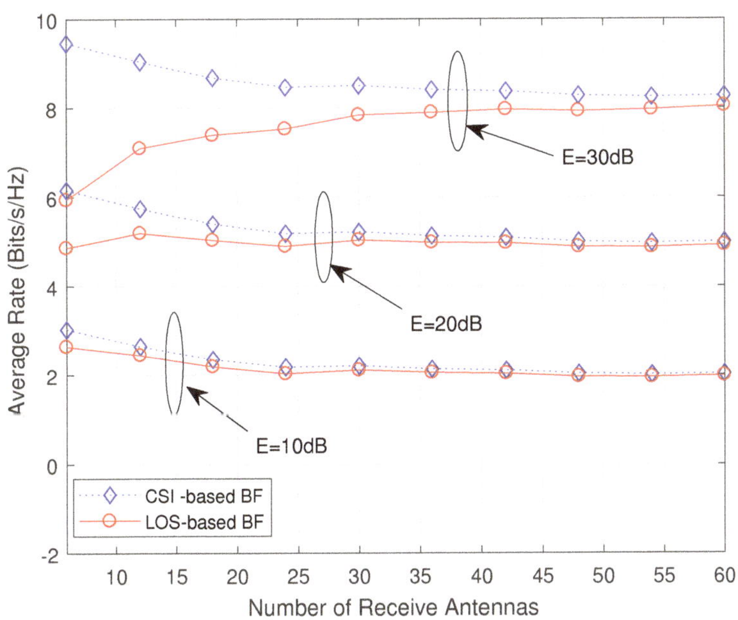

Figure 2. The ergodic achievable rate versus the number of receive antennas for comparing the LOS-based scheme with the whole CSI-based scheme.

Next, we consider validating Proposition 2. For that, we need to compare the ergodic achievable rate of the proposed LOS-based EGT/EGC scheme with that of the perfect CSI-based MRT/MRC scheme. We set $\mu = 1/2$ when the numbers of antennas at the transmitter and receiver grow large. For the parameter $E = 10, 20, 30$ dB, as M increases from 6 to 60, Figure 2 plots the two ergodic achievable rates, R_S and R_P. It can be found from Figure 2 that both of the ergodic achievable rates can tend to the same limit results for the given values of E. However, with an increase of E, the speed of rate convergence appears to be slower and slower.

Finally, we pay our attention to the classical DF cooperative relay system consisting of the source-relay and relay-destination links and set the identical parameters mentioned above in the two links. When $N_1 = N_2 = 6$, as M increases from 6 to 60, Figure 3 plots the two average rates R_S and R_P for $\kappa = 5, 15$ dB. It can be found from Figure 3 that, with an increasing κ, both R_S and R_P improve and R_S is closer to R_P. It should be noticed that R_P denotes the average rate for the traditional linear processing scheme based on the whole CSI as considered in [6]. For obtaining a comprehensive comparison with the scheme based on the whole CSI in Rayleigh fading discussed in [24], Figure 3 also includes a rate curve which corresponds to $\kappa = -\infty$ dB. Interestingly, with $\kappa = 5$ dB, the LOS-based scheme always obviously outperforms the scheme based on the whole CSI in Rayleigh fading.

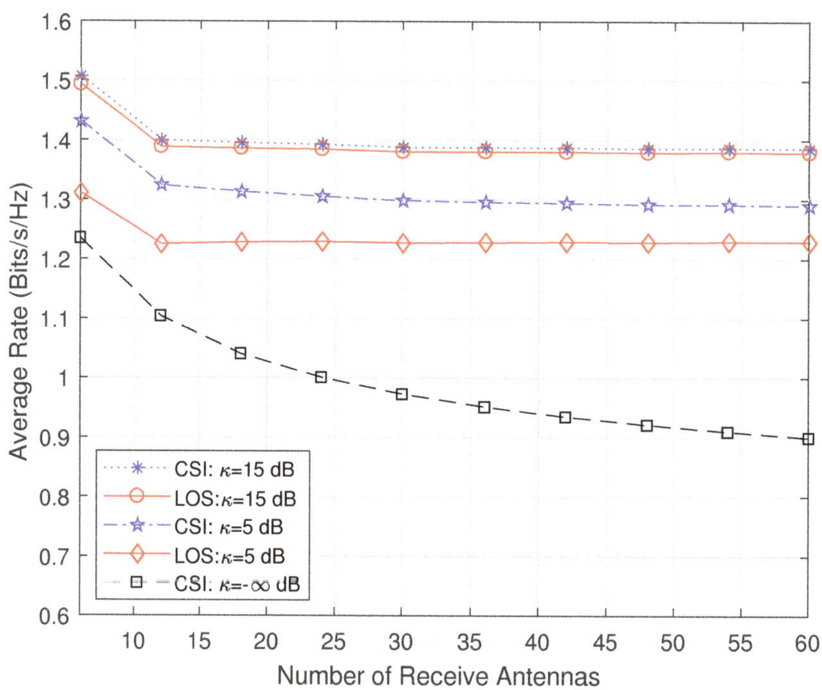

Figure 3. The ergodic achievable rate versus the number of relay antennas for comparing the case with Ricean fading and the case with Rayleigh fading.

6. Conclusions

In this paper, we have developed the transmission scheme of LOS-based EGT/EGC for point-to-point massive-MIMO systems in frequency-selective Ricean fading channels without and with spatial correlation. In particular, we have derived expressions of the system achievable rate and determined several power scaling laws. In addition, we have also generalized our analysis to the cooperative relaying scenarios with DF and AF protocols, respectively. It is shown by our simulation results that the spatial correlation can improve the system performance and thus is an advantage, which is contrary to the traditional point of view. Compared to the Rayleigh fading environments, deployment of large scale antenna arrays in Ricean fading environments would be more suitable. For instance, massive MIMO can be applied in microwave backhaul links [26].

Author Contributions: Formal Analysis, Q.Y.; Validation, Y.S.; Writing—Original Draft Preparation, Q.Y., Y.S.; Writing—Review and Editing, D.-W.Y.; Supervision, D.-W.Y.

Funding: This research was funded by the open research fund of State Key Laboratory of Integrated Services Networks, the Fundamental Research Funds for the Central Universities Grant No. 3132016347, and the Natural Science Foundation of Liaoning Province Grant No. 201602086.

Conflicts of Interest: The authors declare no conflict of interest.

References

1. Shafi, M.; Molisch, A.F.; Smith, P.J.; Haustein, T.; Zhu, P.; Silva, P.D.; Tufvesson, F.; Benjebbour, A.; Wunder, G. 5G: A tutorial overview of standards, trials, challenges, deployment, and practice. *IEEE J. Sel. Areas Commun.* **2017**, *35*, 1201–1221. [CrossRef]
2. Marzetta, T.L. Noncooperative cellular wireless with unlimited numbers of base station antennas. *IEEE Trans. Wirel. Commun.* **2010**, *9*, 3590–3600. [CrossRef]
3. Ngo, H.Q.; Larsson, E.G.; Marzetta, T.L. Energy and spectral efficiency of very large multiuser MIMO systems. *IEEE Trans. Commun.* **2013**, *61*, 1436–1449.
4. Liu, W.; Han, S.; Yang, C. Energy efficiency scaling law of massive MIMO systems. *IEEE Trans. Commun.* **2017**, *65*, 107–121. [CrossRef]
5. Yang, H.; Marzetta, T.L. Performance of conjugate and zero-forcing beamforming in large-scale antenna systems. *IEEE J. Sel. Areas Commun.* **2013**, *31*, 172–179. [CrossRef]
6. Zhang, Q.; Jin, S.; Wong, K.-K.; Zhu, H.; Matthaiou, M. Power scaling of uplink massive MIMO systems with arbitrary-rank channel means. *IEEE J. Sel. Top. Signal Process.* **2014**, *62*, 966–981. [CrossRef]
7. Rana, M.M.; Kim, J. *Fundamentals of Channel Estimations for Mobile Communications-Existing and New Techniques of a LTE SC-FDMA System*; LAMBERT Academic Publishing: Saarbrucken, Germany, 2012; ISBN 978-3-8454-3137-6.
8. Rusek, F.; Persson, D.; Lau, B.K.; Larsson, E.G.; Marzetta, T.L.; Edfors, O.; Tufvesson, F. Scaling up MIMO: Opportunities and challenges with very large arrays. *IEEE Signal Proc. Mag.* **2013**, *30*, 40–60. [CrossRef]
9. Yue, D.-W.; Zhang, Y.; Jia, Y.-N. Beamforming based on specular component for massive MIMO systems in Rician fading. *IEEE Wirel. Commun. Lett.* **2015**, *4*, 197–200. [CrossRef]
10. Yue, D.-W.; Yan, Q. Beamforming only based on specular component for downlink massive MIMO systems. *Wirel. Pers. Commun.* **2017**, *95*, 1799–1810. [CrossRef]
11. Yue, D.-W.; Meng, Z. LOS component-based equal gain combining for Ricean links in uplink massive MIMO. In Proceedings of the 2016 Sixth International Conference on Instrumentation & Measurement, Computer, Communication and Control (IMCCC), Harbin, China, 21–23 July 2016; pp.476–481.
12. Yue, D.-W.; Zhang, Q.T. Generic approach to the performance anaysis of correlated transmit/receive diversity MIMO systems with/without co-channel interference. *IEEE Trans. Inf. Theory* **2010**, *56*, 1147–1157. [CrossRef]
13. Yue, D.-W. Specular component-based beamforming for broadband massive MIMO systems with doubly-ended correlation. *Electron. Lett.* **2016**, *52*, 1082–1084. [CrossRef]
14. McKay, M.R.; Collings, I.B. On the capacity of frenquency-flat and frequency-selective Rician MIMO channels with single-ended correlation. *IEEE Trans. Wirel. Commun.* **2006**, *5*, 2038–2043. [CrossRef]
15. Dinis, R.; Montezuma, P. Iterative receiver based on the EGC for massive MIMO schemes using SC-FDE modulations. *Electron. Lett.* **2016**, *52*, 972–974. [CrossRef]
16. Love, D.J.; Heath, R.W. Equal gain transmission in multiple-input multiple-output wireless systems. *IEEE Trans. Commun.* **2003**, *51*, 1102–1110. [CrossRef]
17. Zheng, X.; Xie, Y.; Li, J.; Stoica, P. MIMO transmit beamforming under uniform elemental power constraint. *IEEE Trans. Signal Process.* **2007**, *55*, 5395–5406. [CrossRef]

18. Fu, H.; Crussière, M.; Hélard, M. BER analysis for equal gain transmission in downlink multiuser MIMO systems. *IEEE Wirel. Commun. Lett.* **2015**, *4*, 533–536. [CrossRef]
19. Jin, S.; Gao, X.; You, X. On the ergodic capacity of rank-1 Ricean fading MIMO channels. *IEEE Trans. Inf. Theory* **2003**, *53*, 502–517. [CrossRef]
20. Bölcskei, H.; Borgmann, M.; Paulraj, A.J. Impact of the propagation environment on the the performance of space-frequency coded MIMO-OFDM. *IEEE J. Sel. Areas Commun.* **2003**, *21*, 427–439. [CrossRef]
21. Chen, J. When does asymptotic orthogonality exist for very large arrays? In Proceedings of the Global Communications Conference (GLOBECOM), Atlanta, GA, USA, 9–13 December 2013; pp. 4251–4255.
22. Tulino, A.M.; Verdu, S. Random matrix theory and wireless communications. Found. *Trends Commun. Inf. Theory* **2004**, 1, 1–182. [CrossRef]
23. Wang, C.-X.; Ge, X.; Cheng, X.; Zhang, G.; Thompson, J. Cooperative MIMO channel models: A survey. *IEEE Commun. Mag.* **2010**, *48*, 80–87. [CrossRef]
24. Ngo, H.Q.; Suraweera, H.A.; Matthaiou, M.; Larsson, E.G. Multipair full-duplex relaying with massive arrays and linear processing. *IEEE J. Sel. Areas Commun.* **2014**, *32*, 1721–1736. [CrossRef]
25. Suraweera, H.A.; Krikidis, I.; Yuen, C. Antenna selection in the full-duplex multi-antenna relay channel. In Proceedings of the 2013 IEEE International Conference on Communications (ICC), Budapest, Hungary, 9–13 June 2013; pp. 4823–4828.
26. Durisi, G.; Tarable, A.; Camarda, C.; Devassy, R.; Montorsi, G. Capacity bounds for MIMO microwave backhaul links affected by phase noise. *IEEE Trans. Commun.* **2014**, 62, 920–929. [CrossRef]

electronics

MDPI

Article

Hybrid Beamforming for Millimeter-Wave Heterogeneous Networks

Mostafa Hefnawi

Department of Electrical and Computer Engineering, Royal Military College of Canada, Kingston, ON K7K 7B4, Canada; hefnawi@rmc.ca

Received: 23 December 2018; Accepted: 23 January 2019; Published: 28 January 2019

Abstract: Heterogeneous networks (HetNets) employing massive multiple-input multiple-output (MIMO) and millimeter-wave (mmWave) technologies have emerged as a promising solution to enhance the network capacity and coverage of next-generation 5G cellular networks. However, the use of traditional fully-digital MIMO beamforming methods, which require one radio frequency (RF) chain per antenna element, is not practical for large-scale antenna arrays, due to the high cost and high power consumption. To reduce the number of RF chains, hybrid analog and digital beamforming has been proposed as an alternative structure. In this paper, therefore, we consider a HetNet formed with one macro-cell base station (MBS) and multiple small-cell base stations (SBSs) equipped with large-scale antenna arrays that employ hybrid analog and digital beamforming. The analog beamforming weight vectors of the MBS and the SBSs correspond to the the best-fixed multi-beams obtained by eigendecomposition schemes. On the other hand, digital beamforming weights are optimized to maximize the receive signal-to-interference-plus-noise ratio (SINR) of the effective channels consisting of the cascade of the analog beamforming weights and the actual channel. The performance is evaluated in terms of the beampatterns and the ergodic channel capacity and shows that the proposed hybrid beamforming scheme achieves near-optimal performance with only four RF chains while requiring considerably less computational complexity.

Keywords: hybrid beamforming; massive MIMO; HetNets; mmWaves

1. Introduction

Recently, heterogeneous networks (HetNets) that use massive multiple-input multiple-output (MIMO) and millimeter-wave (mmWave) technologies has emerged as a promising solution to enhance the network capacity and coverage of next-generation 5G cellular networks [1–6]. Small cell deployment in HetNets can achieve high signal to interference plus noise ratio (SINR) and dense spectrum reuse, mmWave can address the current challenge of bandwidth shortage, and the large number of antenna arrays [7–10] are essential for mmWaves to compensate for channel attenuation. In Reference [11] we applied the concept of massive multiuser (MU)-MIMO to enhance both the access and the backhaul links in HetNets, and it was shown that such a concept could significantly improve the system performance in terms of link reliability, spectral efficiency, and energy efficiency. Traditional MIMO-beamforming systems require a dedicated radio frequency (RF) chain for each antenna element, which becomes impractical with massive MIMO systems due to either cost or power consumption. To reduce the number of RF chains, hybrid beamforming (HBF), which combines RF analog and baseband digital beamformers, has been proposed as a promising solution [12–17]. Figure 1 shows a general hybrid configuration that connects N_a antenna elements to N_d RF chains, where $N_d < N_a$, using an analog RF beamforming matrix built from only phase-shifters. Two widely-used analog beamformer architectures for hybrid beamforming are shown in Figure 2. The fully-connected hybrid beamforming structure of Figure 2a provides a full beamforming gain per transceiver—but with high complexity—by connecting each RF chain to all antennas through a network of $N_d \times N_a$ phase shifters [12–15]. Figure 2b, on the other hand, shows a partially-connected structure, where each RF chain is connected to N_a/N_d number of sub-arrays. Such a structure has a lower hardware complexity at the price of reduced beamforming gain.

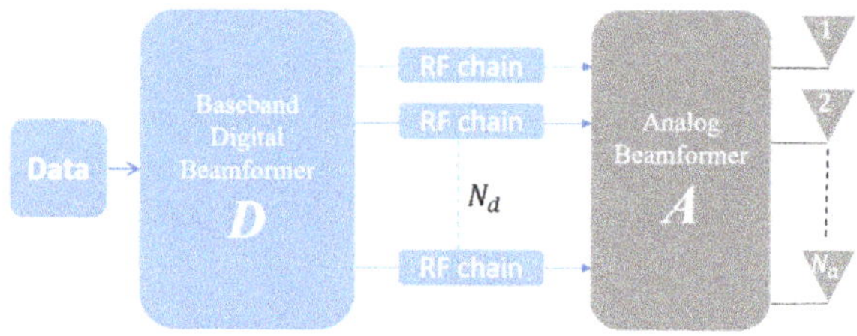

Figure 1. Hybrid beamforming.

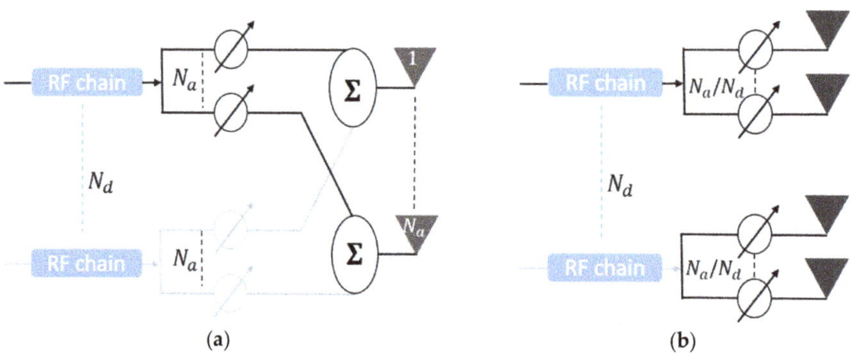

Figure 2. The architecture of analog beamformers: (**a**) Fully-connected structure; (**b**) partially-connected structure.

Previous studies on massive hybrid MIMO mainly focused on single-user systems [12–14]. On the other hand, MU-MIMO cases were studied in References [15–17]. In Reference [15] a scheme called Joint Spatial Division Multiplexing (JSDM) was proposed to create multiple virtual sectors which reduce the overhead and computational complexity of downlink training and uplink feedback. In References [16,17] it was shown that the required number of RF chains only needs to be twice the number of data streams to achieve the same performance of any fully-digital beamforming scheme. These studies, however, did not consider HBF in the context of HetNets and focused primarily on macro-cellular systems. In this paper, we consider a HetNet where both the macro-cell base stations (MBSs) and small-cell base stations (SBSs) are equipped with fully-connected massive hybrid antenna arrays, while all mobile users have a single antenna. We propose a low-complexity HBF that is fully-based on eigenbeamforming. The MBSs and the SBSs select the best-fixed multi-beams by eigendecomposition of the access and backhaul channels. The selected beams are then used by the digital beamformers, which are based on the maximization of the receive SINR of the effective channels consisting of the cascade of the analog beamforming weights and the actual channel [18,19].

2. System Model

We consider the access and backhaul uplinks in the HetNet of Figure 3, where K cognitive small cells and their L_s small-cell users (SUs) are concurrently sharing the same frequency band with one MBS and their L_p macro-cell primary users (PUs). It is assumed that both the MBS and SBSs are equipped with massive hybrid antenna arrays while the SUs and PUs are equipped with a single antenna. The SBSs are acting as smart relays between the SUs and the MBS with N_a- element transmitting/receiving antenna arrays and N_d RF chains. The MBS is equipped with M_a- element antenna arrays and M_d RF chains. It is also assumed that both the SBS and the MBS perform an OFDM-based transmission and that the analog beamformers are identical for all subcarriers while adapting digital beamformers in each subcarrier.

Let $\boldsymbol{x}_s[f_n] = \left\{x_1^s, x_2^s, \cdots, x_{L_s}^s\right\}$ and $\boldsymbol{x}_p[f_n] = \left\{x_1^p, x_2^p, \cdots, x_{L_p}^p\right\}$ denote, respectively, the set of L_s SUs signals and L_p PUs signals transmitted on each subcarrier, and f_n, $n = 1, \cdots, N_c$, where N_c denotes the number of subcarriers per OFDM symbol in the system. The analysis is done separately on each subcarrier. For brevity therefore, we drop the frequency index f_n.

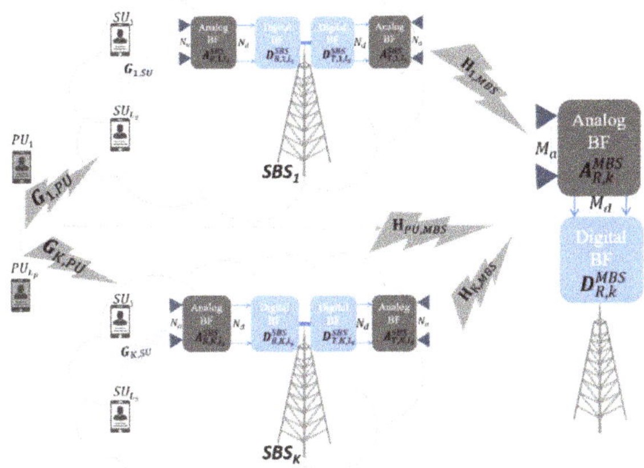

Figure 3. System model: hybrid beamforming-based HetNet with one macro-cell and K small-cells.

2.1. Access Link

The $N_a \times 1$ received signal vector $\boldsymbol{y}_{k,SBS}$ at the kth SBS is given by

$$\boldsymbol{y}_{k,SBS} = \boldsymbol{G}_{k,SU}\boldsymbol{x}_s + \boldsymbol{G}_{k,PU}\boldsymbol{x}_p + \boldsymbol{n}_{k,SBS}, \tag{1}$$

where $\boldsymbol{G}_{k,SU} \in \mathbb{C}^{N_a \times L_s}$ is the channel matrix between the kth SBS and its L_s users, $\boldsymbol{G}_{k,PU} \in \mathbb{C}^{N_a \times L_p}$ is the channel matrix between the kth SBS and the L_p PUs, $\boldsymbol{x}_s \in \mathbb{C}^{L_s \times 1}$ is the transmitted signal vector of L_s users in the kth small-cell, $\boldsymbol{x}_p \in \mathbb{C}^{L_p \times 1}$ is the transmitted signal vector of L_p users in the HetNet, and $\boldsymbol{n}_{k,SBS} \in \mathbb{C}^{N_a \times 1}$ is the received complex additive white Gaussian noise (AWGN) vector at the kth SBS.

It should be noted that in Equation (1), the interference between small-cells was neglected. This was justified by the fact that small-cell base stations are using a large number of antennas, which enables sharp beamforming towards their users without harming neighboring small-cells.

The kth SBS received signal, $\boldsymbol{y}_{k,SBS}$, is first applied to an $N_a \times N_d$ receive analog beamforming weight matrix, $\boldsymbol{A}_{R,k,l_s}^{SBS}$, whose output is directed to an $N_d \times N_d$ receive digital beamforming weight vector $\boldsymbol{D}_{R,k,l_s}^{SBS}$. If we denote the combined digital-analog receive beamformer for the l_s^{th} user as $\mathbf{w}_{R,k,l_s} = \boldsymbol{A}_{R,k,l_s}^{SBS}\boldsymbol{D}_{R,k,l_s}^{SBS}$, then the detection of the l_s^{th} user signal by its kth SBS can be expressed as

$$\begin{aligned} r_{k,l_s} &= \mathbf{w}_{R,k,l_s}^{\mathrm{H}}\boldsymbol{y}_{k,SBS} = \mathbf{w}_{R,k,l_s}^{H}\boldsymbol{G}_{k,SU}\boldsymbol{x}_s + \mathbf{w}_{R,k,l_s}^{H}\boldsymbol{G}_{k,PU}\boldsymbol{x}_p + \mathbf{w}_{R,k,l_s}^{H}\boldsymbol{n}_{k,SBS} \\ &= \mathbf{w}_{R,k,l_s}^{H}\boldsymbol{g}_{k,l_s}x_{l_s}^{s} + \mathbf{w}_{R,k,l_s}^{H}\sum_{i=1,i\neq l_s}^{L_s}\boldsymbol{g}_{k,i}x_i^s + \mathbf{w}_{R,k,l_s}^{H} + \boldsymbol{G}_{k,PU}\boldsymbol{x}_p + \mathbf{w}_{R,k,l_s}^{H}\boldsymbol{n}_{k,SBS} \end{aligned} \tag{2}$$

where g_{k,l_s} is the l_s^{th} column of $G_{k,SU}$ that represents the channel between the kth SBS and its l_s^{th} user.

If we denote $\mathbf{H}_{AL,k,l_s} = \left(A_{R,k,l_s}^{SBS}\right)^H g_{k,l_s}$ as the effective access channel between the kth SBS and its l_s^{th} user, then for a set of selected beams, i.e. known $\left(A_{R,k,l_s}^{SBS}\right)^H$, the SINR can be expressed in terms of the digital beamformer, D_{R,k,l_s}^{SBS}, as

$$\gamma_{k,l_s}^{SBS} = \frac{\left(D_{R,k,l_s}^{SBS}\right)^H \mathbf{H}_{AL,k,l_s} x_{l_s}^s \left(x_{l_s}^s\right)^H \mathbf{H}_{AL,k,l_s}^H D_{R,k,l_s}^{SBS}}{\left(D_{R,k,l_s}^{SBS}\right)^H \mathbf{B}_{AL} D_{R,k,l_s}^{SBS}}, \tag{3}$$

where $\mathbf{B}_{AL}$ is the covariance matrix of the interference-plus-noise given by

$$\mathbf{B}_{AL} = \underbrace{\sum_{i=1,i\neq l_s}^{L_s} (A_{R,k,i}^{SBS})^H g_{k,i} x_i^s (x_i^s)^H g_{k,i}^H A_{R,k,i}^{SBS}}_{\text{Interference from } L_s-1 \text{ SUs}} + \underbrace{(A_{R,k,l_s}^{SBS})^H G_{k,PU} x_p x_p^H G_{k,PU}^H A_{R,k,l_s}^{SBS}}_{\text{Interference from } L_p \text{ PUs}} + \sigma_n^2 (A_{R,k,l_s}^{SBS}) A_{R,k,l_s}^{SBS}, \tag{4}$$

2.2. Backhaul Link

The hybrid beamforming weights at the backhaul link are obtained based on orthogonal pilot signals transmitted from each SBS to the MBS. The kth SBS applies its pilot signal $s_k^p \in \mathbb{C}^{N_d \times 1}$ to an $N_d \times N_d$ transmit digital beamforming weight vector $D_{T,k}^{SBS}$ followed by an $N_a \times N_d$ transmit analog beamforming matrix $A_{T,k}^{SBS}$. If we denote the combined digital-analog transmit beamformer for the kth SBS as $\mathbf{w}_{T,k} = A_{T,k}^{SBS} D_{T,k}^{SBS}$, then the array output of the MBS can be written as

$$y_{\text{MBS}}^p = \sum_{k=1}^{K} \mathbf{H}_{k,MBS} \mathbf{w}_{T,k} s_k^p + \mathbf{H}_{PU,MBS} x_p + n_{MBS}, \tag{5}$$

where y_{MBS}^p is the $M_a \times 1$ vector containing the outputs of the M_a- element antenna array of the MBS, $\mathbf{H}_{k,MBS}$ is the $N_a \times M_a$ channel matrix representing the transfer functions from the N_a- element antenna array of the kth SBS to the M_a- element antenna array of the MBS, $\mathbf{H}_{PU,MBS}$ is the $M_a \times L_p$ channel matrix from the L_p PUs to the MBS's M_a- element antenna array, and $\mathbf{n}_{MBS}$ is the received $M_a \times 1$ complex AWGN vector at the MBS.

The MBS detects the kth SBS signal by applying the output of the array y_{MBS}^p to the $M_a \times M_d$ receive analog weight matrix $A_{R,k}^{MBS}$ followed by the $M_d \times M_d$ receive digital beamforming weight vector $D_{k,R}^{MBS}$. If we denote the combined digital-analog receive beamformer for the kth SBS as $c_k = A_{k,R}^{MBS} D_{k,R}^{MBS}$, then the detection of the kth SBS signal by the MBS can be expressed as

$$\hat{r}_k = c_k^H y_{\text{MBS}}^p = S_k^p + S_{I_k} + S_{I_p} + c_k^H n_{MBS}, \tag{6}$$

where $S_k^p = c_k^H \mathbf{H}_{k,MBS} \mathbf{w}_{T,k} s_k^p$ is the kth SBS signal, $S_{I_k} = c_k^H \sum_{i=1,i\neq k}^{K} \mathbf{H}_{k,MBS} \mathbf{w}_{T,k} s_k^p$ is the interference from $K-1$ other SBSs, and $S_{I_p} = c_k^H \mathbf{H}_{PU,MBS} x_p$ is the interference from L_p PUs.

Assuming that s_k^p are complex-valued random variables with unit power, i.e., $E\left[\|s_k^p\|^2\right] = 1$, and denoting $\mathbf{H}_{BL,k} = \left(A_{R,k}^{MBS}\right)^H \mathbf{H}_{k,MBS} \left(A_{T,k}^{SBS}\right)^H$ as the effective channel between the kth SBS and the MBS, the SINR at the MBS for the kth SBS can be expressed as

$$\gamma_k^{MBS} = \frac{\left(D_{R,k}^{MBS}\right)^H \mathbf{H}_{BL,k} \left(D_{T,k}^{SBS}\right)^H D_{T,k}^{SBS} \mathbf{H}_{BL,k}^H D_{R,k}^{MBS}}{c_k^H \mathbf{B}_{BL} c_k}, \tag{7}$$

where $\mathbf{B}_{BL} = \sum_{i=1,i\neq k}^{K} \mathbf{H}_{i,MBS}\mathbf{w}_{T,i}\mathbf{w}_{T,i}^{H}\mathbf{H}_{i,MBS}^{H} + \mathbf{H}_{PU,MBS}x_p x_p^H \mathbf{H}_{PU,MBS}^{H} + \sigma_n^2 \mathbf{I}_{M_a}$ is the covariance matrix of the interference-plus-noise at the backhaul link.

2.3. End-to-End SINR and Channel Capacity

Once the hybrid beamforming weights of the backhaul link are obtained, they can be used to forward the SUs signals to the MBS. The kth SBS applies the received l_s^{th} user signal, $\mathbf{w}_{R,k,l_s}^{H} g_{k,l_s} x_{l_s}^{s}$, to the hybrid beamformer, $\mathbf{w}_{T,k}$. The $N_{T,a} \times 1$ transmitted signal s_{k,l_s} at the output of the antenna array can then be expressed as

$$s_{k,l_s} = \mathbf{w}_{T,k}\mathbf{w}_{R,k,l_s}^{H} g_{k,l_s} x_{l_s}^{s}, \tag{8}$$

and the expression for the array output of the MBS can be written as

$$y_{MBS} = \mathbf{H}_{PU,MBS}x_p + y_{k,MBS} + \sum_{i=1,\, i\neq k}^{K} y_{i,MBS} + n_{MBS}, \tag{9}$$

where $y_{k,MBS} = \mathbf{H}_{k,MBS}\mathbf{w}_{T,k}\mathbf{w}_{R,k,l_s}^{H} y_k$ is the array output of the MBS from the kth SBS.

Using Equation (2), $y_{k,MBS}$ can be expressed in terms of the l_s^{th} user signal, $x_{l_s}^{s}$, as follows:

$$\begin{aligned} y_{k,MBS} = &\mathbf{H}_{k,MBS}\mathbf{w}_{T,k}\mathbf{w}_{R,k,l_s}^{H} g_{k,l_s} x_{l_s}^{s} \\ &+ \mathbf{H}_{k,MBS}\mathbf{w}_{T,k}\mathbf{w}_{R,k,l_s}^{H} \textstyle\sum_{i=1,i\neq l_s}^{L_s} g_{k,i} x_i^s \\ &+ \mathbf{H}_{k,MBS}\mathbf{w}_{T,k}\mathbf{w}_{R,k,l_s}^{H} G_{k,PU} x_p \\ &+ \mathbf{H}_{k,MBS}\mathbf{w}_{T,k}\mathbf{w}_{R,k,l_s}^{H} n_{k,SBS}, \end{aligned} \tag{10}$$

When the MBS applies the output of the array, y_{MBS}, to the hybrid weight, c_k^H, the detection of the l_s^{th} user signal of the kth SBS by the MBS can be expressed as

$$\hat{x}_{k,l_s} = c_k^H y_{\text{MBS}} = c_k^H \left(\mathbf{S}_{k,l_s} + \mathbf{S}_I^{SBSs} + \mathbf{S}_{k,I}^{SU} + \mathbf{S}_I^{\text{PU}} + \mathbf{N}_{MBS} \right), \tag{11}$$

where

$\mathbf{S}_{k,l_s} = \mathbf{H}_{k,MBS}\mathbf{w}_{T,k}\mathbf{w}_{R,k,l_s}^{H} \mathbf{g}_{k,l_s} x_{l_s}^{s}$ is the l_s^{th} user signal of the kth SBS,
$\mathbf{S}_{k,I}^{\text{SU}} = \mathbf{H}_{k,MBS}\mathbf{w}_{T,k}\mathbf{w}_{R,k,l_s}^{H} \sum_{i=1,i\neq l_s}^{L_s} g_{k,i} x_i^s$ is the interference from the $L_s - 1$ other SUs of kth SBS,
$\mathbf{S}_I^{SBS} = \sum_{i=1,\, i\neq k}^{K} \left(\mathbf{H}_{i,MBS}\mathbf{w}_{T,i}\mathbf{w}_{R,i,l_s}^{H} y_{i,SBS} \right)$ is the interference from the $K-1$ other SBSs.
$\mathbf{S}_I^{\text{PU}} = \mathbf{H}_{k,MBS}\mathbf{w}_{T,k}\mathbf{w}_{R,k,l_s}^{H} G_{k,PU} x_p + c_k^H \mathbf{H}_{PU,MBS} x_p$
$\mathbf{N}_{MBS} = \left(\mathbf{H}_{k,MBS}\mathbf{w}_{T,k}\mathbf{w}_{R,k,l_s}^{H} n_{k,SBS} + n_{MBS} \right)$

The end-to-end SINR at the MBS for the l_s^{th} user of the kth SBS can be expressed as

$$\gamma_{k,l_s}^{MBS} = \frac{c_k^H \mathbf{H}_{k,MBS}\mathbf{w}_{T,k}\mathbf{w}_{R,k,l_s}^{H} g_{k,l_s} x_{l_s}^{s} \left(x_{l_s}^{s} \right)^H g_{k,l_s}^{H} \mathbf{w}_{R,k,l_s}\mathbf{w}_{T,k}^{H}\mathbf{H}_{k,MBS}^{H} c_k}{c_k^H \mathbf{B}_{AL\ \ BL} c_k}, \tag{12}$$

where $\mathbf{B}_{AL-BL} = \mathbf{S}_{k,I}^{\text{SU}}\left(\mathbf{S}_{k,I}^{\text{SU}}\right)^H + \mathbf{S}_I^{SBS}\left(\mathbf{S}_I^{SBS}\right)^H + \mathbf{S}_I^{\text{PU}}\left(\mathbf{S}_I^{\text{PU}}\right)^H + \mathbf{N}_{MBS}(\mathbf{N}_{MBS})^H$ is the covariance matrix of the interference-plus-noise for the l_s^{th} user end-to-end link.

The ergodic channel capacity for each user, l_s, is given by [19]

$$C = \mathbb{E}\left(log_2 \left\{ 1 + \frac{c_k^H \mathbf{H}_{k,MBS}\mathbf{w}_{T,k}\mathbf{w}_{R,k,l_s}^{H} g_{k,l_s} x_s x_s^H g_{k,l_s}^{H} \mathbf{w}_{R,k,l_s}\mathbf{w}_{T,k}^{H}\mathbf{H}_{k,MBS}^{H} c_k}{c_k^H \mathbf{B}_{AL-BL} c_k} \right\} \right), \tag{13}$$

where $\mathbb{E}(.)$ denotes the expectation operator.

2.4. Channel Model

For the access and backhaul links, we consider mmWave propagation channels with limited scattering which can be modelled at each subcarrier by the clustered channel representation [13]. We assume a scattering environment with N_{cl} scattering clusters randomly distributed in space and within each cluster, there are N_{ray} closely located scatterers.

For the backhaul link, the channel matrix at subcarrier f_n between the kth SBS and the MBS can be expressed as

$$\mathbf{H}_{k,MBS,f_n} = \sqrt{\frac{N_a M_a}{N_{cl} N_{ray}}} \sum_{i}^{N_{cl}} \sum_{l=1}^{N_{ray}} \alpha_{il,f_n} \boldsymbol{a}_{MBS,f_n}\left(\varnothing^r_{i,l}, \theta^r_{i,l}\right) \boldsymbol{a}^*_{k,SBS,f_n}\left(\varnothing^t_{i,l}, \theta^t_{i,l}\right), \tag{14}$$

where $\alpha_{il,f_n} = \widetilde{\alpha}_{il} e^{-j2\pi i f_n / N_c}$ are the complex gains of the jth ray in the ith scattering cluster and $\widetilde{\alpha}_{il}$ are assumed i.i.d $\mathrm{CN}(0, \sigma^2_{\widetilde{\alpha},i})$. With $\sigma^2_{\widetilde{\alpha},i}$ representing the average power of the ith cluster, $\varnothing^r_{i,l}$ and $\varnothing^t_{i,l}$ are the azimuth angles of arrival and departure, respectively, $\theta^r_{i,j}$ *and* $\theta^t_{i,j}$ are the elevation angles of arrival and departure, respectively, and $\boldsymbol{a}_{MBS,f_n}\left(\varnothing^r_{i,l}, \theta^r_{i,l}\right)$ and $\boldsymbol{a}_{k,SBS,f_n}\left(\varnothing^t_{i,l}, \theta^t_{i,l}\right)$ represent, respectively, the normalized receive and transmit array response vectors of the MBS and the kth SBS.

For the access link, the channel matrix at subcarrier f_n between the kth SBS and its L_s users can be written as

$$G_{k,SU,f_n} = \sqrt{\frac{L_s M_a}{N_{cl} N_{ray}}} \sum_{i}^{N_{cl}} \sum_{l=1}^{N_{ray}} \alpha_{il,f_n} \boldsymbol{a}_{SBS,f_n}\left(\varnothing^r_{i,l}, \theta^r_{i,l}\right) \boldsymbol{a}^*_{k,SU,f_n}\left(\varnothing^t_{i,l}, \theta^t_{i,l}\right), \tag{15}$$

where $\boldsymbol{a}_{k,SU,f_n}\left(\varnothing^t_{i,l}, \theta^t_{i,l}\right)$ represents the spatial signature vector of the L_s single antenna users.

The N_{ray} azimuth and elevation angles, $\varnothing^{r,t}_{i,l}$ and $\theta^{r,t}_{i,l}$, within the cluster i are assumed to be randomly distributed with a uniformly-random mean cluster angle of $\varnothing^{r,t}_i$ *and* $\theta^{r,t}_i$, respectively, and a constant angular spread of $\sigma_{\varnothing^{r,t}}$ *and* $\sigma_{\theta^{r,t}}$, respectively.

For simplicity, the access links between the MBS and its L_p users ($\mathbf{H}_{PU,MBS}$) and between PUs and the kth SBS ($\boldsymbol{G}_{k,PU}$) are modeled by convetional i.i.d MIMO channels.

Note that in this per-subcarrrier representation, it is assumed that for each subcarrier f_n, the carrier frequency f_c is much larger than $f_c \pm f_n$ and that $\boldsymbol{a}_{MBS,f_n}\left(\varnothing^r_{i,l}, \theta^r_{i,l}\right)$ and $\boldsymbol{a}_{k,SBS,f_n}\left(\varnothing^t_{i,l}, \theta^t_{i,l}\right)$ can approximately be considered equal for all subcarriers. Consequently, the channel covariance matrices are approximately similar with the same set of eigenvectors for all subcarriers and can be replaced by the average of the covariance matrices, i.e., $\mathbf{H}^H_{AL,k,l_s}\mathbf{H}_{AL,k,l_s} = \frac{1}{N_c}\sum_{n=1}^{N_c} \mathbf{H}^H_{AL,k,l_s,f_n}\mathbf{H}_{AL,k,l_s,f_n}$, $\mathbf{H}^H_{BL,k}\mathbf{H}_{BL,k} = \frac{1}{N_c}\sum_{n=1}^{N_c} \mathbf{H}^H_{BL,k,f_n}\mathbf{H}_{BL,k,f_n}$, and $\mathbf{H}^H_{PU,MBS}\mathbf{H}_{PU,MBS} = \frac{1}{N_c}\sum_{n=1}^{N_c} \mathbf{H}^H_{PU,MBS,f_n}\mathbf{H}_{PU,MBS,f_n}$.

3. Proposed Hybrid Beamforming

3.1. Access Link

The kth SBS communicates with each SU through a set of selected beams that corresponds to a set of weight vectors. These weight vectors are obtained using the eigenbeamforming scheme and can be expressed as

$$\begin{gathered} \boldsymbol{A}^{SBS}_{R,k,l_s} = \left[\boldsymbol{a}^{SBS}_{R,k,l_s,1}, \boldsymbol{a}^{SBS}_{R,k,l_s,2}, \cdots, \boldsymbol{a}^{SBS}_{R,k,l_s,N_d}\right] \\ subject\ to \left|\boldsymbol{A}^{SBS}_{R,k,l_s}(i,j)\right|^2 = 1 \end{gathered}, \tag{16}$$

where $\boldsymbol{a}^{SBS}_{R,k,l_s,i}$ denote the ith selected $M_a \times 1$ eigenvector corresponding to the ith maximum eigen value of $\boldsymbol{g}^H_{k,l_s}\boldsymbol{g}_{k,l_s}$.

Since the analog beamforming matrix A^{SBS}_{R,k,l_s} is realized using phase shifters only, its elements, $A^{SBS}_{R,k,l_s}(i,j)$, satisfy $\left|A^{SBS}_{R,k,l_s}(i,j)\right|^2 = 1$. It should be noted that each SBS is using a different analog matrix for each user and that the system model shown in Figure 1 focuses on the detection of the l_s^{th} user of the kth SBS and shows the analog beamformer and the RF chains for one user only. The analog beamformer can be implemented using the Butler matrix as shown in Figure 4, where four users $(L_s = 4)$ and four RF chains per user $(N_d = 4)$ are assumed. Depending upon which 4 ports are activated, 4 beams are produced in specified directions. Since we are assuming 4 different channels, we should expect 4 different ports for each user.

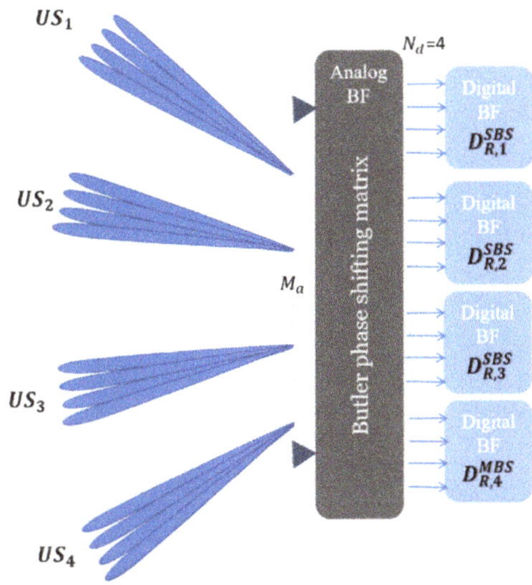

Figure 4. Hybrid beamforming based on Butler matrix for the access link.

Once the analog beams are selected, the received optimal digital weights, D^{SBS}_{R,k,l_s}, are obtained based on the maximization of the access link receive SINR, γ^{SBS}_{k,l_s}, given by Equation (3):

$$D^{SBS}_{R,k,l_s} = B^{-1}_{AL,k,l_s} V_{AL,k,l_s}\,, \tag{17}$$

where V_{AL,k,l_s} denote the eigen vector corresponding to the maximum eigenvalue of the effective access channel, $H^H_{AL,k,l_s} H_{AL,k,l_s}$.

3.2. Backhaul Link

The transmit analog weights of the kth SBS are based on the eigenbeamforming scheme and are given by

$$\begin{array}{c} A^{SBS}_{T,k} = \left[a^{SBS}_{T,k,1}, a^{SBS}_{T,k,2}, \cdots, a^{SBS}_{T,k,N_d}\right] \\ \textit{subject to } \left|A^{SBS}_{T,k}(i,j)\right|^2 = 1 \end{array}, \tag{18}$$

where $a^{SBS}_{T,k,i}$ denote the ith selected $N_a \times 1$ eigenvector corresponding to the ith maximum eigenvalue of $H^H_{k,\text{MBS}} H_{k,\text{MBS}}$.

Assuming channel reciprocity with $N_a = M_a$, the receive analog weight vectors of the MBS are given by $A^{MBS}_{R,k} = A^{SBS}_{T,k}$. It should also be noted that the MBS is using a different analog matrix for each SBS, which can be implemented using the Butler matrix of Figure 4, where mobile users are replaced by SBSs.

For fixed analog beamforming weights, $A^{SBS}_{T,k}$ and $A^{MBS}_{R,k}$, the transmit optimal digital weight vector of the kth SBS, $D^{SBS}_{T,k}$, and the receive optimal digital weight vector of the MBS, $D^{MBS}_{R,k}$, are obtained base on the maximization of the backhaul link receive SINR and are given by

$$D^{SBS}_{T,k} = D^{MBS}_{R,k} = B^{-1}_{BL,k} \boldsymbol{H}_{\boldsymbol{BL,k}} V_{BL}\,, \tag{19}$$

where V_{BL} is the eigenvector corresponding to the maximum eigenvalue of $\mathbf{H}^{H}_{BL,k}\mathbf{H}_{BL,k}$, with $H_{BL,k}$ representing the effective channel given by $H_{BL,k} = \left(A^{MBS}_{R}\right)^{H}_{k} \mathbf{H}_{k,MBS}\left(A^{SBS}_{T}\right)^{H}_{k}$.

4. Simulation Results

In our simulation setups, we considered a HetNet organized into four SBSs ($K = 4$) and one macro-cell. The SBSs and the MBS used the same number of antennas, $N_a = M_a = 64$, and the same number of RF chains, $N_d = M_d = 2$ *or* 4. Each SBS is serving $L_s = 4$ users and the macro-cell is serving 4 users, each transmitting with a single antenna. We assumed QPSK modulation. For the OFDM configurations, we assumed the 256-OFDM system ($N_c = 256$), which is widely deployed in broadband wireless access services.

Figure 5 shows the beampattern of the proposed HBF with four RF chains and the optimal fully-digital one for the access link. It is noted that the optimal beamformer has about five dominant beams, three of which are similar to the selected beams of the proposed HBF. This beampattern means that the data streams can be successfully transmitted through those three beams using the proposed HBF and that near optimal performance could be achieved if we were to bring the number of RF chains close to the number of dominant beams of the optimal beamformer. For the backhaul link, Figure 6 shows very similar beampatterns with more dominant beams.

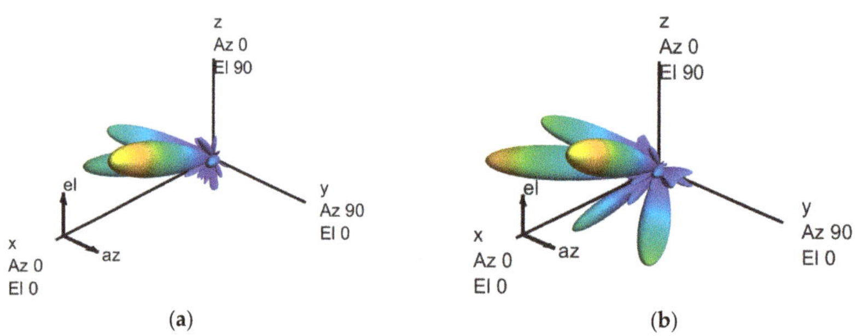

Figure 5. Beampattern of the access link: (**a**) Proposed HBF, 4 RF chains; (**b**) fully-digital beamforming (optimal).

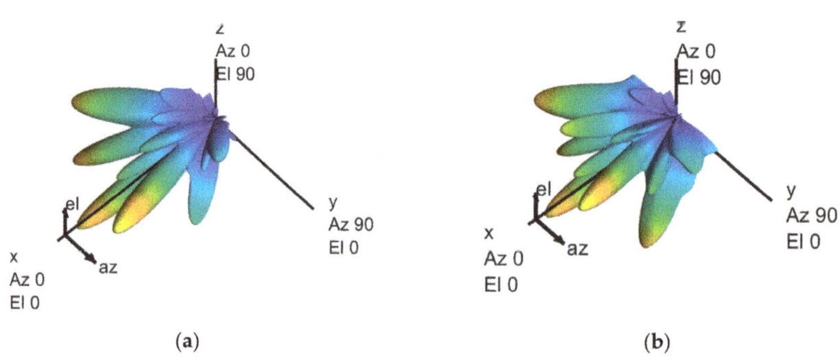

Figure 6. Beampattern of the backhaul link: (**a**) Proposed HBF, 4 RF chains; (**b**) fully-digital beamforming (optimal).

Figure 7, on the other hand, compares the ergodic channel capacity of the proposed HBF and the optimal fully-digital one. It is observed that for both cases the optimal beamformer is outperforming the proposed HBF. However, as we increase the number of RF chains, the performance gap between the two schemes was reduced, and a near-optimal solution was achieved by the proposed HBF using four RF chains. On the other hand, for the single cell MU-MIMO case presented in References [12–14], near optimal performance was achieved with only five RF chains, and for the MU-MIMO case in [16,17], it was shown that the required number RF chains could be reduced to two to achieve fully digital beamforming performance. However, unlike our case, where we have assumed a HetNet with a macro cell and multiple small cognitive cells, these studies focused primarily on macro-cellular systems and did not consider HBF in the context of HetNets.

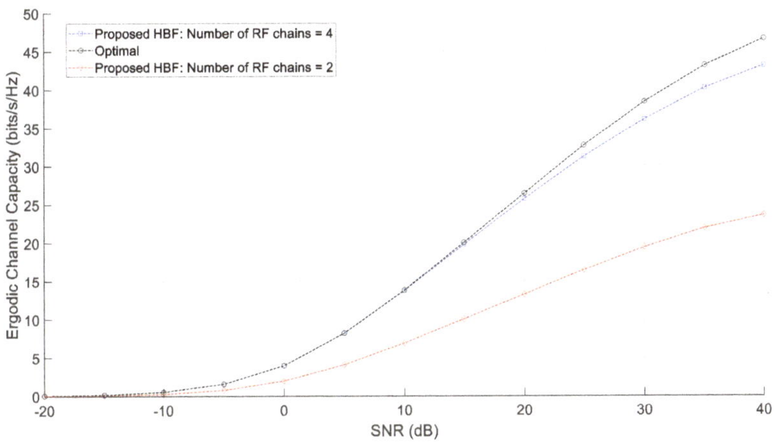

Figure 7. Ergodic channel capacity of the proposed HBF for different number of RF chains.

5. Conclusions

In this paper, we employed hybrid beamforming at the access and backhaul links of a mmWave HetNet system. We proposed a low-complexity HBF that was fully-based on MRT/MRC Eigen-beamforming schemes. The performance evaluation in terms of the beam patterns and the ergodic channel capacity showed that the proposed HBF scheme achieved near-optimal performance with only four RF chains and required considerably less computational complexity.

Funding: This research received no external funding.

Acknowledgments: The author would like to thank the Canadian Microelectronics Corporation (CMC) for providing the Heterogeneous Parallel Platform to run the computationally-intensive Monte-Carlo Simulations.

Conflicts of Interest: The authors declare no conflict of interest.

References

1. Siddique, U.; Tabassum, H.; Hossain, E.; Kim, D.I. Wireless backhauling of 5G small cells: Challenges and solution approaches. *IEEE Wirel. Commun.* **2015**, *22*, 22–31. [CrossRef]
2. Gao, Z.; Dai, L.; Mi, D.; Wang, Z.; Imran, M.A.; Shakir, M.Z. MmWave Massive MIMO Based Wireless Backhaul for 5G Ultra-Dense Network. *IEEE Wirel. Commun.* **2015**, *22*, 13–21. [CrossRef]
3. Tabassum, H.; Hamdi, S.A.; Hossai, E. Analysis of massive MIMO-enabled downlink wireless backhauling for full-duplex small cells. *IEEE Trans. Commun.* **2016**, *64*, 2354–2369. [CrossRef]
4. Shariat, M.; Pateromichelakis, E.; Quddus, A.; Tafazolli, R. Joint TDD Backhaul and Access Optimization in Dense Small-Cell Networks. *IEEE Trans. Veh. Technol.* **2015**, *64*, 5288–5299. [CrossRef]
5. ElSawy, H.; Hossain, E.; Kim, D.I. HetNets with cognitive small cells: User offloading and Distributed channel allocation techniques. *IEEE Commun. Mag.* **2013**, *51*, 28–36. [CrossRef]

6. Yan, Z.; Zhou, W.; Chen, S.; Liu, H. Modeling and Analysis of Two-Tier HetNets with Cognitive Small Cells. *IEEE Access.* **2016**, *5*, 2904–2912. [CrossRef]
7. Marzetta, T.L. Noncooperative cellular wireless with unlimited numbers of base station antennas. *IEEE Trans. Wirel. Commun.* **2010**, *9*, 3590–3600. [CrossRef]
8. Rusek, F.; Persson, D.; Lau, B.; Larsson, E.; Marzetta, T.L.; Edfors, O.; Tufvesson, F. Scaling up MIMO: Opportunities and challenges with very large arrays. *IEEE Signal Process. Mag.* **2013**, *30*, 40–60. [CrossRef]
9. Ngo, H.Q.; Larsson, E.G.; Marzetta, T.L. Energy and spectral efficiency of very large multiuser MIMO systems. *IEEE Trans. Commun.* **2013**, *61*, 1436–1449.
10. Hoydis, J.; ten Brink, S.; Debbah, M. Massive MIMO in the UL/DL of cellular networks: How many antennas do we need? *IEEE J. Sel. Areas Commun.* **2013**, *31*, 160–171. [CrossRef]
11. Hefnawi, M. Capacity-Aware Multi-User Massive MIMO for Heterogeneous Cellular Network. In Proceedings of the IEEE International Conference on Selected Topics in Mobile and Wireless Networking, Tangier, Morocco, 20–22 June 2018.
12. Sohrabi, F.; Yu, W. Hybrid beamforming with finite-resolution phase shifters for large-scale MIMO systems. In Proceedings of the IEEE Workshop Signal Processing Advances in Wireless Communications, Stockholm, Sweden, 28 June–1 July 2015.
13. El Ayach, O.; Rajagopal, S.; Abu-Surra, S.; Pi, Z.; Heath, R. Spatially sparse precoding in millimeter wave MIMO systems. *IEEE Trans. Wirel. Commun.* **2014**, *13*, 1499–1513. [CrossRef]
14. Alkhateeb, A.; El Ayach, O.; Leus, G.; Heath, R. Channel estimation and hybrid precoding for millimeter wave cellular systems. *IEEE J. Sel. Top. Signal Process.* **2014**, *8*, 831–846. [CrossRef]
15. Sohrabi, F.; Yu, W. Hybrid digital and analog beamforming design for large-scale MIMO systems. In Proceedings of the IEEE International Conference on Acoustics, Speech, Signal Process (ICASSP), Brisbane, Australia, 19–24 April 2015.
16. Liang, L.; Dai, Y.; Xu, W.; Dong, X. How to approach zero-forcing under RF chain limitations in large mmwave multiuser systems? In Proceedings of the IEEE/CIC International Conference on Communications in China, Shanghai, China, 13–15 October 2014.
17. Liang, L.; Xu, W.; Dong, X. Low-complexity hybrid precoding in massive multiuser MIMO systems. *IEEE Wirel. Commun.* **2014**. [CrossRef]
18. Kang, M. A comparative study on the performance of MIMO MRC systems with and without cochannel interference. *IEEE Trans. Commun.* **2004**, *52*, 1417–1425. [CrossRef]
19. Sulyman, A.I.; Hefnawi, M. Adaptive MIMO Beamforming Algorithm Based on Gradient Search of the Channel Capacity in OFDMSDMA System. *IEEE Commun. Lett.* **2008**, *12*, 642–644. [CrossRef]

electronics

MDPI

Article

Spatio-Radio Resource Management and Hybrid Beamforming for Limited Feedback Massive MIMO Systems

Hedi Khammari [1], Irfan Ahmed [2,*], Ghulam Bhatti [3] and Masoud Alajmi [1]

1 Department of Computer Engineering, College of Computers and Information Technology, Taif University, Taif 21974, Saudi Arabia; hkhamari@tu.edu.sa (H.K.); ms.alajmi@tu.edu.sa (M.A.)

2 Department of Electrical Engineering, Higher Colleges of Technology, Ruwais 12389, UAE

3 Department of Computer Science, College of Computers and Information Technology, Taif University, Taif 21974, Saudi Arabia; gbhatti@tu.edu.sa

* Correspondence: iahmed3@hct.ac.ae

Received: 5 August 2019; Accepted: 17 September 2019; Published: 20 September 2019

Abstract: In this paper, a joint spatio–radio frequency resource allocation and hybrid beamforming scheme for the massive multiple-input multiple-output (MIMO) systems is proposed. We consider limited feedback two-stage hybrid beamformimg for decomposing the precoding matrix at the base-station. To reduce the channel state information (CSI) feedback of massive MIMO, we utilize the channel covariance-based RF precoding and beam selection. This beam selection process minimizes the inter-group interference. The regularized block diagonalization can mitigate the inter-group interference, but requires substantial overhead feedback. We use channel covariance-based eigenmodes and discrete Fourier transforms (DFT) to reduce the feedback overhead and design a simplified analog precoder. The columns of the analog beamforming matrix are selected based on the users' grouping performed by the K-mean unsupervised machine learning algorithm. The digital precoder is designed with joint optimization of intra-group user utility function. It has been shown that more than 50% feedback overhead is reduced by the eigenmodes-based analog precoder design. The joint beams, users scheduling and limited feedbacK-based hybrid precoding increases the sum-rate by 27.6% compared to the sum-rate of one-group case, and reduce the feedback overhead by 62.5% compared to the full CSI feedback.

Keywords: hybrid beamforming; massive MIMO; resource allocation

1. Introduction

The scarcity of available frequency band for wireless communications has led to the inclusion of millimeter Wave (mmWave) frequencies in cellular communications. This has opened the doors for massive multiple-input multiple-output (MIMO) systems. Due to high transmission frequencies, fabrication of large number of antennas with small form factor has become possible. MmWave band has inherent hindrances, like, high path-loss and absorption-loss. It has been known that MIMO systems advantages (spatial multiplexing or diversity gain) are scaled-up with the number of antennas. In summary, one can enjoy the benefits of the large bandwidth available at mmWave frequencies by combating high path and absorption losses with massive MIMO directional beamforming. Future mmWave massive MIMO-based cellular networks will be as shown in Figure 1. Due to the high pathloss on one hand and high directional gain on the other hand, the inter-cell interference and cell boundaries will become meaningless. The fixed area size cell boundaries of traditional cellular will probably no longer exist in the future mmWave massive MIMO systems. Narrow beams can serve distant user equipment (UE) without interfering other UEs provided that there is no obstacle between BS and intended UE, whereas a closely located UE may deprive of connection due to the obstacles.

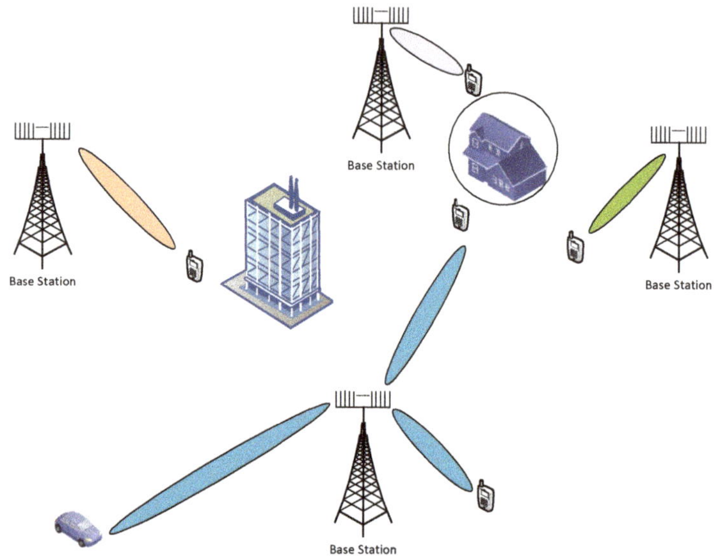

Figure 1. MmWave massive MIMO based cellular system.

The cost of massive MIMO is in terms of excessive feedback overhead for channel estimation along with the hardware complexity of RF chains (increased number of radio-frequency (RF) chains). The feedback overhead has been tackled separately for frequency division duplex (FDD) and time division duplex (TDD) systems. In FDD systems, the uplink channel estimation consists of fewer overheads compared to the downlink channel estimation, because generally, the number of transmit antennas N_t is larger than the number of users K, and the number of receive antennas per user n_k ($N_t \gg K$ and $N_t \gg n_k$). The most common technique to reduce the downlink channel estimation overhead is joint spatial division multiplexing (JSDM) [1]. The JSDM uses two-stage precoding: second order channel statistics (covariance)-based user grouping and the traditional MU-MIMO linear precoding (zero-forcing) for the inter-user interference mitigation based on the low-dimensional effective channel. In TDD, only uplink channel estimation is done and the downlink channel estimates are obtained by the transpose of the uplink channel using the channel reciprocity principle. The TDD massive MIMO systems suffer from pilot contamination when the BS receives non-orthogonal pilot signals from the neighboring cells. This pilot contamination degrades the channel estimation and hence, affects both uplink combining and downlink precoding.

In traditional MIMO systems, a separate RF chain (analog-to-digital converter/digital-to-analog converter, serial-to-parallel/parralel-to-serial converter, up/down converter etc) is required for each antenna, but the high power consumption makes it infeasible for the case of massive MIMO systems. Hybrid beamforming technique resolves this problem by dividing the precoding/combining into baseband digital processing and RF analog processing. The hybrid precoding and combining offer extra degrees of freedom in space domain with a large number of antennas and analog beamforming [2]. The hybrid beamforming can be realized by using MU-MIMO precoding as baseband digital precoding and the statistical channel state information-based pre-beamforming as RF analog precoding. This limited feedback (due to average CSI) configuration is particularly suited for massive MIMO mmWave systems with a large number of antennas but relatively small number of RF chains [3]. It has been shown [4], that the covariance-based limited feedback works well for mmWave massive MIMO systems, where the number of users is small with respect to the number of BS antennas and the channels are formed by a few multi path components (MPCs) with small angular spread.

Limited work has been done on the joint multiuser massive MIMO resource allocation and hybrid beamforming design. Although mmWave massive MIMO system has a potential of tremendous increase of spectral efficiency. However, the cost and power consumption of power-hungry radio frequency chains (analog-to-digital converter (ADC)/digital-to-analog converter (DAC), parallel to serial converter, serial to parallel converter, up converter/down converter) make it impractical to build a complete RF chain for each antenna. A promising solution to this problem is hybrid beamforming, where the precoder at the transmitter is divided into two parts: analog precoder and digital precoder. The analog precoder (usually a network of phase shifters) at the RF stage reduces the number of RF chains required for the digital precoder. In order to configure these precoders, the transmitter requires channel state information in the form of uplink feedback from users, but in the presence of massive antennas, this feedback becomes a huge load on the wireless uplink, especially in FDD mode. JSDM [4] is a technique used to reduce the feedback overhead. It uses slowly varying average channel statistics to implement the analog precoder; then, the digital precoder is realized by using a low-dimensional effective channel. Till now, different variants of the JSDM have been proposed. Li et al. [5] generalize the JSDM scheme to support non-orthogonal virtual sectorization and with multiple RF chains at both link ends. It uses the Kronecker channel model to decouple the transmit and receive beamforming. Under this channel, the analog beamformer is obtained by stacking strongest eigenbeams of the channel covariance matrix and then the digital beamformer is based on a weighted minimum mean squared error (MMSE) with effective channel. However, the Kronecker model does not characterize the mmWave channel where transmitter and receiver have coupling effects due to highly directional transmission. In [6], the authors apply JSDM using a geometrical channel model and find hybrid precoder and combiner at transmitter and receivers, respectively. Hybrid beamforming with switches (HBwS) has been introduced in [7], where, $L \times N_t$ analog beamformer is controlled by $N_{RF} \times L$ instantaneous CSI based switches. N_t is the number of transmit antennas, N_{RF} is the number of RF chains, and $N_{RF} < L < N_t$. Another switch-based analog beamforming is proposed in [8] but it requires instantaneous CSI for both switching network and the phase shifter network. Also it contains $L = N_t$. The JSDM implementation also requires the training in the downlink to estimate the channel covariance matrix. Most of the work assumes that the CSI is known at both ends. In [9], authors consider the joint optimization of the training resource allocation and channel-statistics-based analog beamformer design by using user centric virtual sectorization. There are different structures for the phase shifter-based analog beamformer, namely, fully connected, sub-connected, and dynamically connected [10]. Park [11] investigate JSDM with these analog beamformer architectures. The dynamic architecture gives better result at the cost of added complexity. In [12], authors propose a hybrid beamforming method with unified analog beamformer by Subspace Construction (SC) based on partial CSI in massive MIMO OFDM system. In [13], statistical CSI based analog beamformer uses regularized block diagonalization to mitigate the inter-group interference and instantaneous CSI based digital beamformer utilizes the weighted MMSE to suppress intra-group interference. Jiang et al. [14] jointly optimize the user selection and beam selection during analog beamforming design. They use Lyapunov-drift optimization framework to obtain the optimal solution. Their work only focuses on the design of statistical CSI based analog precoder and user/beam selection. Our previous work [15] on resource allocation for transmit beamforming develops digital and analog precoders which maximize the sum rate with total power and desired number of RF chains constraints. The provided solutions require full instantaneous CSI at the transmitter and receiver, which, in case of the massive MIMO, consists of large number of pilot transmission in downlink and channel information feedback in the uplink. In this work we exploit the channel similarities by grouping (K-Mean machine learning) the users based on the location information. Low complexity DFT matrix based analog precoder is derived using statistical CSI. This greatly reduces the feedback overhead for the design of zero-forcing digital precoder.

Machine learning (ML) applications for the physical layer of wireless communication systems have been widely reported in [16]. Most of the conventional transmitter and receiver blocks can be replaced by an ML-based auto encoder as suggested by the authors. The large number of antennas in massive MIMO leads to the challenging issue of channel estimation in mmWave communications. A common practice in TDD massive MIMO systems is to utilize the channel reciprocity to get the downlink CSI from uplink channel information estimates. However, in FDD, the channel reciprocity is not applicable and the downlink CSI estimation is very difficult. The downlink channel estimation is known to be hampered by the pilot contamination effect (user to base-station). The quality of channel estimates is deteriorated by the mutual interference caused by the non-orthogonal pilots in a cell. In [17], a supervised learning-based pilot decontamination scheme for massive MIMO uplink is reported. In the proposed ML-based solution, the users' locations in all cells and the pilot assignments stand for the input features and output labels, respectively. In [18], a deep learning network CsiNet is used to learn the CSI-to-codeword transformation (codebook approach is usually adopted to reduce the feedback overhead) at users' terminals and inverse CsiNet at base-station. The authors of [19], suggest a learning-based antenna selection for massive MIMO systems. It uses a multiclass K-NN and support vector machine (SVM) for data-driven optimal antenna selection. Wang et al. [20] employs K-nearest neighbor $(K-NN)$ supervised learning for the N beams allocation among K users. In [21], a reinforcement learning based framework for radio resource management in radio access networks has been proposed. In our previous work [22], we used neural networks to reduce the execution time of the computationally intensive resource allocation part of the joint resource allocation and hybrid beamforming design in [15]. However, in this work, we use K-mean based unsupervised machine learning scheme to group the users based on their spatial locations. To the best of our knowledge, there is no research work that jointly consider the spatio–radio resource management and the hybrid beamforming in massive MIMO systems.

In this work, we use spatial channel covariance matrices for the analog beamforming design. We also consider the users to RF beam mapping. This mapping requires channel state information and a search over all possible beam combinations at the base-station. This search is exponential in the number of users [23]. Due to this exponential increase in complexity, we use DFT-based eigenmode beams with RF switches.

Contribution: In this paper, we develop joint spatio–radio resource and hybrid precoding algorithms for limited feedback wideband massive MIMO systems. The contributions of this paper are summarized as follow.

- First, we consider the problem of joint hybrid precoder design with limited feedback and user-beam selection to maximize the sum proportional rate under the total power constraint. The formulated mixed integer programming problem is then transformed to the relaxed-convex optimization problem.
- Second, a low complexity suboptimal solution is provided for the optimization problem. The algorithm generates the analog beamforming matrix, digital beamforming matrix, and the set of users in each group. The DFT/eigenmodes-based analog beamforming is formed using limited statistical CSI feedback from the users. Then, the digital precoder design with users selection is done iteratively.
- Finally, we develop a K-Mean algorithm based unsupervised machine learning scheme for users grouping. These users groups are used to form the limited feedback (statistical channel state information) based analog beamforming matrix. The proposed machine learning based analog beamforming along with the zero-forcing digital precoding and user scheduling gives better performance than the DFT/eigenmodes-based solution.

The rest of the paper is organized as follows. System, signal, and channel model along with the problem formulation are described in Section 2. Section 3 introduce the relaxed-convex transformation of the formulated mixed integer optimization problem. Suboptimal solution to the joint resource allocation and hybrid beamforming based on eigenmodes and discrete Fourier transform is given in Section 4. Section 5 proposes machine learning based users grouping and beam selection for joint optimization problem. Simulation results are given in Section 6, followed by the conclusions in Section 7.

Notations: Bold upper and lower case letters denote vectors and matrices, respectively. The notations $\mathbf{X}^{-1}$, $\mathbf{X}^{\dagger}$, $\mathbf{X}^{T}$, $\mathbf{X}^{H}$, and $tr(\mathbf{X})$ denote the inverse, pseudo-inverse, transpose, Hermitian transpose, and trace of a matrix $\mathbf{X}$. $vec\{\cdot\}$ is a vector operator, $diag\{x_1, ..., x_n\}$ is diagonal matrix, and $\otimes$ is the Kronecker product. $\|\cdot\|_F$ denotes the Frobenius norm. The $n \times n$ identity matrix is denoted by $\mathbf{I}_n$. $\mathbb{E}\{\cdot\}$ represents the expectation with respect to the random variable within the brackets.

2. System Model

Consider a FDD MU-MIMO downlink system where a base station (BS) with N_t antennas is located at the cell center and transmits to K single antenna users as shown in Figure 2. There are G groups of users such that the group $g \in \mathcal{G} = \{1, ..., G\}$. Each group contains K_g users.

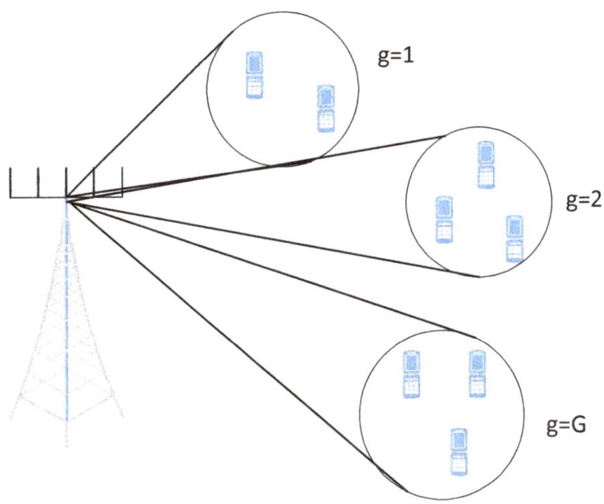

Figure 2. System Model.

Assume that the BS and users have the knowledge of the channel. We consider multi-carrier OFDM transmission with narrow-band blocK-fading channel. The BS is equipped with N_t antennas in linear antenna array (ULA) configuration. The information signal block $\mathbf{S} \in \mathbb{C}^{K \times N_f}$ at the input of the BS transceiver for the user k is given as

$$\mathbf{s}_k = [s_1, s_2, \ldots, s_{N_f}], \quad \forall k, \tag{1}$$

and for the subchannel n,

$$\mathbf{s}_n = [s_1, s_2, \ldots, s_K]^T, \quad \forall n, \tag{2}$$

where N_f and N_s are the number of subchannels and the number of symbols per subchannel, respectively. In a subchannel n, the information symbol vector is $\mathbf{s} \in \mathbb{C}^{N_s \times 1}$. We assume $N_s = K$, such that the transmit signal per subchannel n satisfying $\mathbb{E}\{\mathbf{s}_n \mathbf{s}_n^H\} = \frac{P_n}{K}\mathbf{I}_K$, where $P_n = P_T / N_f$ is the transmit power per subchannel and P_T is the total transmit power of the BS. The transmit signal vector $\mathbf{X}$ is obtained from $\mathbf{F}^B\mathbf{S}$, where $\mathbf{F}^B \in \mathbb{C}^{N_t \times N_s}$ is the precoding matrix. The hybrid beamforming divides the precoding matrix into baseband digital precoding matrix $\mathbf{F}^{DB} \in \mathbb{C}^{N_{RF} \times N_s}$ and RF analog precoding matrix $\mathbf{F}^{AB} \in \mathbb{C}^{N_t \times N_{RF}}$, where N_{RF} is the number of RF chains as shown in Figure 3. The transmit signal $\mathbf{X} \in \mathbb{C}^{N_t \times N_f}$ is given by

$$\begin{aligned} \mathbf{X} &= \mathbf{F}^{B}\mathbf{S} \\ &= \mathbf{F}^{AB}\mathbf{F}^{DB}\mathbf{S} \end{aligned} \tag{3}$$

Also, the precoding matrix must satisfy

$$\begin{aligned} \mathbb{E}\{tr(\mathbf{X}\mathbf{X}^{H})\} &\leq P_T \\ \mathbb{E}\{tr(\mathbf{F}^{AB}\mathbf{F}^{DB}\mathbf{S}\mathbf{S}^{H}\mathbf{F}^{DB^H}\mathbf{F}^{AB^H})\} &\leq P_T \\ tr(\mathbf{F}^{AB}\mathbf{F}^{DB}\mathbb{E}\{\mathbf{S}\mathbf{S}^{H}\}\mathbf{F}^{DB^H}\mathbf{F}^{AB^H}) &\leq P_T \end{aligned} \tag{4}$$

since $\mathbb{E}\{\mathbf{S}\mathbf{S}^{H}\} = \frac{P_T}{N_f K}\mathbf{I}_{KN_f}$, therefore,

$$tr(\mathbf{F}^{AB}\mathbf{F}^{DB}\mathbf{F}^{DB^H}\mathbf{F}^{AB^H}) \leq N_f K \tag{5}$$

The transmit signal in subchannel n is $\mathbf{x}_n \in \mathbb{C}^{N_t \times 1}$. Thus, the received signal vector $\mathbf{y}_n \in \mathbb{C}^{K \times 1}$ at K users in subchannel n is given by

$$\begin{aligned} \mathbf{y}_n &= \mathbf{H}_n^{H}\mathbf{x}_n + \mathbf{w}_n \\ &= \mathbf{H}_n^{H}\mathbf{F}^{AB}\mathbf{F}_n^{DB}\mathbf{s}_n + \mathbf{w}_n \end{aligned} \tag{6}$$

where $\mathbf{H}_n \triangleq [\mathbf{h}_{1,n}, ..., \mathbf{h}_{K,n}] \in \mathbb{C}^{N_t \times K}$ is the channel matrix with $\mathbf{h}_{k,n} = [h_{1,k}, ..., h_{N_t,k}]^H$ being the channel vector from BS to user k in subchannel n, $\mathbf{x}_n = \mathbf{F}_n^{DB}\mathbf{F}^{AB}\mathbf{s}_n$, and $\mathbf{w}_n \sim \mathcal{CN}(\mathbf{0}, \sigma^2\mathbf{I}_K)$ be the additive white Gaussian noise (AWGN) in subchannel n at the users. The RF beamforming $\mathbf{F}^{RF}$ is performed in time domain and the same beamforming is applied on all subchannels, whereas, the digital beamforming $\mathbf{F}_n^{DB}$ is performed in frequency domain on the per subchannel basis [11]. In the nth subchannel, the jth UE receives the sum of all transmitted signals for K UEs over its MIMO channel $\mathbf{H}_{j,i}$ as

$$y_{j,n} = \sum_{k=1}^{K} \mathbf{h}_{j,n}^{H}\mathbf{x}_{k,n} + w_{j,n} \tag{7}$$

where $\mathbf{h}_{j,n}$ is the $N_t \times 1$ channel vector. We denote the rank of the channel matrix $\mathbf{H}_{j,n}$ by $r_{j,n}$, where $0 \leq r_{j,n} \leq \min(K, N_t)$, $\forall n$. In matrix form, the above equation is given as

$$\mathbf{y}_{j,n} = \mathbf{h}_{j,n}^{H}\mathbf{x}_n + \mathbf{w}_{j,n} \tag{8}$$

The $1 \times N_f$ received signal at the kth UE is given by

$$\mathbf{y}_k = [\mathbf{h}_{k,1}^{H}\mathbf{F}^{AB}\mathbf{F}_1^{DB}\mathbf{s}_1, ..., \mathbf{h}_{k,N_f}^{H}\mathbf{F}^{AB}\mathbf{F}_{N_f}^{DB}\mathbf{s}_{N_f}] + \mathbf{w}_k, \tag{9}$$

Combining the signals for all UEs in a K dimensional received signal vector $\mathbf{y} = [\mathbf{y}_1, ..., \mathbf{y}_K]^H$, we get the system equation as

$$\mathbf{Y} = \mathbf{H}^{H}\mathbf{F}^{AB}\mathbf{F}^{DB}\mathbf{S} + \mathbf{W}, \tag{10}$$

where $\mathbf{Y}, \mathbf{W} \in \mathbb{C}^{K \times N_f}$.

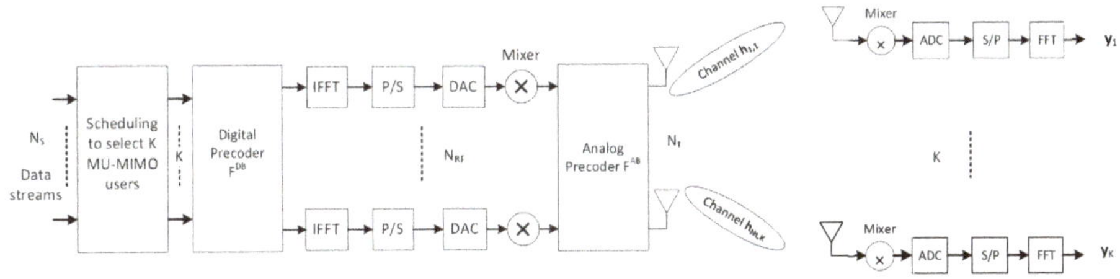

Figure 3. Block diagram of mmWave massive MIMO BS and UEs.

2.1. Channel Model

Generally, massive MIMO channel models are categorized in two types (i) analytical models and (ii) physical models [4]. Analytical models are commonly used for the theoretical analysis of wireless communication systems. The most commonly used analytical model is Kronecker channel model. It is a correlation-based model and characterizes the MIMO channel matrix in terms of the separate transmit and receive side spatial correlation matrices [24],

$$\begin{aligned}\mathbf{R}_H &= \mathbb{E}\{vec\{\mathbf{H}\}vec\{\mathbf{H}^H\}\} \\ &= \mathbf{R}_{tx} \otimes \mathbf{R}_{rx}\end{aligned} \tag{11}$$

under the above assumptions, the channel model **H** is simplified to Kronecker model,

$$\mathbf{H} = \mathbf{R}_{rx}^{1/2}\mathbf{K}\mathbf{R}_{tx}^{1/2} \tag{12}$$

where $\mathbf{K} \sim \mathcal{CN}(0,1)$ is an i.i.d. unit variance MIMO channel matrix, $\mathbf{R}_{tx}$ and $\mathbf{R}_{rx}$ are the transmit and receive corrrelation matrices, respectively. The transmit and receive correlation matrices are given as [24],

$$\mathbf{R}_{tx} = \mathbb{E}\{\mathbf{H}^H\mathbf{H}\}, \qquad \mathbf{R}_{rx} = \mathbb{E}\{\mathbf{H}\mathbf{H}^H\} \tag{13}$$

The physical models explicitly model wave propagation parameters like the complex amplitude, DoD, DoA, and delay of an MPC [24,25]. MmWave propagation leads to limited spatial scattering due to the high free-space pathloss. In addition, the large tightly packed antenna arrays lead to high levels of antenna correlation. The sparse scattering and antennas spatial correlation makes many of the commonly used statistical fading distributions inaccurate for mmWave channel modeling. Therefore, we use extended Saleh-Valenzuela model, which accurately describes the mathematical structure present in mmWave channels [26,27]. For simplicity, we assume that each scattering cluster around the transmitter and receiver contributes a single propagation path [28].

In general, the mmWave MIMO channel matrix between the BS with N_t transmit antennas and a user k with n_r receive antennas in subchannel n, can be modeled as double directional channel,

$$\mathbf{H}_{k,n} = \sqrt{\frac{N_t n_r}{\rho_{k,n} L}} \sum_{l=0}^{L} \alpha_{k,n,l} \mathbf{a}(\phi_{k,n,l}) \mathbf{b}^H(\theta_{k,n,l}), \tag{14}$$

where L is the total number of multipaths, $\alpha_{k,n,l}$ is the complex gain of the lth path with i.i.d. $\mathcal{CN}(0,1)$, and $\rho_{k,n}$ is the distance dependent pathloss between the BS and user k [29]. The LOS path is included with $l = 0$. Moreover, **a** and **b** are the receive and transmit steering vectors, respectively. The variables $\phi_{k,n,l} \in [0, 2\pi)$ and $\theta_{k,n,l} \in [0, 2\pi)$ are the lth path's azimuth angles (boresight angles in the receive array and transmit array) of arrival and departure, respectively. The steering vectors are given by

$$\mathbf{a}(\phi_{k,n,l}) = \frac{1}{\sqrt{N_r}}[a_1(\phi_l), ..., a_{n_r}(\phi_{k,n,l})]^T \tag{15}$$

$$\mathbf{b}(\theta_{k,n,l}) = \frac{1}{\sqrt{N_t}}[b_1(\theta_l), ..., b_{N_t}(\theta_{k,n,l})]^T \tag{16}$$

The elements of transmit and receive steering vectors are given by

$$b_i(\theta_{k,n,l}) = e^{-j\omega\tau_{i,l}} = e^{-j2\pi(i-1)\frac{d_t}{\lambda}\sin(\theta_{k,n,l})},\ i = 1, 2, ..., N_t \tag{17}$$

$$a_i(\phi_{k,n,l}) = e^{-j\omega\tau_{i,l}} = e^{-j2\pi(i-1)\frac{d_r}{\lambda}\sin(\phi_{k,n,l})},\ i = 1, 2, ..., n_r \tag{18}$$

where λ is the wavelength, $\omega = \frac{2\pi}{\lambda}$, τ_i is the beamforming delay, and d_t and d_r are the antenna spacing at the transmitter and receiver, respectively.

The channel matrix in (14) can also be written in more compact form as

$$\mathbf{H}_{k,n} = \sqrt{\nu}\mathbf{A}_{k,n}\mathbf{D}_{k,n}\mathbf{B}_{k,n}^H \tag{19}$$

where $\nu = \frac{N_t n_r}{\rho L}$ and, $\mathbf{A}_{k,n}$ and $\mathbf{B}_{k,n}$ consist of stacked steering vectors of AoA and AoD, respectively, i.e., $\mathbf{A}_{k,n} = [\mathbf{a}(\phi_{k,n,1}), \mathbf{a}(\phi_{k,n,2}), \ldots, \mathbf{a}(\phi_{k,n,L})]$ and $\mathbf{B}_{k,n} = [\mathbf{b}(\theta_{k,n,1}), \mathbf{b}(\theta_{k,n,2}), \ldots, \mathbf{b}(\theta_{k,n,L})]$. The matrix $\mathbf{D}_{k,n}$ is a diagonal matrix, given as $\mathbf{D}_{k,n} = diag\{\alpha_{k,n,1}, \alpha_{k,n,2}, \ldots, \alpha_{k,n,L}\}$. The small scale fading at user k in subchannel n in multipath component (MPC) is given by $\alpha_{k,n,l}$ with zero mean and variance $\sigma^2_{k,n,l}$. Assume that each MPC is i.i.d. such that $\sum_{l=1}^{L}\sigma^2_{k,n,l} = 1$. We can express the channel model in (19) as

$$\mathbf{H}_{k,n} = \sqrt{\nu}\mathbf{A}_{k,n}\mathbf{\Sigma}_{k,n}\bar{\mathbf{D}}_{k,n}\mathbf{B}_{k,n}^H, \tag{20}$$

where $\mathbf{\Sigma}_{k,n} = diag\{\sigma^2_{k,n,1}, \sigma^2_{k,n,2} \ldots, \sigma^2_{k,n,L}\}$ and $\bar{\mathbf{D}}_{k,n} = diag\{\bar{\alpha}_{k,n,1}, \bar{\alpha}_{k,n,2} \ldots, \bar{\alpha}_{k,n,L}\}$ with $\bar{\alpha}_{k,n,l} = \frac{\alpha_{k,n,l}}{\sigma_{k,n,l}}$ such that $E\{\bar{\alpha}_{k,n,l}\} = 0$ and $E\{\bar{\alpha}^2_{k,n,l}\} = 1$.

Substituting (20) in (13) and averaging over small scale fading, we get the transmit and receive correlation matrices for user k in the subchannel n as

$$\mathbf{R}_{tx,k,n} = \nu\mathbf{B}_{k,n}\mathbf{\Sigma}^2\mathbf{B}_{k,n}^H, \tag{21}$$

$$\mathbf{R}_{rx,k,n} = \nu\mathbf{A}_{k,n}\mathbf{\Sigma}^2\mathbf{A}_{k,n}^H, \tag{22}$$

For mmWave massive MIMO systems with large number of antennas, the steering vectors are asymptotically orthogonal to each other [6]:

$$\begin{aligned} &\mathbf{a}^H(\phi_{k,n})\mathbf{a}(\phi_{k,n'}) \approx 0, \\ &\mathbf{b}^H(\theta_{k,n})\mathbf{b}(\theta_{k,n'}) \approx 0. \end{aligned} \tag{23}$$

Moreover, in mmWave massive MIMO, acquisition of the instantaneous full CSI is not practical. Instead, an average CSI in terms of $[\mathbf{A}_{k,n}]$, $[\mathbf{B}_{k,n}]$, and $[\Sigma_{k,n}]$ is a practical solution for the beamforming design because the coherence time of the channel statistics based CSI is of the order of few seconds or more as compared to the small scale of the order of milli-second [6].

2.2. Problem Formulation

The hybrid beamforming divides the beamforming matrix into two parts: covariance-based pre-beamforming matrix $\mathbf{F}^{AB}$ realized by analog beamformers and the reduced dimension MU-MIMO digital precoding based on the effective channel $\mathbf{H}^H\mathbf{F}^{AB}$ (omitting the subchannel subscript for simplicity). We assume that K users are divided into G groups, such that, the group g contains K_g number of users. Since users are near the ground level and surrounded by the scatterers compared to the scatterer-free elevated base-station, we assume one-ring model [1] and all users in group g

experience the same azimuth center angle (θ_g) and angular spread (Δ_g). In this case, $\mathbf{R}_{rx} = \mathbf{I}$ in (12), therefore, the channel covariance matrix of each user in group g is given by [30]

$$\mathbf{R}_g = \mathbb{E}\{\mathbf{h}_{g_k}\mathbf{h}_{g_k}^H\} \tag{24}$$

for which the eigenvalue decomposition gives

$$\mathbf{R}_g = \mathbf{U}_g\mathbf{\Lambda}_g\mathbf{U}_g^H \tag{25}$$

where $\mathbf{U}_g \in \mathbb{C}^{N_t \times r_g}$ is a tall unitary matrix ($\mathbf{U}_g\mathbf{U}_g^H = \mathbf{I}$) comprises the eigenvectors of $\mathbf{R}_g$ and $\mathbf{\Lambda}_g \in \mathbb{R}^{r_g \times r_g}$ is diagonal matrix with r_g nonzeros positive eigenvalues along the diagonal. The $i,j-th$ element of covariance matrix $\mathbf{R}_g$ represents the correlation between the channel coefficients antenna element i and j as

$$[\mathbf{R}_g]_{i,j} = \frac{1}{2\Delta_g}\int_{\theta_g-\Delta_g}^{\theta_g+\Delta_g} e^{-j2\pi\frac{d}{\lambda}(i-j)\sin\theta}d\theta, \tag{26}$$

where d is the distance between antenna elements of ULA and λ is the wavelength of carrier frequency. Using the Karhunen-Loeve representation, the channel vector of user k in group g is given as

$$\mathbf{h}_{g_k} = \mathbf{U}_g\mathbf{\Lambda}_g^{1/2}\mathbf{z}_g = \mathbf{U}_g\tilde{\mathbf{h}}_{g_k} \tag{27}$$

where $\mathbf{z}_g \in \mathbb{C}^{N_t \times 1} \sim \mathcal{CN}(\mathbf{0}, \mathbf{I}_{r_g})$ and $\tilde{\mathbf{h}}_{g_k}$ is beam domain channel. For large N_t, $\mathbf{U}_g$ tends to discrete Fourier transform (DFT) matrix $\Delta_{N_t} \in \mathbb{C}^{N_t \times N_t}$ [31]. Each column of $\mathbf{U}_g$ represents one direction of angle-of-departure (AoD), i.e., a *beam*.

Alternatively, for the case, when dominant eigenvalues $\hat{r}_g \leq r_g$, then, the channel matrix can be written as ([13], Equation (5))

$$\mathbf{h}_{g_k} = N_t^{1/2}\mathbf{R}_g^{1/2}\mathbf{z}_{g_k} \tag{28}$$

The limited feedback-based hybrid beamforming consists of analog pre-beamforming matrix $\mathbf{F}_g^{AB} \in \mathbb{C}^{N_t \times N_{RF,g}}$ responsible for spatial group formation and inter-group interference mitigation; and the digital multi-users precoding matrix $\mathbf{F}_g^{DB} \in \mathbb{C}^{N_{RF,g} \times S_g}$ for spatial multiplexing inside the group and inter-user interference mitigation. Here, $N_{RF,g}$ is the number of RF chains for group g such that $S_g < N_{RF,g} < \hat{r}_g$ and S_g is the number of multi-carrier information symbols vectors for group g with $N_{RF} = \sum_{g=1}^{G} N_{RF,g}$ and $S = \sum_{g=1}^{G} S_g$. The overall analog pre-beamforming matrix $\mathbf{F}^{AB} \in \mathbb{C}^{N_t \times N_{RF}}$ is given by

$$\mathbf{F}^{AB} = [\mathbf{F}_1^{AB}, ..., \mathbf{F}_G^{AB}], \tag{29}$$

and the overall digital beamforming matrix $\mathbf{F}^{DB} \in \mathbb{C}^{N_{RF} \times N_s}$ is given by

$$\mathbf{F}^{DB} = diag[\mathbf{F}_1^{DB}, ..., \mathbf{F}_G^{DB}] \tag{30}$$

and the overall channel matrix

$$\mathbf{H} = [\mathbf{H}_1, ..., \mathbf{H}_G], \tag{31}$$

where the channel matrix of group g is defined as $\mathbf{H}_g \triangleq [\mathbf{h}_{g_1}, ..., \mathbf{h}_{gK_g}]$.

The analog pre-beamforming $\mathbf{F}_g^{AB}$ is based on the slowly varying channel covariance matrix $\mathbf{R}_g$ and can be implemented by the DFT matrix (when N_t is large), whereas, the digital beamformer $\mathbf{F}_g^{DB}$ is based on the instantaneous channel information of the reduced dimension effective channel $\mathbf{H}_g^H\mathbf{F}_g^{AB}$. The overall effective channel is given by

$$\mathbf{H}_{eff}^{H} = \mathbf{H}^{H}\mathbf{F}^{AB} = \begin{bmatrix} \mathbf{H}_1^H\mathbf{F}_1^{AB} & \mathbf{H}_1^H\mathbf{F}_2^{AB} & \cdots & \mathbf{H}_1^H\mathbf{F}_G^{AB} \\ \mathbf{H}_2^H\mathbf{F}_1^{AB} & \mathbf{H}_2^H\mathbf{F}_2^{AB} & \cdots & \mathbf{H}_2^H\mathbf{F}_G^{AB} \\ \vdots & \vdots & \ddots & \vdots \\ \mathbf{H}_G^H\mathbf{F}_1^{AB} & \mathbf{H}_G^H\mathbf{F}_2^{AB} & \cdots & \mathbf{H}_G^H\mathbf{F}_G^{AB} \end{bmatrix} \tag{32}$$

The excessive pilot transmission in downlink and feedback in uplink of FDD system can be reduced by only sending the group-wise average CSI based channel estimates in uplink. This is accomplished by using the diagonal elements $\mathbf{H}_g^H\mathbf{F}_g^{AB}$ as feedback information with the size of $K_g \times N_{RF,g}$ for $g = 1, ...G$. The analog pre-beamforming is designed in such a way that the other elements of matrix (32) $\mathbf{H}_g^H\mathbf{F}_{g'}^{AB} \approx 0$ for all $g' \neq g$. This group-wise division creates virtual sectors, each group corresponds to a virtual sector [30].

The second order channel statistics-based RF beamformer $\mathbf{F}^{AB}$ remains the same across multiple coherence blocks which gives the effective instantaneous channel between BS and user k as

$$\mathbf{h}_{n,g_k,eff}^{H} \triangleq \mathbf{h}_{n,g_k}^{H}\mathbf{F}_g^{AB}, \tag{33}$$

with $\mathbf{h}_{n,g_k,eff} \in \mathbb{C}^{N_{RF,g}\times 1}$. Therefore, channel statistics-based CSI sufficiently reduces the feedback overhead on each user, otherwise, for instantaneous CSI, each user have to send the $N_t \times 1$ size of channel estimate on the uplink channel. The covariance of effective channel $\mathbf{h}_{n,g_k,eff}^{H}$ is given by using (13) as,

$$\begin{aligned} \mathbb{E}\{\mathbf{h}_{n,g_k,eff}\mathbf{h}_{n,g_k,eff}^{H}\} &= \mathbb{E}\{\mathbf{F}_g^{AB^H}\mathbf{h}_{n,g_k}\mathbf{h}_{n,g_k}^{H}\mathbf{F}_g^{AB}\} \\ &= \nu\mathbf{F}_g^{AB^H}\mathbf{B}_{k,n}\Sigma^2\mathbf{B}_{k,n}^{H}\mathbf{F}_g^{AB} \end{aligned} \tag{34}$$

The analog beamformer consists of columns of the DFT matrix, which can be easily implemented by phase shifter network. Therefore, $\mathbf{F}_{n,g}^{AB}$ can be obtained by eigenvalue decomposition of channel covariance matrix. With the group-wise hybrid beamforming, the received signal $\mathbf{y}_{g,n}$ for group g in subchannel n becomes

$$\mathbf{y}_{g,n} = \mathbf{H}_{g,n}^{H}\mathbf{F}_g^{AB}\mathbf{F}_{g,n}^{DB}\mathbf{s}_{g,n} + \mathbf{H}_{g,n}^{H}\sum_{g'\neq g}\mathbf{F}_{g'}^{AB}\mathbf{F}_{g',n}^{DB}\mathbf{s}_{g',n} + \mathbf{w}_{g,n} \tag{35}$$

and the received signal of user k in group g in subchannel n is given by

$$\begin{aligned} y_{g_k,n} = {} & \mathbf{h}_{g_k,n}^{H}\mathbf{F}_g^{AB}\mathbf{f}_{g_k,n}^{DB}\mathbf{s}_{g_k,n} + \underbrace{\mathbf{h}_{g_k,n}^{H}\sum_{k'\neq k}\mathbf{F}_g^{AB}\mathbf{f}_{g_{k'},n}^{DB}\mathbf{s}_{g_{k'},n}}_{\text{Inter-user interference}} \\ & + \underbrace{\mathbf{h}_{g_k,n}^{H}\sum_{g'\neq g}\sum_{j}\mathbf{F}_{g'}^{AB}\mathbf{f}_{g'_j,n}^{DB}\mathbf{s}_{g'_j,n}}_{\text{Inter-group interference}} + w_{g_k,n} \end{aligned} \tag{36}$$

The received signal to interference and noise ratio (SINR) at the user k in group g and subchannel n is given by

$$SINR_{g_k,n} = \frac{|\mathbf{h}_{g_k,n}^{H}\mathbf{F}_g^{AB}\mathbf{f}_{g_k,n}^{DB}|^2}{|\mathbf{h}_{g_k,n}^{H}|^2\left(\sum_{k'\neq k}|\mathbf{F}_g^{AB}\mathbf{f}_{g_{k'},n}^{DB}|^2 + \sum_{g'\neq g}\sum_j|\mathbf{F}_{g'}^{AB}\mathbf{f}_{g'_j,n}^{DB}|^2\right) + \sigma^2} \tag{37}$$

The spectral efficiency of user k in group g and subchannel n is expressed as

$$R_{g_k,n} = \Psi_{g_k,n}\log_2(1 + SINR_{g_k,n}), \tag{38}$$

where $\Psi_{g_k,n}$ is the binary variable such that it is equal to 1 if user k is selected in group g in the subchannel n. In order to achieve balance tradeoff between throughput and fairness [32], we use proportional fairness (PF) based throughput maximization. We define per user proportional fairness metric as

$$U(\mathbf{F}_g^{AB}, \mathbf{f}_{g_k,n}^{DB}) = \frac{R_{g_k,n}(t)}{\bar{R}_{g_k,n}(t)}, \forall n, g_k, \tag{39}$$

where $\bar{R}_{g_k,n}(t)$ is average throughput (moving average) over a past window of length $T_w = 1/\alpha$ [33], as

$$\bar{R}_{g_k,n}(t) = \alpha R_{g_k,n}(t-1) + (1-\alpha)\bar{R}_{g_k,n}(t-1), \tag{40}$$

The large number of antennas in massive MIMO systems enable the use of the eigenmodes of the channel covariance matrix, i.e., $\mathbf{B}_{k,n}$ comprises of the columns of the DFT matrix [6]. DFT-based beams with $N_t = 16$ and $N_t = 64$ are shown in Figure 4a,b, respectively.

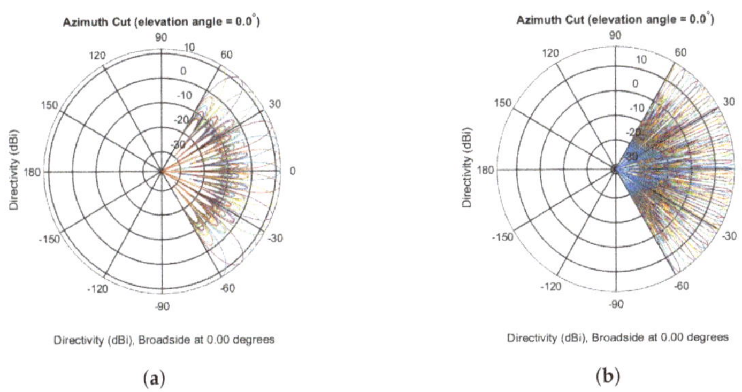

Figure 4. DFT-based beams in a 120 sector. (**a**) DFT-based beams in a 120 sector with $N_t = 16$; (**b**) DFT-based beams in a 120 sector with $N_t = 64$.

The beam steering matrix $\mathbf{B}_{k,n}$ consists of selected columns of $N_t \times N_t$ DFT matrix Δ_{N_t} such that

$$\mathbf{F}_g^{AB} = \mathbf{B}_{k,n} = \Delta_{N_t}\mathbf{Y}_n \tag{41}$$

where $\Delta_{N_t} = [\mathbf{b}_1, \mathbf{b}_2, ..., \mathbf{b}_{N_t}]$ consisting of all eigenmodes and $\mathbf{Y}_n$ is an $N_t \times r_R$ binary beam selection matrix, with r_R is the rank of the channel covariance matrix. The selection matrix $\mathbf{Y}_n \in \mathbb{C}^{N_t \times r_R}$ with only a single one on each row and column such that $\sum_i [\mathbf{Y}_n]_{i,j} = 1 \quad \forall j$. Now we formulate our optimization problem for joint spatio–radio resource allocation and precoders design with the objective to maximize the utility function as

$$\begin{aligned}
\max_{\mathbf{F}_g^{AB}, \mathbf{f}_{g_k,n}^{DB}} \quad & \sum_{g=1}^{G}\sum_{n=1}^{N_f}\sum_{k=1}^{K_g} U\left(\mathbf{F}_g^{AB}, \mathbf{f}_{g_k,n}^{DB}\right) \\
\text{subject to} \quad & \\
C1: \quad & tr(\mathbf{F}_n^{DB^H}\mathbf{F}^{AB^H}\mathbf{F}^{AB}\mathbf{F}_n^{DB}) \leq P_n, \quad \forall n \\
C2: \quad & rank(\mathbf{F}^{AB}\mathbf{F}^{DB}) \leq N_{RF}, \\
C3: \quad & \sum_{k=1}^{K_g} \Psi_{k,n} \leq K_g, \quad \forall g, n, \\
C4: \quad & tr(\mathbf{Y}_n) \leq G, \quad \forall n, \\
C5: \quad & [\Psi_{k,n}], [\mathbf{Y}_n] \in \{0,1\}.
\end{aligned} \tag{42}$$

The above optimization problem is a mixed integer programming (MIP) problem with coupling between the digital and RF precoders in the power constraint. This MIP problem is NP-hard [14].

3. Relaxed-Convex Transformation

Though the above MIP optimization problem is NP-hard, it can be transformed to a relaxed convex optimization problem by (i) relaxing the binary integer constraints to real number between 0 and 1 [14], and (ii) decoupling the digital and analog precoders. For decoupling purpose, we make use of change of variables $\mathbf{F}_n^{DB} = (\mathbf{F}^{AB^H}\mathbf{F}^{AB})^{-\frac{1}{2}}\tilde{\mathbf{F}}_n^{DB}$, where $\tilde{\mathbf{F}}_n^{DB}$ is the *equivalent* digital precoder [34]. Thus, the problem in (42) can be written as

$$\begin{aligned}
\max_{\mathbf{F}_g^{AB},\tilde{\mathbf{f}}_{g_k,n}^{DB}} \quad & \sum_{g=1}^{G}\sum_{n=1}^{N_f}\sum_{k=1}^{K_g} \tilde{U}_{g_k,n}\left(\mathbf{F}_g^{AB},\mathbf{f}_{g_k,n}^{DB}\right) \qquad (43)\\
\text{subject to} \quad & \\
C1: \quad & tr(\tilde{\mathbf{F}}_n^{DB^H}\tilde{\mathbf{F}}_n^{DB}) \le P_n, \quad \forall n\\
C2: \quad & rank(\tilde{\mathbf{F}}^{DB}) \le N_{RF},\\
C3: \quad & \sum_{k=1}^{K_g}\Psi_{k,n} \le K_g, \quad \forall g,n,\\
C4: \quad & tr(\Upsilon_n) \le G, \quad \forall n,\\
C5: \quad & 0 \le [\Psi_{k,n}] \le 1, \quad 0 \le [\Upsilon_n] \le 1,
\end{aligned}$$

For a given RF precoder $\mathbf{F}^{AB}$ and the knowledge of perfect CSI at the base-station, the digital precoder can be obtained by conventional MU-MIMO techniques, e.g., the *zero-forcing* and *block diagonalization* [15].

For the digital precoder, we adopt the ZF precoder for no multiuser interference among the users in each groups. The beamforming vector of user k is chosen to be orthogonal to the effective channel vectors of all the other users in the group. Zero-forcing is a suboptimal but low complexity approach within the linear precoders' class. ZF precoder is asymptotically optimal among all downlink beamforming techniques in high SNR region. It guarantees high spectral efficiency for large-scale antennas with low-complexity linear processing [35]. For $N_t \gg N_r$, it has shown that zero-forcing beamforming can achieve up to 98% of the non-linear dirty paper coding (DPC) capacity [36]. In order to make this paper self-contained, we describe the block diagonalization briefly. Since digital precoder is used to mitigate the multiuser interference within a groups and all groups are independent, we omit the subscript g. First we consider the downlink transmission over one subchannel n with the general case of BS with N_t antennas and K_n users with n_k antennas each, such that $\sum_{k=1}^{K} n_k = N_r$. The downlink channel on the subchannel n is expressed as $N_r \times N_t$ matrix,

$$\mathbf{H}_{n,eff} = [\mathbf{H}_{1,n,eff}^T, ..., \mathbf{H}_{K_n,n,eff}^T]^T \qquad (44)$$

For user k, we define the following $(N_r - n_k) \times N_t$ channel matrix

$$\mathbf{H}'_{k,n,eff} = [\mathbf{H}_{1,n,eff}^T, ..., \mathbf{H}_{K-1,n,eff}^T, \mathbf{H}_{k+1,n,eff}^T, ..., \mathbf{H}_{K_n,n,eff}^T]^T \qquad (45)$$

Let the rank of $\mathbf{H}'_{k,n,eff}$ be denoted by $r'_{k,n}$, then the nullspace of $\mathbf{H}'_{k,n,eff}$ has dimension $N_t - r'_{k,n} \ge n_k$. Performing the SVD of each user's channel matrix in subchannel n leads to the following

$$\mathbf{H}'_{k,n,eff} = \mathbf{U}'_{k,n}\boldsymbol{\Sigma}'_{k,n}\mathbf{V}'^H_{k,n} = \mathbf{U}'_{k,n}\boldsymbol{\Sigma}'_{k,n}[\mathbf{V}'^{(1)}_{k,n}\ \mathbf{V}'^{(0)}_{k,n}]^H, \qquad (46)$$

where $\mathbf{U}'_{k,n}$ and $\mathbf{V}'_{k,n}$ are the unitary matrices. The columns of $\mathbf{U}'_{k,n}$ are the left singular vectors of $\mathbf{H}'_{k,n,eff}$, the columns of $\mathbf{V}'_{k,n}$ are the right singular vectors of $\mathbf{H}'_{k,n,eff}$, and $\boldsymbol{\Sigma}'_{k,n}$ is a diagonal matrix in which the diagonal entries are the singular values of $\mathbf{H}'_{k,n,eff}$. In the last equality of (46), $\mathbf{V}^{'(1)}_{k,n}$ holds the first $r'_{k,n}$ right singular vectors of $\mathbf{H}'_{k,n,eff}$ and $\mathbf{V}^{'(0)}_{k,n}$ contains the $N_t - r'_{k,n}$ singular vectors of $\mathbf{H}'_{k,n,eff}$ which are in the nullspace of $\mathbf{H}'_{k,n,eff}$. The columns of $\mathbf{V}^{'(0)}_{k,n}$ are best suited for user k beamforming matrix $\mathbf{F}^{DB}_{k,n}$, because they will provide zero interference at other UEs. Usually $\mathbf{V}^{'(0)}_{k,n}$ contains more number of columns than the n_k, therefore we use some linear combinations of the columns of $\mathbf{V}^{'(0)}_{k,n}$ to make at most n_k columns.

$$\mathbf{H}_{k,n,eff}\mathbf{V}^{'(0)}_{k,n} = \mathbf{U}_{k,n}\begin{bmatrix} \mathbf{\Sigma}_{k,n} & \mathbf{0} \\ \mathbf{0} & \mathbf{0} \end{bmatrix}\begin{bmatrix} \mathbf{V}^{(1)}_{k,n} & \mathbf{V}^{(0)}_{k,n} \end{bmatrix}^H, \tag{47}$$

where $\mathbf{H}_{k,n,eff}\mathbf{V}^{'(0)}_{k,n}$ gives the matrix with columns as the linear combinations of the columns of $\mathbf{V}^{'(0)}_{k,n}$. The right hand side of the equation is the SVD of $\mathbf{H}_{k,n,eff}\mathbf{V}^{'(0)}_{k,n}$, where $\mathbf{\Sigma}_{k,n}$ is $r_{k,n} \times r_{k,n}$ diagonal matrix and $\mathbf{V}^{(1)}_{k,n}$ represents the $r_{k,n}$ singular vectors with nonzero singular values of $\mathbf{H}_{k,n,eff}\mathbf{V}^{'(0)}_{k,n}$. The Equation (47) can also be written as,

$$\begin{aligned} \mathbf{H}_{k,n,eff}\mathbf{V}^{'(0)}_{k,n} &= \mathbf{U}_{k,n}\mathbf{\Sigma}_{k,n}\mathbf{V}^{(1)^H}_{k,n} \\ \mathbf{H}_{k,n,eff} &= \mathbf{U}_{k,n}\mathbf{\Sigma}_{k,n}\mathbf{V}^{(1)^H}_{k,n}\mathbf{V}^{'(0)^H}_{k,n} \end{aligned} \tag{48}$$

The transmit beamforming matrix that maximizes the user k throughput without any inter-user interference is obtained as,

$$\mathbf{F}^{DB}_{k,n} = \mathbf{V}^{'(0)}_{k,n}\mathbf{V}^{(1)}_{k,n} \tag{49}$$

The transmit digital beamforming matrix for subchannel n is defined as

$$\mathbf{F}^{DB}_{n} = [\mathbf{F}^{DB}_{1,n}, ..., \mathbf{F}^{DB}_{K_n,n}]\mathbf{P}^{1/2}_{n}, \tag{50}$$

where $\mathbf{F}^{B}_{k,n}{}^{H}\mathbf{F}^{B}_{k,n} = \mathbf{I}, \quad 1 \leq k \leq K_n$ and $\mathbf{P}_n$ is a block diagonal matrix whose elements scale the power allocated to each interference-free virtual subchannel for all UEs. The receive combining matrix for this user is $\mathbf{U}_{k,n}$ [37].

In the case of single antenna users, complete diagonalization is achieved entirely at the BS by channel inversion, i.e., $\mathbf{F}^{DB}_{n} = (\mathbf{H}^H_{n,eff})^{\dagger}$, where $(\mathbf{H}^H_{n,eff})^{\dagger}$ is the pseudo-inverse of $\mathbf{H}^H_{n,eff}$ [38].

$$\begin{aligned} \mathbf{F}^{DB}_{n} &= \beta_n \left(\mathbf{H}^H_{n,eff}\right)^{\dagger} \\ [\mathbf{f}^{DB}_{1,n}, ..., \mathbf{f}^{DB}_{K,n}] &= \beta_n \left[[\mathbf{h}_{1,n}, ..., \mathbf{h}_{K,n}]^H\mathbf{F}^{AB}\right]^{\dagger} \end{aligned} \tag{51}$$

where β_n is a normalization factor chosen to satisfy the power constraint and is given by

$$\beta^2_n = \frac{K}{||\mathbf{F}^{AB}\mathbf{F}^{DB}_{n}||^2_F} \tag{52}$$

Using the definition of the pseudo-inverse, we get,

$$\mathbf{F}^{DB}_{n} = \beta_n \left(\mathbf{H}_{n,eff}\mathbf{H}^H_{n,eff} + N_{RF}\varrho\mathbf{I}_{N_{RF}}\right)^{-1}\mathbf{H}_{n,eff} \tag{53}$$

where ϱ is the regularization parameter, $\varrho = 0$ for ZF precoding and $\varrho = \frac{N_s}{N_{RF}\eta}$ for regularized ZF, with $\eta = P_{T,n}/\sigma^2$. Lastly, introducing the group subscript again, the SINR of user g_k is given by

$$SINR_{g_k,n} = \frac{\frac{P_{T,n}}{K_g}\beta^2|\mathbf{h}_{g_k,n}^H \mathbf{F}_g^{AB}\mathbf{F}_{g,n}^{DB}\mathbf{F}_{g,n}^{DB^H}\mathbf{F}_g^{AB^H}\mathbf{h}_{g_k,n}|^2}{|\mathbf{h}_{g_k,n}^H|^2\left(\sum_{k'\neq k}|\mathbf{F}_g^{AB}\mathbf{f}_{g_{k'},n}^{DB}|^2 + \sum_{g'\neq g}\sum_j |\mathbf{F}_{g'}^{AB}\mathbf{f}_{g'_j,n}^{DB}|^2\right) + \sigma^2} \tag{54}$$

and the PF sum rate is calculated as

$$\hat{U} = \sum_{g=1}^{G}\sum_{n=1}^{N_f}\sum_{k=1}^{K_g}\tilde{U}_{g_k,n}. \tag{55}$$

4. Suboptimal Solution

Joint optimization of analog and digital beamformers is challenging because they use different channel information for the design of analog and digital beamformers. Hybrid beamforming methods consider decoupled designs of analog and digital beamforming to reduce the complexity of joint optimization, but the main challenge remains the use of different channel information. To approximate the optimal solution to this mixed integer programming problem, we summarize our proposed algorithm below:

The analog precoder is formed by selecting K_g columns of DFT matrix of eigenvectors of channel covariance matrix $\mathbf{R}_g$ of users' group g in (41) to minimize the inter-group interference $\mathbf{I}_g$,

$$\begin{aligned} \min_{\mathbf{Y}} \quad & \mathbf{I}_g(\mathbf{Y}) \\ \text{subject to} \quad & \\ & tr(\mathbf{Y}_n) \leq G, \quad \forall n, \end{aligned} \tag{56}$$

where $\mathbf{I}_g = \sum_{i=1}^{G}\sum_{\substack{j=1 \\ j\neq i}}^{G} |\mathbf{H}_i^H \mathbf{F}_j^{AB}|^2$. To solve the MIP problem, we divide the solution into two parts. In the first part, we get the analog precoder using the selected columns of the DFT matrix which maximize the PF sum-rate. The inherent benefit of the DFT matrix is its constant modulus which enables the use of analog phase shifters and RF switches to realize the analog beamforming. In the second part, for the given analog precoder, intra-group users scheduling is performed and a ZF digital precoder is designed to maximize the sum-rate utility function. The decoupling of the analog and digital precoders design makes the solution suboptimal but tractable [34]. The joint hybrid beamforming and user scheduling Algorithm 1 takes K_g, N_f, N_{RF}, N_t, and K. It generates the analog beamforming matrix $\mathbf{F}^{AB}$, digital beamforming matrix $\mathbf{F}^{DB}$, and the set of users in each group. The first part of the algorithm (line 9 to line 19) forms DFT/eigenmodes-based analog beamforming using limited statistical CSI feedback from the users. Beam and users pairing within each group is taken place in this part of the algorithm. The *while* loop at line 12 executes till all the binary combinations in $N_t \times N_t$ are exhausted with the condition that each column contain exactly one binary 1 and total number of 1s are equal to the number of streams (or number of RF chains). The second part (starts from line 39) assigns the radio resources to users to maximize the utility function.

Algorithm 1 Joint Resource Allocation and Hybrid Beamforming Design Algorithm.

Inputs
K_g: Number of users per group
N_f: Number of subchannels
K_n: Number of UEs to be scheduled in subchannel n
N_{RF}: Number of RF chains
N_t: Number of transmit antennas
Initialization
$\mathcal{K}_g = \{1, ..., K_g\}$, $U(\mathbf{F}_n^{DB})^{last} = 0$, $\mathbf{Y}_g^{last} = \mathbf{0}$, $\mathcal{K}_n = \{\}$, $\forall n$
{Average CSI-based Beam Selection and RF Precoder Design}
Evaluate channel covariance matrix $\mathbf{R}_g$ from CSI at users: $\mathbf{R}_g = \mathbb{E}\{\mathbf{h}_{g_k}\mathbf{h}_{g_k}^H\}$
Find transmit steering $\mathbf{B}_{k,n}$ by Eigenvector based unitary matrix $\mathbf{U}_g$ of $\mathbf{R}_g$ or DFT matrix $\mathbf{\Delta}_{N_t}$.
while $g \leq G$ **do**
    while $tr(\mathbf{Y}_g) \leq N_{RF,g}$, set $\{[Y_g]_{i,i}, ..., [Y_g]_{N_t,N_t}\} = \{0, 1\}$ **do**
        Calculate $U(\mathbf{Y}_g)$, employing zero forcing $\mathbf{F}_{n,g}^{DB}$
        if $U(\mathbf{Y}_g) \geq U(\mathbf{Y}_g^{last})$ **then**
            $\mathbf{Y}_g^{last} \leftarrow \mathbf{Y}_g$
        end if
    end while
end while
Get $\mathbf{F}_g^{AB} = \Delta_{N_t}\mathbf{Y}_g^{last}$
{Instantaneous CSI-based Users Selection and Digital Precoder Design}
while $n \leq N_f$ **do**
    while $b \leq N_{RF,g}$ **do**
        while $k \leq K_g$ **do**
            Compute $U(\mathbf{F}_n^{DB}) = U(\mathbf{F}_n^{DB}) + U(\mathbf{f}_{k',n}^{DB}), \quad \forall k' \in \mathcal{K}_g$
            $k'^* = \arg\max_{k'}\{U(\mathbf{F}_n^{DB})\}$
            $\Psi_{k'^*,n} = 1$
            Update $\mathbf{F}_n^{DB}$ and $U(\mathbf{F}_n^{DB})^{update}$ with user k'^*
            if $U(\mathbf{F}_n^{DB})^{updated} \geq U(\mathbf{F}_n^{DB})^{last}$ **then**
                $\mathcal{K}_n = \mathcal{K}_n \cup \{k'^*\}$
                $\mathcal{K}_g = \mathcal{K}_g - \{k'^*\}$
                $U(\mathbf{F}_n^{DB})^{last} = U(\mathbf{F}_n^{DB})^{updated}$
                $k{+}{+}$
            else
                break
            end if
        end while
        $b{+}{+}$
    end while
    $n{+}{+}$
end while
Stack the beamforming matrices $\mathbf{F}^{DB} = [\mathbf{F}_1^{DB}, ..., \mathbf{F}_{N_f}^{DB}]$
Output
$\mathbf{F}^{AB} = [\mathbf{F}_1^{AB}, ..., \mathbf{F}_G^{AB}]$
$\mathbf{F}^{DB} = [\mathbf{F}_1^{DB}, ..., \mathbf{F}_{N_f}^{DB}]$
$[\mathcal{K}_1, ..., \mathcal{K}_{N_f}]$

The Algorithm 1 is illustrated in flowchart Figure 5.

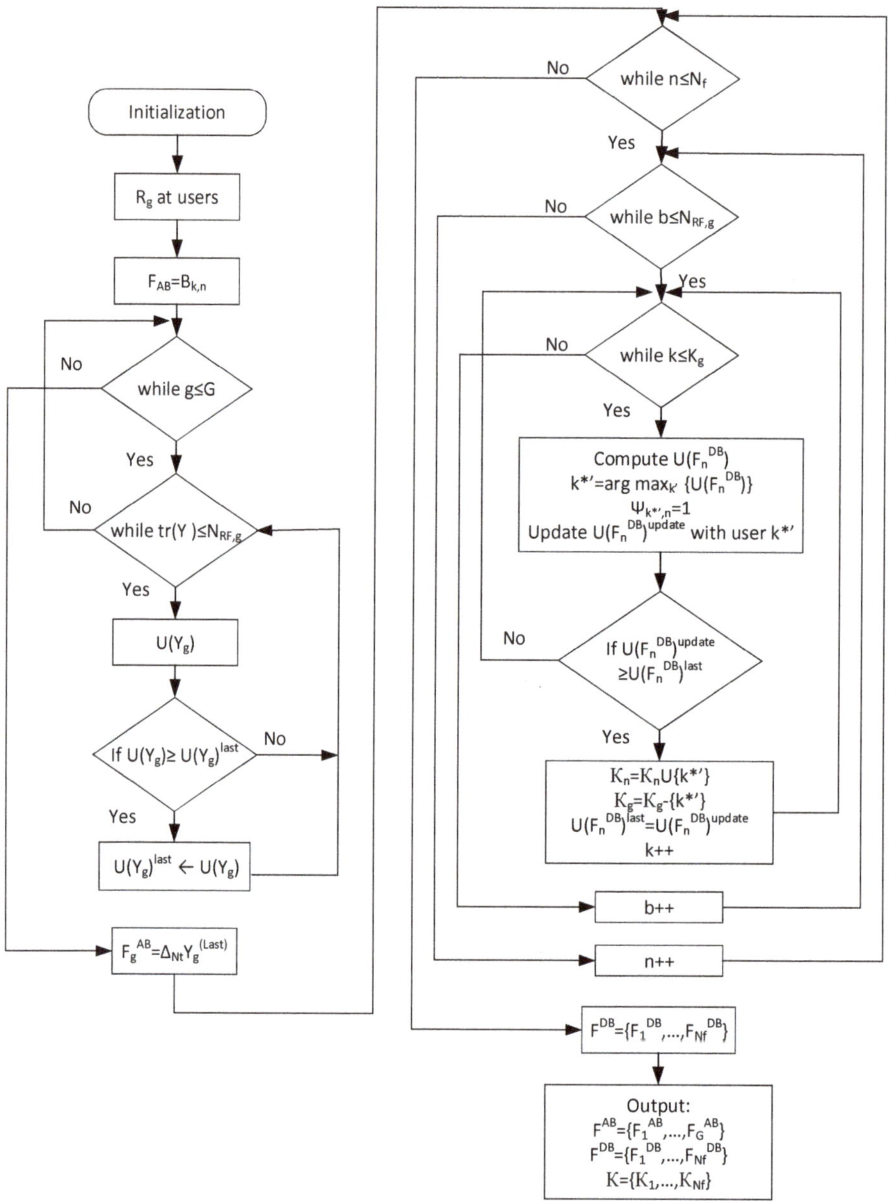

Figure 5. Flowchart illustration of Algorithm 1.

5. Machine Learning: K-Means Based Optimal Users Grouping for Analog Beamforming

In this section, we use machine learning technique to group the users. Then, the DFT based fixed switched-beams are used to realize the analog beamforming matrix. The joint users scheduling and hybrid beamforming architecture with ML-based users grouping is shown in Figure 6.

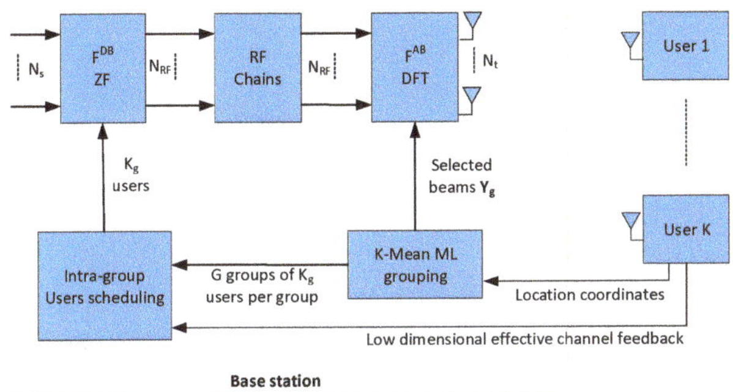

Figure 6. Joint users scheduling and hybrid beamforming architecture with ML-based users grouping.

Machine learning algorithms can broadly be divided into two main categories, namely supervised learning and unsupervised learning algorithms. The former class of algorithms learn by training on the input labeled examples, called training dataset, $\{(x^{(1)}, y^{(1)}), (x^{(2)}, y^{(2)}), (x^{(3)}, y^{(3)}), ..., (x^{(m)}, y^{(m)})\}$, where the ith example $(x^{(i)}, y^{(i)})$ consists of the ith instance of feature vector $x^{(i)}$ and the corresponding label $y^{(i)}$. Given a labeled training dataset, these algorithms try to find the decision boundary that separates the positive and negative labeled examples by fitting a hypothesis to the input dataset. Unsupervised machine learning algorithms, on the other hand, are given an unlabeled input dataset. These algorithms are used for extracting information or features from the dataset. These features might be related, but not confined, to the underlying structures or patterns in the input data, relationships in data items, grouping/clustering of data items, etc. Discovered features are meant to provide a deeper insight into the input dataset that can subsequently be exploited for achieving specific goals. Clustering algorithms make an important part of unsupervised learning where the input examples are grouped into two or more separate clusters based on some features. The K-Means (KM) algorithm, is probably the most popular clustering algorithm. It is an iterative algorithm that starts with a set of initial centroids given to it as input. During each iteration, it performs the following two steps.

1. **Assign Cluster:** For every user, the algorithm computes the distance between the user and every centroid. The user is then associated to the cluster with the closest centroid. During this step, a user might change its association from one cluster to another one.
2. **Recompute centroids:** Once all users have been associated to their respective cluster, the new position of centroid for every cluster is then calculated.

Figure 7a depicts how the cluster centroids keep moving across iterations until the system stabilizes for an example network consisting of thirty users being grouped in five clusters. The system becomes stable in only five iterations and the final cluster layout is shown in Figure 7b.

Let us define the following notations to be used later in this section.

$$K = \text{Total number of clusters being formed.}$$

$$x^{(i)} = \text{Location coordinates of user } u^{(i)}. \text{ In our case, } x^{(i)} \in \mathbb{R}^2$$

$$c^{(i)} = \text{Cluster to which the user } u^{(i)} \text{ is currently associated.}$$

$$\mu_k = \text{Centroid of } k^{th} \text{ cluster, } \mu_k \in \mathbb{R}^2$$

$$\mu_{c^{(i)}} = \text{Centroid of the cluster to which the user } u^{(i)} \text{ is currently associated.}$$

Now the cost function J can be defined as

$$J(c^{(1)}, c^{(2)}, ..., c^{(m)}, \mu_1, \mu_2, \ldots, \mu_K) = \frac{1}{m}\sum_{i=1}^{m} ||x^{(i)} - u_{(c^i)}||^2 \tag{57}$$

with the following optimization objective function.

$$\min_{c^{(1)},,...,c^{(m)},\mu_1,...,\mu_K} \quad J(c^{(1)}, c^{(2)}, ..., c^{(m)}, \mu_1, \mu_2, \ldots, \mu_K)$$

It may be pointed out that Equation (57) allows us to compare multiple clustering layouts based on their cost and select the one with the lowest cost. The above optimization objective function constitutes a non-convex and NP-hard problem because it has many possible local minima and integer optimization variable $c^{(i)}$. The KM algorithm heuristically optimize this function by alternate minimization method. It iterates between two steps (Assign cluster and Recompute centroids) as described above.

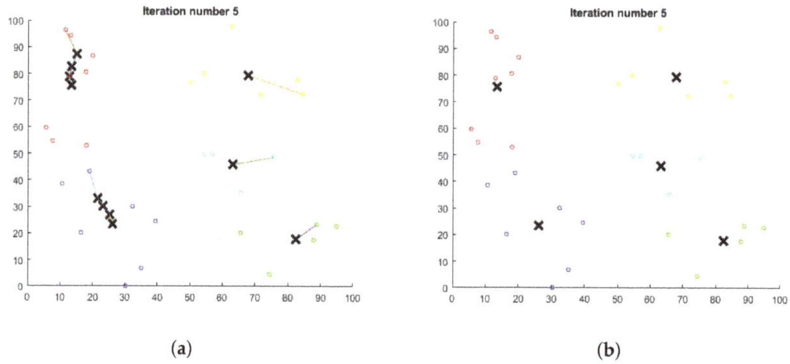

Figure 7. Change in position of centroids as K-Means clustering algorithm progresses. (**a**) shows the transition of cluster centroids (shown as crosses) up to iteration 5, whereas, (**b**) shows only the final stable state after iteration 5. In the figures, cross-sign represents cluster centroids and the colored-circle-sign represents the user associated with the same group or cluster.

In this section, we use the KM algorithm for optimal clustering of m users competing for resources in a particular cell. The clustering is performed based on their geographic location, thus our input dataset $\{u^{(1)}, u^{(2)}, u^{(3)}, ..., u^{(m)}\}$ has m vectors $u^{(i)}, 1 \leq i \leq m$, consisting of location coordinates, of ith user. For the sake of simplicity, we assume these users are deployed in a two dimensional area, i.e., a plane and so $u^{(i)} = (x_1^{(i)}, x_2^{(i)})$, i.e., an ordered pair of location coordinates. Our clustering algorithm is summarized in Algorithm 2.

The proposed algorithm takes the location coordinates of m users as input. It also takes two numbers min_k and max_k as additional input. The algorithm outputs the best number of clusters, k, such that $min_k \leq k \leq max_k$, and corresponding members of each cluster. It starts with $k = min_k$ and randomly selects k user locations as the initial centroids (line 6). It assigns the closest centroid to each user (line 8) and then computes new centroids by calculating the center/average location of all nodes in each cluster (line 11). So, in effect, the location of centroids keeps moving in successive iterations. It repeats the above two steps until the change in centroid positions is zero or negligible. We repeat the test max_t times with a new set of randomly chosen initial centroids every time. During every test, the discovered centroids, corresponding centroid assignment to users, and the cost are saved (lines 14–16) for later comparison. After running the loop for max_t times, we select and store the best k centroids resulting from the test with the lowest cost while discarding the remaining (lines 19–21).

The same is repeated for the next value of k, i.e., $k = k + 1$, until $k > max_k$. At the end we have $cnt = max_k - max_k$ vectors μ_k, one for each value of k, the corresponding assignment vector $\mathbf{a}_k$ and cost c_k. Finally, we choose the vector $\boldsymbol{\mu}$ having the lowest cost and corresponding assignment vector $\mathbf{a}$ among cnt stored cases. That is the best number of clusters and corresponding centroids that the algorithm found.

Algorithm 2 K-Means based users grouping algorithm.

$cnt = 0$
for $k = min_k max_k$ **do**
    $cnt = cnt + 1$
    for $t = 1 : max_t$ **do**
        repeat
            Randomly choose initial k centroids $\mu_1, \mu_2, \mu_3, ..., \mu_k$
            for $i = 1 : m$ **do**
                $a^{(i)} = j, \quad 1 \le j \le k$, such that μ_j is the centroid closest to $u^{(i)}$
            end for
            for $l = 1 : k$ **do**
                $\mu_l =$ mean of all users/points $u^{(i)}$ assigned to lth centroid
            end for
        until converges
        $\boldsymbol{\mu}^{(t)} = (\mu_1, \mu_2, \mu_3, \ldots, \mu_k)$
        $\mathbf{a}^{(t)} = (a^{(1)}, a^{(2)}, a^{(3)}, \ldots, a^{(m)})$
        $c^{(t)} = cost(\mu_1, \mu_2, \mu_3, \ldots, \mu_k)$
    end for
    $idx = \arg_{min}\{c(t), 1 \le t \le max_t\}$
    $\boldsymbol{\mu}_{\mathbf{k}}{}^{(k)} = \boldsymbol{\mu}^{(idx)}, 1 \le idx \le max_t$
    $\mathbf{a}_{\mathbf{k}}{}^{(k)} = \mathbf{a}^{(idx)}, 1 \le idx \le max_t$
    $c_k^{(k)} = c^{(idx)}, 1 \le idx \le max_t$
end for
$index = \arg_{min}\{c_k^{(k)}, 1 \le k \le cnt\}$
$\boldsymbol{\mu} = \boldsymbol{\mu}_{\mathbf{k}}{}^{(index)}, 1 \le index \le cnt$
$\mathbf{a} = \mathbf{a}_{\mathbf{k}}{}^{(index)}, 1 \le index \le cnt$
$n = index$

After the groups formation, BS sends this information to all users, where users use this information to form reduced average statistical CSI. For example, a user in a group of 5, needs to send the average statistical CSI only after 1/5 of regular feedback interval time.

6. Simulation Results

Consider the downlink of a multiuser massive MIMO single cell with three 120 degree sectors. We neglect inter-sector interference and focus on a single 120 degree sector served by a ULA of $N_t = 64$ isotropic antennas at BS. The users grouping forms virtual sectors inside 120 physical sector.

In simulation, the results are obtained by averaging over 100 drops. In each drop we randomly generate spatial correlation matrices $\mathbf{R}_g$. For each realization of spatial correlation matrix $\mathbf{R}_g$, we simulate 1000 realizations of instantaneous channel $\mathbf{H}$.

The joint spatio–radio scheduling and hybrid precoder scheme first forms the users groups and then selects the beams that maximizes the sum-rate through downlink training process. Secondly, it calculates the ZF based digital precoder using low dimensional effective channel feedback from the users.

Figure 8 shows the CDF of the non-zero eigenvalues of channel covariance matrix. Notice that approximately 50% of the non-zero eigenvalues are close to zero. The sum-rate increases as the number of groups increases at the cost of increased feedback overhead as shown in Figure 9. Using machine learning technique in Section 5 we can get optimal number of groups from channel covariance feedback. This results in increased sum-rate with substantial reduced feedback. The optimal $G = 3$ gives 27.6% increase in sum-rate compared to when $G = 1$ and 62.5% decrease in feedback overhead compared to $G = 8$. The comparison of performance of ML-based users grouping with previous work cannot be provided because there is no previous work that uses ML-based technique to reduce the CSI feedback overhead in massive MIMO systems. Many papers use users grouping in massive MIMO hybrid beamforming [3,5,39,40], but they do not utilize ML-based users grouping. Therefore, we have compared our proposed solution with two benchmarks of full-CSI ($G = K$) and coarse-CSI ($G = 1$). Figure 10 shows sum-rate with number of users at $10dB$ SNR. For a fixed number of groups $G = 3$, the increase in number of users, increases number of users per group. Due to the fixed number of groups, the feedback overhead remains the constant. Sum-rate is increasing with users because we assumed $N_{RF} = N_s = K$. If we fix the number of RF chains to some hardware limit, then the sum-rate will saturate at specific number of users. It can be seen in Figure 10, that increasing number of users per group decreases the slop of the sum-rate for limited CSI schemes. This decrease is due the increase in intra-group interference.

Sum-rate also depends on the number of RF chains but this dependence is not linear as shown in Figure 11. This figure shows sum-rate variation with number of RF chains N_{RF} when $N_s = 8$, $K = 8$, $N_t = 64$, and $SNR = 10$ dB. Sum rate increases with number of RF chains because it yields better conditioned effective channel matrix. It can be seen that the spectral efficiency does not increase monotonically with N_{RF} and saturates at $N_{RF} = N_t$ where hybrid precoding is turned to the pure digital precoding. The increase in spectral efficiency with the number of RF chains comes at the cost of higher dimensional effective channel feedback overhead and power consumption in RF chains.

The spectral efficiency of the proposed scheme also varies with number of transmit antennas as shown in Figure 12. In the figure, $N_{RF} = N_s = K = 8$, $SNR = 10$ dB, and BS has 16, 64, 128 or 256 ULA antennas. The performance gain increases with the increase in number of transmit antennas because large antennas array increases the resolution of the transmit beams (also depicts in figure 4) and, hence, decreases the potential of inter-beams interference.

In general, the spectral efficiency is a function of SNR and for the $SNR = 10$ dB, our ML-based users grouping and hybrid beamforming scheme gives 27.6% increased sum-rate at the cost of 33.3% extra feedback overhead as compared to the coarse-CSI case (G = 1). Our proposed scheme incurs 62.5% reduced feedback at the cost of 25.2% reduction in sum-rate as compared to the full-CSI case (G = K).

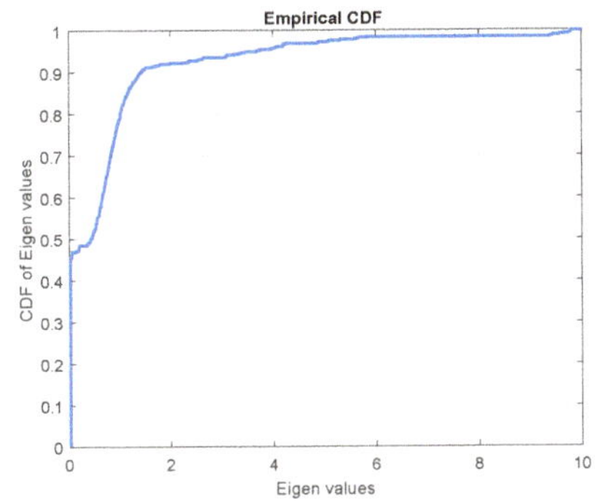

Figure 8. CDF of non-zero eigenvalues of channel covariance matrix $\mathbf{R}_g$ for $N_t = 64$.

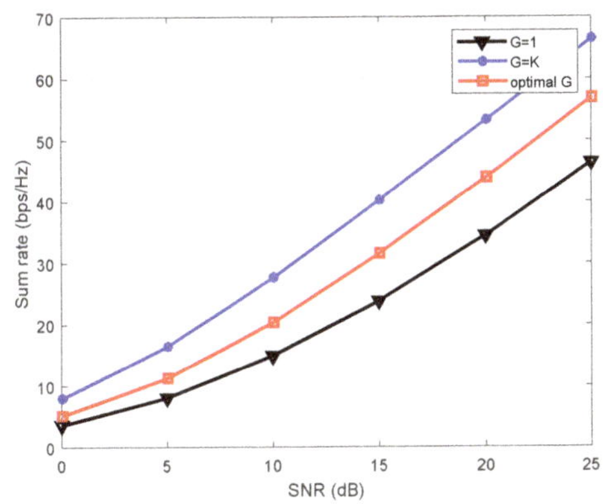

Figure 9. Sum-rate Vs SNR with different number of groups, $K = 8$.

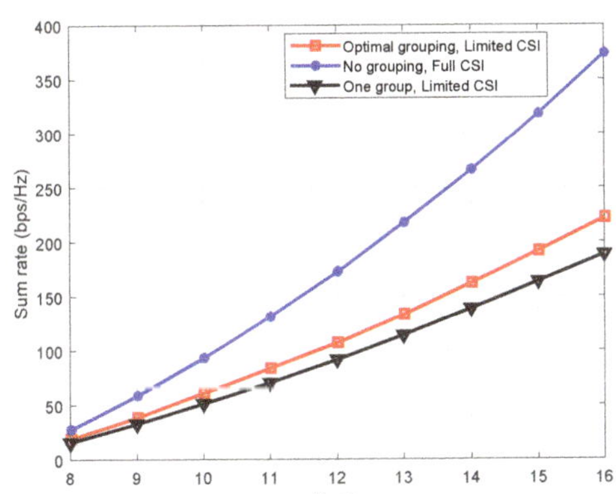

Figure 10. Sum-rate Vs number of users with different number of groups and CSI, $SNR = 10$ dB.

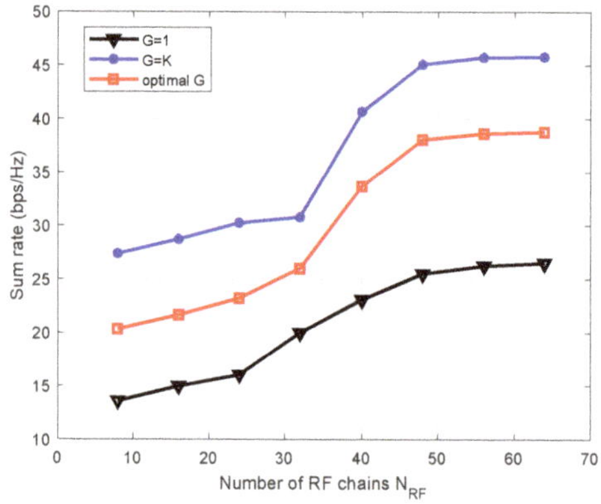

Figure 11. Sum-rate Vs number of RF chains. Number of users $K = 8$, number of transmit antennas $N_t = 64$, and $SNR = 10$ dB.

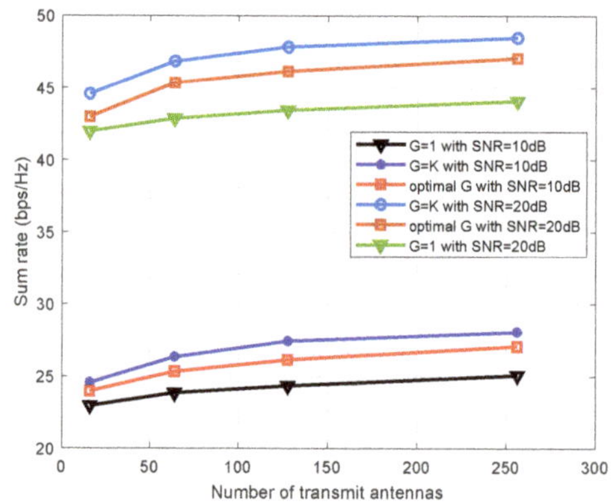

Figure 12. Sum-rate Vs number of transmit antennas. The number of RF chains at BS is fixed at 8, number of users is 8 and the results are obtained for $SNR = 10$ dB and $SNR = 20$ dB.

7. Conclusions

This paper studied the limited feedback two-stage hybrid beamformimg for decomposing the precoding matrix at the base-station. The huge channel state information feedback of massive MIMO has been reduced by the channel covariance-based RF precoding and beam selection. The well-known regularized block diagonalization can mitigate the inter-group interference, but requires substantial feedback. We used K-mean algorithm based unsupervised machine learning technique for users grouping and channel covariance-based eigenmodes/discrete Fourier transforms to reduce the feedback overhead and designed a simplified analog precoder. The digital precoder is designed with joint optimization of intra-group user utility function. It has been shown that more than 50% feedback overhead is reduced by the eigenmodes-based analog precoder design. The spatio–radio resources scheduling and limited feedback-based hybrid precoding increases the sum-rate by 27.6% compared to the sum-rate of one-group case at the cost of 33.3% extra feedback overhead, and reduces the feedback overhead by 62.5% at the cost of 25.2% reduction in sum-rate, compared to the full CSI feedback.

Author Contributions: H.K. and I.A., contributed the key idea and defined the main system model. G.B. and M.A. assisted with the system model and the mathematical analysis. I.A. developed the problem formulation.

G.B. and H.K. developed the model for users grouping. H.K., I.A., G.B. and M.A. analysed all the results and added the relevant discussions. All authors contributed to the paper write up.

Funding: This research was supported by King Abdul Aziz City for Science and Technology Project under Grant PC-37-66.

Conflicts of Interest: The authors declare no conflict of interest.

Abbreviations

The following abbreviations are used in this manuscript:

ADC	Analog-to-digital converter
AWGN	Additive white Gaussian noise
BS	Base station
CDF	Cumulative distribution function
CSI	Channel state information
DAC	Digital-to-analog converter
DFT	Discrete Fourier transforms
DoA	Direction of arrival
DoD	Direction of departure
DPC	Dirty paper coding
FDD	Frequency division duplex
HBwS	Hybrid beamforming with switches
LOS	Line-of-sight
MIMO	Multiple-input multiple-output
MIP	Mixed integer programming
ML	Machine learning
MMSE	Minimum mean squared error
MPCs	Multi path components
MRT/MRC	Maximum ratio transmission/Maximum ratio combining
MU-MIMO	Multi-User MIMO
OFDM	Orthogonal frequency division multiplexing
RBD	Regularized block diagonalization
RF	Radio frequency
SC	Subspace construction
SINR	Signal-to-interference-plus-noise ratio
SVD	Singular value decomposition
SVM	Support vector machine
TDD	Time division duplex
UE	User equipment
ULA	Uniform linear array

References

1. Adhikary, A.; Nam, J.; Ahn, J.Y.; Caire, G. Joint spatial division and multiplexing-The large-scale array regime. *IEEE Trans. Inf. Theory* **2013**, *59*, 6441–6463. [CrossRef]
2. Kutty, S.; Member, S.; Sen, D. Beamforming for millimeter wave communications: An inclusive survey. *IEEE Commun. Surv. Tutor.* **2016**, *18*, 949–973. [CrossRef]
3. Nam, J.; Adhikary, A.; Ahn, J.Y.; Caire, G. Joint Spatial Division and Multiplexing: Opportunistic Beamforming, User Grouping and Simplified Downlink Scheduling. *IEEE J. Sel. Top. Signal Process.* **2014**, *8*, 876–890. [CrossRef]
4. Adhikary, A.; Al Safadi, E.; Samimi, M.; Wang, R.; Caire, G.; Rappaport, T.; Molisch, A. Joint Spatial Division and Multiplexing for Wave Channels. *IEEE J. Sel. Areas Commun.* **2014**. [CrossRef]
5. Li, Z.; Han, S.; Sangodoyin, S.; Wang, R.; Molisch, A.F. Joint Optimization of Hybrid Beamforming for Multi-User Massive MIMO Downlink. *IEEE Trans. Wirel. Commun.* **2018**, *1276*, 1–14. [CrossRef]

6. Li, Z.; Han, S.; Molisch, A.F. Optimizing Channel-Statistics-Based Analog Beamforming for Millimeter-Wave Multi-User Massive MIMO Downlink. *IEEE Trans. Wirel. Commun.* **2017**, *16*, 4288–4303. [CrossRef]
7. Ratnam, V.; Molisch, A.; Bursalioglu, O.Y.; Papadopoulos, H.C. Hybrid Beamforming with Selection for Multi-user Massive MIMO Systems. *IEEE Trans. Signal Process.* **2018**, *66*, 4105–4121. [CrossRef]
8. Liu, A.; Lau, V. Phase only RF precoding for massive MIMO systems with limited RF chains. *IEEE Trans. Signal Process.* **2014**, *62*, 4505–4515. [CrossRef]
9. Li, Z.; Han, S.; Molisch, A.F. User-Centric Virtual Sectorization for Millimeter-Wave Massive MIMO Downlink. *IEEE Trans. Wirel. Commun.* **2018**, *17*, 445–460. [CrossRef]
10. Ahmed, I.; Khammari, H.; Shahid, A.; Musa, A.; Kim, K.S.; Poorter, E.D.; Moerman, I. A Survey on Hybrid Beamforming Techniques in 5G: Architecture and System Model Perspectives. *IEEE Commun. Surv. Tutor.* **2018**, *20*, 3060–3097. [CrossRef]
11. Park, S.; Alkhateeb, A.; Heath, R.W. Dynamic Subarrays for Hybrid Precoding in Wideband mmWave MIMO Systems. *IEEE Trans. Wirel. Commun.* **2017**, *16*, 2907–2920. [CrossRef]
12. Zhu, D.; Li, B.; Liang, P. A Novel Hybrid Beamforming Algorithm with Unified Analog Beamforming by Subspace Construction Based on Partial CSI for Massive MIMO-OFDM Systems. *IEEE Trans. Commun.* **2017**, *65*, 594–607. [CrossRef]
13. Jeon, Y.; Song, C.; Lee, S.R.; Maeng, S.; Jung, J.; Lee, I. New Beamforming Designs for Joint Spatial Division and Multiplexing in Large-Scale MISO Multi-User Systems. *IEEE Trans. Wirel. Commun.* **2017**, *16*, 3029–3041. [CrossRef]
14. Jiang, Z.; Chen, S.; Zhou, S. Joint User Scheduling and Beam Selection Optimization for Beam-Based Massive MIMO Downlinks. *IEEE Trans. Wirel. Commun.* **2018**, *17*, 2190–2204. [CrossRef]
15. Ahmed, I.; Khammari, H.; Shahid, A. Resource allocation for transmit hybrid beamforming in decoupled millimeter wave multiuser-MIMO downlink. *IEEE Access* **2017**, *5*, 170–182. [CrossRef]
16. O'Shea, T.J.; Hoydis, J. An Introduction to Deep Learning for the Physical Layer. *IEEE Trans. Cogn. Commun. Netw.* **2017**, *3*, 563–575. [CrossRef]
17. Kim, K.; Lee, J.; Choi, J. Deep Learning Based Pilot Allocation Scheme (DL-PAS) for 5G Massive MIMO System. *IEEE Commun. Lett.* **2018**, *22*, 828–831. [CrossRef]
18. Wen, C.K.; Shih, W.T.; Jin, S. Deep Learning for Massive MIMO CSI Feedback. *IEEE Wirel. Commun. Lett.* **2018**, 1–4. [CrossRef]
19. Joung, J. Machine Learning-Based Antenna Selection in Wireless Communications. *IEEE Commun. Lett.* **2016**, *20*, 2241–2244. [CrossRef]
20. Wang, J.B.; Wang, J.; Wu, Y.; Wang, J.Y.; Zhu, H.; Lin, M.; Wang, J. A Machine Learning Framework for Resource Allocation Assisted by Cloud Computing. *IEEE Netw.* **2018**, *32*, 144–151. [CrossRef]
21. Calabrese, F.D.; Wang, L.; Ghadimi, E.; Peters, G.; Soldati, P. Learning Radio Resource Management in 5G Networks: Framework, Opportunities and Challenges. *arXiv* **2016**, arXiv:1611.10253.
22. Ahmed, I.; Khammari, H. Joint Machine Learning based Resource Allocation and Hybrid Beamforming Design for Massive MIMO Systems. In Proceedings of the IEEE Globecom Workshops (GC Wkshps), Abu Dhabi, UAE, 9–13 December 2018; pp. 1–6.
23. Rahman, M.S.; Josiam, K. Low complexity RF beam search algorithms for millimeter-wave systems. In Proceedings of the 2014 IEEE Global Communications Conference (GLOBECOM 2014), Austin, TX, USA, 8–12 December 2014; pp. 3815–3820. [CrossRef]
24. Almers, P.; Bonek, E.; Burr, A.; Czink, N.; Debbah, M.; Degli-Esposti, V.; Hofstetter, H.; Kyölsti, P.; Laurenson, D.; Matz, G.; et al. Survey of channel and radio propagation models for wireless MIMO systems. *Eurasip J. Wirel. Commun. Netw.* **2007**, *2007*, 019070. [CrossRef]
25. Andrews, J.G.; Bai, T.; Kulkarni, M.; Alkhateeb, A.; Gupta, A.; Heath, R.W. Modeling and Analyzing Millimeter Wave Cellular Systems. *IEEE Trans. Commun.* **2017**, *65*, 403–430. [CrossRef]
26. Ayach, O.E.; Rajagopal, S.; Abu-Surra, S.; Pi, Z.; Heath, R.W. Spatially sparse precoding in millimeter wave MIMO systems. *IEEE Trans. Wirel. Commun.* **2014**, *13*, 1499–1513. [CrossRef]
27. Sohrabi, F.; Yu, W. Hybrid digital and analog beamforming design for large-scale MIMO systems. In Proceedings of the IEEE International Conference on Acoustics, Speech and Signal Processing (ICASSP), Brisbane, Australia, 19–24 April 2015; pp. 2929–2933. [CrossRef]

28. Ayach, O.E.; Heath, R.W.; Abu-Surra, S.; Rajagopal, S.; Pi, Z. Low complexity precoding for large millimeter wave MIMO systems. In Proceedings of the IEEE International Conference on Communications (ICC), Ottawa, ON, Canada, 10–15 June 2012; pp. 3724–3729. [CrossRef]
29. Rappaport, T.S.; Gutierrez, F.; Ben-Dor, E.; Murdock, J.N.; Qiao, Y.; Tamir, J.I. Broadband millimeter-wave propagation measurements and models using adaptive-beam antennas for outdoor Urban cellular communications. *IEEE Trans. Antennas Propag.* **2013**, *61*, 1850–1859. [CrossRef]
30. Nam, J.; Ahn, J.Y.; Adhikary, A.; Caire, G. Joint spatial division and multiplexing: Realizing massive MIMO gains with limited channel state information. In Proceedings of the 2012 46th Annual Conference on Information Sciences and Systems (CISS 2012), Princeton, NJ, USA, 21–23 March 2012; Volume 1. [CrossRef]
31. Lu, C.; Wang, W.; Zhong, W.; Gao, X. User scheduling and beam allocation for massive MIMO systems with two-stage precoding. In Proceedings of the IEEE International Symposium on Personal, Indoor and Mobile Radio Communications (PIMRC), Valencia, Spain, 4–8 September 2016. [CrossRef]
32. Andrews, M. A survey of scheduling theory in wireless data networks. In *Wireless Communications*; Springer: New York, NY, USA, 2007; pp. 1–17.
33. Escudero-Garzas, J.J.; Devillers, B.; Garcia-Armada, A. Fairness-Adaptive Goodput-Based Resource Allocation in OFDMA Downlink with ARQ. *IEEE Trans. Veh. Technol.* **2014**, *63*, 1178–1192. [CrossRef]
34. Alkhateeb, A., Jr.; Heath , R.W. Frequency Selective Hybrid Precoding for Limited Feedback Millimeter Wave Systems. *IEEE Trans. Commun.* **2016**, *64*, 1801–1818. [CrossRef]
35. Kang, J.; Lee, J.H.; Choi, W. Dimension-optimized Two-Stage Precoder Design for Massive MIMO Systems with Limited Feedback. In Proceedings of the IEEE Military Communications Conference MILCOM, Los Angeles, CA, USA, 29–31 October 2018; pp. 530–535. [CrossRef]
36. Tataria, H.; Smith, P.J.; Greenstein, L.J.; Dmochowski, P.A.; Shafi, M. Performance and analysis of downlink multiuser MIMO systems with regularized zero-forcing precoding in Ricean fading channels. In Proceedings of the IEEE International Conference on Communications, Kuala Lumpur, Malaysia, 22–27 May 2016; pp. 1–7.
37. Ho, W.; Liang, Y. Optimal Resource Allocation for Multiuser MIMO-OFDM Systems With User Rate Constraints. *IEEE Trans. Veh. Technol.* **2009**, *58*, 1190–1203. [CrossRef]
38. Spencer, Q.; Swindlehurst, A.; Haardt, M. Zero-forcing methods for downlink spatial multiplexing in multiuser MIMO channels. *IEEE Trans. Signal Process.* **2004**, *52*, 461–471. [CrossRef]
39. Li, Z.; Han, S.; Molisch, A.F. Hybrid beamforming design for millimeter-wave multi-user massive MIMO downlink. In Proceedings of the 2016 IEEE International Conference on Communications (ICC 2016), Kuala Lumpur, Malaysia, 22–27 May 2016. [CrossRef]
40. Kang, J.; Lee, J.H.; Choi, W. Machine Learning-Based Dimension Optimization for Two-Stage Precoder in Massive MIMO Systems with Limited Feedback. *Appl. Sci.* **2019**, *9*, 2894. [CrossRef]

electronics

MDPI

Article

Evaluation of Multi-Beam Massive MIMO Considering MAC Layer Using IEEE802.11ac and FDD-LTE

Fumiya Muramatsu [1], Kentaro Nishimori [1,*], Ryotaro Taniguchi [1], Takefumi Hiraguri [2] and Tsutomu Mitsui [2]

1 Graduate School of Science and Technology, Niigata University, Niigata-shi, Niigata 950-2102, Japan; muramatsu@gis.ie.niigata-u.ac.jp (F.M.); taniguchi@gis.ie.niigata-u.ac.jp (R.T.)

2 Faculty of Fundamental Engineering, Nippon Institute of Technology, 4-1 Gakuendai, Miyashiro-machi, Minamisaitama-gun, Saitama 345-8501, Japan; hira@nit.ac.jp (T.H.); ben321@outlook.jp (T.M.)

* Correspondence: nishimori@ie.niigata-u.ac.jp; Tel.: +81-25-262-7431

Received: 26 January 2019; Accepted: 14 February 2019; Published: 18 February 2019

Abstract: Massive multiple-input multiple-output (MIMO) transmission has attracted attention as a key technology for use in fifth-generation mobile communication systems. Multi-beam massive MIMO systems that apply beam selection in analog components and blind algorithms in digital components to eliminate the requirement for channel state information have been proposed as a method for reducing overhead. In this study, we developed an adaptive modulation scheme for implementing multi-beam massive MIMO and used computer simulation to compare it with digital and analog–digital hybrid beam-forming methods. The effectiveness of the proposed system was verified in a medium access control layer based on the IEEE802.11ac and frequency division duplex-LTE representative wireless communication standards.

Keywords: massive MIMO; analog multi-beam; hybrid beam-forming; PHY layer; MAC layer

1. Introduction

Cellular network data traffic volumes have increased significantly with the advent of smart devices. This trend will continue to grow as Internet of Things (IoT) equipment and big-data services become more common. To meet this demand, recent research and development have focused on achieving a 20-Gbps or more standard for future wireless communication [1,2].

Multiple-input multiple-output (MIMO) systems can be used to improve the transmission rate within a limited frequency band [3]. To this end, multi-user MIMO (MU-MIMO) has been developed to enable MIMO transmission to multiple users [4]. MU-MIMO technologies have been standardized for LTE-advanced and IEEE802.11ac [5,6]. Accordingly, massive MIMO is viewed as a fifth-generation (5G) technology, and it is expected to play an important role in achieving 5G targets [1,7].

In a MIMO system, a base station (BS) acquires channel state information (CSI) between itself and a user terminal (UT) [8]. As CSI acquisition under massive MIMO involves numerous UT–BS antenna pair channels, one of the most important challenges in implementing it is the acquisition of large amounts of CSI with a small overhead. To achieve this, it is necessary to evaluate not only the physical (PHY) layer but also the medium access control (MAC) layer. The MAC layer is a communication protocol of a part of the data link layer corresponding to the second layer of the OSI reference model in IEEE802, and it is located in a layer that is one layer above the physical layer.

Implicit beamforming (BF) has been proposed as an approach for eliminating CSI feedback in MU-MIMO systems [9]. However, the communication efficiency of short-packet communication decreases even if implicit BF is applied in a massive MIMO system [10].

CSI estimation and initial UT tracking, which require a significant amount of overhead, are necessary in hybrid BF massive MIMO systems [1]. A hybrid BF approach referred to as multi-beam massive MIMO transmission has been proposed to solve this problem. CSI estimation is unnecessary in this approach. The effectiveness of multi-beam massive MIMO has been demonstrated through computer simulation [11–13]. In multi-beam massive MIMO systems, a number of analog multi-beams are created for initial UT tracking and a subset of these beams with high received power are selected for use. As multi-beams have narrow beam widths, interference signals can be mitigated and residual interference can be cancelled by applying a blind array algorithm to digital beam components using only the information pertaining to a received signal [14–16]. This paper proposes a multi-beam massive MIMO approach that achieves high transmission efficiency and flexible communication using asynchronous UT transmission. As the proposed system does not require timing synchronization among UTs, all UTs can freely transmit signals while avoiding collision between their signals. The diminished signal transmission time resulting from this configuration reduces the total power consumption by the UTs.

In our previous studies, we proposed multi-beam massive MIMO beam-selection methods [13] that utilized received-signal information such as the power difference and amplitude correlation between beams. This approach was capable of appropriately performing beam-selection at a high signal-to-interference noise ratio (SINR). Furthermore, this method could be performed through signal processing using a simple configuration and was highly suitable for hybrid analog-digital massive MIMO.

Channel quality indicator (CQI) values, which can serve as an index of modulation scheme determination, cannot be obtained under multi-beam massive MIMO because it does not perform CSI estimation. In this paper, we propose a simple adaptive modulation scheme for multi-beam massive MIMO transmission based on amplitude correlation and received power. Under this method, an appropriate modulation scheme can be simply determined based on the relationship between amplitude correlation and the SINR. We assessed the proposed method by evaluating the transmission rates it achieved using the IEEE802.11ac and frequency division duplex (FDD)-LTE standards. In addition, we evaluated the throughput of the method under the MAC layer of each standard to consider overhead.

The rest of this paper is organized as follows: The general massive MIMO transmission approach is described in Section 2. In Section 3, we describe the use of amplitude correlation for modulation and propose a simple adaptive modulation scheme for multi-beam massive MIMO. In Section 4, we present our simulation model and describe its application in evaluating simple adaptive modulation schemes. Then, we describe the results of the performance evaluation of the proposed multi-beam massive MIMO method through the computer simulation of the IEEE802.11ac and FDD-LTE standard environments.

2. Massive MIMO Transmission and Proposed System

The bandwidth required by 5G systems is achieved using frequencies of 20 GHz or higher. Massive MIMO utilizes BF technology to resolve propagation loss, which is one of the more serious problems encountered under high-frequency bands. Typically, a technique known as digital BF (DBF) is adopted, as shown on the right-hand side of Figure 1. In this technique, weight values are calculated through digital signal processing. However, the massive MIMO systems that apply DBF experience problems with respect to power consumption and implementation cost [1,17,18].

The left-hand side of Figure 1 shows the configuration of a typical hybrid BF massive MIMO with sub-arrays [19]. In the figure, N_L and N_K indicate the number of elements and receivers in a sub-array, respectively. This configuration produces a hybrid analog-digital BF that tracks a desired signal through analog control while removing interference using digital signal processing, and it has attracted significant interest [20–22]. Massive MIMO is known to be highly effective in addressing the power consumption and implementation cost problems owing to its hybridization of analog and digital approaches.

Multi-beam massive MIMO eliminates the requirement for CSI estimation by forming numerous narrow directional beams and spatially separating the signals of each UT. As a result, the synchronization between a BS and a UT, which is required for CSI estimation, becomes unnecessary. Under standard MU-MIMO, UTs must transmit signals to a BS simultaneously to employ CSI feedback prior to communication. As multi-beam massive MIMO does not require CSI feedback, a UT can communicate with a BS at any time.

Figure 2 shows the configuration of analog multi-beam massive MIMO under uplink communication, along with its 16 multi-beam pattern. The proposed system generates N orthogonal multiple beams in the analog component and uses a butler matrix circuit to achieve multi-beam formation [23,24]. The butler matrix circuit, which was introduced in [24], forms an orthogonal beam using a Fast Fourier Transform (FFT) on a spatially arranged array. This system is also effective from the viewpoint of the pilot contamination problem in the multi-cell scenario because of the signals by multiple UTs can be received without the CSI estimation by the multi-beam circuit.

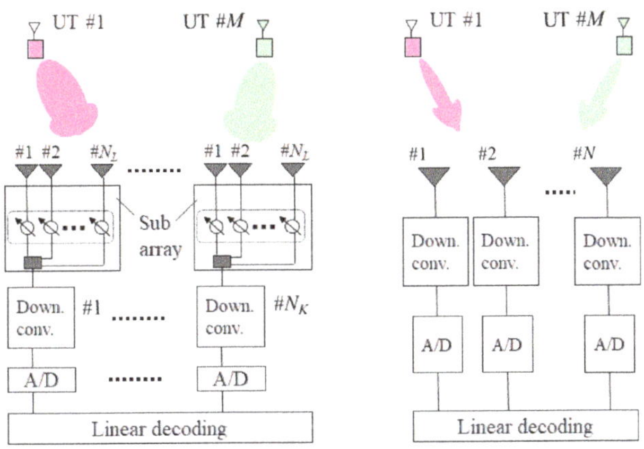

Figure 1. Uplink configurations of typical analog–digital hybrid (**left**) and full-digital (**right**) beamforming (BF) massive multiple-input multiple-output (MIMO) systems.

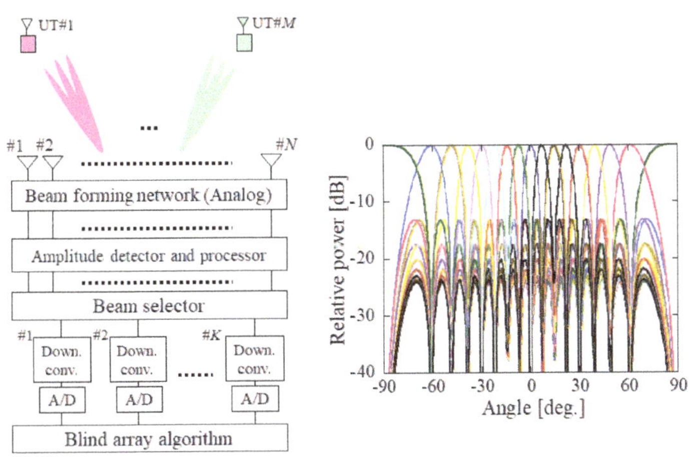

Figure 2. Uplink configuration of multi-beam massive MIMO in uplink and 16 multi-beam pattern.

The proposed method performs beam selection using an amplitude detector and a processor. Through beam selection, it is possible to spatially separate a specific UT using beam directivity. However, interference signals from the other UTs, which are also received by side lobes, cannot be rejected through only the multi-beam forming network. A method that uses only amplitude information, such as amplitude correlation, must be used to achieve beam selection with low interference power.

The system developed in this study applies robust independent component analysis (ICA) using a blind adaptive algorithm [16,25]. ICA is a blind signal processing technique that is commonly used in image processing and medicine. In this technique, an observed random vector is decomposed into statistically independent variables [16]. ICA does not require CSI because it utilizes only received signals. The use of ICA allows for a hybrid configuration for efficient transmission in massive MIMO systems in which multi-beams are applied to analog components and a blind algorithm is applied to digital components.

3. Adaptive Modulation for Multi-Beam Massive MIMO

As discussed in the previous section, multi-beam massive MIMO does not require CSI estimation, thereby eliminating the requirement for a process to obtain channel information at the BS side. As a result, it is not possible to use a modulation and coding scheme (MCS) index to combine the modulation scheme and coding rate from the CQI value, as is done under current wireless communication standards [5,6]. Therefore, we developed an adaptive modulation scheme based on the received power and amplitude correlation obtained from performing beam selection during uplink.

Here, we explain the beam-selection method and amplitude correlation in multi-beam massive MIMO. Beam-selection is one of the major challenges in multi-beam massive MIMO. As a directivity peak cannot be directed to a given UT, it is possible for one beam to receive signals from multiple UTs. It has been confirmed that in such situations, signal separation cannot be perfectly performed in digital components, even by a blind algorithm. To address this, the authors previously proposed a beam-selection method that utilizes the power differences between the multi-beams and correlation values obtained from amplitude information [13]. By employing this method, it is possible to achieve beam selection with a high SINR by setting thresholds.

Under the proposed beam-selection method, amplitude correlation is calculated based on the covariance between the amplitudes of received and adjacent beams. The amplitude of a received signal is defined as a vector, $\boldsymbol{x}$, whose number of elements is equivalent to the lengths of all transmission-signal symbols. The correlation coefficient, $\rho(\boldsymbol{x},\boldsymbol{y})$, with respect to the signal amplitudes, $\boldsymbol{x}$ and $\boldsymbol{y}$, is denoted as

$$\rho(\boldsymbol{x},\boldsymbol{y}) = \frac{\sum_{i=1}^{N}(x_i-\overline{x})(y_i-\overline{y})}{\sqrt{\sum_{i=1}^{N}(x_i-\overline{x})^2}\sqrt{\sum_{i=1}^{N}(y_i-\overline{y})^2}}, \tag{1}$$

where $\overline{x}$ is the mean of all elements in $\boldsymbol{x}$.

Figure 3 shows the relationship between amplitude correlation and the SINR. The SINR of the received signal tends to be high when the amplitude correlation is high. Based on this finding, we examined a method of determining a simple adaptive modulation approach on the downlink using the amplitude correlation and received power obtained on the uplink.

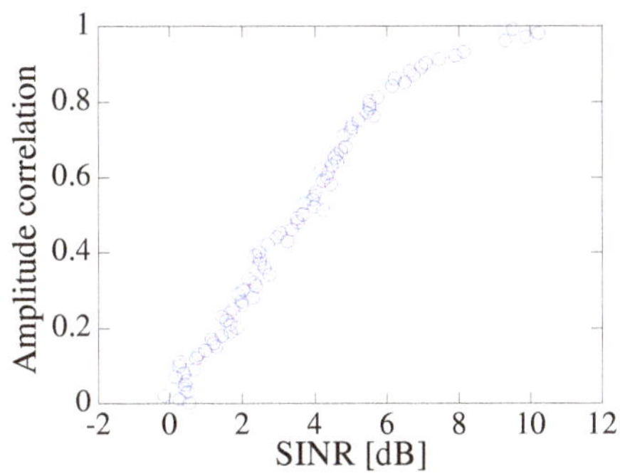

Figure 3. Relationship between amplitude correlation and signal-to-interference noise ratio (SINR).

4. Computer Simulation

The results of the computer simulation of the proposed system were verified in terms of the Rayleigh fading and angular spread of quadrature amplitude modulation (QAM) signals [26,27]. The simulation conditions are listed in Table 1. Figure 4 shows the simulation model, which is a scattering ring model [27] with a specific angular spread. The model applies 101 paths per UT with an angular spread of 1.0 degree. We assume flat fading in narrowband signals and do not consider delay spread. Sixty-four elements are arranged in the horizontal direction at intervals of 0.5 wavelengths, and beam width is approximately 1.60 degrees. As the assumed transmission method involves a single carrier for narrowband signals, delay waves have no influence on signal separation. The signals for each sub-carrier are regarded as narrowband signals to apply them as actual broadband signals through orthogonal frequency division multiplexing. In the UT distribution, the center direction of the BS is set as 0 degrees and individual UTs are placed at random angles within a range of -60 to 60 degrees. A constant distance is set between the BS and all UTs so that a signal-to-noise ratio (SNR) of 20 [dB] per UT is received at the BS.

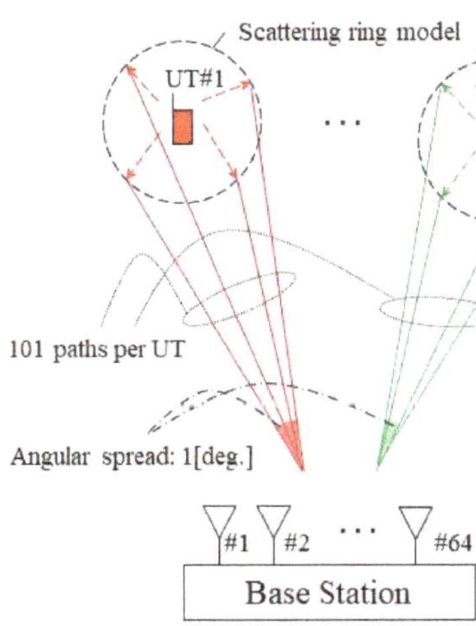

Figure 4. Simulation model.

Table 1. Simulation conditions.

Number of receiving antennas	64
Number of UTs	20
Number of demodulated UTs	8
SNR per UT, before array gain acquisition	20 [dB]
Modulation scheme	4QAM (QPSK) – 1024QAM
Trial number	3000
Data length	3000
Data smoothing size	3000
Number of iterations	30

First, the processing of the uplink was simulated transmitting QPSK signals from all 20 UTs to the BS, which recorded the amplitude correlation and received power for each beam. On the downlink, the BS transmitted QPSK 1024-QAM signals to eight users with high reception power. Then, the UTs received signals from the BS calculated bit error rate (BER). We assumed that a modulation scheme yielding a BER of less than 10^{-2} could be applied [28,29]. We evaluated the applicable modulation schemes in terms of the SINR and received power recorded on the uplink to simulate an adaptive modulation scheme.

4.1. Adaptive Modulation

The results produced by the simulation procedure described in Section 4 are shown in Figure 5, in which the left-hand and right-hand figures show the relationship between the applicable modulation method and amplitude correlation and received power, respectively. Even though amplitude correlation increases with the modulation order, it is not appropriate to select a modulation method using only amplitude correlation because the standard deviation of the results is large. Therefore, the received power relationship shown in the right-hand figure was used to develop an indicator for determining the modulation method (Table 2). In the table, ρ and P represent amplitude correlation and received power, respectively.

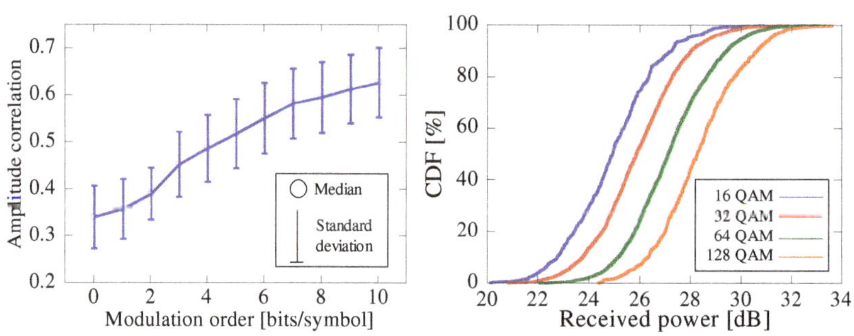

Figure 5. Relationships between applicable modulation scheme and amplitude correlation (**left**) and received power (**right**).

Table 2. Adaptive modulation scheme for eight user terminals (UTs).

Modulation Scheme	Conditions
None	$\rho < 0.4$
16-QAM	$\rho \geq 0.4$ and $P < 20$ [dB]
32-QAM	$\rho \geq 0.4$ and $20 \leq P < 24$ [dB]
64-QAM	$\rho \geq 0.4$ and $24 \leq P < 28$ [dB]
128-QAM	$\rho \geq 0.4$ and $P \geq 28$ [dB]

4.2. IEEE802.11ac

We evaluated the indicator shown in Table 2 and the downlink effectiveness of multi-beam massive MIMO transmission based on the IEEE802.11ac procedures [5,30]. The simulation conditions and model were the same as those used in Section 4. Equation (2) was used as an index of evaluation based on the BER calculated on the UTs.

$$R = \begin{cases} M\,(1 - \mathrm{BER}) & (\mathrm{BER} \leq 10^{-2}) \\ 0 & (\text{otherwise}) \end{cases} \quad [\text{bit/symbol}], \tag{2}$$

where M denotes the number of bits per symbol and R denotes the number of bits per symbol obtained independently of the coding rate.

The assessment was divided into parts, i.e., the evaluation of the PHY layer without considering the overhead arising from the control signal and the evaluation of the MAC layer considering overhead. Table 3 shows the relationship between the MCS index and the transmission rate under the IEEE802.11ac 20-[MHz] operation [5]. The values of R' are obtained by multiplying the modulation order by the coding rate, and they are equivalent to R, which is the number of bits per symbol. Accordingly, it was assumed that an MCS index that satisfies $R \geq R'$ could be applied to the R obtained on the UTs. Based on this, we could evaluate the transmission rate corresponding to the MCS index.

Table 3. Relationship between MCS index and transmission rate under IEEE802.11ac at 20 MHz.

I_{MCS}	Modulation	Coding Rate	R' [bits/symbol]	Transmission Rate [Mbps]
0	BPSK	1/2	0.5	7.2
1	QPSK	1/2	1.0	14.4
⋮	⋮	⋮	⋮	⋮
9	256QAM	3/4	6.0	86.7

In the MAC layer, throughput is calculated according to the downlink MU-MIMO procedure shown in Figure 6 [5]. As seen in the figure, the BS first transmits request signals to all UTs to obtain channel information. Once the BS has obtained channel feedback from each UT data transmission signal, data reception processing is performed for all UTs. Throughput is obtained by dividing the data size of frame aggregation, which was set as 63,000 bytes in this evaluation, by the time required by the processes. As multi-beam massive MIMO does not require procedures such as CSI estimation or feedback, we assumed that no time other than that necessary for data transmission was required.

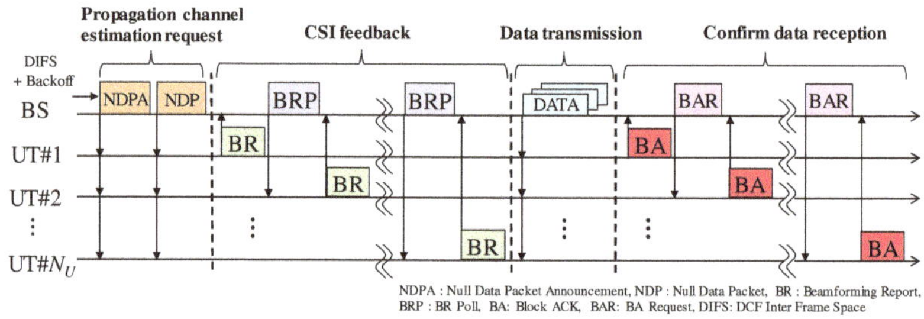

Figure 6. Downlink multi-user (MU)-MIMO procedure under IEEE802.11ac.

We evaluated the performance of a typical hybrid BF massive MIMO and compared it with those of the sub-array and full-digital BF configurations discussed in Section 2. Table 4 shows the precoding and decoding methods used under each configuration in the downlink. For the hybrid BF, we assumed that maximum ratio combining (MRC) was performed in the analog component using a phase shifter [31]. In the digital component, the minimum mean square error (MMSE) weight calculated in the uplink was used as a precoder [31]. Eight sub-arrays (N_K = 8) and eight sub-array elements (N_L = 8) were utilized, resulting in 64 elements. In the full-digital BF, transmission precoding was performed using a block diagonalization (BD) method with 64 elements [32].

Table 4. Precoding and decoding on downlink.

Massive MIMO Configuration	Precoding	Decoding
Multi-beam massive MIMO (Figure 2)	Analog multi-beam	Robust ICA
Hybrid BF massive MIMO (Figure 1)	MRC and MMSE	MMSE
Full-digital BF massive MIMO (Figure 1)	BD method	MMSE

Figure 7 shows the cumulative distribution function (CDF) characteristics of transmission rate and throughput. The adaptive modulation results show the characteristics obtained for each trial when the highest modulation order is selected in the modulation scheme in which the BER is 10^{-2} or less. The proposed modulation results are obtained using the characteristics based on Table 2. At the assumed transmission rate, the full-digital BF configuration achieves 86.7 [Mbps], which is the maximum transmission rate obtainable under IEEE802.11ac, because it scans a sharp beam width using 64 elements. The median value of multi-beam massive MIMO is approximately 70 [Mbps], and it is asymptotic to the ideal characteristic obtained by utilizing the proposed simple adaptive modulation. The hybrid BF configuration characteristic achieves a low modulation order because beam width is widened when there are eight elements per sub-array. The full-digital configuration can achieve a throughput of only approximately 40 [Mbps] because of the overhead of channel estimation and feedback obtained when there are 64 antenna elements. In hybrid BF, overhead occurs owing to eight elements, which is the number of sub-arrays. As shown in Figure 6, throughput decreases when the transmission rate of even one UT is slow because IEEE802.11ac requires the synchronization of all UTs, as shown in Figure 6. This reduces the throughput characteristic of the hybrid BF. The multi-beam massive MIMO characteristic is the best among the three configurations because it does not require synchronization and overhead and can effectively utilize the transmission rate obtained in the PHY layer.

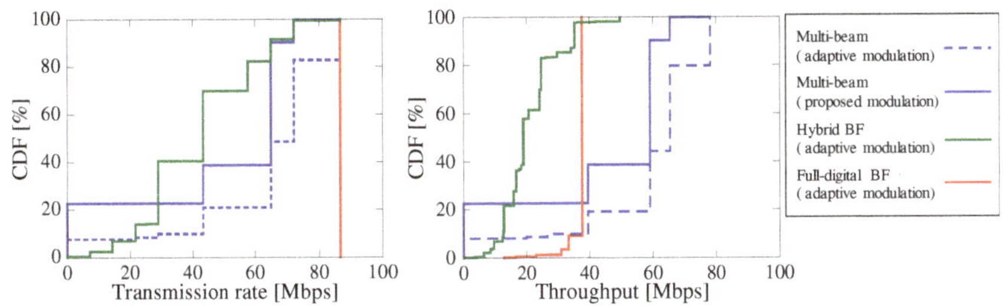

Figure 7. Characteristics of transmission rate and throughput under IEEE802.11ac.

4.3. FDD-LTE

We evaluated the proposed method under FDD-LTE (Release 15) [6].We assumed a bandwidth of 20 [MHz] per UT and 100 resource blocks (RBs). Table 5 shows the relationship between the MCS index and the transmission rate in LTE when there is one antenna. The transport block size (TBS) is based on [6], and the transmission rate is calculated based on the efficiency per bandwidth. The TBS corresponds to an actual data component divided into RBs and transmitted, while user data indicate the resource size for a given control signal. As in the IEEE802.11ac simulation, we compared the values of R and R' produced by the respective approaches and evaluated the results in terms of the corresponding transmission rates.

Table 5. Relationship between MCS index and transmission rate under frequency division duplex (FDD)-LTE (100 RBs, 20 MHz).

I_{MCS}	Modulation	TBS	User Data	Coding Rate	R' [bits/symbol]	Transmission Rate [Mbps]
0	QPSK	2792	30000	0.094	0.188	7.22
1	QPSK	4548	30000	0.154	0.308	9.61
2	QPSK	7224	30000	0.242	0.484	21.7
⋮	⋮	⋮	⋮	⋮	⋮	⋮
27	256QAM	84760	120000	0.707	5.656	127.0

The throughput in LTE, which is denoted here by T, can be calculated as follows: [6,33].

$$\begin{aligned} T &= TBS \times N_{\text{str}} \\ &\fallingdotseq N_{\text{subc}} \times N_{\text{slot}} \times N_{\text{sym}} \times M \times N_{\text{str}} \times CR \ [\text{bps}], \end{aligned}$$

where N_{subc} is the number of sub-carriers, N_{slot} is the number of slots per second, N_{sym} is the number of symbols per slot, N_{str} is the number of streams, and CR is the coding rate.

We apply Equation (3) to consider control signals because Equation (3) cannot express the number of control signals that can be eliminated under multi-beam massive MIMO.

$$T = TBS \times \frac{Userdata'}{Userdata} \times N_{\text{subf}} \times N_{\text{str}} \ [\text{bps}], \tag{3}$$

where N_{subf} is the number of sub-frames per second. The value of $Userdata'$ is obtained by increasing or decreasing the number of control signals from secured $Userdata$, which is defined in [6]. For example, when I_{MCS} = 27 in Table 1, $Userdata$ is obtained as 120,000 [bits]. This corresponds to a $Userdata'$ of 128,000 [bits] in the multi-beam massive MIMO, which is obtained by removing the control signal from $Userdata$. As there is no complete definition of a control signal in the case of 64 elements, we assumed that one antenna port would be added to handle the additional control signal required each time the number of antennas was doubled. Therefore, $Userdata'$ was 83,200 [bits] for 64 antennas under the full-digital configuration.

Figure 8 shows the relationship between the transmission rate and throughput based on Equation (3). We used this figure to evaluate throughput based on overhead for each configuration.

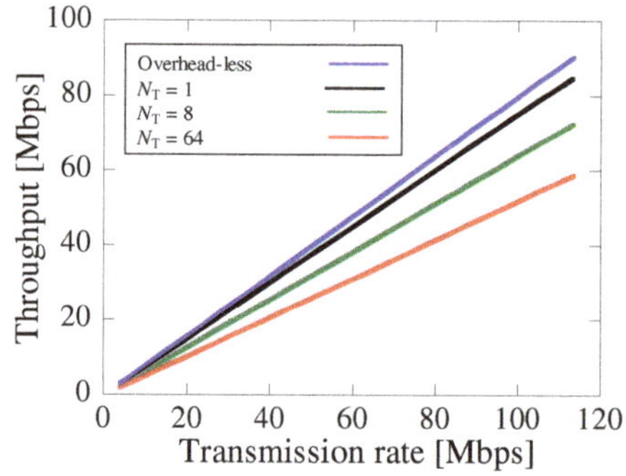

Figure 8. Transmission rate versus throughput from Equation (3).

Figure 9 shows the CDF characteristics of the transmission rate and throughput under FDD-LTE. The modulation scheme for each characteristic is determined in the same manner as that used in the evaluation under IEEE802.11ac. The transmission rate follows the same trend as that under IEEE802.11ac. However, unlike IEEE 802.11ac, a high transmission rate can be achieved even with a low modulation order. This suggests that the performance of the hybrid BF approaches that of multi-beam massive MIMO. An evaluation of throughput based on Figure 8 reveals that it increases linearly with transmission rate under all configurations. Given that multi-beam massive MIMO has the lowest number of control signals of all configurations, it achieves the highest throughput at a given transmission rate. This verifies the effectiveness of multi-beam massive MIMO, which can effectively utilize the transmission rate in the PHY layer.

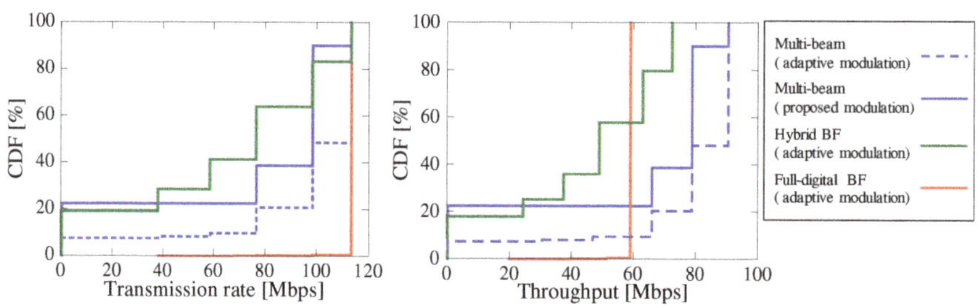

Figure 9. Characteristics of transmission rate and throughput under FDD-LTE.

5. Conclusions

In this paper, we proposed a simple adaptive modulation method for multi-beam massive MIMO and evaluated its performance when implemented in a MAC layer. The effectiveness of the proposed transmission method was validated through computer simulation. The performance of the method asymptotically approached the ideal characteristic because of the implementation of simple adaptive modulation using amplitude correlation and received power. The proposed method was also shown to achieve higher throughput characteristics compared to a hybrid approach with sub-arrays and a full-digital beam-forming configuration under the IEEE802.11ac and FDD-LTE environments. Future works include the further theoretical analysis and experimental evaluation using the analog multi-beam by Butler matrix circuit.

Author Contributions: This study was led by F.M. while K.N., R.T., T.H. and T.M. assisted with the computer simulations.

Funding: This work was partially supported by the SCOPE #165004002, #185004002 and KAKENHI, Grant-in-Aid for Scientific Research (B) (17H01738, 17H03262).

Conflicts of Interest: The authors declare no conflict of interest.

References

1. Papadopoulos, H.; Wang, C.; Bursalioglu, O.; Hou, X.; Kishiyama, Y. Massive MIMO technologies and challenges towards 5G. *IEICE Trans. Commun.* **2015**, *E99-B*, 602–621. [CrossRef]
2. Obara, T.; Suyama, S.; Shen, J.; Okumura, Y. Joint Processing of Analog Fixed Beamforming and CSI-Based Precoding for Super High Bit Rate Massive MIMO Transmission Using Higher Frequency Bands. *IEICE Trans. Commun.* **2015**, *E98-B*, 1474–1481. [CrossRef]
3. Foschini, G.J.; Gans, M.J. On limits of wireless communications in a fading environment when using multiple antennas. *Wirel. Pers. Commun.* **1998**, *6*, 331–335. [CrossRef]
4. Gesbert, D.; Kountouris, M.; Heath, R.W., Jr.; Chae, C.B. ; Salzer, T. Shifting the MIMO paradigm. *IEEE Signal Process. Mag.* **2007**, *24*, 36–46. [CrossRef]
5. *IEEE P802.11ac./D5.0, Part 11: Wireless LAN Medium Access Control (MAC and Physical Layer (PHY) Specifications;* IEEE Standard for Information Technology; IEEE: New York, NY, USA, 2013.
6. *3rd Generation Partnership Project (3GPP). TS 36.213 Physical Layer Procedures;* V10.2.0; 3GPP: Sophia Antipolis, France, 2018.
7. Lu, L.; Li, G.Y.; Swindlehurst, A.L.; Ashikhmin, A.; Zhang, R. An overview of massive MIMO: Benefits and challenges. *IEEE J Sel. Top. Signal Process.* **2014**, *8*, 742–758. [CrossRef]
8. Spencer, Q.H.; Peel, C.B.; Swindlehurst, A.L.; Haardt, M. An introduction to the multi-user MIMO downlink. *IEEE Commun. Mag.* **2004**, *42*, 60–67. [CrossRef]
9. Murakami, T.; Fukuzono, H.; Takatori, Y.; Mizoguchi, M. Multiuser MIMO with implicit channel feedback in massive antenna systems. *IEICE Commun. Express* **2013**, *2*, 336–342. [CrossRef]
10. Hiraguri, T.; Nishimori, K. Survey of transmission methods and efficiency using MIMO technologies for wireless LAN systems. *IEICE Trans. Commun.* **2015**, *E98-B*, 1250–1267. [CrossRef]
11. Kameyama, K.; Nishimori, K.; Hiraguri, T.; Yamada, H.; Makino, H. Analog multi-beam massive MIMO transmission eliminating CSI estimation. *IEICE Trans. Commun.* **2016**, *J99-B*, 743–752. (In Japanese)
12. Taniguchi, R.; Nishimori, K.; Makino, H. Multi-beam massive MIMO using constant modulus algorithm for QAM signals employing amplitude and phase offset compensation. *IEICE Trans. Commun.* **2017**, *E100-B*, 262–268. [CrossRef]
13. Muramatsu, F.; Nishimori, K.; Taniguchi, R.; Hiraguri, T. Multi-beam massive MIMO with beam-selection using only amplitude information in uplink channel. *IEICE Trans. Commun.* **2018**, *E101-B*, 1544–1551. [CrossRef]
14. Treichler, J.R.; Agee, B.G. A new approach to multipath correction of constant modulus signals. *IEEE Trans. Acoust. Speech Signal Process.* **1983**, *31*, 459–472. [CrossRef]
15. Agee, B. The least-Square CMA: A new technique for rapid correction of constant modulus signals. *Proc. IEEE ICASSP* **1986**, 953–956.
16. Hyvarinen, A. Fast and robust fixed-point algorithms for independent component analysis. *IEEE Trans. Neural Netw.* **1999**, *10*, 626–634. [CrossRef]
17. Sohrabi, F.; Yu, W. Hybrid digital and analog beamforming design for large-scale antenna arrays. *IEEE J. Sel. Top. Signal Process.* **2016**, *10*, 501–513. [CrossRef]
18. Bogale, T.E.; Le, L.B. Beamforming for multiuser massive MIMO systems: Digital versus hybrid analog-digital. In Proceedings of the 2014 IEEE Global Communications Conference, Austin, TX, USA, 8–12 December 2014; pp. 4066–4071.
19. Molisch, A.F.; Ratnam, V.; Han, S.; Li, Z.; Nguyen, L.H.; Li, L.; Haneda, K. Hybrid beamforming for massive MIMO: A survey. *IEEE Commun. Mag.* **2017**, *55*, 134–141. [CrossRef]
20. Ngo, H.Q.; Larsson, E.G.; Marzetta, T.L. Energy and spectral efficiency of very large multiuser MIMO systems. *IEEE Trans. Commun.* **2013**, *61*, 1436–1449.
21. Rusek, F.; Persson, D.; Lau, B.K.; Larsson, E.G.; Marzetta, T.L.; Edfors, O.; Tufvesson, F. Scaling up MIMO: Opportunities and challenges with very large arrays. *IEEE Signal Process. Mag.* **2013**, *30*, 40–60. [CrossRef]

22. Hoydis, J.; Brink, S.T.; Debbah, M. Massive MIMO in the UL/DL of cellular networks: How many antennas do we need? *IEEE J. Sel. Areas Commun.* **2013**, *31*, 160–171. [CrossRef]
23. Butler, J.; Lowe, R. Beamforming matrix simplifies design of electronically scanned antennas. *Electron. Des.* **1961**, *9*, 170–173.
24. Yamamoto, S.; Hirokawa, J.; Ando, M. A single-layer hollow-waveguide 8-way Butler matrix with modified phase shifters. In Proceedings of the ISAP2015, Hobart, Australia, 9–12 November 2015; pp. 937–940.
25. Zarzoso, V.; Comon, P. Robust independent component analysis by iterative maximization of the kurtosis contrast with algebraic optimal step size. *IEEE Trans. Neural Netw.* **2010**, *21*, 248–261. [CrossRef]
26. Gesbert, D.; Bolcskei, H.; Gore, D.A.; Paulraj, A.J. Outdoor MIMO wireless channels: Models and performance prediction. *IEEE Trans. Commun.* **2002**, *50*, 1926–1934. [CrossRef]
27. Oestges, C.; Erceg, V.; Paulraj, A.J. A physical scattering model for MIMO macrocellular broadband wireless channels. *IEEE J. Sel. Areas Commun.* **2003**, *21*, 721–729. [CrossRef]
28. Chung, S.T.; Goldsmith, A.J. Degrees of freedom in adaptive modulation: A unified view. *IEEE Trans. Commun.* **2001**, *49*, 1561–1571. [CrossRef]
29. Nishimori, K.; Yomo, H.; Popovski, P. Distributed interference cancellation for cognitive radios using periodic signals of the primary system. *IEEE Trans. Wirel. Commun.* **2011**, *10*, 2791–2981. [CrossRef]
30. Kikuma, N.; Nishimori, K.; Hiraguri, T. Effect of user antenna selection on block beamforming algorithms for suppressing inter-user interference in multiuser MIMO system. *IEICE Trans. Commun.* **2018**, *E101-B*, 1523–1535. [CrossRef]
31. Tanbourgi, R.; Dhillon, H.S.; Andrews, J.G.; Jondral F.K. Effect of spatial interference correlation on the performance of maximum ratio combining. *IEEE Trans. Wirel. Commun.* **2014**, *13*, 3307–3316. [CrossRef]
32. Stankovic, V.; Haardt, M. Generalized design of multi-user MIMO precoding matrices. *IEEE Trans. Wirel. Commun.* **2008**, *7*, 953–961. [CrossRef]
33. Rathi, S.; Malik, N.; Chahal, N.; Malik, S. Throughput for TDD and FDD 4G LTE systems. *Int. J. Innov. Technol. Explor. Eng. (IJITEE)* **2014**, *3*, 73–77.

electronics

MDPI

Article

Computationally Efficient Channel Estimation in 5G Massive Multiple-Input Multiple-output Systems

Imran Khan [1], Mohammad Haseeb Zafar [1], Majid Ashraf [1] and Sunghwan Kim [2,*]

1 Department of Electrical Engineering, University of Engineering & Technology, Peshawar P.O.B. 814, Pakistan; imran_khan@uetpeshawar.edu.pk (I.K.); haseeb@uetpeshawar.edu.pk (M.H.Z.); majid@uetpeshawar.edu.pk (M.A.)

2 School of Electrical Engineering, University of Ulsan, Ulsan 44610, Korea

* Correspondence: sungkim@ulsan.ac.kr; Tel.: +82-052-259-1401

Received: 9 November 2018; Accepted: 28 November 2018; Published: 3 December 2018

Abstract: Traditional channel estimation algorithms such as minimum mean square error (MMSE) are widely used in massive multiple-input multiple-output (MIMO) systems, but require a matrix inversion operation and an enormous amount of computations, which result in high computational complexity and make them impractical to implement. To overcome the matrix inversion problem, we propose a computationally efficient hybrid steepest descent Gauss–Seidel (SDGS) joint detection, which directly estimates the user's transmitted symbol vector, and can quickly converge to obtain an ideal estimation value with a few simple iterations. Moreover, signal detection performance was further improved by utilizing the bit log-likelihood ratio (LLR) for soft channel decoding. Simulation results showed that the proposed algorithm had better channel estimation performance, which improved the signal detection by 31.68% while the complexity was reduced by 45.72%, compared with the existing algorithms.

Keywords: 5G; massive MIMO; computational efficiency; precoding algorithms; channel estimation

1. Introduction

Multiple-input multiple-output (MIMO) technology is becoming more and more mature, especially when combined with orthogonal frequency division multiplexing (OFDM) [1–5], which has been successfully applied in multiple wireless communications fields such as Long-Term Evolution (LTE) and LTE-Advanced. However, traditional MIMO technology can only achieve a 4×4 or 8×8 scale system [6], which makes it difficult to meet the explosive growth in mobile data services. Therefore, in recent years, massive MIMO has been proposed based on traditional MIMO technology [7]. Massive MIMO systems configure up to hundreds of antenna arrays at the base station to serve multiple single-antenna end-users simultaneously [8], which can improve spectrum utilization and power utilization in wireless communications systems by two to three orders of magnitude [9–11]. This has become one of the most promising enabling technologies and one of the hottest research directions in 5G [12]. The maximum likelihood (ML) algorithm is the optimal algorithm in massive MIMO detection algorithms, but its computational complexity increases exponentially with the number of system antennas and the modulation order of baseband signals. It is difficult for it to be fast, effective, and realized in practical applications. Linear detection methods, such as the zero-forcing (ZF) algorithm and minimum mean square error (MMSE) algorithm, can achieve near-optimal detection performance in massive MIMO systems. The complexity in this kind of detection algorithm is greatly reduced, compared with the complexity of the ML algorithm, but introduces a complex high-dimensional matrix inversion operation, so a low-cost and efficient engineering implementation is still a problem to be solved. Aimed at this problem, many simplified algorithms based on the MMSE detection scheme have been proposed in recent years, and can be roughly divided into three types: The series expansion class-approximation method [13,14], the iterative class-approximation method [15,16] and a gradient-based search for an approximate solution [17–20]. The authors in [13] proposed a method of using Neumann series expansion to approximate the inverse MMSE filter matrix, but when the number of expansion stages was gradually increased $(i > 2)$, the computational complexity was still high, even equal to or exceeding the exact MMSE. The complexity of the filter matrix inversion algorithm also loses a large degree of detection performance. The authors in [14] applied the Newton algorithm derived from the first-order Taylor series expansion (similar to Neumann series expansion) to massive MIMO signal detection, and used the iterative method to improve the estimation accuracy of the MMSE filter matrix inversion. However, from the aspects of detection performance and computational complexity, the algorithm based on the Newton iteration was not dominant. Different from the two series expansion-based algorithms above, it is necessary to estimate the signal vector sent by the user by inverting the approximate matrix. Some iterative algorithms based on solving linear equations, such as the Richardson iterative (RI) algorithm [15] and the successive over-relaxation (SOR) algorithm [16], use the special properties of the MMSE filter's symmetric positive definite matrix. Through the method of solving linear equations, they directly estimate the transmission vector, thus avoiding the inversion of high-dimensional matrices.

The RI and SOR algorithms mentioned above have lower computational complexity at a fixed number of iterations, but RI convergence is slower and requires a higher number of iterations to achieve certain detection performance requirements. In SOR, although the detection performance is close to excellent, its internal iterative structure means the algorithm cannot be implemented in parallel in practical applications. The third type of algorithm is mainly designed and implemented based on the idea of a matrix gradient, including the conjugate-gradient (CG) method [17] and the steepest descent (SD) method [18]. This type of algorithm uses the matrix gradient search method and avoids the high-dimensional matrix inversion problem. However, compared to the method of series expansion, the CG and SD algorithms bring about great improvement in detection performance, but calculation of the matrix gradient after each iteration also causes higher complexity.

In this paper, a low-complexity joint detection algorithm was proposed. The SD algorithm had a good convergence direction at the beginning of the iteration, and the Gauss–Seidel (GS) algorithm with low complexity mentioned in [19] was combined with the SD method (called SDGS), which provided an effective search direction for GS iterations, speeding up convergence and improving the detection performance. Furthermore, applying it to soft output detection gave an approximate calculation method for the bit log-likelihood ratio (LLR) of the channel decoder input. A good compromise between detection performance and computational complexity was achieved.

The rest of this paper is organized as follows: Section 2 discusses the system model and analytical derivations, Section 3 explains signal detection, while Section 4 explains the mixed iterative algorithm and the proposed algorithm. Section 5 provides the simulation results, while Section 6 concludes the paper.

2. System Model

The research object considered in this paper was the uplink for a massive MIMO system consisting of a base station equipped with N antennas and K single-antenna users where $N \gg K$, as shown in Figure 1. Let $s_c = [s_1, s_2, \ldots, s_K]^T$ denote the $K \times 1$ dimensional symbol vector sent by all users simultaneously, where $s_k \in \varepsilon$ was the transmitted symbol from the kth user, and ε was the modulation symbol set.

Let $H_c \in \mathbb{C}^{N \times K}$ represent the Rayleigh fading channel matrix; then, the $N \times 1$ dimensional signal vector received by the base station could be recorded as:

$$y_c = H_c s_c + n_c \tag{1}$$

where $\boldsymbol{n}_c$ represented an additive white Gaussian noise (AWGN) vector with an $\boldsymbol{N} \times \boldsymbol{1}$ dimensional mean of 0 and a covariance matrix of $\sigma^2 \boldsymbol{I}_N$. Converting the complex model of Equation (1) into an equivalent real model gave:

$$\boldsymbol{y} = \boldsymbol{H}\boldsymbol{s} + \boldsymbol{n} \tag{2}$$

Among these terms, $\boldsymbol{s} \in \mathbb{R}^{2K}$, $\boldsymbol{H} \in \mathbb{R}^{2N \times 2K}$, $\boldsymbol{y} \in \mathbb{R}^{2N}$, and $\boldsymbol{n} \in \mathbb{R}^{2N}$, which were:

$$\begin{gathered} \boldsymbol{H} = \begin{bmatrix} \Re(\boldsymbol{H}_c) & -\Im(\boldsymbol{H}_c) \\ \Im(\boldsymbol{H}_c) & \Re(\boldsymbol{H}_c) \end{bmatrix}, \boldsymbol{y} = \begin{bmatrix} \Re(\boldsymbol{y}_c) \\ \Im(\boldsymbol{y}_c) \end{bmatrix} \\ \boldsymbol{s} = \begin{bmatrix} \Re(\boldsymbol{s}_c) \\ \Im(\boldsymbol{s}_c) \end{bmatrix}, \boldsymbol{n} = \begin{bmatrix} \Re(\boldsymbol{n}_c) \\ \Im(\boldsymbol{n}_c) \end{bmatrix} \end{gathered} \tag{3}$$

Among those, $\Re(\cdot)$ and $\Im(\cdot)$ indicated the real part and imaginary part, respectively.

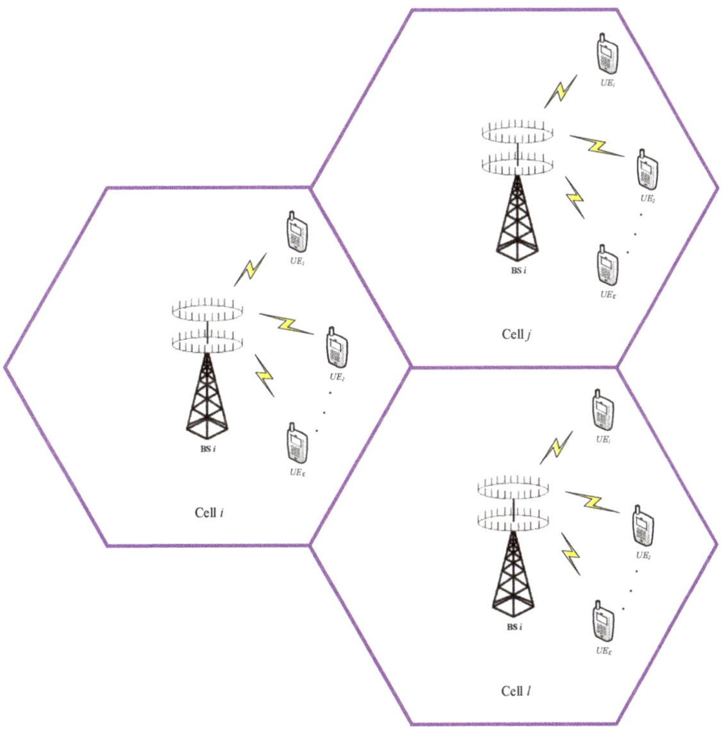

Figure 1. Massive multiple-input multiple-output system model.

2.1. Minimum Mean Square Error Signal Detection

The main task of signal detection was to accurately determine user transmission vector s at the base station through received signal vector y. The transmitted signal vector $\hat{s}$ detected by the MMSE algorithm could be expressed as:

$$\hat{s} = \left(H^H H + \sigma^2 I_{2K}\right)^{-1} H^H y = W^{-1}\hat{y} \tag{4}$$

where $\hat{y} = H^H y$. The filter matrix W of the MMSE detector could be expressed as:

$$W = G + \sigma^2 I_{2K} \tag{5}$$

where $G = H^H H$ was the Gram matrix. In massive MIMO systems, the computational complexity of W^{-1} is $O(K^3)$, which makes the implementation of the MMSE algorithm very complex.

2.2. Log Likelihood Ratio Calculation

Various channel coding techniques are commonly employed in wireless communication systems to improve their error performance, since channel reliability can be used to improve system stability. Conventional MIMO system signal detection generally uses a hard decision method to directly execute symbol decisions on the estimated value of the user-transmitted signal vector, i.e., $\hat{s}$ in Equation (4). In order to output the soft detection information to the back end of the detector, after the MMSE detector estimates $\hat{s}$, the LLR soft information used for channel decoding could be calculated with the following method. First, we needed to restore the estimated $\hat{s}$ and the calculated W^{-1} to the equivalent complex field to get $\hat{s}_c$ and W_c^{-1}. Let $U = W_c^{-1}G_c = W_c^{-1}H_c^H H_c$ denote the equalized channel matrix. The equalized signals obtained by the MMSE filter matrix could be obtained from Equations (2) and (4) as follows:

$$\begin{aligned}\hat{s}_c &= W_c^{-1}G_c s_c + W_c^{-1}H_c^H n_c \\ &= U s_c + W_c^{-1}H_c^H n_c\end{aligned} \tag{6}$$

Then, the estimated value of the symbol transmitted by the ith user is $\hat{s}_{c,i} = \mu_i s_{c,i} + e_i$, where $\mu_i = [U]_{ii} = U_{ii}$ represented the effective channel gain after equalization, and e_i represented the noise plus interference (NPI) term contained in the $\hat{s}_{c,i}$. The noise variance was $v_i^2 = \sum_{j\neq i}^{K}|U_{ji}|^2 + E_{ii}\sigma^2$, where U_{ji} and E_{ii} represented the (j,i)th element of the matrix U and the ith diagonal of the matrix E, respectively, where $E = W_c^{-1}H_c^H\left(W_c^{-1}H_c^H\right)^H = W_c^{-1}G_c W_c^{-1}$. Using the max-log approximation representation given in [11], the LLR $L_{i,b}$ corresponding to the bth bit transmitted by the ith user was expressed as:

$$L_{i,b} = \Upsilon_i\left(\underset{a\,\in\,O_b^0}{min}\left|\frac{\hat{s}_{c,i}}{\mu_i}a\right|^2 - \underset{a'\,\in\,O_b^1}{min}\left|\frac{\hat{s}_{c,i}}{\mu_i}a\right|^2\right) \tag{7}$$

where $\Upsilon_i = \mu_i^2/v_i^2$ represented the signal-to-interference plus noise ratio (SINR), and O_b^0 and O_b^1 represented the modulation symbol set with the bth bit being 0 and 1, respectively.

3. Low Complexity Signal Detection

3.1. Neumann Series Expansion

In a massive MIMO system, the MMSE signal detection algorithm involves a high-dimensional matrix inversion, W^{-1}, with a computational complexity of $O(K^3)$. In order to reduce the computational complexity of W^{-1}, the authors in [11] proposed using Neumann series expansion to approximate matrix inversion results. When W approximates the invertible matrix X and satisfies

$$\lim_{n\to\infty}(I \quad WX)^n = 0 \tag{8}$$

then, the Neumann series can be expressed as

$$W^{-1} = \Sigma_{n=0}^{\infty}\left(X^{-1}(X-W)\right)^n X \tag{9}$$

The decomposition matrix is $W = D + E$, where D is the diagonal matrix of W, and E is the hollow matrix corresponding to W. Since the number of antennas equipped at the base station was much larger than the number of single-antenna users $(N \gg K)$, matrix W has a diagonal dominant characteristic [3]; that is, $W \approx D$. Substituting D for X in Equation (9) gives:

$$W^{-1} = \Sigma_{n=0}^{\infty}\left(-D^{-1}E\right)^n D^{-1} \tag{10}$$

when $\lim_{n \to \infty} \left(-D^{-1}E\right)^n = 0$, the progression of Equation (10) converges. If we only expand the first i term of the Neumann series, we can get:

$$W_i^{-1} = \Sigma_{n=0}^{i-1} \left(-D^{-1}E\right)^n D^{-1} \tag{11}$$

when the value of i is small, the Neumann series expansion can approximate W^{-1} with lower complexity. For example, when $i = 2$, $W_2^{-1} = D^{-1} - D^{-1}ED^{-1}$, which is computationally complex, and the complexity is $O(K^2)$.

3.2. Gauss–Seidel Algorithm

In the Neumann series expansion algorithm, when the number of expansion terms $i \geq 3$, the computational complexity is still $O(K^3)$, which is equal to or even exceeds the complexity of the exact inverse calculation of the MMSE filter matrix. Unlike the Neumann series expansion, which approximates W, the GS algorithm [19] can solve N-dimensional linear equations of the form $Ax = b$ without inverting the matrix, where matrix A is an $N \times N$ dimensional symmetric positive definite matrix, x is the $N \times 1$ dimensional solution vector, and b is the $N \times 1$ dimensional measurement vector. Decomposing matrix A into a diagonal element matrix, D_A, a strict lower triangular element matrix, L_A, and a strict upper triangular element matrix, L_A^H, the GS algorithm can estimate x by the following iterative method:

$$\hat{x}^{(i)} = (D_A + L_A)^{-1} \left(b - L_A^H \hat{x}^{(i-1)}\right) \tag{12}$$

where $i = 1, 2, \ldots$ represents the number of iterations of the GS algorithm. In a massive MIMO system, as the number of base station antennas increases substantially, when it is much larger than the number of single-antenna users $(N \gg K)$, the individual column vectors of the uplink channel matrix H are progressively orthogonal [20], and $W = G + \sigma^2 I_{2K}$ is a symmetric positive definite matrix. Similarly, W can be decomposed into:

$$W = \left(D + L + L^H\right) \tag{13}$$

Among those terms, D, L, and L^H, respectively, is the diagonal element matrix of W, the strict lower triangular element matrix, and the strict upper triangular matrix. The GS algorithm can be used to avoid inverting the high-dimensional matrix, which directly estimates the transmitted signal vector $\hat{s}$:

$$\hat{s}^{(i)} = (D + L)^{-1} \left(\hat{y} - L^H \hat{s}^{(i-1)}\right) \tag{14}$$

where $\hat{s}^{(0)}$ represents the initial solution and is usually set to a zero vector.

4. Proposed Algorithm

4.1. Hybrid Iterative Algorithm Structure

The SD algorithm based on matrix gradient search can have a good convergence direction at the beginning of the iteration [18], while the GS iterative algorithm has lower complexity. Therefore, using the above characteristics, this paper proposed a hybrid iteration of the joint SD and GS algorithm. The joint algorithm (called the SDGS algorithm) speeds up convergence of the iterative effect of the algorithm without increasing the complexity, and achieves error performance close to the MMSE ideal matrix inversion detection method. The steps are in SDGS Algorithm.

SDGS Algorithm

- **Step 1:** For the diagonal approximation's initial value setting, Equation (4) can be converted to $W\hat{s} = \hat{y}$; W is a symmetric positive definite matrix and a diagonally dominant matrix, so W^{-1} is also a diagonally dominant matrix.
- **Step 2:** Determine the initial solution using D^{-1} instead of W^{-1}:

$$s^{(0)} = D^{-1}\hat{y} \tag{15}$$

Since D is a diagonal matrix, it is obvious that calculating D^{-1} requires only low complexity, and the initial value, $s^{(0)}$, is set to the initial value of the SD algorithm according to Equation (15).

- **Step 3:** The iterative results of the first two GS algorithms are represented by the SD algorithm, and the second GS iteration result can be expressed as:

$$\begin{aligned} s^{(2)} &= (D+L)^{-1}\left(\hat{y} - L^H s^{(1)}\right) = (D+L)^{-1}\left[((D+L) - W)s^{(1)} + \hat{y}\right] \\ &= s^{(1)} + (D+L)^{-1}\left(\hat{y} - Ws^{(1)}\right) = s^{(1)} + (D+L)^{-1} r^{(1)} \end{aligned} \tag{16}$$

$$\text{where } r^{(1)} = \hat{y} - W\left(s^{(0)} + ur^{(0)}\right) = \hat{y} - Ws^{(0)} - uWr^{(0)} = r^{(0)} - up^{(0)};\ u = \frac{\left(r^{(0)}\right)^H r^{(0)}}{\left(p^{(0)}\right)^H r^{(0)}} \tag{17}$$

- **Step 4:** Combine single SD and GS iterations into one hybrid iteration by substituting Equation (17) and

$$s^{(1)} = s^{(0)} + ur^{(0)} \rightarrow s^{(2)} = s^{(0)} + ur^{(0)} + (D+L)^{-1}\left(r^{(0)} - up^{(0)}\right) \tag{18}$$

This represents the first two GS iterations as Equation (18); update the mixed iteration value $\hat{s}^{(1)} = s^{(2)}$, and then perform the next GS iteration.

- **Step 5:** Using the $(i-1)$th GS iteration using Equation (14), ideal estimated value $\hat{s}^{(i)}$ of the transmitted signal vector s can be obtained by setting the appropriate number of iterations, i:

$$\hat{s}^{(i)} = (D+L)^{-1}\left(\hat{y} - L^H \hat{s}^{(i-1)}\right) \tag{19}$$

Then, $\hat{s}^{(i)}$ is related to the complex domain for the next soft decision, so the hybrid iterative algorithm can converge very quickly after a small number of iteration.

4.2. Approximate Log-Likelihood Ratio Calculation

The low-complexity MMSE signal detection algorithm described in [13–16] directly estimates the transmitted signal vector $\hat{\mathbf{s}}$ without calculating $\boldsymbol{W}^{-1}$. The exact calculation of the LLR for the channel decoder input is described in Section 1 (i.e., using the exact $\boldsymbol{W}^{-1}$ matrix inversion information), which is not difficult to find with Equation (7). When the LLR of the first bit transmitted by the $\boldsymbol{i}$**th** user is $\boldsymbol{L}_{i,b}$, the inverse $\boldsymbol{W}^{-1}$ of the MMSE detector filter matrix $\boldsymbol{W}$ needs to be used again to calculate the SINR of the $\boldsymbol{i}$**th** user. Consider using the $\boldsymbol{W}$ approximation of the diagonal property to replace $\boldsymbol{W}^{-1}$ and $\boldsymbol{D}^{-1}$, that is, $\widetilde{\boldsymbol{W}}^{-1} \approx \boldsymbol{D}^{-1}$, and then convert it to the complex domain to get $\widetilde{\boldsymbol{W}}_c^{-1}$, in order to obtain the approximate channel gain and NPI variance, expressed as:

$$\widetilde{\mu}_i = \widetilde{\boldsymbol{U}}_{ii} \tag{20}$$

$$\widetilde{v}_i^2 = \sum_{j \neq i}^{K} \left|\widetilde{\boldsymbol{U}}_{ji}\right|^2 + \widetilde{\boldsymbol{E}}_{ii}\sigma^2 \tag{21}$$

where $\widetilde{\boldsymbol{U}} \approx \widetilde{\boldsymbol{W}}_c^{-1} \boldsymbol{G}_c = \boldsymbol{D}_c^{-1}\boldsymbol{G}_c$, and $\widetilde{\boldsymbol{E}} \approx \widetilde{\boldsymbol{W}}_c^{-1}\boldsymbol{G}_c\widetilde{\boldsymbol{W}}_c^{-1} = \widetilde{\boldsymbol{U}}\widetilde{\boldsymbol{W}}_c^{-1} = \widetilde{\boldsymbol{U}}\boldsymbol{D}_c^{-1}$, so we can calculate $\boldsymbol{Y}_i = \widetilde{\mu}_i^2/\widetilde{v}_i^2$.

4.3. Complexity Analysis

According to the number of real multiplications required in the algorithm, the computational complexity of the SDGS detection algorithm proposed in this paper was analyzed. Since all linear MMSE detection algorithms and the proposed algorithm must calculate the filter matrix, $W = G + I_{2k}$, and the matched filter signal, $\hat{y} = H^H y$, then only the other parts were analyzed for complexity, mainly using the following three parts of the composition.

4.3.1. Initial Value and First Iteration Calculation

Equation (15) requires $2K$ multiplications. The first iteration is mainly to calculate $r^{(0)} = \hat{y} - Ws^{(0)}$, $p^{(0)} = Wr^{(0)}$, and scalar. Obviously, the respective $4K^2$, $4K^2$, and $4K$ sub-multiplications are required. Combining the first iteration of Equation (18), a total of $(2K^2 + 10K)$ multiplications are required.

4.3.2. GS iteration

Equation (19) can be expressed as $(D + L)\hat{s}^{(i)} = \hat{y} - L^H\hat{s}^{(i-1)} = c$. After i iterations, the calculation of $\hat{s}^{(i)}$ mainly comes from the following two steps: First, c is a $2K \times 2K$ strictly lower triangular element matrix; $2K \times 2K$ and the $2K \times 1$ vector $\hat{s}^{(i-1)}$ are multiplied, and c must be multiplied $(2K^2 - K)$ times. Second, in Equation (19), the mth element, $\hat{s}_m^{(i)}$, can be expressed as:

$$s_m^{(i)} = \begin{cases} \frac{c_1}{L_{11}}, & m = 1 \\ \frac{c_m - \sum_{k=1}^{m-1} s_k^{(i)} L_{mk}}{L_{mm}}, & m = 2, \ldots, 2K \end{cases} \quad (22)$$

where c_m represents the mth element of c, and L_{mk} represents the mth row and kth column element of the lower triangular matrix $(D + L)$. When $m = 1$, it is obvious that $\hat{s}_1^{(i)}$ requires $2K$ multiplications, and all $\hat{s}_m^{(i)}$ $(m = 2, \ldots, 2K)$ require $(2K^2 - K)$ multiplications, so a total of $2K^2$ multiplications are required for each iteration.

4.3.3. LLR calculation

The computational complexity of this part mainly came from the calculation of the effective channel gain and the NPI variance after equalization. It can be known from Equations (20) and (21) that all the elements of the matrix $\widetilde{U}$ and the pair of matrices $\widetilde{E}$ need to be calculated. Obviously, the former requires $2K^2$ multiplications, while the latter only requires $2K$ multiplications. Therefore, a total of $(2K^2 + 2K)$ multiplications were required for this step.

In summary, the total complexity required for the joint iterations to be applied to the soft decision was $2K^2(i + 2) + 12Ki$, which reduced the computational complexity by an order of magnitude, compared to the traditional MMSE algorithm. The complexity of the number of iterations was kept at $O(K^2)$. In addition, considering the application scenarios of hard decision detection, Table 1 also gave a comparison of the computational complexity in the four detection algorithms.

Table 1. Complexity comparison of four kinds of detection algorithms for hard decision calculation.

Algorithm	Real Multiplications	Performance
NS [11]	$(8K^3 - 8K^2 + 2K)(i + 2) + (4K^2 - 4K)$ $(i > 1)$	General
CG [15]	$4K^2(i + 1) + 10Ki$	Better
GS [17]	$4K^2 i + 2K$	Better
SDGS	$2K^2(i + 2) + 12Ki$	Optimal

5. Simulation Results

We deployed Matlab (R2017a, Mathworks, Natick, MA, USA) for performing analysis and experimentation. In order to verify the soft and hard detection performance of the SDGS algorithm proposed in this paper, this section presents Monte Carlo simulation results based on Matlab. The main simulation parameters configured are in Table 2.

Table 2. Simulation parameters.

Parameter	Value
Number of transmitter antennas (N)	64–128
Number of receiver antennas (K)	16
Baseband modulation mode	16 QAM
Signal-to-noise ratio (SNR)	10–12 dB
Code rate	1/2
Channel characteristics	Uncorrelated Rayleigh fading
Number of iterations (i)	1000

Figure 2 compares the bit error rate (BER) based on Neumann series (NS) expansion, the conjugate gradient (CG) detection algorithm, the Gauss–Seidel iterative detection algorithm, the MMSE exact inversion detection algorithm, and the proposed SDGS joint algorithm under different antenna configurations. The decision mode is a hard decision; that is, estimated signal vector $\hat{s}$ is directly judged. The simulation results showed that the detection performance of the various algorithm increased with the number of iterations or the number of items expanded by the Neumann series. For example, when the number of iterations $i = 2$, the BER performance of the SDGS algorithm was much better than the BER when the number of items expanded by the Neumann series was 2. By comparing the performance in Figure 2a,b, it can be seen that with the increase in the ratio of the number of base station antennas to the number of users (N/K), the BER performance of the various algorithms was greatly improved. For example, if the BER was to reach 10^{-3}, the MMSE algorithm and the proposed algorithm require an SNR of about 13 dB when the antenna configuration is 64 $\times$ 16, and only 8 dB when the configuration is 128 $\times$ 16.

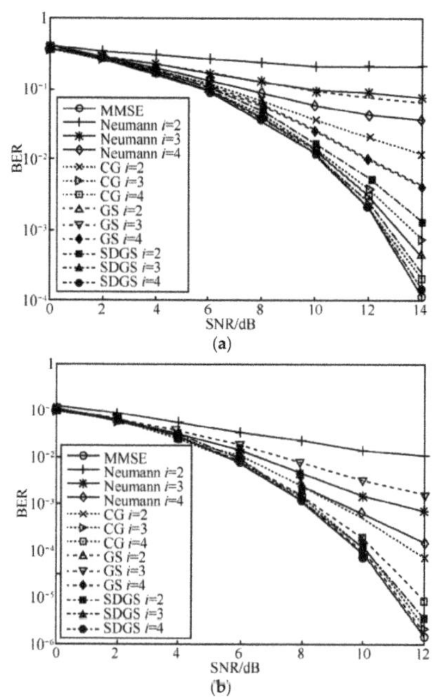

Figure 2. Hard decision bit error rate (BER) performance comparison. (**a**) Analysis at 64 $\times$ 16 antenna configuration; (**b**) Analysis at 128 $\times$ 16 antenna configuration.

Figure 3 shows the soft decision simulation results for the two antenna configurations. With BER based on the NS, CG, and GS iterative detection algorithms, the MMSE exact inversion detection algorithm and the SDGS joint algorithm were compared. We set the system's convolutional code rate to 1/2, and the LLR calculation used the approximate calculation method described in this paper. Simulation results showed that no matter what kind of MMSE receiver was used, the soft decision was checked. The measured performance was much better than the hard decision. For example, when the BER reached 10^{-4}, when the antenna was configured, the MMSE algorithm and the proposed algorithm required an SNR of 10 dB for hard decisions and only 5 dB for soft decisions. In addition, for the same number of iterations, the BER performance of the SDGS algorithm proposed in this paper was better than the other three simplified algorithms, and after a few iterations, the detection performance could quickly approach the detection performance of the ideal MMSE filter matrix inversion.

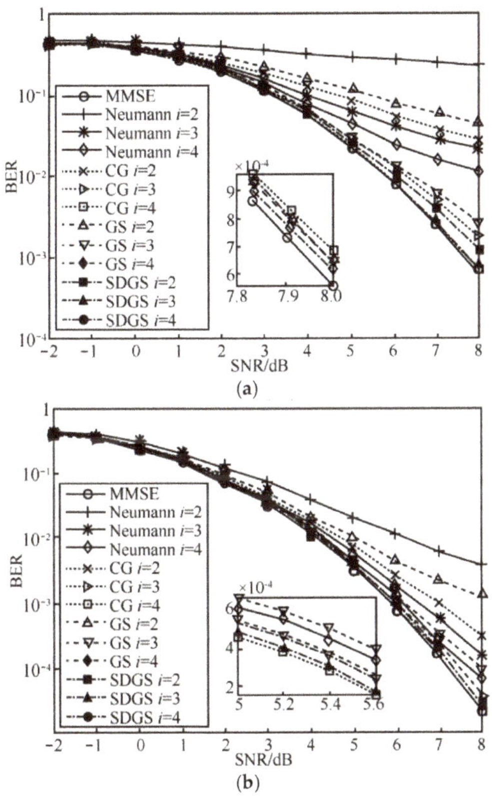

Figure 3. Soft decision bit error rate (BER) performance comparison. (**a**) Analysis at 64 × 16 antenna configuration; (**b**) Analysis at 128 × 16 antenna configuration.

Figure 4 shows the hard decision BER comparison of the proposed SDGS algorithm with NS, CG, GS and MMSE under a high fading scenario with 128 × 16 antenna configuration. As can be seen from Figure 4 the BER of the proposed SDGS algorithm was better and followed the MMSE performance with increasing SNR and number of iterations. Moreover, due to high fading impact on the SNR, there was a gap between the proposed SDGS algorithm and MMSE algorithm at a high SNR level.

Figure 5 shows the BER comparison of the proposed SDGS algorithm with NS, CG, GS and MMSE under a low fading level and 128 × 16 antenna configuration. It can be seen from Figure 5 that all the algorithms showed lower BER and better performance as compared with the hard decision BER performance in Figures 2 and 4. Therefore, to keep the system performance in a suitable level, the fading level and number of iterations should be considered, which has an obvious impact on the system's overall performance. Furthermore, the proposed SDGS algorithm in Figure 5 had a close BER performance with MMSE which indicated that the SDGS algorithm showed better performance in the low fading level.

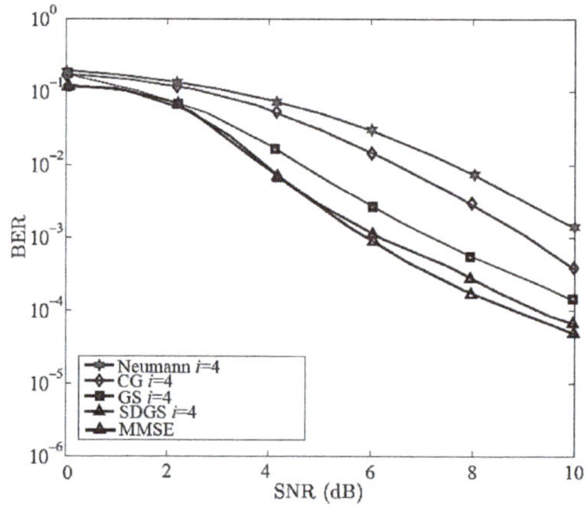

Figure 4. Hard decision bit error rate (BER) performance comparison with high fading level at 128 × 16 antenna configuration.

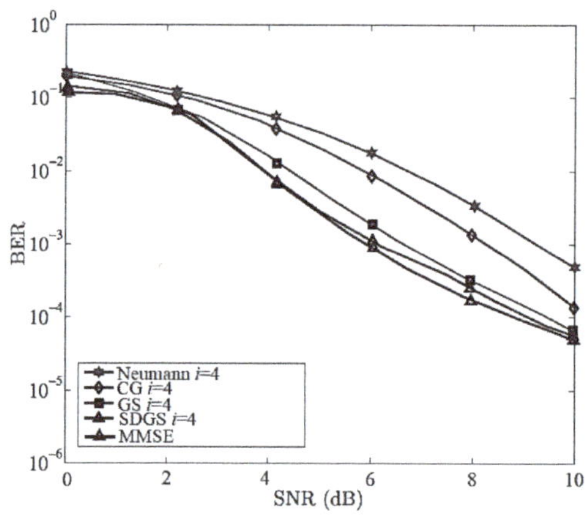

Figure 5. Soft decision bit error rate (BER) performance comparison with slow fading level at 128 × 16 antenna configuration.

6. Conclusions

Signal detection methods based on MMSE filtering in massive MIMO systems are widely used, but matrix inversion with higher complexity makes it more difficult to implement them in practical applications. Some methods of approximate inversion, such as Neumann series expansion, has reduced the detection complexity, but due to a large degree of detection performance loss; others avoid the complex matrix inversion and directly estimate the signal vector. Although, computational complexity is reduced by orders of magnitude, detection performance needs to be improved. Based on the MMSE criterion, this paper proposes a low-complexity, hybrid, iterative SDGS joint detection algorithm, which directly estimates the user's transmitted symbol vector and can quickly converge to obtain an ideal estimation value with a few simple iterations. The matrix inversion operation is avoided, and algorithm complexity is kept at $O(K^2)$. In addition, in order to make full use of soft information, the algorithm is applied to the soft decision, and an approximate calculation method of the LLR for channel decoding is given, which further improves the signal detection performance. Theoretical derivation and simulation results show that the SDGS algorithm can be used as one of the most effective solutions for signal detection in massive MIMO systems.

Author Contributions: I.K. conceived and designed the presented idea and developed the theory, performed the simulations, and wrote the paper. M.H.Z. analyzed the research and performed experimentations, and M.A. provided extensive technical support throughout the research. S.K. provided extensive support in the theoretical analysis and provided funding support.

Funding: This research was supported by the Basic Science Research Program through the National Research Foundation of Korea (NRF-2016R1D1A1B03934653).

Conflicts of Interest: The authors declare no conflicts of interest.

References

1. Chang, Y.K.; Ueng, F.B.; Jhang, Y.W. Turbo MIMO-OFDM Receiver in Time-Varying Channels. *KSII Trans. Internet Inf. Syst.* **2018**, *12*, 3704–3724.
2. Tseng, S.M.; Chen, Y.F. Average PSNR Optimized Cross Layer User Grouping and Resource Allocation for Uplink MU-MIMO OFDMA Video Communications. *IEEE Access* **2018**, *6*, 50559–50571. [CrossRef]
3. Ghosh, A.; Ratasuk, R.; Mondal, B.; Mangalvedhe, N.; Thomas, T. LTE-advanced: Next-generation wireless broadband technology. *IEEE Wirel. Commun.* **2010**, *17*, 10–22. [CrossRef]
4. Khan, I.; Zafar, M.H.; Jan, M.T.; Lloret, J.; Basheri, M.; Singh, D. Spectral and Energy Efficient Low-Overhead Uplink and Downlink Channel Estimation for 5G Massive MIMO Systems. *Entropy* **2018**, *20*, 92. [CrossRef]
5. Rusek, F.; Persson, D.; Lau, B.K.; Larsson, E.G.; Marzetta, T.L.; Edfors, O.; Tufvesson, F. Scaling up MIMO: Opportunities and challenges with very large arrays. *IEEE Signal Process. Mag.* **2012**, *30*, 40–60. [CrossRef]
6. Larsson, E.G.; Edfors, O.; Tufvesson, F.; Marzetta, T.L. Massive MIMO for next-generation wireless systems. *IEEE Commun. Mag.* **2014**, *52*, 186–195. [CrossRef]
7. Marzetta, T.L. Noncooperative cellular wireless with unlimited numbers of base station antennas. *IEEE Trans. Wirel. Commun.* **2010**, *9*, 3590–3600. [CrossRef]
8. Khan, I.; Singh, D. Efficient compressive sensing based sparse channel estimation for 5G massive MIMO systems. *AEU-Int. J. Electron. Commun.* **2018**, *89*, 181–190. [CrossRef]
9. Ngo, H.Q.; Larsson, E.G.; Marzetta, T.L. Energy and spectral efficiency of very large multiuser MIMO systems. *IEEE Trans. Commun.* **2011**, *61*, 1436–1449.
10. Dai, L.; Wang, Z.; Yang, Z. Spectrally efficient time-frequency training OFDM for mobile large-scale MIMO systems. *IEEE J. Sel. Areas Commun.* **2013**, *31*, 251–263. [CrossRef]
11. Benmimoune, M.; Driouch, E.; Ajib, W. Joint antenna selection in grouping in Massive MIMO Systems. In Proceedings of the 10th International Symposium on Communication Systems, Networks and Digital Signal Processing (CSNDSP), Prague, Czech Republic, 20–22 July 2016; pp. 1–6.
12. Khan, I. A Robust Signal Detection Schemefor 5G Massive Multiuser MIMO Systems. *IEEE TVT.* **2018**, *67*, 9567–9604.
13. Wu, M.; Yin, B.; Wang, G.; Dick, C.; Cavallaro, J.R.; Studer, C. Large-scale MIMO detection for 3GPP LTE: Algorithms and FPGA implementations. *IEEE J. Sel. Top. Signal Process.* **2014**, *85*, 916–929. [CrossRef]
14. Tang, C.; Liu, C.; Yuan, L.; Xing, Z. High precision low complexity matrix inversion based on Newton iteration for data detection in the massive MIMO. *IEEE Commun. Lett.* **2016**, *20*, 490–493. [CrossRef]
15. Gao, X.; Dai, L.; Yuen, C.; Zhang, Y. Low-complexity MMSE signal detection based on Richardson method for large-scale MIMO systems. In Proceedings of the Vehicular Technology Conference, Vancouver, BC, Canada, 14–17 September 2014; pp. 1–5.
16. Gao, X.; Dai, L.; Hu, Y.; Wang, Z.; Wang, Z. Matrix inversion-less signal detection using SOR method for uplink large-scale MIMO systems. In Proceedings of the IEEE Global Communications Conference, Qingdao, China, 8–12 December 2014; pp. 3291–3295.

17. Hu, Y.; Wang, Z.; Gaol, X.; Ning, J. Low-complexity signal detection using the CG method for uplink large-scale MIMO systems. In Proceedings of the IEEE International Conference on Communications, Macau, China, 19–21 November 2014; pp. 477–481.
18. Hageman, L.A. *The Iterative Solution for Large Linear Systems*; Academic Press: Manhattan, NY, USA, 1971; pp. 9–22.
19. Dai, L.; Gao, X.; Su, X.; Han, S.; Chih-Lin, I.; Wang, Z. Low-complexity soft-output signal detection based on a Gauss-Seidel method for uplink multiuser large-scale MIMO systems. *IEEE Trans. Veh. Technol.* **2014**, *64*, 4839–4845. [CrossRef]
20. Wu, M.; Yin, B.; Vosoughi, A.; Studer, C.; Cavallaro, J.R.; Dick, C. Approximate matrix inversion for high-throughput data detection in the large-scale MIMO uplink. In Proceedings of the International Symposium on Circuits & Systems, Beijing, China, 19–23 May 2013; pp. 12155–12158.

electronics

MDPI

Article

A Novel Iterative Discrete Estimation Algorithm for Low-Complexity Signal Detection in Uplink Massive MIMO Systems

Hui Feng [1], Xiaoqing Zhao [2], Zhengquan Li [2,*] and Song Xing [3]

1 Department of Electronic Science and Technology, Tongji University, Shanghai 201804, China
2 Jiangsu Provincial Engineering Laboratory of Pattern Recognition and Computational Intelligence, Jiangnan University, Wuxi 214122, China
3 Department of Information Systems, California State University, Los Angeles, CA 90032, USA
* Correspondence: lzq722@jiangnan.edu.cn; Tel.: +86-0510-8591-0633

Received: 1 August 2019; Accepted: 29 August 2019; Published: 2 September 2019

Abstract: In this paper, a novel iterative discrete estimation (IDE) algorithm, which is called the modified IDE (MIDE), is proposed to reduce the computational complexity in MIMO detection in uplink massive MIMO systems. MIDE is a revision of the alternating direction method of multipliers (ADMM)-based algorithm, in which a self-updating method is designed with the damping factor estimated and updated at each iteration based on the Euclidean distance between the iterative solutions of the IDE-based algorithm in order to accelerate the algorithm's convergence. Compared to the existing ADMM-based detection algorithm, the overall computational complexity of the proposed MIDE algorithm is reduced from $O\left(N_t^3\right) + O\left(N_r N_t^2\right)$ to $O\left(N_t^2\right) + O\left(N_r N_t\right)$ in terms of the number of complex-valued multiplications, where Ntand Nr are the number of users and the number of receiving antennas at the base station (BS), respectively. Simulation results show that the proposed MIDE algorithm performs better in terms of the bit error rate (BER) than some recently-proposed approximation algorithms in MIMO detection of uplink massive MIMO systems.

Keywords: massive MIMO systems; MIDE algorithm; low computational complexity; BER

---

1. Introduction

With the development of the mobile Internet and the Internet of Things, much high data rate communication is required in the new generation of cellular networks like 5G [1]. By equipping hundreds of antennas at the base station (BS) serving tens of users, the massive multiple-input multiple-output (MIMO) is deemed one key technology for meeting the 5G requirements due to its improvements in data throughput, link reliability, higher spectral efficiency, and better communication quality compared with the traditional MIMO usage [2–4].

However, when applying the massive MIMO, a major computational challenge is the data detection in uplink MIMO systems due to the large increase in the system dimensions [5]. The maximum likelihood (ML) is the optimal detection approach on data detection, but its computational complexity grows exponentially with the number of user equipment (UE) and the modulation order [6,7]. Some suboptimal detection alternatives are proposed to reduce the computational complexity while obtaining a good bit error rate (BER) performance. For example, the linear minimum mean squared error (LMMSE) algorithm is one of the widely-used suboptimal detection algorithms with near-optimal BER performance and reduced computational complexity [8]. However, the LMMSE algorithm still involves the computation of the Gram matrix, as well as matrix inversion, where their corresponding computational complexity is $O\left(N_r N_t^2\right)$ and $O\left(N_t^3\right)$, respectively, with N_t denoting the number of single-antenna UE and N_r denoting the number of antennas at the BS.

It is worth noting that some approaches for approximating the matrix inversion have been proposed to reduce the computational complexity [9–19], among which, for example, the Neumann series (NS) approximation is used to approximate the matrix inversion by a series of truncated NS expansions [9,10]. However, only marginal reduction in the computational complexity can be reached with the increased terms of the NS expansion. Hence, various classical iterative algorithms have been provided to approximate the inverse matrix in LMMSE detection to achieve low computational complexity, which include the Richardson (RI) algorithm [11], the Jacobi algorithm [12–15], the successive over relaxation (SOR) algorithm [16], the symmetric successive over relaxation (SSOR) algorithm [17], and the Gauss–Seidel (GS) algorithm [18,19]. The computational complexity of the matrix inverse is reduced by these approximation-based detection algorithms from $O\left(N_t^3\right)$ to $O\left(N_t^2\right)$. It is noted that these algorithms involving the inverse-matrix approximation in LMMSE detection (also known as the approximated LMMSE) achieve a near-LMMSE performance, but with lower computational complexity.

It is well known that the approximated LMMSE detection algorithms achieve a substantial performance loss when $N_r > N_t$ in multiuser massive MIMO systems. Various algorithms have been proposed to obtain a better BER performance than that of LMMSE detection in multiuser massive MIMO systems, which include the non-convex and the convex optimization algorithms [20–22].

For example, the alternating minimization (AltMin) algorithm is one of the non-convex optimization algorithms, which is applied to the data detection in a multiuser massive MIMO system [20]. Specifically, the AltMin algorithm converts the ML detection problem into a sum of convex functions by decomposing the received vector into multiple sub-vectors. Hence, the non-convex problem is transformed into the convex problem in the AltMin algorithm. The AltMin algorithm shows better BER performance than that of the LMMSE detector in overloaded network scenarios with relatively low computational complexity. However, it shows near-LMMSE performance with even higher computational complexity when the ratio of the number of BS antennas to the number of single-antenna users is larger.

Similarly, some convex optimization algorithms are used to solve the non-convex optimization problems, which include, for example, the alternating direction method of multipliers (ADMM) detection algorithm [21,22]. In the multiuser massive MIMO system, ADMM demonstrates better BER performance than the LMMSE detection algorithm with the relatively low computational complexity of the iterative procedure. However, the computational complexity of the preprocessing in the ADMM algorithm includes the calculation complexity of the Gram matrix and LDLdecomposition [23,24], which results in very high computational complexity for massive MIMO systems.

To make a tradeoff between the performance and the computational complexity with different antenna configurations, the iterative discrete estimation (IDE) algorithm is integrated into the ADMM algorithm [25], which presents low computational complexity due to the avoidance of the calculation of the Gram matrix and LDL decomposition. Motivated by the aforementioned algorithms, we propose a modified IDE (MIDE) algorithm to present a better BER performance and lower computational complexity than the ADMM algorithms. To summarize, the main contributions of this work are listed as follows.

- A novel iterative data detection algorithm for uplink multiuser massive MIMO systems is designed by exploiting the IDE-based algorithmic framework. The proposed MIDE algorithm refactors the detection algorithm as a series of simpler subproblems with closed-form solutions.
- A heuristic damping factor is defined based on the Euclidean distance instead of a fixed factor. Compared with the fixed damping factor, this self-updated damping factor contributes to a faster convergence in the proposed MIDE algorithm.
- The computational complexity analysis indicates that the proposed algorithm has a lower computational complexity than the traditional approximated detection approximation algorithms (LMMSE, AltMin, and ADMM), under the same BER performance. Specifically, the complexity of the novel MIDE detection algorithm is only $O\left(N_t^2\right) + O\left(N_r N_t\right)$.

• Simulation results reveal that for the typical independent and identically distributed (i.i.d.) frequency flat Rayleigh fading channel in massive MIMO systems, the proposed MIDE detection algorithm performs better than the ADMM and AltMin-based detection algorithms and the LMMSE detection algorithm in terms of BER performance with various system configurations.

The rest of the paper is organized as follows. In Section 2, we briefly introduce the system model. Section 3 specifies the proposed low-complexity signal detection based on the IDE algorithm and performs the computational complexity analysis of the algorithms. In Section 4, the numerical simulation results of the BER performance are presented. Finally, Section 5 concludes the paper.

Notation: Throughout the paper, the lowercase and uppercase boldface type is used for vectors (e.g., **a**) and matrices (e.g., **A**). The superscripts $(\cdot)^{-1}$ and $(\cdot)^{H}$ denote the matrix inversion and the conjugate transpose, respectively. The L_2 norms of the vectors are represented by $\|\cdot\|_2$. $\Re$ and $\Im$ denote the real part and the imaginary part of the complex-valued signal, respectively.

The typical uplink massive MIMO system is considered in this work, as shown in Figure 1, where there are N_t single-transmitting antenna UE devices and N_r receiving antennas at the BS [26]. In general, N_r is larger than N_t for an uplink massive MIMO communication system [27].

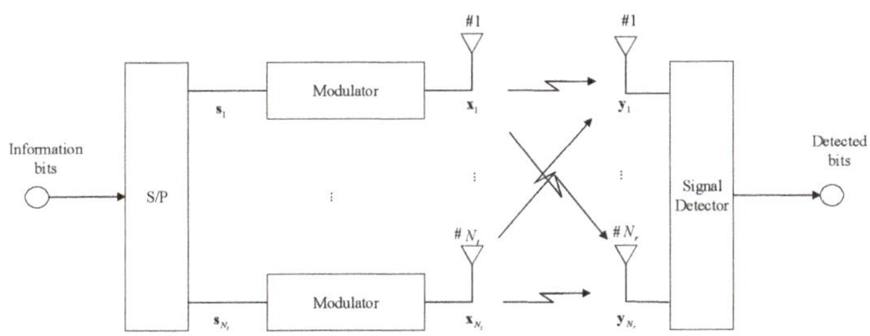

Figure 1. MIMO system signal detection structure diagram.

2. System Model

At the transmitter side, the source information $\mathbf{s} = [\mathbf{s}_1, \cdots, \mathbf{s}_i, \cdots \mathbf{s}_{N_t}]^T$ where each symbol s is mapped to constellation symbols by taking symbols from a set of the constellation alphabet Ω. The transmitted signal $\mathbf{x} = [\mathbf{x}_1, \cdots, \mathbf{x}_i, \cdots \mathbf{x}_{N_t}]^T$ is constructed by the modulated symbol $\mathbf{s}$, where $\mathbf{x}_i$ denotes a signal transmitted by the ith UE device. The vector $\mathbf{y} = [\mathbf{y}_1, \cdots, \mathbf{y}_i, \cdots \mathbf{y}_{N_r}]^T$ represents the receiving signal at the BS, where $\mathbf{y}_i$ denotes a signal received by the ith receive antenna, and:

$$\mathbf{y} = \mathbf{H}\mathbf{x} + \mathbf{w} \tag{1}$$

where $\mathbf{w}$ is the N_r-by-one additive white Gaussian noise (AWGN) vector following $CN(0, \sigma^2)$ with σ^2 representing the average power of the noise. In Equation (1), the matrix $\mathbf{H}$ denotes the N_r-by-N_t flat fading channel gain, and $\mathbf{H}$ can be expressed as:

$$\mathbf{H} = \begin{bmatrix} h_{11} & h_{12} & \cdots & h_{1N_t} \\ h_{21} & h_{22} & \cdots & h_{2N_t} \\ \vdots & \vdots & \ddots & \vdots \\ h_{N_r1} & h_{N_r2} & \cdots & h_{N_rN_t} \end{bmatrix} \tag{2}$$

where the element $h_{i,j}$, $i = \{1, 2, \cdots, N_r\}$,$j = \{1, 2, \cdots, N_t\}$ denotes the channel impulse response between the receiving antenna j and the user i. It is noted that h_{ij} follows an i.i.d. Gaussian distribution with zero mean and unit variance. In addition, the channel matrix $\mathbf{H}$ is assumed to be known perfectly at the BS [28,29].

3. Proposed MIDE Algorithm

We first propose a low-complexity signal detection algorithm, which converts the ML problem into the constrained non-convex problem and utilizes the IDE-based algorithm to solve the problem iteratively. Then, a self-updating method is proposed to update the damping factor based on the Euclidean distance, which accelerates the convergence in the proposed MIDE algorithm. Finally, the computational complexity analysis of the proposed algorithm is provided with the comparison to the conventional algorithms.

3.1. ML Problem Formulation and IDE-Based Algorithm

Detecting the transmitted symbol vector $\mathbf{x}$ at the BS can be done by minimizing the squared Euclidean distance between the received signal vector $\mathbf{y}$ and the hypothesized received signal $\mathbf{Hx}$ with the vector $\mathbf{x}$ constrained to the modulation constellation Ω^{N_t}, which can be represented as:

$$\hat{\mathbf{x}} = \arg\min_{x \in \Omega^{N_t}} \|\mathbf{y} - \mathbf{Hx}\|_2^2 \tag{3}$$

It is noted that in Equation (3), the finite-alphabet constraint $x \in \Omega^{N_t}$ can be converted into the indicator function $\mathbf{I}_\Omega(\hat{\mathbf{x}})$, which is given by:

$$I_\Omega(\hat{\mathbf{x}}) = \begin{cases} 0, \text{if} \hat{\mathbf{x}} \in \Omega^{N_t} \\ \infty, \text{otherwise} \end{cases} \tag{4}$$

By combining Equation (3) and (4), the signal detection problem can be converted into the constrained optimization problem, which is given by:

$$\begin{aligned} &\underset{\mathbf{z},\hat{\mathbf{x}}}{\text{minimize}} \ \|\mathbf{y} - \mathbf{Hz}\|_2^2 + \mathbf{I}_\Omega(\hat{\mathbf{x}}) \\ &\text{s.t. } \mathbf{z} - \hat{\mathbf{x}} = \mathbf{0} \end{aligned} \tag{5}$$

where $\mathbf{z}$ is the least-squares solution of the least-squares and $\hat{\mathbf{x}}$ is the estimated transmitted symbol. Since the optimization problem is defined over complex-valued variables, the Lagrangian objective function for the optimization problem in Equation (5) can be remodeled as:

$$L_\gamma(\mathbf{z}, \hat{\mathbf{x}}, \mathbf{u}) = \|\mathbf{y} - \mathbf{Hz}\|_2^2 + \mathbf{I}_\Omega(\hat{\mathbf{x}}) + \gamma\|\mathbf{z} - \hat{\mathbf{x}}\| + \mathbf{u}^H(\mathbf{z} - \hat{\mathbf{x}}) \tag{6}$$

where $\gamma > 0$ is the penalty parameter and $\mathbf{u}$ is the dual vector. In order to solve this problem efficiently, we decompose it into three sub-problems. First, we solve $\mathbf{z}$ while holding $\hat{\mathbf{x}}$ and $\mathbf{u}$ fixed; then, we solve $\hat{\mathbf{x}}$ while holding $\mathbf{z}$ and $\mathbf{u}$ fixed; finally, we solve $\mathbf{u}$ while holding $\hat{\mathbf{x}}$ and $\mathbf{z}$ fixed. Specifically, the following procedure is repeated with iterations.

$$\mathbf{z}^k = \arg\min_{\mathbf{z}} L_\gamma\left(\mathbf{z}, \hat{\mathbf{x}}^{k-1}, \mathbf{u}^{k-1}\right) \tag{7}$$

$$\hat{\mathbf{x}}^k = \arg\min_{\hat{\mathbf{x}}} L_\gamma\left(\mathbf{z}^k, \hat{\mathbf{x}}^k, \mathbf{u}^{k-1}\right) \tag{8}$$

$$\mathbf{u}^k = \mathbf{u}^{k-1} + \gamma\left(\mathbf{z}^k - \hat{\mathbf{x}}^k\right) \tag{9}$$

Note that the $\mathbf{z}$-minimization procedure is convex, but the $\hat{\mathbf{x}}$-update is projected onto a convex set Ω^{N_t}. To make the iterative procedure converge, the IDE-based algorithm is applied to remove the dual vector $\mathbf{u}$ at each iterative and turn the $\mathbf{z}$-update and $\hat{\mathbf{x}}$-update to reach a consensus. After manipulation, the $\hat{\mathbf{x}}$-update involves solving a linearly-constrained minimum Euclidean-norm problem, and the $\mathbf{z}$-update in IDE is given by:

$$\mathbf{z}^k = \hat{\mathbf{x}}^{k-1} + \left[diag\left(\mathbf{H}^H\mathbf{H}\right)\right]^{-1}\mathbf{H}^H\left(\mathbf{y} - \mathbf{H}\hat{\mathbf{x}}^{k-1}\right) \tag{10}$$

Hence, the $\hat{\mathbf{x}}$-update step can be represented as:

$$\hat{\mathbf{x}}^{k+1} = \Pi_\Omega\left(\mathbf{z}^{k+1}\right) \tag{11}$$

where $\Pi_\Omega(\cdot)$ is the projection onto Ω^{N_t}, i.e., the elements of $\mathbf{z}^{k+1}$ can be implemented through simple rounding of each component to the closest element in Ω.

Additionally, to make the iteration converge, a fixed damping factor α is employed in the IDE-based algorithm to update the iterative solution $\hat{\mathbf{x}}$, i.e.,

$$\hat{\mathbf{x}}_d^{k+1} = \left(1 - \alpha^k\right)\hat{\mathbf{x}}_d^k + \alpha^k\hat{\mathbf{x}}^{k+1} \tag{12}$$

where $\hat{\mathbf{x}}_d^{k+1}$ denotes the solution after updating. By applying $\hat{\mathbf{x}}_d$ to the $\mathbf{z}$-update, the IDE-based algorithm for detection can be expressed as:

$$\mathbf{z}^k = \hat{\mathbf{x}}_d^{k-1} + \left[diag\left(\mathbf{H}^H\mathbf{H}\right)\right]^{-1}\mathbf{H}^H\left(\mathbf{y} - \mathbf{H}\hat{\mathbf{x}}_d^{k-1}\right) \tag{13}$$

$$\hat{\mathbf{x}}^k = \Pi_\Omega\left(\mathbf{z}^k\right) \tag{14}$$

$$\hat{\mathbf{x}}_d^k = \left(1 - \alpha^k\right)\hat{\mathbf{x}}_d^{k-1} + \alpha^k\hat{\mathbf{x}}^k \tag{15}$$

3.2. Modified IDE-Based Detection Algorithm with Self-Update Damping

The performance of the IDE-based detection algorithm is influenced by the choice of the damping factor. In early studies, this parameter was fixed, e.g., 0.05 in [25]. However, a fixed damping factor is not applicable for all cases, and an optimal damping factor is not easy to obtain. We propose an MIDE herein to decide a proper damping factor α, by which the $\hat{\mathbf{x}}_d$-update step is analyzed as follows.

$$\begin{aligned}
\hat{\mathbf{x}}_d^1 &= \left(1 - \alpha^0\right)\hat{\mathbf{x}}_d^0 + \alpha^0\hat{\mathbf{x}}^1, \\
\hat{\mathbf{x}}_d^2 &= \left(1 - \alpha^1\right)\hat{\mathbf{x}}_d^1 + \alpha^1\hat{\mathbf{x}}^2 \\
&= \left(1 - \alpha^1\right)\left(\left(1 - \alpha^0\right)\hat{\mathbf{x}}_d^0 + \alpha^0\hat{\mathbf{x}}^1\right) + \alpha^1\hat{\mathbf{x}}^2 \\
&= \left(1 - \alpha^1\right)\left(1 - \alpha^0\right)\hat{\mathbf{x}}_d^0 + \left(1 - \alpha^1\right)\alpha^0\hat{\mathbf{x}}^1 + \alpha^1\hat{\mathbf{x}}^2 \\
&\vdots \\
\hat{\mathbf{x}}_d^{k+1} &= \left(1 - \alpha^k\right)\hat{\mathbf{x}}_d^k + \alpha^k\hat{\mathbf{x}}^{k+1} \\
&= \left(1 - \alpha^{k-1}\right)\cdots\left(1 - \alpha^0\right)\hat{\mathbf{x}}_d^0 + \left(1 - \alpha^{k-1}\right)\cdots\left(1 - \alpha^1\right)\alpha^0\hat{\mathbf{x}}^1 + \cdots + \alpha^{k-1}\hat{\mathbf{x}}^{k+1}
\end{aligned} \tag{16}$$

Since there is no prior information of the final result $\hat{\mathbf{x}}_d$, the initial value of $\hat{\mathbf{x}}_d^0$ can be set as a zero vector. Hence, the expression for $\hat{\mathbf{x}}_d$ in Equation (16) is composed of the solution of $\hat{\mathbf{x}}$ with different values of the damping factor. Based on the expression in Equation (16), the convergence of the iterations can be measured by the difference between $\hat{\mathbf{x}}_d^k$ and $\hat{\mathbf{x}}^{k+1}$. Specifically, the Euclidean distance, which is one of the widely-used approaches for measuring the distance between two vectors [30], is defined as:

$$d^k\left(\hat{\mathbf{x}}_d^k, \hat{\mathbf{x}}^{k+1}\right) = \sqrt{\sum_{i=1}^{N_t}\left(\hat{\mathbf{x}}_d^k(i,1) - \hat{\mathbf{x}}^{k+1}(i,1)\right)^2} \tag{17}$$

We can notice from Equation (17) that the smaller the distance d, the closer the two vectors, implying the convergence of the IDE iterations. Based on this discussion, the following heuristic damping factor at the kth iteration is defined as:

$$\alpha^k = \frac{d^k}{d^k + q} \tag{18}$$

where q is a positive constant. Obviously, a higher value of d leads to a higher α, and vice versa. In other words, when $d^k \to 0$, $\alpha^k \to 0$, and when $d^k \to 1$, $\alpha^k \to 1$. We utilize the first iteration result d^1 to obtain the value of p. According to Equation (16), d^1 can be calculated as:

$$\begin{aligned} d^1 &= \sqrt{\sum_{i=1}^{N_t} \left(\hat{\mathbf{x}}_d^0(i,1) - \hat{\mathbf{x}}^1(i,1)\right)^2} \\ &= \sqrt{\sum_{i=1}^{N_t} \left(\hat{\mathbf{x}}^1(i,1)\right)^2} \\ &= \sqrt{\sum_{i=1}^{N_t} \left(\Re\left(\hat{\mathbf{x}}^1(i,1)\right)\right)^2 + \left(\Im\left(\hat{\mathbf{x}}^1(i,1)\right)\right)^2} \end{aligned} \tag{19}$$

It is noticed that the value of d^1 is varying and decided based on several factors such as the modulation method, the noise in the system, etc. For the ease of calculation, the expectation of d^1 is computed instead of the direct calculation of d^1. Since the real part and the imaginary part of the vector $\hat{\mathbf{x}}^1$ have the same uniform distribution, the expectation for d^1 is obtained as:

$$E\left(d^1\right) = 2E\left(\sqrt{\sum_{i=1}^{N_t} \left(\Re\left(\hat{\mathbf{x}}^1(i,1)\right)\right)^2}\right) \tag{20}$$

The expectation of d^1 is based on the constellation points of the modulation scheme. With 16-QAM, for example, $\Re\left(\hat{\mathbf{x}}^1(i,1)\right)$ can be $\{-3,-1,+1,+3\}$, and the probability p of each possible value of the point is the same without the prior information, i.e., $p = 0.25$ in this case. Hence, Equation (19) can be rewritten as:

$$E\left(d^1\right) = 2E\left(\sqrt{\sum_{i=1}^{N_t} \left(\Re\left(\hat{\mathbf{x}}^1(i,1)\right)\right)^2}\right) = 2N_t\sqrt{\frac{\sum cand^2}{\sqrt{M}}} \tag{21}$$

where $cand$ represents the candidate value set of $\Re\left(\hat{\mathbf{x}}^1(i,1)\right)$ and M represents the modulation cardinality. For the ease of description, we still take the 16-QAM as an example. Then, $cand$ is denoted as the candidate values $\{-3,-1,+1,+3\}$ and $\sum cand^2 = (-3)^2 + (-1)^2 + 1^2 + 3^2$. M represents the modulation cardinality. Based on experience, the value of $\alpha^1 = 0.8$ is set as 0.8. Then, substituting Equation (22) into Equation (18) yields the estimation of q as:

$$q = \frac{E\left(d^1\right)}{4} = \frac{N_t}{2}\sqrt{\frac{\sum cand^2}{\sqrt{M}}} \tag{22}$$

The simulation results in Section 4 show that the proposed MIDE algorithm can improve the BER performance significantly compared to the IDE-based algorithm, which employs the fixed damping factor. The procedure of the proposed MIDE detection algorithm is illustrated in Algorithm 1.

Algorithm 1: The pseudocode of the MIDE detection algorithm.

Input: 1) $\mathbf{H}$, the channel matrix
2) $\mathbf{y}$, the received signal matrix
3) K, the number of iterations
Output: the detected signal $\tilde{\mathbf{x}} \leftarrow \hat{\mathbf{x}}^K$
Preprocessing
$\mathbf{D} \leftarrow \left[diag\left(\mathbf{H}^H\mathbf{H}\right)\right]^{-1}$;
$\mathbf{W} \leftarrow \mathbf{D}\mathbf{H}^H$;
$q \leftarrow \frac{N_t}{2}\sqrt{\frac{\sum cand^2}{\sqrt{M}}}$;
Initialization
$\hat{\mathbf{x}}_d^0 = \mathbf{0}$;
Iterative process (repeat K times)
for $k = 1$–K (iteration times) **do**
$\mathbf{z}^k \leftarrow \hat{\mathbf{x}}_d^{k-1} + \mathbf{W}\left(\mathbf{y} - \mathbf{H}\hat{\mathbf{x}}_d^{k-1}\right)$
$\hat{\mathbf{x}}^k \leftarrow \Pi_\Omega\left(\mathbf{z}^k\right)$
$\mathbf{r}^k \leftarrow \hat{\mathbf{x}}_d^{k-1} - \hat{\mathbf{x}}^k$
$d^k \leftarrow \sqrt{\sum_{i=1}^{N_t}\left(\Re\left(\mathbf{r}^k(i,1)\right)\right)^2 + \left(\Im\left(\mathbf{r}^k(i,1)\right)\right)^2}$
$\alpha^k \leftarrow \frac{d^k}{d^k+q}$;
$\hat{\mathbf{x}}_d^k \leftarrow \left(1-\alpha^k\right)\hat{\mathbf{x}}_d^{k-1} + \alpha^k\hat{\mathbf{x}}^k$;
end for

3.3. Analysis of the Complexity of the Algorithm

In this subsection, we analyze the computational complexity of the proposed MIDE algorithm, which is dominated by the multiplications operations. Hence, we compute the number of complex-valued multiplications as the measurement of the computational complexity of the algorithm [31].

It can be found in Algorithm 1 that the computational complexity is composed of three parts including (1) preprocess, (2) the $\mathbf{x}$-update procedure, and (3) the $\mathbf{x}_d$-update procedure.

(1) Preprocess: The first part comes from the related computation before the iterative process. The main factors affecting the computational complexity of the preprocess are the computation of $\mathbf{D}$ and the multiplication of the $N_t \times N_t$ diagonal matrix and the $N_t \times N_r$ matrix. Let $\mathbf{h}_i$ be the ith column of the complex-valued channel matrix $\mathbf{H}$. Then, the diagonal calculation can be presented as:

$$\mathbf{D}_{ij} = \begin{cases} \frac{1}{\mathbf{h}_i^H\mathbf{h}_j}, & i = j \\ 0, & i \neq j \end{cases} \tag{23}$$

Therefore, the complexity of the preprocessing is counted as $N_t^2 + N_rN_t$.

(2) $\mathbf{x}$-update procedure: The second part comes from the x-update procedure, which involves the computation of two multiplications of the $N_t \times N_t$ matrix and the $N_t \times 1$ vector. Thus, the complexity is counted as $2N_rN_t$.

(3) $\mathbf{x}_d$-update procedure: The third part originates from updating the value of $\mathbf{x}_d$. As can be seen in the expression of $\mathbf{x}_d^k$, the computation of this part includes the update of the Euclidean distance d^k

and two scalar multiplications with $N_t \times 1$ vectors. Then, the complexity in this part is counted as $3N_t$.

Therefore, the overall required number of complex multiplications by the MIDE algorithm is $N_t N_r + N_t^2 + K\left(2N_r N_t + 3N_t\right)$. For comparison, the similar calculations of the complexity of the typical detection algorithms (e.g., LMMSE with full matrix inversion, AltMin, ADMM) are given, as well as the one of the proposed MIDE in Table 1.

Table 1. Complexity comparison in terms of complex multiplication operations. LMMSE, linear minimum mean squared error; AltMin, alternating minimization; ADMM, alternating direction method of multipliers; MIDE, modified iterative discrete estimation.

Algorithm	Complexity
LMMSE [8]	$\frac{5}{6}N_t^3 + \frac{3}{4}N_t^2 + \frac{4}{3}N_t + N_r N_t^2 + N_r N_t$
AltMin [18]	$(4K+2)\, N_r N_t$
ADMM [20]	$\frac{1}{2}N_t^3 + N_t^2 + N_r N_t^2 + N_r N_t + K\left(N_t^2 + N_r N_t\right)$
MIDE	$N_t N_r + N_t^2 + K\left(2N_r N_t + 3N_t\right)$

Note that all these algorithms utilize the approximation approaches to solve the ML problem. It is obvious that the proposed MIDE algorithm and AltMin algorithm have lower computational complexity at each iteration among the compared iterative approaches. The numerical analysis will be provided for further analysis of the computational complexity, which depends on the number of iterations K and is presented in Section 4.

4. Simulation Results

4.1. BER Performance Evaluation

The BER performance of the proposed algorithm was evaluated and compared with the ones of other detection algorithms by numerical simulations. The simulation parameters are listed in Table 2. Several typical detection algorithms were selected for comparison, which were introduced in Section 1 and listed as LMMSE, AltMin, and ADMM.

We first considered the number of iterations for the antenna configurations of $N_t \times N_r = 16 \times 128$, 32×128 and 64×128. Figure 2 illustrates the BER performance of the proposed MIDE detection algorithm against the number of iterations, where the SNR was set as 3 dB. It is observed that with the increase of N_r/N_t, the convergence number of the iterations required by the proposed MIDE was almost the same and a very small one, e.g., 10 in all simulations. This demonstrated a reliable BER performance and a fast detection convergence in our proposed MIDE algorithm. Moreover, with the increase of the transmitting antennas, the diversity gain of the MIMO system decreased, leading to a degradation of the BER performance of the proposed algorithm and the compared MIMO detectors.

Table 2. Simulation parameters.

Channel Model	Uncorrelated Rayleigh Flat Fading
Modulation scheme	16-QAM
Number of transmitting antennas (N_t)	16, 32, 64
Number of receiving antennas (N_r)	128

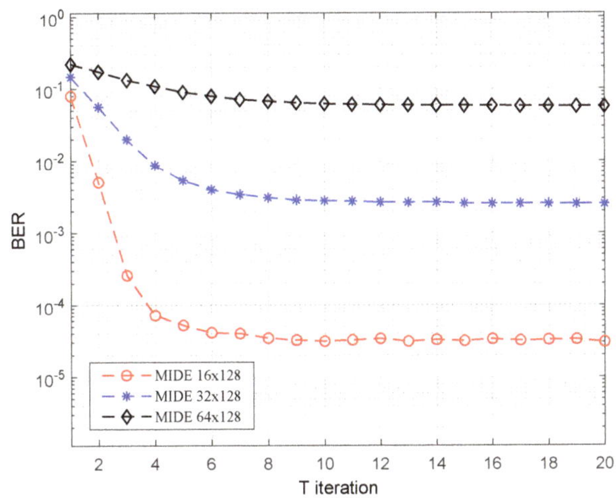

Figure 2. BER performance versus proposed iterations with SNR = 3 dB.

Figure 3 depicts the BER performance comparison of the IDE-based detection algorithm with the fixed damping factor $\alpha = 0.05$ and the proposed MIDE detection with the self-updated damping factor. It is clear that the proposed MIDE-based detection with the self-updated damping factor showed better BER performance than the conventional IDE-based detection algorithm with all antenna configurations.

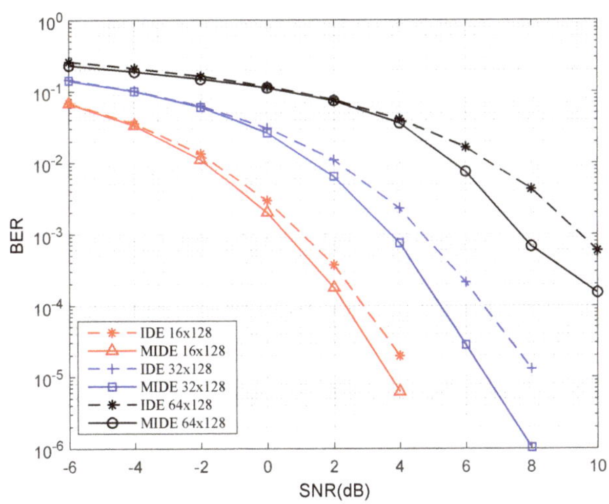

Figure 3. BER performance of the conventional damping factor and the proposed damping factor.

Moreover, Figure 4 compares the BER performances of the proposed MIDE algorithm and the conventional ADMM algorithm. From the figure, it is clear that the BER performance of both algorithms degraded when the number of users increased. However, it is observed that the proposed MIDE-based detection performed better than the conventional ADMM detection in terms of BER performance with all antenna configurations. Furthermore, we can observe from Figure 4 that when the target of BER was set as 10^{-3}, the SNR required by the proposed algorithm was at least 0.5 dB less than the one of the conventional ADMM algorithm.

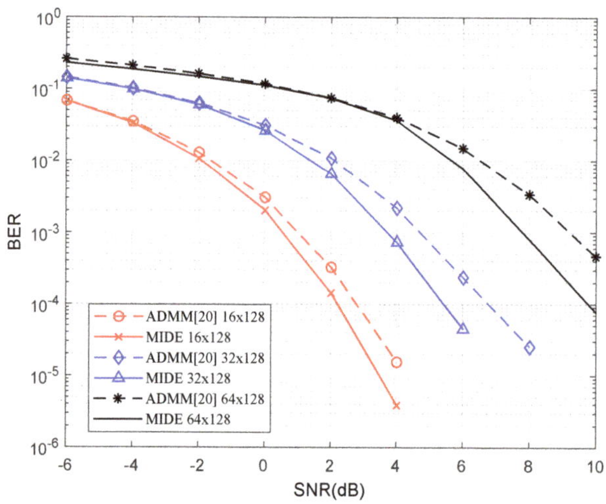

Figure 4. BER performance comparison of the conventional ADMM and the proposed algorithm.

Finally, we show the BER performance comparison of the conventional AltMin algorithm, the LMMSE algorithm, and the proposed MIDE algorithm in Figure 5.

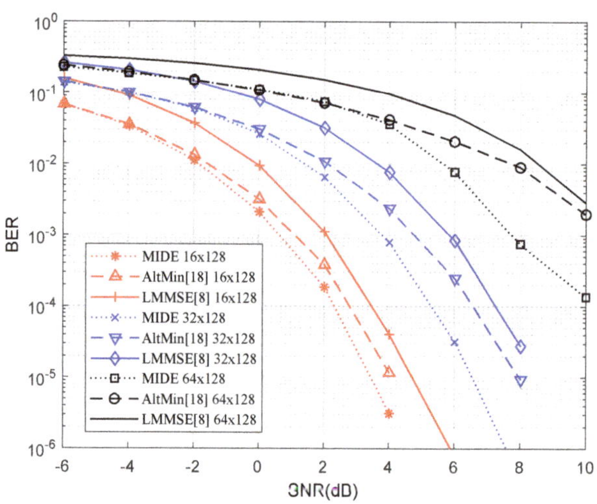

Figure 5. BER performance comparison of the AltMin algorithm, the LMMSE algorithm, and the proposed algorithm.

It is clear that with any ratio of N_r/N_t, the proposed MIDE algorithm performed better than all the compared algorithms in terms of BER performance.

4.2. Computational Complexity Comparison

The computational complexity of ADMM, AltMin, and the proposed MIDE algorithms depends on the number of iterations K. The compared algorithms had a different number of iterations to reach convergence with different antenna configurations. We fixed the number of receiving antennas to $N_r = 128$, and the number of transmitting antennas N_t was increased from 16 to 84. Further, we set K = 5, 14, and 10 for the conventional ADMM algorithm, the AltMin algorithm [20], and the proposed

MIDE algorithm, respectively, based on the convergence simulations. Figure 6 illustrates the total number of multiplications vs. the number of transmitting antennas, which was based on the analysis of the computational complexity of the algorithms in Section 3.3.

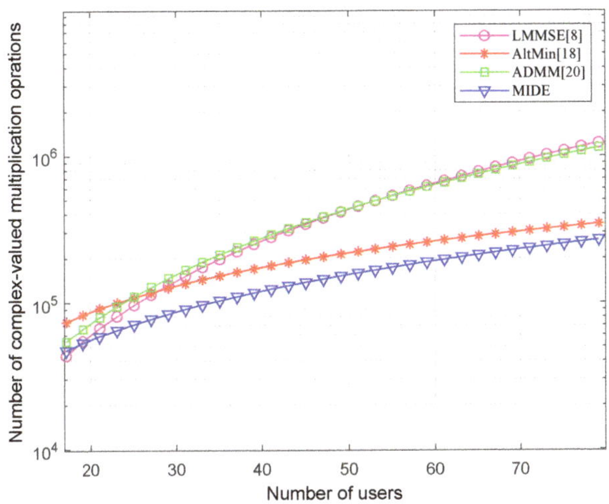

Figure 6. Computational complexity comparison against the number of users.

From Figure 6, we can see that the computational complexity increased with the number of users in all compared algorithms. However, the proposed MIDE algorithm achieved the lowest computational complexity among all compared algorithms under various antenna configurations. Specifically, the MIDE algorithm showed a relatively lower computational complexity than the AltMin algorithm, which was proven to have a low complexity detection for uplink massive systems in [20]. In addition, from Figure 6, the proposed MIDE algorithm achieved much lower computational complexity than the LMMSE and ADMM detection algorithms when the dimension of the MIMO system became larger. As a consequence, the proposed MIDE detection is much more applicable for massive MIMO systems with its low computational complexity.

5. Conclusions

In this paper, we proposed a low-complexity, IDE-based detection algorithm in uplink massive MIMO systems. The proposed MIDE algorithm avoided the calculation of the Gram matrix, the matrix inversion, and LDL decomposition to reach a low computational complexity. In addition, a self-updating damping method was provided with the damping factor estimated and updated at each iteration based on the Euclidean distance between the latest two detection solutions, which accelerated the convergence of the IDE-based detection algorithms. Simulation results showed that the proposed MIDE algorithm performed better than the conventional LMMSE, AltMin, and ADMM detection algorithms in terms of the BER performance and the computational complexity.

Author Contributions: H.F. and X.Z. contributed the idea generation; S.X. and X.Z. conducted the study design and paper writing; Z.L. helped the analysis of the simulation results.

Funding: This work is supported in part by the National Natural Science Foundation of China (No. 61571108, No. 61701197, and No. 61801193), the open research fund of National Mobile Communications Research Laboratory of millimeter wave, Southeast University (No. 2018D15), the open research fund of the National Key Laboratory of millimeter wave, Southeast University (No.K201918), the Open Foundation of Key Laboratory of Wireless Communication, Nanjing University of Posts and Telecommunication (No. 2017WICOM01), and the Wuxi Science and Technology Development Fund (No. H20191001).

Conflicts of Interest: The authors declare no conflict of interest.

References

1. Xiang, W.; Zheng, K.; Shen, X.S. *5G Mobile Communications;* Springer International Publishing: Basel, Switzerland, 2017.
2. Lu, L.; Li, G.Y.A.; Swindlehurst, L.; Ashikhmin, A.; Zhang, R. An overview of massive MIMO: Benefits and challenges. *IEEE J. Sel. Topics Signal Process.* **2014**, *8*, 742–758. [CrossRef]
3. Wu, M.; Yin, B.; Wang, G.; Dick, C.; Cavallaro, J.R.; Studer, C. Large-scale MIMO detection for 3GPP LTE: algorithms and FPGA implementations. *IEEE J. Sel. Topics Signal Process.* **2014**, *8*, 916–929. [CrossRef]
4. Ngo, H.; Larsson, E.; Marzetta, T. Energy and spectral efficiency of very large multiuser MIMO systems. *IEEE Trans. Commun.* **2012**, *61*, 1436–1449.
5. Rusek, F.; Persson, D.; Lau, B.K.; Larsson, E.G.; Marzetta, T. L.; Edfors, O.; Tufvesson, F. Scaling up MIMO: Opportunities and challenges with very large arrays. *IEEE Signal Process. Mag.* **2013**, *30*, 40–60. [CrossRef]
6. Chockalingam, A.; Rajan, B.S. *Large MIMO Systems;* Cambridge University Press: Cambridge, UK, 2014.
7. Kuchi, K.; Ayyar, A.B. Performance analysis of ML detection in MIMO systems with co-channel interference. *IEEE Commun. Lett.* **2011**, *15*, 786–788. [CrossRef]
8. Mandloi, M.; Bhatia, V. Low-complexity near-optimal iterative sequential detection for uplink massive MIMO systems. *IEEE Commun. Lett.* **2017**, *21*, 568–571. [CrossRef]
9. Gustafsson, O.; Bertilsson, E.; Klasson, J.; Ingemarsson, C. Approximate Neumann Series or exact matrix inversion for massive MIMO? In Proceedings of the 2017 IEEE 24th Symposium on Computer Arithmetic (ARITH), London, UK, 24–26 July 2017.
10. Fang, L.; Huang, D. Neumann Series Expansion Based LMMSE Channel Estimation for OFDM Systems. *IEEE Commun. Lett.* **2016**, *20*, 748–751. [CrossRef]
11. Gao, X.; Dai, L.; Ma, Y.; Wang, Z. Low-complexity near-optimalsignal detection for uplink large-scale MIMO systems. *Electron. Lett.* **2014**, *50*, 1326–1328. [CrossRef]
12. Qin, X.; Yan, Z.; He, G. A near-optimal detection scheme based on joint steepest descent and Jacobi method for uplink massive MIMO systems. *IEEE Commun. Lett.* **2016**, *20*, 276–279. [CrossRef]
13. Lee, Y. Decision-aided Jacobi iteration for signal detection in massive MIMO systems. *Electron. Lett.* **2017**, *53*, 1552–1554. [CrossRef]
14. Kong, B.Y.; Park, I. Low-complexity symbol detection for massive MIMO uplink based on Jacobi method. In Proceedings of the IEEE 27th Annual International Symposium on Personal, Indoor, and Mobile Radio Communications (PIMRC), Valencia, Spain, 1–5 September 2016.
15. Minango, J.; de Almeida, C.; Daniel Altamirano, C. Low-complexity MMSE detector for massive MIMO systems based on Damped Jacobi method. In Proceedings of the 2017 IEEE 28th Annual International Symposium on Personal, Indoor, and Mobile Radio Communications (PIMRC), Montreal, QC, Canada, 1–5 October 2017.
16. Deng, Q.; Guo, L.; Dong, C.; Lin, J.; Meng, D.; Chen, X. High-throughput signal detection based on fast matrix inversion updates for uplink massive multiuser multiple-input multi-output systems. *IET Commun.* **2017**, *11*, 2228–2235. [CrossRef]
17. Sun, Y.; Li, Z.; Zhang, C.; Zhang, R.; Yan, F.; Shen, L. Low complexity signal detector based on SSOR iteration for large-scale MIMO systems. In Proceedings of the 9th International Conference on Wireless Communications and Signal Processing (WCSP), Nanjing, China, 1–6 October 2017.
18. Dai, L.; Gao, X.; Su, X.; Han, S.; Chih-Lin, I.; Wang, Z. Low-complexity soft-output signal detection based on Gauss–Seidel method for uplink multiuser large-scale MIMO systems. *IEEE Trans. Veh. Technol.* **2015**, *64*, 4839–4845. [CrossRef]
19. Zeng, J.; Lin, J.; Wang, Z. An improved Gauss–Seidel algorithm and its efficient architecture for massive MIMO systems. *IEEE Trans. Circuits Syst. II-Express Briefs.* **2018**, *65*, 1194–1198. [CrossRef]
20. Elgabli, A.; Elghariani, A.; Aggarwal, V.; Bell, M.R. A low-complexity detection algorithm for uplink massive MIMO systems based on Alternating Minimization. *IEEE Wirel. Commun. Lett.* **2019**, *8*, 917–920. [CrossRef]
21. Souto, N.; Dinis, R. MIMO detection and equalization for single-carrier systems using the Alternating Direction Method of Multipliers. *IEEE Signal Process. Lett.* **2016**, *23* , 1751–1755. [CrossRef]
22. Shahabuddin, S.; Juntti, M.; Studer, C. Admm-based infinity normdetection for large MU-MIMO: Algorithm and vlsi architecture. In Proceedings of the IEEE International Symposium on Circuits and Systems (ISCAS), Baltimore, MD, USA,1–4 May 2017.

23. Zhang, K.; Zhang, Y. Recursive algorithm for calculating eigenvalues of real symmetric matrix based on LDLT decomposition. *Computer Engineering and Applications*. **2008**, *44*, 78–80.
24. Chou, C.; Wu, J. Low-complexity MIMO precoder design with LDL^H channel decomposition. *IEEE Trans. Veh. Technol.* **2011**, *60*, 2368–2372. [CrossRef]
25. Wang, C.; Wen, C.; Jin, S.; Tsai, S. Finite-Alphabet precoding for massive MU-MIMO with low-resolution DACs. *IEEE Trans. Wirel. Commun.* **2018**, *17*, 4706–4720. [CrossRef]
26. Chen, J. A low complexity data detection algorithm for uplink multiuser massive MIMO systems. *IEEE J. Sel. Areas Commun.* **2017**, *35*, 1701–1714. [CrossRef]
27. Jin, F.; Liu, Q.; Liu, H.; Wu, P. A low complexity signal detection scheme based on improved Newton iteration for massive MIMO systems. *IEEE Commun. Lett.* **2019**, *23*, 748–751. [CrossRef]
28. Wu, Z.; Ge, L.; You, X.; Zhang, C. Efficient near-MMSE detector for large-scale MIMO systems. In Proceedings of the 2017 IEEE International Workshop on Signal Processing Systems (SiPS), Lorient, France, 1–6 October 2017.
29. Wu, C.; Huang, W.; Chung, W. Robust update algorithms for Zero-Forcing detection in uplink large-scale MIMO systems. *IEEE Commun. Lett.* **2018**, *22*, 424–427. [CrossRef]
30. Monteiro, F.A.; Wassell, I.J. Euclidean distances in quantized spaces with pre-stored components for MIMO detection. In Proceedings of the European Conference on Wireless Technologies, Munich, Germany, 8–10 October 2007.
31. Hunger, R. *Floating Point Operations in Matrix-Vector Calculus*; Technical Report; Munich University of Technology: Munich, Germany, 2007.

electronics

MDPI

Article

Energy Efficiency Optimization for Massive MIMO Non-Orthogonal Unicast and Multicast Transmission with Statistical CSI

Wenjin Wang *, Yufei Huang, Li You, Jiayuan Xiong, Jiamin Li and Xiqi Gao

National Mobile Communications Research Laboratory, Southeast University, Nanjing 210096, China
* Correspondence: wangwj@seu.edu.cn; Tel.: +86-025-8379-0506

Received: 16 July 2019; Accepted: 30 July 2019; Published: 1 August 2019

Abstract: We study the energy efficiency (EE) optimization problem in non-orthogonal unicast and multicast transmission for massive multiple-input multiple-output (MIMO) systems with statistical channel state information of all receivers available at the transmitter. Firstly, we formulate the EE maximization problem. We reduce the number of variables to be solved and simplify this large-dimensional-matrix-valued problem into a real-vector-valued problem. Next, we lower the computational complexity significantly by replacing the objective with its deterministic equivalent to avoid the high-complex expectation operation. With guaranteed convergence, we propose an iterative algorithm on beam domain power allocation using the minorize maximize algorithm and Dinkelbach's transform and derive the locally optimal power allocation strategy to achieve the optimal EE. Finally, we illustrate the significant EE performance gain of our EE maximization algorithm compared with the conventional approach through conducting numerical simulations.

Keywords: energy efficiency; non-orthogonal unicast and multicast transmission; statistical channel state information; massive MIMO; beam domain

1. Introduction

As mobile data expands rapidly, it is expected that global wireless data traffic will surpass 100 exabytes per month by 2023 [1]. A considerable proportion of the data traffic, such as massive software updating and sports broadcasting, is of common interest, which stimulates the demand for services that can deliver the same data to a group of user terminals (UTs) efficiently. Since physical layer multicasting can provide efficient point-to-multipoint wireless transmission, it has great potential for future mobile communication systems [2–4].

Recently, non-orthogonal unicast and multicast (NOUM) transmission has been gaining increasing interest [5–7]. At the transmitter, the unicast and multicast signals are precoded and then sent out to the receivers simultaneously, sharing the same time-frequency resources. Compared with the conventional orthogonal unicast and multicast (OUM) transmission, NOUM transmission is more spectrum-efficient, and more suitable for scenarios where both multicast and unicast signals are needed by a UT. Massive multiple-input multiple-output (MIMO) has become one of the core technologies of the fifth generation wireless system for its significant performance in energy efficiency (EE) and spectral efficiency [8,9]. Therefore, there has been considerable research on the combination of multicast transmission and massive MIMO systems [6,10,11]. Please note that mutual coupling is a major concern in massive MIMO because it can weaken the system performance [12–16]. We assume perfect isolation between the antennas without loss of generality.

EE has become a significant design criterion for wireless communication systems [17–19]. The broad-scale antenna arrays equipped at the base station (BS) cause the power consumption to increase in massive MIMO systems, and the energy consumed by wireless communications is responsible for greenhouse gas emissions [20], which motivates the need to design energy-efficient systems [8,21,22]. EE of a massive MIMO system was considered in [8]. However, it ignores the power consumed by the BS circuit, while in [21], research on maximizing the EE and power transfer efficiency for wireless-powered systems was analyzed, taking the circuit power consumption into account. In [22], how the system parameters (number of antennas, transmitted power and number of UTs) affect the EE of a multi-user MIMO system was investigated.

There are also previous works that studied energy-efficient NOUM transmission in massive MIMO systems [23–25]. In [23], energy-efficient NOUM beamforming in multi-cell multi-user MIMO scenario was studied. An optimization beamforming algorithm was proposed in [24] to optimize the EE in the multi-cell multicast system. The extension of the problem was investigated in [25], which takes antenna selection into consideration.

Please note that most of the previous works made the assumption that the UTs' instantaneous channel state information (CSI) is available at the BS. However, in realistic systems, obtaining good estimates of instantaneous CSI is a challenging job [26–28]. Compared with obtaining instantaneous CSI, the acquisition of statistical CSI is easier and more precise. In [11], rate maximization problem for NOUM massive MIMO transmission was considered, and the EE maximization problem for physical-layer multicast transmission was investigated in [29], both assumed that the BS only has access to the UTs' statistical CSI.

To our knowledge, the research on EE optimization of NOUM transmission for massive MIMO systems with statistical CSI at the transmitter has not been studied yet. We investigate this problem in our work, and the major contributions we provide in this paper are listed as follows:

- With statistical CSI, we formulate the EE maximization problem for NOUM transmission in the massive MIMO scenario.
- We determine the optimal transmit directions of the multicast and unicast transmission in closed-form, respectively, and then simplify the large-scale complex-matrix-valued precoding design problem into a real-vector-valued power allocation problem in the beam domain.
- We reduce the computational complexity of the EE optimization problem significantly by replacing the objective function with its deterministic equivalent (DE).
- With guaranteed convergence, we propose an algorithm on beam domain power allocation using the minorize maximize (MM) algorithm and Dinkelbach's transform. We deal with the EE optimization problem by iteratively solving a series of related convex optimization problems.

The remainder of the paper is constructed as follows. The channel model is introduced in Section 2. The EE maximization problem is formulated and investigated in Section 3. Numerical simulations are conducted in Section 4. Section 5 summarizes the paper.

Column vectors and matrices are represented by lower and upper case boldface letters, respectively, whereas italic letters stand for scalars, and the following are other notations used in this paper.

- We adopt $\mathbb{R}^{M\times N}$ to represent $M \times N$ real-valued vector space and $\mathbb{C}^{M\times N}$ to denote $M \times N$ complex-valued vector space.
- $\mathbf{I}_M$ represents the identity matrix of size $M \times M$.
- $\mathbf{X} \succeq \mathbf{0}$ indicates that matrix $\mathbf{X}$ is positive semidefinite.
- $\mathsf{E}\{.\}$ represents the expectation operation.
- $\odot$ denotes the Hadamard product.
- Denote $\mathrm{tr}\{.\}$ as the trace operation, $(.)^T$ as the transpose operation, $(.)^*$ as the conjugate operation, $(.)^H$ as the conjugate-transpose operation, and $\det\{.\}$ as the determinant operation.
- $\sim$ stands for "be distributed as", and $\triangleq$ stands for "be defined as".

2. System Model

Consider a single cell massive MIMO system with an M-antenna BS, jointly serves K UTs. Denote by $\mathcal{K} \triangleq \{1, 2, \ldots, K\}$ the UT set, where the kth UT is equipped with N_k antennas. The multicast and unicast services are carried out with the same time-frequency resources. The BS sends a multicast signal that is of common interest to all the UTs in the cell while delivering unique messages to UTs according to each UT's demand during the downlink transmission, as shown in Figure 1.

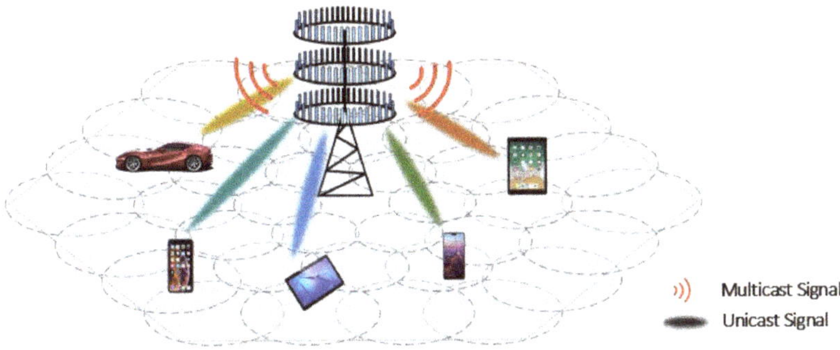

Figure 1. System model of NOUM.

Assume the downlink signal sent by the BS is denoted by

$$\mathbf{x} = \mathbf{x}^{\mathrm{m}} + \sum_{k \in \mathcal{K}} \mathbf{x}_k^{\mathrm{u}} \in \mathbb{C}^{M \times 1}, \tag{1}$$

where $\mathbf{x}^{\mathrm{m}} \in \mathbb{C}^{M \times 1}$ represents the multicast signal and $\mathbf{x}_k^{\mathrm{u}} \in \mathbb{C}^{M \times 1}$ denotes the unicast signal sent to the kth UT. Assume that $\mathbf{x}^{\mathrm{m}}$ and $\mathbf{x}_k^{\mathrm{u}}$ are mutually uncorrelated, zero-mean, and their covariance matrices are $\mathbf{Q}^{\mathrm{m}}$ and $\mathbf{Q}_k^{\mathrm{u}}$, respectively. Define $\mathrm{tr}\{\mathbf{Q}^{\mathrm{m}}\}$ as the multicast transmission power and $\mathrm{tr}\{\mathbf{Q}_k^{\mathrm{u}}\}$ as the unicast transmission power. At the kth UT, the received signal is denoted by

$$\mathbf{y}_k = \mathbf{H}_k \mathbf{x} + \mathbf{n}_k \in \mathbb{C}^{N_k \times 1}, \tag{2}$$

where $\mathbf{H}_k$ is the downlink channel matrix of size $N_k \times M$, and $\mathbf{n}_k \sim \mathcal{CN}\left(\mathbf{0}, \sigma^2 \mathbf{I}_{N_k}\right)$ represents the additive circularly symmetric complex-valued Gaussian noise with the variance being σ^2.

We adopt Weichselberger's channel model [30,31] in our work because the correlation properties between the transmit and receive ends of Weichselberger's channel model are jointly considered rather than separately characterized in the Kronecker model. Then, we can write the downlink channel matrix in (2) as

$$\mathbf{H}_k = \mathbf{U}_k \mathbf{G}_k \mathbf{V}_k^H \in \mathbb{C}^{N_k \times M}, \tag{3}$$

where $\mathbf{U}_k \in \mathbb{C}^{N_k \times N_k}$ and $\mathbf{V}_k \in \mathbb{C}^{M \times M}$ are deterministic unitary matrices. $\mathbf{G}_k \in \mathbb{C}^{N_k \times M}$ represents the downlink channel matrix in the beam domain [26,27,32], and the elements of $\mathbf{G}_k$ are independently distributed random variables with zero-mean. Denote $\mathbf{\Omega}_k$ as the beam domain channel power matrix

$$\mathbf{\Omega}_k = \mathsf{E}\{\mathbf{G}_k \odot \mathbf{G}_k^*\} \in \mathbb{R}^{N_k \times M}, \tag{4}$$

where the average power of $[\mathbf{G}_k]_{i,j}$ is represented by $[\mathbf{\Omega}_k]_{i,j}$. As $\mathbf{\Omega}_k$ has the property of remaining approximately constant while the frequency changes widely, the statistical CSI can be obtained accurately and efficiently [32].

The vast number of antenna arrays employed at the BS brings about new channel properties for massive MIMO systems. For example, as the BS antenna number M tends to infinity, the eigenvector matrices of the transmit correlation matrices between the BS and all UTs tend to be the same and are only affected by the BS array topology [32,33]. Denote the corresponding deterministic unitary matrix as $\mathbf{V}$, and then in the massive MIMO scenario, the downlink channel matrix becomes

$$\mathbf{H}_k \overset{M\to\infty}{=} \mathbf{U}_k \mathbf{G}_k \mathbf{V}^H. \tag{5}$$

Please note that many of the previous works on massive MIMO adopted the channel model mentioned in (5) such as [26,29,34], and it can achieve quite accurate performance [34].

3. NOUM Transmission in Massive MIMO

3.1. Problem Formulation

Consider a NOUM massive MIMO system. We assume that there is only one multicast group without loss of generality. Consider the case when the kth ($\forall k$) UT knows its own instantaneous CSI with proper pilot design [33], while the BS only has access to the statistic CSI of all UTs.

Rewrite the received signal at the kth UT by inserting (1) into (2) as follows

$$\mathbf{y}_k = \mathbf{H}_k \mathbf{x}^{\mathrm{m}} + \mathbf{H}_k \mathbf{x}_k^{\mathrm{u}} + \sum_{k' \neq k} \mathbf{H}_k \mathbf{x}_{k'}^{\mathrm{u}} + \mathbf{n}_k. \tag{6}$$

Each UT will decode the common multicast signal and its desired unicast signal in order by applying successive interference cancellation (SIC) method.

During the process of multicast decoding, the kth UT regards the term $\mathbf{H}_k \mathbf{x}^{\mathrm{m}}$ in (6) as the desired message while treating the others as interference. For the covariance matrix of the interference and noise, we have

$$\mathbf{K}_k^{\mathrm{m}} = \underbrace{\sigma^2 \mathbf{I}_{N_k}}_{\text{noise}} + \underbrace{\sum_{k' \in \mathcal{K}} \mathsf{E}\left\{\mathbf{H}_k \mathbf{Q}_{k'}^{\mathrm{u}} \mathbf{H}_k^H\right\}}_{\text{interference}} \in \mathbb{C}^{N_k \times N_k}. \tag{7}$$

Since UT k has the knowledge of its own instantaneous CSI and the covariance matrix $\mathbf{K}_k^{\mathrm{m}}$, during the multicast transmission, we denote by R_k^{m} the ergodic rate of the kth UT

$$R_k^{\mathrm{m}} = \mathsf{E}\left\{\log\det\left\{\mathbf{K}_k^{\mathrm{m}} + \mathbf{H}_k \mathbf{Q}^{\mathrm{m}} \mathbf{H}_k^H\right\}\right\} - \log\det\left\{\mathbf{K}_k^{\mathrm{m}}\right\}. \tag{8}$$

Denote multicast ergodic rate as $\min_k R_k^{\mathrm{m}}$. By inserting the massive system model in (5) and $\det\{\mathbf{I} + \mathbf{MN}\} = \det\{\mathbf{I} + \mathbf{NM}\}$, the Sylvester's determinant identity, into (8), the multicast rate R_k^{m} in (8) becomes

$$R_k^{\mathrm{m}} = \mathsf{E}\left\{\log\det\left\{\overline{\mathbf{K}}_k^{\mathrm{m}} + \mathbf{G}_k \mathbf{V}^H \mathbf{Q}^{\mathrm{m}} \mathbf{V} \mathbf{G}_k^H\right\}\right\} - \log\det\left\{\overline{\mathbf{K}}_k^{\mathrm{m}}\right\}, \tag{9}$$

where $\overline{\mathbf{K}}_k^{\mathrm{m}}$ is defined as

$$\begin{aligned}\overline{\mathbf{K}}_k^{\mathrm{m}} &\triangleq \mathbf{U}_k^H \mathbf{K}_k^{\mathrm{m}} \mathbf{U}_k \\ &= \sigma^2 \mathbf{I}_{N_k} + \sum_{k' \in \mathcal{K}} \mathsf{E}\left\{\mathbf{G}_k \mathbf{V}^H \mathbf{Q}_{k'}^{\mathrm{u}} \mathbf{V} \mathbf{G}_k^H\right\} \in \mathbb{C}^{N_k \times N_k}.\end{aligned} \tag{10}$$

Define a matrix-valued function $\mathbf{A}_k(\mathbf{X})$ by $\mathbf{A}_k(\mathbf{X}) \triangleq \mathsf{E}\left\{\mathbf{G}_k \mathbf{X} \mathbf{G}_k^H\right\}$. Since all the elements of $\mathbf{G}_k$ are zero-mean and independently distributed, the off-diagonal elements of $\mathbf{A}_k(\mathbf{X})$ are zero, so $\mathbf{A}_k(\mathbf{X})$ is a diagonal matrix-valued function of size $N_k \times N_k$, and its ith diagonal element is

$$[\mathbf{A}_k]_{i,i} = \mathrm{tr}\left\{\mathrm{diag}\left\{\left([\boldsymbol{\Omega}_k]_{i,:}\right)^T\right\} \mathbf{X}\right\}. \tag{11}$$

Then the terms $\mathsf{E}\left\{\mathbf{G}_k \mathbf{V}^H \mathbf{Q}_{k'}^{\mathrm{u}} \mathbf{V} \mathbf{G}_k^H\right\}$ in (10) can be rewritten as $\mathbf{A}_k\left(\mathbf{V}^H \mathbf{Q}_{k'}^{\mathrm{u}} \mathbf{V}\right)$.

For the unicast signal decoding, with SCI, the multicast signal is removed, so the interference only contains the unicast signal meant for other UTs. For the covariance matrix of the interference and noise at the kth UT, we have

$$\mathbf{K}_k^{\mathrm{u}} = \underbrace{\sigma^2 \mathbf{I}_{N_k}}_{\text{noise}} + \underbrace{\sum_{k' \neq k} \mathsf{E}\left\{\mathbf{H}_k \mathbf{Q}_{k'}^{\mathrm{u}} \mathbf{H}_k^H\right\}}_{\text{interference}} \in \mathbb{C}^{N_k \times N_k}. \tag{12}$$

Then we denote by R_k^{u} the ergodic rate of the kth UT during the unicast transmission

$$R_k^{\mathrm{u}} = \mathsf{E}\left\{\log\det\left\{\mathbf{K}_k^{\mathrm{u}} + \mathbf{H}_k \mathbf{Q}_k^{\mathrm{u}} \mathbf{H}_k^H\right\}\right\} - \log\det\left\{\mathbf{K}_k^{\mathrm{u}}\right\}. \tag{13}$$

By inserting the massive system model in (5) and the Sylvester's determinant identity into (13), the unicast rate R_k^{u} at the kth UT becomes

$$R_k^{\mathrm{u}} = \mathsf{E}\left\{\log\det\left\{\overline{\mathbf{K}}_k^{\mathrm{u}} + \mathbf{G}_k \mathbf{V}^H \mathbf{Q}_k^{\mathrm{u}} \mathbf{V} \mathbf{G}_k^H\right\}\right\} - \log\det\left\{\overline{\mathbf{K}}_k^{\mathrm{u}}\right\}, \tag{14}$$

where $\overline{\mathbf{K}}_k^{\mathrm{u}}$ is defined as

$$\overline{\mathbf{K}}_k^{\mathrm{u}} \triangleq \sigma^2 \mathbf{I}_{N_k} + \sum_{k' \neq k} \mathbf{A}_k\left(\mathbf{V}^H \mathbf{Q}_{k'}^{\mathrm{u}} \mathbf{V}\right) \in \mathbb{C}^{N_k \times N_k}, \tag{15}$$

and the definition of $\mathbf{A}_k(\mathbf{X})$ is given in (11).

Next, we consider the system power consumption. Apply the power consumption model the same as the one used in [29,35] as follows

$$P = \mu\left(\mathrm{tr}\left\{\mathbf{Q}^{\mathrm{m}}\right\} + \sum_{k \in \mathcal{K}} \mathrm{tr}\left\{\mathbf{Q}_k^{\mathrm{u}}\right\}\right) + MP_{\mathrm{c}} + P_{\mathrm{s}}, \tag{16}$$

where the constant-coefficient $\mu \geq 1$ accounts for the reciprocal of the transmit amplifier drain efficiency. $\mathrm{tr}\left\{\mathbf{Q}^{\mathrm{m}}\right\}$ means the multicast transmit power, and $\sum_{k \in \mathcal{K}} \mathrm{tr}\left\{\mathbf{Q}_k^{\mathrm{u}}\right\}$ denotes the total unicast transmit power. P_{c} stands for the constant circuit power consumption per antenna and is unaffected by the actual transmit power. P_{s} represents the BS static power consumption and is irrelevant to the number of antennas.

In the following, we formulate the EE optimization problem for NOUM massive MIMO system. We aim at identifying the optimal transmit covariance matrices $\mathbf{Q}^{\mathrm{m}}$ and $\mathbf{Q}_k^{\mathrm{u}}$ for multicast and unicast transmission that can maximize the system EE, respectively. We define a weight matrix $\mathbf{u} = [u_0, u_1, \ldots, u_K]$ with u_0 being the weight of multicast rate and u_k being the weight of kth unicast rate. Then we can denote by R the weighted sum rate as follows:

$$R \triangleq u_0 K(\min_k R_k^{\mathrm{m}}) + \sum_{k \in \mathcal{K}} u_k R_k^{\mathrm{u}}, \tag{17}$$

and the EE of the considered system with bandwidth W is given by

$$EE = \frac{WR}{P} = \frac{W\left(u_0 K(\min_k R_k^{\mathrm{m}}) + \sum_{k \in \mathcal{K}} u_k R_k^{\mathrm{u}}\right)}{\mu\left(\mathrm{tr}\left\{\mathbf{Q}^{\mathrm{m}}\right\} + \sum_{k \in \mathcal{K}} \mathrm{tr}\left\{\mathbf{Q}_k^{\mathrm{u}}\right\}\right) + MP_{\mathrm{c}} + P_{\mathrm{s}}}. \tag{18}$$

Therefore, the EE maximization problem is stated as

$$\begin{aligned} \max_{\mathbf{Q}^{\mathrm{m}},\mathbf{Q}_k^{\mathrm{u}},\forall k\in\mathcal{K}} \quad & \frac{W\left(u_0 K(\min_k R_k^{\mathrm{m}}) + \sum_{k\in\mathcal{K}} u_k R_k^{\mathrm{u}}\right)}{\mu\left(\mathrm{tr}\left\{\mathbf{Q}^{\mathrm{m}}\right\} + \sum_{k\in\mathcal{K}} \mathrm{tr}\left\{\mathbf{Q}_k^{\mathrm{u}}\right\}\right) + MP_{\mathrm{c}} + P_{\mathrm{s}}}, \\ \text{s.t.} \quad & \mathrm{tr}\left\{\mathbf{Q}^{\mathrm{m}}\right\} + \sum_{k\in\mathcal{K}} \mathrm{tr}\left\{\mathbf{Q}_k^{\mathrm{u}}\right\} \le P_{\max}, \\ & \mathbf{Q}^{\mathrm{m}} \succeq \mathbf{0},\ \mathbf{Q}_k^{\mathrm{u}} \succeq \mathbf{0}\ (\forall k\in\mathcal{K}), \end{aligned} \tag{19}$$

where $P_{\max}$ is the power budget at the BS.

3.2. Optimal Transmit Directions

The problem in (19) aims at designing large-dimensional complex matrices $\mathbf{Q}^{\mathrm{m}}$ and $\mathbf{Q}_k^{\mathrm{u}}$ $(\forall k)$, and the computational complexity can be very high. To simplify this problem, first, we decompose the transmit covariance matrices as $\mathbf{Q}^{\mathrm{m}} = \mathbf{\Phi}^{\mathrm{m}}\mathbf{\Lambda}^{\mathrm{m}}\left(\mathbf{\Phi}^{\mathrm{m}}\right)^H$ and $\mathbf{Q}_k^{\mathrm{u}} = \mathbf{\Phi}_k^{\mathrm{u}}\mathbf{\Lambda}_k^{\mathrm{u}}\left(\mathbf{\Phi}_k^{\mathrm{u}}\right)^H$, respectively. $\mathbf{\Phi}^{\mathrm{m}}$ and $\mathbf{\Phi}_k^{\mathrm{u}}$ are constituted by the eigenvectors of $\mathbf{Q}^{\mathrm{m}}$ and $\mathbf{Q}_k^{\mathrm{u}}$, respectively, which represent the directions of the transmitted signals. Meanwhile, $\mathbf{\Lambda}^{\mathrm{m}}$ and $\mathbf{\Lambda}_k^{\mathrm{u}}$ are diagonal matrices with their diagonal elements constituted by the eigenvalues of $\mathbf{Q}^{\mathrm{m}}$ and $\mathbf{Q}_k^{\mathrm{u}}$, respectively, which denote the allocated power over the corresponding directions.

The following theorem determines the values of the eigenvectors of $\mathbf{Q}^{\mathrm{m}}$ and $\mathbf{Q}_k^{\mathrm{u}}$.

Theorem 1. *The optimal multicast and unicast transmit covariance matrices of problem (19) is*

$$\mathbf{Q}^{\mathrm{m,opt}} = \mathbf{V}\mathbf{\Lambda}^{\mathrm{m}}\mathbf{V}^H, \quad \mathbf{Q}_k^{\mathrm{u,opt}} = \mathbf{V}\mathbf{\Lambda}_k^{\mathrm{u}}\mathbf{V}^H, \forall k, \tag{20}$$

where $\mathbf{\Lambda}^{\mathrm{m}}$ and $\mathbf{\Lambda}_k^{\mathrm{u}}$ $(\forall k)$ are both diagonal matrices, and the matrix $\mathbf{V}$ equals to the eigenvector matrices of the correlation matrices between the BS and all UTs and only depends on the BS array topology. The eigenvectors of $\mathbf{Q}^{\mathrm{m}}$ and $\mathbf{Q}_k^{\mathrm{u}}$ are given by the columns of the matrix $\mathbf{V}$,

Proof. Please refer to the Appendix A. □

Theorem 1 above indicates that when solving problem (19), since the eigenvectors are deterministic, we only have to determine the power allocation matrix denoted by $\mathbf{\Lambda} \triangleq \{\mathbf{\Lambda}^{\mathrm{m}}, \mathbf{\Lambda}_1^{\mathrm{u}}, \mathbf{\Lambda}_2^{\mathrm{u}}, \ldots, \mathbf{\Lambda}_K^{\mathrm{u}}\}$, which reduces the number of variables to be optimized and the computational complexity significantly. Therefore, the large-dimensional complex-matrix-valued EE maximization problem can be transformed into a real-vector-valued power allocation problem in the beam domain.

Rewrite $\overline{\mathbf{K}}_k^{\mathrm{m}}$ and $\overline{\mathbf{K}}_k^{\mathrm{u}}$ as follows

$$\overline{\mathbf{K}}_k^{\mathrm{m}}(\mathbf{\Lambda}) \triangleq \sigma^2\mathbf{I}_{N_k} + \sum_{k'\in\mathcal{K}} \mathbf{A}_k\left(\mathbf{\Lambda}_{k'}^{\mathrm{u}}\right), \tag{21}$$

$$\overline{\mathbf{K}}_k^{\mathrm{u}}(\mathbf{\Lambda}) \triangleq \sigma^2\mathbf{I}_{N_k} + \sum_{k'\neq k} \mathbf{A}_k\left(\mathbf{\Lambda}_{k'}^{\mathrm{u}}\right), \tag{22}$$

and without loss of optimality, we can simplify the problem in (19) into the problem below

$$\begin{aligned} \max_{\mathbf{\Lambda}} \quad & \frac{W\left(u_0 K\left(\min_k R_k^{\mathrm{m}}(\mathbf{\Lambda})\right) + \sum_{k\in\mathcal{K}} u_k R_k^{\mathrm{u}}(\mathbf{\Lambda})\right)}{\mu\left(\mathrm{tr}\left\{\mathbf{\Lambda}^{\mathrm{m}}\right\} + \sum_{k\in\mathcal{K}} \mathrm{tr}\left\{\mathbf{\Lambda}_k^{\mathrm{u}}\right\}\right) + MP_{\mathrm{c}} + P_{\mathrm{s}}}, \\ \text{s.t.} \quad & \mathrm{tr}\left\{\mathbf{\Lambda}^{\mathrm{m}}\right\} + \sum_{k\in\mathcal{K}} \mathrm{tr}\left\{\mathbf{\Lambda}_k^{\mathrm{u}}\right\} \le P_{\max}, \\ & \mathbf{\Lambda}^{\mathrm{m}} \succeq \mathbf{0},\ \mathbf{\Lambda}^{\mathrm{m}}\ \text{diagonal},\ \mathbf{\Lambda}_k^{\mathrm{u}} \succeq \mathbf{0},\ \mathbf{\Lambda}_k^{\mathrm{u}}\ \text{diagonal}\ (\forall k\in\mathcal{K}), \end{aligned} \tag{23}$$

with

$$R_k^{\mathrm{m}}(\mathbf{\Lambda}) = \underbrace{\mathsf{E}\left\{\log\det\left\{\overline{\mathbf{K}}_k^{\mathrm{m}}(\mathbf{\Lambda}) + \mathbf{G}_k\mathbf{\Lambda}^{\mathrm{m}}\mathbf{G}_k^H\right\}\right\}}_{\triangleq s_k^+(\mathbf{\Lambda})} - \underbrace{\log\det\left\{\overline{\mathbf{K}}_k^{\mathrm{m}}(\mathbf{\Lambda})\right\}}_{\triangleq s_k^-(\mathbf{\Lambda})}, \tag{24}$$

$$R_k^{\mathrm{u}}(\mathbf{\Lambda}) = \underbrace{\mathsf{E}\left\{\log\det\left\{\overline{\mathbf{K}}_k^{\mathrm{u}}(\mathbf{\Lambda}) + \mathbf{G}_k\mathbf{\Lambda}_k^{\mathrm{u}}\mathbf{G}_k^H\right\}\right\}}_{\triangleq t_k^+(\mathbf{\Lambda})} - \underbrace{\log\det\left\{\overline{\mathbf{K}}_k^{\mathrm{u}}(\mathbf{\Lambda})\right\}}_{\triangleq t_k^-(\mathbf{\Lambda})}. \tag{25}$$

Denote the lower bound of $R_k^{\mathrm{m}}(\mathbf{\Lambda})$ $(\forall k)$ as an auxiliary variable η, the problem in (23) can be equivalently expressed as

$$\begin{aligned} \max_{\mathbf{\Lambda}} \quad & \frac{W\left(u_0K\eta + \sum_{k\in\mathcal{K}} u_k R_k^{\mathrm{u}}(\mathbf{\Lambda})\right)}{\mu\left(\mathrm{tr}\left\{\mathbf{\Lambda}^{\mathrm{m}}\right\} + \sum_{k\in\mathcal{K}}\mathrm{tr}\left\{\mathbf{\Lambda}_k^{\mathrm{u}}\right\}\right) + MP_{\mathrm{c}} + P_{\mathrm{s}}}, \\ \text{s.t.} \quad & R_k^{\mathrm{m}}(\mathbf{\Lambda}) \geq \eta \ (\forall k \in \mathcal{K}), \\ & \mathrm{tr}\left\{\mathbf{\Lambda}^{\mathrm{m}}\right\} + \sum_{k\in\mathcal{K}}\mathrm{tr}\left\{\mathbf{\Lambda}_k^{\mathrm{u}}\right\} \leq P_{\max}, \\ & \mathbf{\Lambda}^{\mathrm{m}} \succeq \mathbf{0},\ \mathbf{\Lambda}^{\mathrm{m}}\ \text{diagonal},\ \mathbf{\Lambda}_k^{\mathrm{u}} \succeq \mathbf{0},\ \mathbf{\Lambda}_k^{\mathrm{u}}\ \text{diagonal}\ (\forall k \in \mathcal{K}). \end{aligned} \tag{26}$$

3.3. Energy-Efficient Power Allocation for NOUM Transmission

By observing problem (26), we can conclude that the numerator of the objective function is a difference of concave functions. We adopt the MM algorithm to deal with the problem. It is an iteration optimization process, where during each iteration, we replace the objective function with its lower bound function.

In this problem, we substitute $s_k^-(\mathbf{\Lambda})$ in (24) and $t_k^-(\mathbf{\Lambda})$ in (25) with their first-order Taylor expansions, respectively, to transfer the numerator of the objective function into a concave function, which leads to a concave-linear fractional program. We can solve problem (26) by solving a series of substitution problems iteratively. Then at the pth iteration, $\mathbf{\Lambda}_{(p)} = \left\{\mathbf{\Lambda}_{(p)}^{m}, \mathbf{\Lambda}_{1,(p)}^{\mathrm{u}}, \ldots, \mathbf{\Lambda}_{K,(p)}^{\mathrm{u}}\right\}$, and the sub-problem is

$$\begin{aligned} \mathbf{\Lambda}_{(p+1)} = & \\ \arg\max_{\mathbf{\Lambda}} \quad & \frac{W\left(u_0K\eta + \sum_{k\in\mathcal{K}} u_k\left(t_k^+(\mathbf{\Lambda}) - t_k^-\left(\mathbf{\Lambda}_{(p)}\right) - \sum_{a\neq k}\mathrm{tr}\left\{\left(\frac{\partial t_k^-(\mathbf{\Lambda}_{(p)})}{\partial\mathbf{\Lambda}_a^{\mathrm{u}}}\right)^T\left(\mathbf{\Lambda}_a^{\mathrm{u}} - \mathbf{\Lambda}_{a,(p)}^{\mathrm{u}}\right)\right\}\right)\right)}{\mu\left(\mathrm{tr}\left\{\mathbf{\Lambda}^{\mathrm{m}}\right\} + \sum_{k\in\mathcal{K}}\mathrm{tr}\left\{\mathbf{\Lambda}_k^{\mathrm{u}}\right\}\right) + MP_{\mathrm{c}} + P_{\mathrm{s}}}, \\ \text{s.t.} \quad & s_k^+(\mathbf{\Lambda}) - s_k^-\left(\mathbf{\Lambda}_{(p)}\right) - \sum_{a\in\mathcal{K}}\mathrm{tr}\left\{\left(\frac{\partial s_k^-\left(\mathbf{\Lambda}_{(p)}\right)}{\partial\mathbf{\Lambda}_a^{\mathrm{u}}}\right)^T\left(\mathbf{\Lambda}_a^{\mathrm{u}} - \mathbf{\Lambda}_{a,(p)}^{\mathrm{u}}\right)\right\} - \eta \geq 0\ (\forall k \in \mathcal{K}), \\ & \mathrm{tr}\left\{\mathbf{\Lambda}^{\mathrm{m}}\right\} + \sum_{k\in\mathcal{K}}\mathrm{tr}\left\{\mathbf{\Lambda}_k^{\mathrm{u}}\right\} \leq P_{\max}, \\ & \mathbf{\Lambda}^{\mathrm{m}} \succeq \mathbf{0},\ \mathbf{\Lambda}^{\mathrm{m}}\ \text{diagonal},\ \mathbf{\Lambda}_k^{\mathrm{u}} \succeq \mathbf{0},\ \mathbf{\Lambda}_k^{\mathrm{u}}\ \text{diagonal}\ (\forall k \in \mathcal{K}), \end{aligned} \tag{27}$$

where the gradients of $s_k^-\left(\mathbf{\Lambda}_{(p)}\right)$ and $t_k^-\left(\mathbf{\Lambda}_{(p)}\right)$ with respect to $\mathbf{\Lambda}_a^{\mathrm{u}}$ are defined by $\Delta s_k^{(p)}$ and $\Delta t_k^{(p)}$, respectively, with their diagonal elements being

$$\left[\Delta s_k^{(p)}\right]_{i,i} = \left[\frac{\partial s_k^-\left(\mathbf{\Lambda}_{(p)}\right)}{\partial\mathbf{\Lambda}_a^{\mathrm{u}}}\right]_{i,i} = \sum_{n=1}^{N_k}\frac{[\mathbf{\Omega}_k]_{n,i}}{\sigma^2 + \sum_{k'\in\mathcal{K}}\sum_{m=1}^{M}[\mathbf{\Omega}_k]_{n,m}\left[\mathbf{\Lambda}_{k',(p)}^{\mathrm{u}}\right]_{m,m}}, \tag{28}$$

$$\left[\Delta t_k^{(p)}\right]_{i,i} = \left[\frac{\partial t_k^-\left(\mathbf{\Lambda}_{(p)}\right)}{\partial \mathbf{\Lambda}_a^{\mathrm{u}}}\right]_{i,i} = \begin{cases} \sum\limits_{n=1}^{N_k} \frac{[\mathbf{\Omega}_k]_{n,i}}{\sigma^2 + \sum\limits_{k' \neq k} \sum\limits_{m=1}^{M} [\mathbf{\Omega}_k]_{n,m}\left[\mathbf{\Lambda}_{k',(p)}^{\mathrm{u}}\right]_{m,m}}, & a \neq k, \\ 0, & a = k, \end{cases} \tag{29}$$

respectively.

Since $t_k^-\left(\mathbf{\Lambda}_{(p)}\right)$ and $\mathbf{\Lambda}_{a,(p)}^{\mathrm{u}}$ in (27) are constant in each iteration, we can ignore them and obtain an equivalent optimization problem as

$$\begin{aligned} &\mathbf{\Lambda}_{(p+1)} = \\ &\arg\max_{\mathbf{\Lambda}} \ \frac{W\left(u_0 K\eta + \sum\limits_{k\in\mathcal{K}} u_k \left(t_k^+(\mathbf{\Lambda}) - \sum\limits_{a\neq k} \mathrm{tr}\left\{\left(\frac{\partial t_k^-(\mathbf{\Lambda}_{(p)})}{\partial \mathbf{\Lambda}_a^{\mathrm{u}}}\right)^T \mathbf{\Lambda}_a^{\mathrm{u}}\right\}\right)\right)}{\mu\left(\mathrm{tr}\left\{\mathbf{\Lambda}^{\mathrm{m}}\right\} + \sum\limits_{k\in\mathcal{K}} \mathrm{tr}\left\{\mathbf{\Lambda}_k^{\mathrm{u}}\right\}\right) + MP_{\mathrm{c}} + P_{\mathrm{s}}}, \\ &\quad \text{s.t.} \quad s_k^+(\mathbf{\Lambda}) - s_k^-\left(\mathbf{\Lambda}_{(p)}\right) - \sum_{a\in\mathcal{K}} \mathrm{tr}\left\{\left(\frac{\partial s_k^-\left(\mathbf{\Lambda}_{(p)}\right)}{\partial \mathbf{\Lambda}_a^{\mathrm{u}}}\right)^T \left(\mathbf{\Lambda}_a^{\mathrm{u}} - \mathbf{\Lambda}_{a,(p)}^{\mathrm{u}}\right)\right\} - \eta \geq 0 \ (\forall k \in \mathcal{K}), \\ &\qquad\quad \mathrm{tr}\left\{\mathbf{\Lambda}^{\mathrm{m}}\right\} + \sum_{k\in\mathcal{K}} \mathrm{tr}\left\{\mathbf{\Lambda}_k^{\mathrm{u}}\right\} \leq P_{\max}, \\ &\qquad\quad \mathbf{\Lambda}^{\mathrm{m}} \succeq \mathbf{0},\ \mathbf{\Lambda}^{\mathrm{m}} \text{ diagonal},\ \mathbf{\Lambda}_k^{\mathrm{u}} \succeq \mathbf{0},\ \mathbf{\Lambda}_k^{\mathrm{u}} \text{ diagonal } (\forall k \in \mathcal{K}). \end{aligned} \tag{30}$$

Although the numerator of the objective function and constraint of the transformed sub-problem (30) are concave, the computational complexity can still be quite high if the expectation operation is manipulated using Monte-Carlo methods. Via applying the large-dimensional random matrix theory in [36,37], we further reduce the optimization complexity by substituting the minuends of $R_k^{\mathrm{m}}(\mathbf{\Lambda})$ and $R_k^{\mathrm{u}}(\mathbf{\Lambda})$ with their DEs, respectively.

First, we define a diagonal matrix-valued function $\mathbf{Y}_k(\mathbf{X})$ of size $M \times M$, and its ith diagonal element is

$$[\mathbf{Y}_k(\mathbf{X})]_{i,i} = \mathrm{tr}\left\{\mathsf{diag}\left\{[\mathbf{\Omega}_k]_{:,i}\right\}\mathbf{X}\right\}. \tag{31}$$

Then, we can write the DE of $s_k^+(\mathbf{\Lambda})$ as

$$\begin{aligned} S_k^+(\mathbf{\Lambda}) = &\log\det\left\{\mathbf{I}_M + \mathbf{\Gamma}_k^{\mathrm{m}}\mathbf{\Lambda}^{\mathrm{m}}\right\} \\ &+ \log\det\left\{\widetilde{\mathbf{\Gamma}}_k^{\mathrm{m}} + \overline{\mathbf{K}}_k^{\mathrm{m}}(\mathbf{\Lambda})\right\} - \mathrm{tr}\left\{\mathbf{I}_{N_k} - \left(\widetilde{\mathbf{\Phi}}_k^{\mathrm{m}}\right)^{-1}\right\}, \end{aligned} \tag{32}$$

where $\mathbf{\Gamma}_k^{\mathrm{m}}$, $\widetilde{\mathbf{\Gamma}}_k^{\mathrm{m}}$ and $\widetilde{\mathbf{\Phi}}_k^{\mathrm{m}}$ are given by

$$\begin{aligned} \mathbf{\Gamma}_k^{\mathrm{m}} &= \mathbf{Y}_k\left(\left(\widetilde{\mathbf{\Phi}}_k^{\mathrm{m}}\overline{\mathbf{K}}_k^{\mathrm{m}}(\mathbf{\Lambda})\right)^{-1}\right) \in \mathbb{C}^{M\times M}, \\ \widetilde{\mathbf{\Gamma}}_k^{\mathrm{m}} &= \mathbf{A}_k\left(\mathbf{\Lambda}^{\mathrm{m}}\left(\mathbf{I}_M + \mathbf{\Lambda}^{\mathrm{m}}\mathbf{\Gamma}_k^{\mathrm{m}}\right)^{-1}\right) \in \mathbb{C}^{N_k\times N_k}, \\ \widetilde{\mathbf{\Phi}}_k^{\mathrm{m}} &= \mathbf{I}_{N_k} + \widetilde{\mathbf{\Gamma}}_k^{\mathrm{m}}\left(\overline{\mathbf{K}}_k^{\mathrm{m}}(\mathbf{\Lambda})\right)^{-1} \in \mathbb{C}^{N_k\times N_k}, \end{aligned} \tag{33}$$

and the definition of $\mathbf{A}_k(\mathbf{X})$ is given in (11).

Likewise, we have the DE of $t_k^+(\mathbf{\Lambda})$ as

$$\begin{aligned} T_k^+(\mathbf{\Lambda}) = &\log\det\left\{\mathbf{I}_M + \mathbf{\Gamma}_k^{\mathrm{u}}\mathbf{\Lambda}_k^{\mathrm{u}}\right\} \\ &+ \log\det\left\{\widetilde{\mathbf{\Gamma}}_k^{\mathrm{u}} + \overline{\mathbf{K}}_k^{\mathrm{u}}(\mathbf{\Lambda})\right\} - \mathrm{tr}\left\{\mathbf{I}_{N_k} - \left(\widetilde{\mathbf{\Phi}}_k^{\mathrm{u}}\right)^{-1}\right\}, \end{aligned} \tag{34}$$

where $\boldsymbol{\Gamma}_k^{\mathrm{u}}$, $\widetilde{\boldsymbol{\Gamma}}_k^{\mathrm{u}}$ and $\widetilde{\boldsymbol{\Phi}}_k^{\mathrm{u}}$ are given by

$$\begin{aligned}
\boldsymbol{\Gamma}_k^{\mathrm{u}} &= \mathbf{Y}_k\left(\left(\widetilde{\boldsymbol{\Phi}}_k^{\mathrm{u}}\overline{\mathbf{K}}_k^{\mathrm{u}}\left(\boldsymbol{\Lambda}\right)\right)^{-1}\right) \in \mathbb{C}^{M\times M},\\
\widetilde{\boldsymbol{\Gamma}}_k^{\mathrm{u}} &= \mathbf{A}_k\left(\boldsymbol{\Lambda}_k^{\mathrm{u}}\left(\mathbf{I}_M+\boldsymbol{\Lambda}_k^{\mathrm{u}}\boldsymbol{\Gamma}_k^{\mathrm{u}}\right)^{-1}\right) \in \mathbb{C}^{N_k\times N_k},\\
\widetilde{\boldsymbol{\Phi}}_k^{\mathrm{u}} &= \mathbf{I}_{N_k}+\widetilde{\boldsymbol{\Gamma}}_k^{\mathrm{u}}\left(\overline{\mathbf{K}}_k^{\mathrm{u}}\left(\boldsymbol{\Lambda}\right)\right)^{-1} \in \mathbb{C}^{N_k\times N_k}.
\end{aligned} \tag{35}$$

With the DEs of $s_k^{+}\left(\boldsymbol{\Lambda}\right)$ and $t_k^{+}\left(\boldsymbol{\Lambda}\right)$ defined above, the optimization problem in (30) becomes

$$\begin{aligned}
&\boldsymbol{\Lambda}_{(p+1)} = \\
&\arg\max_{\boldsymbol{\Lambda}} \frac{W\left(u_0K\eta+\sum\limits_{k\in\mathcal{K}}u_k\left(T_k^{+}\left(\boldsymbol{\Lambda}\right)-\sum\limits_{a\neq k}\mathrm{tr}\left\{\left(\frac{\partial t_k^{-}\left(\boldsymbol{\Lambda}_{(p)}\right)}{\partial\boldsymbol{\Lambda}_a^{\mathrm{u}}}\right)^T\boldsymbol{\Lambda}_a^{\mathrm{u}}\right\}\right)\right)}{\mu\left(\mathrm{tr}\left\{\boldsymbol{\Lambda}^{\mathrm{m}}\right\}+\sum\limits_{k\in\mathcal{K}}\mathrm{tr}\left\{\boldsymbol{\Lambda}_k^{\mathrm{u}}\right\}\right)+MP_{\mathrm{c}}+P_{\mathrm{s}}},\\
&\quad\text{s.t.}\quad S_k^{+}\left(\boldsymbol{\Lambda}\right)-s_k^{-}\left(\boldsymbol{\Lambda}_{(p)}\right)-\sum_{a\in\mathcal{K}}\mathrm{tr}\left\{\left(\frac{\partial s_k^{-}\left(\boldsymbol{\Lambda}_{(p)}\right)}{\partial\boldsymbol{\Lambda}_a^{\mathrm{u}}}\right)^T\left(\boldsymbol{\Lambda}_a^{\mathrm{u}}-\boldsymbol{\Lambda}_{a,(p)}^{\mathrm{u}}\right)\right\}-\eta\geq 0\ \left(\forall k\in\mathcal{K}\right),\\
&\quad\mathrm{tr}\left\{\boldsymbol{\Lambda}^{\mathrm{m}}\right\}+\sum_{k\in\mathcal{K}}\mathrm{tr}\left\{\boldsymbol{\Lambda}_k^{\mathrm{u}}\right\}\leq P_{\max},\\
&\quad\boldsymbol{\Lambda}^{\mathrm{m}}\succeq\mathbf{0},\ \boldsymbol{\Lambda}^{\mathrm{m}}\ \text{diagonal},\ \boldsymbol{\Lambda}_k^{\mathrm{u}}\succeq\mathbf{0},\ \boldsymbol{\Lambda}_k^{\mathrm{u}}\ \text{diagonal}\ \left(\forall k\in\mathcal{K}\right).
\end{aligned} \tag{36}$$

We can observe from the optimization problem in (36) that the denominator and numerator of the objective function are linear and concave functions of $\boldsymbol{\Lambda}$, respectively. We invoke Dinkelbach's transform [38] to deal with this concave-linear program. We can obtain the solution to (36) via solving a series of problems below

$$\begin{aligned}
&\left\{\boldsymbol{\Lambda}_{(p)}^{(q+1)},\eta^{(q+1)}\right\} = \\
&\quad\arg\max_{\boldsymbol{\Lambda}}\ W\left(u_0K\eta+\sum_{k\in\mathcal{K}}u_k\left(T_k^{+}\left(\boldsymbol{\Lambda}\right)-\sum_{a\neq k}\mathrm{tr}\left\{\left(\frac{\partial t_k^{-}\left(\boldsymbol{\Lambda}_{(p)}\right)}{\partial\boldsymbol{\Lambda}_a^{\mathrm{u}}}\right)^T\boldsymbol{\Lambda}_a^{\mathrm{u}}\right\}\right)\right)-\chi_{(p)}^{(q)}P\left(\boldsymbol{\Lambda}\right),\\
&\quad\text{s.t.}\quad S_k^{+}\left(\boldsymbol{\Lambda}\right)-s_k^{-}\left(\boldsymbol{\Lambda}_{(p)}\right)-\sum_{a\in\mathcal{K}}\mathrm{tr}\left\{\left(\frac{\partial s_k^{-}\left(\boldsymbol{\Lambda}_{(p)}\right)}{\partial\boldsymbol{\Lambda}_a^{\mathrm{u}}}\right)^T\left(\boldsymbol{\Lambda}_a^{\mathrm{u}}-\boldsymbol{\Lambda}_{a,(p)}^{\mathrm{u}}\right)\right\}-\eta\geq 0\ \left(\forall k\in\mathcal{K}\right),\\
&\quad\mathrm{tr}\left\{\boldsymbol{\Lambda}^{\mathrm{m}}\right\}+\sum_{k\in\mathcal{K}}\mathrm{tr}\left\{\boldsymbol{\Lambda}_k^{\mathrm{u}}\right\}\leq P_{\max},\\
&\quad\boldsymbol{\Lambda}^{\mathrm{m}}\succeq\mathbf{0},\ \boldsymbol{\Lambda}^{\mathrm{m}}\ \text{diagonal},\ \boldsymbol{\Lambda}_k^{\mathrm{u}}\succeq\mathbf{0},\ \boldsymbol{\Lambda}_k^{\mathrm{u}}\ \text{diagonal}\ \left(\forall k\in\mathcal{K}\right),
\end{aligned} \tag{37}$$

where $P\left(\boldsymbol{\Lambda}\right)=\mu\left(\mathrm{tr}\left\{\boldsymbol{\Lambda}^{\mathrm{m}}\right\}+\sum\limits_{k\in\mathcal{K}}\mathrm{tr}\left\{\boldsymbol{\Lambda}_k^{\mathrm{u}}\right\}\right)+MP_{\mathrm{c}}+P_{\mathrm{s}}$, q is the iteration index, and $\chi_{(p)}^{(q)}$ is the auxiliary variable. During each iteration, we update $\chi_{(p)}^{(q)}$ using the following equation

$$\chi_{(p)}^{(q)} = \frac{W\left(u_0K\eta^{(q)}+\sum\limits_{k\in\mathcal{K}}u_k\left(T_k^{+}\left(\boldsymbol{\Lambda}_{(p)}^{(q)}\right)-\sum\limits_{a\neq k}\mathrm{tr}\left\{\left(\frac{\partial t_k^{-}\left(\boldsymbol{\Lambda}_{(p)}\right)}{\partial\boldsymbol{\Lambda}_a^{\mathrm{u}}}\right)^T\boldsymbol{\Lambda}_{a,(p)}^{\mathrm{u},(q)}\right\}\right)\right)}{P\left(\boldsymbol{\Lambda}_{(p)}^{(q)}\right)}. \tag{38}$$

From the analysis above, we can observe that the proposed EE optimization algorithm involves two-layer iterations. During the outer iteration, via invoking the MM algorithm, we replace the numerator of the objective function in (26) with its lower bound function, thus making the numerator concave. The MM-based algorithm is guaranteed to converge to the locally optimal solution [39–41]; in the inner iteration, we transform the fractional problem in (36) into solvable convex optimization problems in (37) via Dinkelbach's transform, which can derive the global optimum solution to (36) with guaranteed convergence [42]. After several iterations, we can obtain the optimal beam domain power allocation matrix $\mathbf{\Lambda}$. Please note that $\mathbf{\Lambda}$ is locally optimal due to the local optimality of MM algorithm. We present our algorithm in Algorithm 1.

Algorithm 1 Energy-Efficient Power Allocation Algorithm in the Beam Domain for Massive MIMO NOUM Transmission

Input: Beam domain channel statistics $\mathbf{\Omega}_k$, initial power allocation matrix $\mathbf{\Lambda}_{(0)}$, outer iteration threshold ϵ_1 and inner iteration threshold ϵ_2

Output: Power allocation matrix $\mathbf{\Lambda}$ in the beam domain

Initialization: $\overline{EE}_{(-1)} = 0, p = 0$

Calculate

$$\overline{EE}_{(p)} = \frac{W\left(u_0 K\left(\min_k\left\{S_k^+\left(\mathbf{\Lambda}_{(p)}\right) - s_k^-\left(\mathbf{\Lambda}_{(p)}\right)\right\}\right) + \sum_{k\in\mathcal{K}} u_k\left(T_k^+\left(\mathbf{\Lambda}_{(p)}\right) - t_k^-\left(\mathbf{\Lambda}_{(p)}\right)\right)\right)}{P\left(\mathbf{\Lambda}_{(p)}\right)} \tag{39}$$

while $\left|\overline{EE}_{(p)} - \overline{EE}_{(p-1)}\right| \geq \epsilon_1$ **do**

  Initialization: $q = 0$, let $\mathbf{\Lambda}_{(p)}^{(q)} = \mathbf{\Lambda}_{(p)}$, calculate $\chi_{(p)}^{(q)}$ with (38)

  while $\left|\chi_{(p)}^{(q)} - \chi_{(p)}^{(q-1)}\right| \geq \epsilon_2$ **do**

    Let $q = q + 1$

    Calculate $\mathbf{\Lambda}_{(p)}^{(q)}$ via solving problem (37) with $\chi_{(p)}^{(q-1)}$

    Calculate $\chi_{(p)}^{(q)}$ using (38)

  end while

  Let $p = p + 1$

  Let $\mathbf{\Lambda}_{(p)} = \mathbf{\Lambda}_{(p-1)}^{(q)}$

  Calculate $\overline{EE}_{(p)}$ with (39)

end while

return $\mathbf{\Lambda} = \mathbf{\Lambda}_{(p)}$

4. Numerical Results

We provide numerical simulation results to demonstrate the performance of the EE optimization algorithm proposed above for NOUM transmission massive MIMO scenario with statistical CSI. Table 1 illustrates how the numerical simulation parameters are set.

First of all, in Figure 2, we illustrate the convergence performance by showing the iteration process of our EE optimization algorithm under different transmit power budgets $P_{\max}$. The horizontal ordinate is the outer iteration index. As we can see, the EE converges after only a few iterations. Also, we can observe that in the lower power budget regime, the EE performance convergences faster than that in the higher power budget regime.

Table 1. Simulation parameters.

Parameter	Value
Scenario	Suburban macro
Channel model	3GPP SCM
Pathloss	-120 dB $(\forall k)$
Array topology	ULA with antenna spacing half wavelength
Noise variance	$\sigma^2 = -131$ dBm
Number of UTs	$K = 8$
Number of BS antennas	$M = 128$
Number of UT antennas	$N_k = 4\ (\forall k)$
Transmission bandwidth	W= 10 MHz
Amplifier drain efficiency	$\mu = 5$
Circuit power consumption per antenna	$P_c = 30$ dBm
Static power consumption	$P_s = 40$ dBm
Weights	$u_0 = 0.7, u_k = 0.3\ (\forall k)$

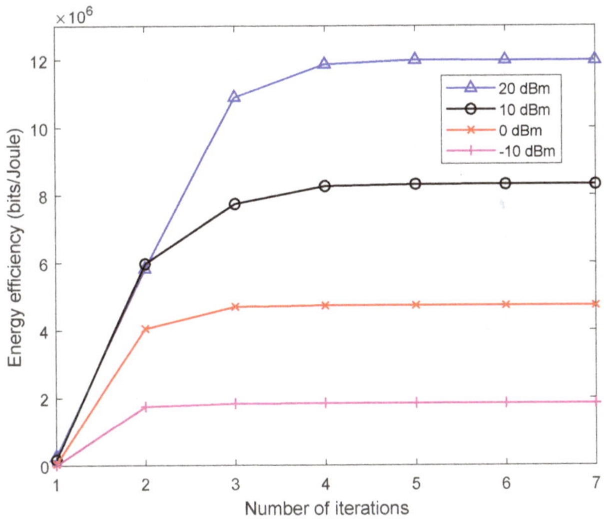

Figure 2. The convergence performance of the proposed EE optimization algorithm for different power budgets $P_{\max}$.

Then, we evaluate the EE of the NOUM transmission versus the power budget $P_{\max}$ under different numbers the antennas M at the BS in Figure 3. As we can see, the EE performance decreases when the BS antenna number M increases for the reason that in the power consumption model we adopted in (16), the total circuit power consumption grows linearly with M, the BS antenna number.

Next, the comparison of the EE performance of the power allocation algorithm proposed above with the rate maximization approach [11] is shown in Figure 4. We notice that the EE performance of the two approaches are similar at low transmit power budget regime. However, when the transmit power budget gets high, the EE performance of the rate maximization approach decreases, while that of our EE maximization approach remains high. This indicates that the rate maximization approach can achieve almost EE optimal when $P_{\max}$ is low. However, our EE maximization approach outperforms the rate maximization one at high transmit power budget regime.

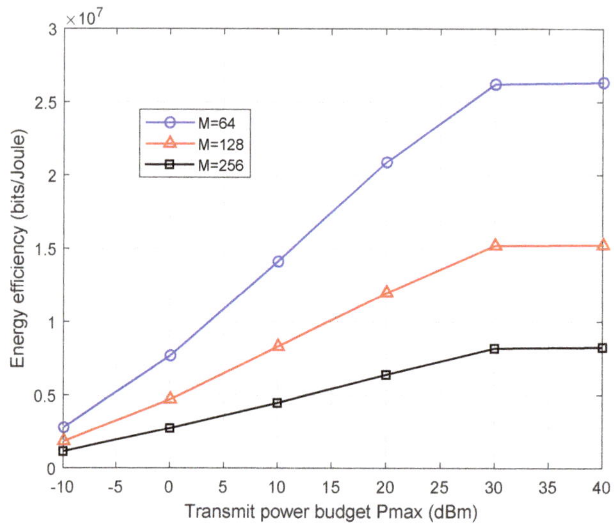

Figure 3. The EE performance of the NOUM transmission versus the power budget $P_{\max}$ for different numbers of BS antennas M.

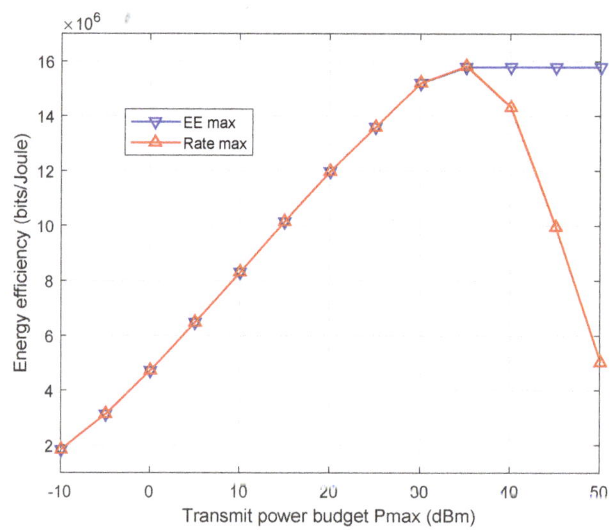

Figure 4. The EE performance of the proposed beam domain power allocation algorithm compared with the rate maximization approach.

Finally, in Figure 5, the EE performance of our power allocation approach and that of full CSI approach, which assumes instantaneous CSI is known at the BS, is compared. Since full CSI is an ideal case, it can achieve better EE performance than other imperfect CSI situation. However, the full CSI case suffers from pilot overhead. As Figure 5 illustrates, our proposed algorithm surpasses the full CSI approach with 3/7 pilot overhead [13] in the EE performance.

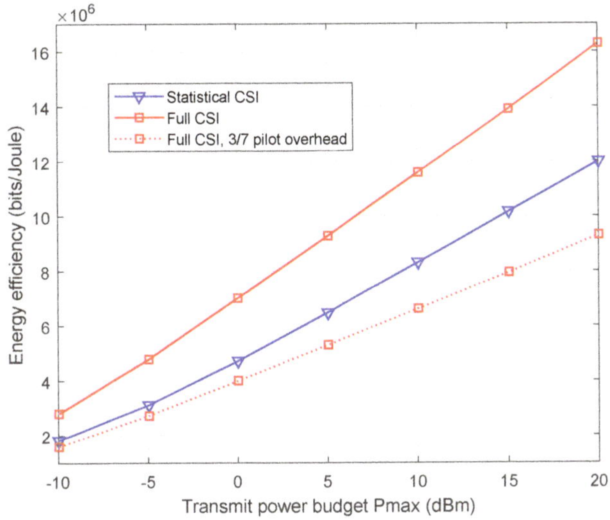

Figure 5. The comparison on the EE performance of proposed algorithm, full CSI case and full CSI with 3/7 overhead.

5. Conclusions

To conclude, we considered the EE optimization problem in NOUM transmission systems with statistical CSI available at the BS. We first formulated the EE maximization problem, and then determined the closed-form optimal eigenvectors of the multicast and unicast transmit covariance matrices for optimal EE, respectively. Next, with guaranteed convergence, we proposed a beam domain power allocation algorithm adopting the MM algorithm, DE and Dinkelbach's transform and derived the locally optimal power allocation strategy to achieve the EE optimization. Finally, with numerical results, we presented the performance gain of our EE maximization algorithm compared with the conventional approach.

Author Contributions: W.W. and Y.H. perceived the idea and wrote the manuscript. L.Y., J.X., J.L. and X.G. gave valuable suggestions on the structuring of the paper and assisted in the revising and proofreading.

Funding: This work was supported by the National Key R&D Program of China under Grant 2018YFB1801103, the National Natural Science Foundation of China under Grants 61631018, 61801114, 61761136016, 61771264, 61871465, and 61501113, the Natural Science Foundation of Jiangsu Province under Grant BK20170688, the Fundamental Research Funds for the Central Universities, and the Huawei Cooperation Project.

Acknowledgments: The authors would like to thank the Editor and the anonymous reviewers for their valuable comments and suggestions.

Conflicts of Interest: The authors declare no conflict of interest.

Appendix A. Proof of Theorem 1

Firstly, we can tell from (11) that the value of $\mathbf{A}_k(\mathbf{X})$ is only affected by the diagonal elements of $\mathbf{X}$, so $\overline{\mathbf{K}}_k^{\mathrm{m}}$ in (10) and $\overline{\mathbf{K}}_k^{\mathrm{u}}$ in (15) are irrelevant to the off-diagonal elements of $\mathbf{V}^H\mathbf{Q}_{k'}^{\mathrm{u}}\mathbf{V}$. Then, the power consumption in (16) has no relationship with the off-diagonal elements of $\mathbf{V}\mathbf{\Lambda}^{\mathrm{m}}\mathbf{V}^H$ or $\mathbf{V}\mathbf{\Lambda}_k^{\mathrm{u}}\mathbf{V}^H$. Moreover, applying the proof method similar to [44], we can conclude that $\mathbf{V}\mathbf{\Lambda}^{\mathrm{m}}\mathbf{V}^H$ and $\mathbf{V}\mathbf{\Lambda}_k^{\mathrm{u}}\mathbf{V}^H$ should be diagonal to maximize R_k^{m} and R_k^{u}, respectively. Therefore, to maximize the objective function in (19), $\mathbf{Q}^{\mathrm{m}}$ and $\mathbf{Q}_k^{\mathrm{u}}$ should be both diagonal matrices. This concludes the proof.

References

1. Lecompte, D.; Gabin, F. Evolved multimedia broadcast/multicast service (eMBMS) in LTE-advanced: Overview and Rel-11 enhancements. *IEEE Commun. Mag.* **2012**, *50*, 68–74. [CrossRef]
2. Wang, W.; Liu, A.; Zhang, Q.; You, L.; Gao, X.; Zheng, G. Robust multigroup multicast transmission for frame-based multi-beam satellite systems. *IEEE Access* **2018**, *6*, 46074–46083. [CrossRef]
3. You, L.; Liu, A.; Wang, W.; Gao, X. Outage constrained robust multigroup multicast beamforming for multi-beam satellite communication systems. *IEEE Wirel. Commun. Lett.* **2019**, *8*, 352–355. [CrossRef]
4. You, L.; Wang, J.; Wang, W.; Gao, X. Secure multicast transmission for massive MIMO with statistical channel state information. *IEEE Signal Process. Lett.* **2019**, *26*, 803–807. [CrossRef]
5. Chen, E.; Tao, M.; Liu, Y. Joint base station clustering and beamforming for non-orthogonal multicast and unicast transmission with backhaul constraints. *IEEE Trans. Wirel. Commun.* **2018**, *17*, 6265–6279. [CrossRef]
6. Sadeghi, M.; Björnson, E.; Larsson, E.G.; Yuen, C.; Marzetta, T. Joint unicast and multi-group multicast transmission in massive MIMO systems. *IEEE Trans. Wirel. Commun.* **2018**, *17*, 6375–6388. [CrossRef]
7. Wang, J.; Xu, H.; Zhu, B.; Fan, L.; Zhou, A. Hybrid beamforming design for mmWave joint unicast and multicast transmission. *IEEE Commun. Lett.* **2018**, *22*, 2012–2015. [CrossRef]
8. Ngo, H.Q.; Larsson, E.G.; Marzetta, T.L. Energy and spectral efficiency of very large multiuser MIMO systems. *IEEE Trans. Commun.* **2013**, *61*, 1436–1449.
9. Hoydis, J.; ten Brink, S.; Debbah, M. Massive MIMO in the UL/DL of cellular networks: How many antennas do we need? *IEEE J. Sel. Areas Commun.* **2013**, *31*, 160–171. [CrossRef]
10. Sadeghi, M.; Björnson, E.; Larsson, E.G.; Yuen, C.; Marzetta, T.L. Max–min fair transmit precoding for multi-group multicasting in massive MIMO. *IEEE Trans. Wirel. Commun.* **2018**, *17*, 1358–1373. [CrossRef]
11. You, L.; Xiong, J.; Li, K.; Wang, W.; Gao, X. Non-orthogonal unicast and multicast transmission for massive MIMO with statistical channel state information. *IEEE Access* **2018**, *6*, 66841–66849. [CrossRef]
12. Alibakhshikenari, M.; Virdee, B.S.; See, C.H.; Abd-Alhameed, R.A.; Falcone, F.; Limiti, E. High-isolation leaky-wave array antenna based on CRLH-metamaterial implemented on SIW with 30o frequency beam-scanning capability at millimetre-waves. *Electronics* **2019**, *8*, 642. [CrossRef]
13. Song, Y.J.; Sarabandi, K. Equivalent circuit model for metamaterial-based electromagnetic band-gap isolator. *IEEE Antennas Wirel. Propag. Lett.* **2012**, *11*, 1366–1369. [CrossRef]
14. Alibakhshikenari, M.; Khalily, M.; Virdee, B.S.; See, C.H.; Abd-Alhameed, R.A.; Limiti, E. Mutual-coupling isolation using embedded metamaterial EM bandgap decoupling slab for densely packed array antennas. *IEEE Access* **2019**, *7*, 51827–51840. [CrossRef]
15. Farsi, S.; Aliakbarian, H.; Schreurs, D.; Nauwelaers, B.; Vandenbosch, G.A.E. Mutual coupling reduction between planar antennas by using a simple microstrip U-section. *IEEE Antennas Wirel. Propag. Lett.* **2012**, *11*, 1501–1503. [CrossRef]
16. Alibakhshikenari, M.; Khalily, M.; Virdee, B.S.; See, C.H.; Abd-Alhameed, R.A.; Limiti, E. Mutual coupling suppression between two closely placed microstrip patches using EM-bandgap metamaterial fractal loading. *IEEE Access* **2019**, *7*, 23606–23614. [CrossRef]
17. Singh, K.; Gupta, A.; Ratnarajah, T. QoS-driven energy-efficient resource allocation in multiuser amplify-and-forward relay networks. *IEEE Trans. Signal Inf. Process. Over Netw.* **2017**, *3*, 771–786. [CrossRef]
18. Singh, K.; Gupta, A.; Ratnarajah, T. QoS-driven resource allocation and EE-balancing for multiuser two-way amplify-and-forward relay networks. *IEEE Trans. Wirel. Commun.* **2017**, *16*, 3189–3204. [CrossRef]
19. Singh, K.; Gupta, A.; Ratnarajah, T. A utility-based joint subcarrier and power allocation for green communications in multi-user two-way regenerative relay networks. *IEEE Trans. Commun.* **2017**, *65*, 3705–3722. [CrossRef]
20. Li, G.Y.; Xu, Z.; Xiong, C.; Yang, C.; Zhang, S.; Chen, Y.; Xu, S. Energy-efficient wireless communications: tutorial, survey, and open issues. *IEEE Wirel. Commun.* **2011**, *18*, 28–35. [CrossRef]
21. Khan, T.A.; Yazdan, A.; Heath, R.W. Optimization of power transfer efficiency and energy efficiency for wireless-powered systems with massive MIMO. *IEEE Trans. Wirel. Commun.* **2018**, *17*, 7159–7172. [CrossRef]
22. Björnson, E.; Sanguinetti, L.; Hoydis, J.; Debbah, M. Optimal design of energy-efficient multi-user MIMO systems: Is massive MIMO the answer? *IEEE Trans. Wirel. Commun.* **2015**, *14*, 3059–3075. [CrossRef]

23. Tervo, O.; Tran, L.; Chatzinotas, S.; Juntti, M.; Ottersten, B. Energy-efficient joint unicast and multicast beamforming with multi-antenna user terminals. In Proceedings of the 2017 IEEE 18th International Workshop on Signal Processing Advances in Wireless Communications (SPAWC), Sapporo, Japan, 3–6 July 2017; pp. 1–5.
24. He, S.; Huang, Y.; Jin, S.; Yang, L. Energy efficient coordinated beamforming design in multi-cell multicast networks. *IEEE Commun. Lett.* **2015**, *19*, 985–988. [CrossRef]
25. Tervo, O.; Tran, L.; Pennanen, H.; Chatzinotas, S.; Ottersten, B.; Juntti, M. Energy-efficient multicell multigroup multicasting with joint beamforming and antenna selection. *IEEE Trans. Signal Process.* **2018**, *66*, 4904–4919. [CrossRef]
26. You, L.; Gao, X.; Xia, X.; Ma, N.; Peng, Y. Pilot reuse for massive MIMO transmission over spatially correlated rayleigh fading channels. *IEEE Trans. Wirel. Commun.* **2015**, *14*, 3352–3366. [CrossRef]
27. You, L.; Gao, X.; Swindlehurst, A.L.; Zhong, W. Channel acquisition for massive MIMO-OFDM with adjustable phase shift pilots. *IEEE Trans. Signal Process.* **2016**, *64*, 1461–1476. [CrossRef]
28. You, L.; Chen, X.; Wang, W.; Gao, X. Coordinated multicast precoding for multi-cell massive MIMO transmission exploiting statistical channel state information. *Electronics* **2018**, *7*, 338. [CrossRef]
29. You, L.; Wang, W.; Gao, X. Energy-efficient multicast precoding for massive MIMO transmission with statistical CSI. *Energies* **2018**, *11*, 3175. [CrossRef]
30. Weichselberger, W.; Herdin, M.; Ozcelik, H.; Bonek, E. A stochastic MIMO channel model with joint correlation of both link ends. *IEEE Trans. Wirel. Commun.* **2006**, *5*, 90–100. [CrossRef]
31. Gao, X.; Jiang, B.; Li, X.; Gershman, A.B.; McKay, M.R. Statistical eigenmode transmission over jointly correlated MIMO channels. *IEEE Trans. Inf. Theory* **2009**, *55*, 3735–3750. [CrossRef]
32. You, L.; Gao, X.; Li, G.Y.; Xia, X.; Ma, N. BDMA for millimeter-wave/terahertz massive MIMO transmission with per-beam synchronization. *IEEE J. Sel. Areas Commun.* **2017**, *35*, 1550–1563. [CrossRef]
33. Sun, C.; Gao, X.; Jin, S.; Matthaiou, M.; Ding, Z.; Xiao, C. Beam division multiple access transmission for massive MIMO communications. *IEEE Trans. Commun.* **2015**, *63*, 2170–2184. [CrossRef]
34. Wen, C.; Jin, S.; Wong, K.; Chen, J.; Ting, P. Channel estimation for massive MIMO using Gaussian-mixture Bayesian learning. *IEEE Trans. Wirel. Commun.* **2015**, *14*, 1356–1368. [CrossRef]
35. Ng, D.W.K.; Lo, E.S.; Schober, R. Energy-efficient resource allocation in OFDMA systems with large numbers of base station antennas. *IEEE Trans. Wirel. Commun.* **2012**, *11*, 3292–3304. [CrossRef]
36. Lu, A.; Gao, X.; Xiao, C. Free deterministic equivalents for the analysis of MIMO multiple access channel. *IEEE Trans. Inf. Theory* **2016**, *62*, 4604–4629. [CrossRef]
37. Couillet, R.; Debbah, M. *Random Matrix Methods for Wireless Communications*; Cambridge University Press: Cambridge, UK, 2011. [CrossRef]
38. Dinkelbach, W. On nonlinear fractional programming. *Manag. Sci.* **1967**, *13*, 492–498. [CrossRef]
39. Lipp, T.; Boyd, S. Variations and extension of the convex-concave procedure. *Optim. Eng.* **2016**, *17*, 263–287. [CrossRef]
40. Yuille, A.L.; Rangarajan, A. The concave-convex procedure. *Neural Comput.* **2003**, *15*, 915–936. [CrossRef]
41. Hunter, D.R.; Lange, K. A tutorial on MM algorithms. *Am. Stat.* **2004**, *58*, 30–37. [CrossRef]
42. Zappone, A.; Jorswieck, E. Energy efficiency in wireless networks via fractional programming theory. *Found. Trends® Commun. Inf. Theory* **2015**, *11*, 185–396. [CrossRef]
43. Marzetta, T.L. Noncooperative Cellular Wireless with Unlimited Numbers of Base Station Antennas. *IEEE Trans. Wirel. Commun.* **2010**, *9*, 3590–3600. [CrossRef]
44. Tulino, A.M.; Lozano, A.; Verdú, S. Capacity-achieving input covariance for single-user multi-antenna channels. *IEEE Trans. Wirel. Commun.* **2006**, *5*, 662–671. [CrossRef]

electronics

MDPI

Article

Artificial Intelligence-Based Discontinuous Reception for Energy Saving in 5G Networks

Mudasar Latif Memon [1], Mukesh Kumar Maheshwari [2], Navrati Saxena [3,*], Abhishek Roy [4] and Dong Ryeol Shin [3,*]

1 College of Information and Communication Engineering, Sungkyunkwan University, Suwon 16419, Korea
2 Department of Electrical Engineering, Bahria University, Karachi 75260, Pakistan
3 College of Software, Sungkyunkwan University, Suwon 16419, Korea
4 Advanced Communication Technology, Wireless System Design, MediaTek USA Inc., San Jose, CA 95134, USA
* Correspondence: navrati@skku.edu (N.S.); drshin@skku.edu (D.R.S.); Tel.: +82-31-299-4676 (N.S.)

Received: 17 June 2019; Accepted: 9 July 2019; Published: 11 July 2019

Abstract: 5G is expected to deal with high data rates for different types of wireless traffic. To enable high data rates, 5G employs beam searching operation to align the best beam pairs. Beam searching operation along with high order modulation techniques in 5G, exhausts the battery power of user equipment (UE). LTE network uses discontinuous reception (DRX) with fixed sleep cycles to save UE energy. LTE-DRX in current form cannot work in 5G network, as it does not consider multiple beam communication and the length of sleep cycle is fixed. On the other hand, artificial intelligence (AI) has a tendency to learn and predict the packet arrival-time values from real wireless traffic traces. In this paper, we present AI based DRX (AI-DRX) mechanism for energy efficiency in 5G enabled devices. We propose AI-DRX algorithm for multiple beam communications, to enable dynamic short and long sleep cycles in DRX. AI-DRX saves the energy of UE while considering delay requirements of different services. We train a recurrent neural network (RNN) on two real wireless traces with minimum root mean square error (RMSE) of 5 ms for trace 1 and 6 ms for trace 2. Then, we utilize the trained RNN model in AI-DRX algorithm to make dynamic short or long sleep cycles. As compared to LTE-DRX, AI-DRX achieves 69% higher energy efficiency on trace 1 and 55% more energy efficiency on trace 2, respectively. The AI-DRX attains 70% improvement in energy efficiency for trace 2 compared with Poisson packet arrival model for $\lambda = 1/20$.

Keywords: discontinuous deception; multiple beam communications; artificial intelligence; energy efficiency; 5G; wireless communications

1. Introduction

The use of cellular gadgets, like smartphones, notebooks, and tablets has comforted our life. The Ericsson mobility report predicts the rise of cellular traffic to 8.8 billion by 2024 [1]. These extensive growing cellular users require improved data rates with heterogeneous services in next generation networks. 5G expects to deal with various types of traffics including periodic and delay tolerant traffic for IoT devices or burst type of traffic for delay intolerant services [2,3]. 3rd Generation Partnership Project (3GPP) planned the standardization of GHz spectrum (mm-wave) to address the users' demand of high bandwidth. However, communication over high-frequency bands of the mm-wave requires directional air interface and narrower beams to reduce the path loss. In directional air interface, UE has to search for best beam pairs and make adequate beam alignment with the next generation nodeB (gNB) [4]. In addition to beam searching process in 5G networks, massive MIMO, higher order modulation schemes and advanced coding techniques also increase UE energy expenses.

Long term evolution (LTE) networks utilize DRX mechanism to reduce the energy consumption of the UE [2,5]. DRX enables a UE to save energy by switching off the radio circuitry part, in case of no incoming data. LTE-DRX turns off radio part during long and short sleep cycles in order to reduce the energy consumption of UE. The sleep cycles in LTE-DRX are of the fixed time period. If any new packet arrives during the fixed time sleep cycle, the packets will be buffered at evolved NodeB (eNB). The eNB serves the stored packets after the completion of each sleep cycle in LTE-DRX. Hence, the LTE-DRX saves the UE energy at the cost of delay [6,7]. LTE-DRX in its current form is not suitable for energy savings in 5G enabled cellular devices [8,9], due to two main reasons.

- The communications in LTE do not consider beamforming, whereas in 5G networks UE has to align the best beam pair before the start of communications.
- The LTE-DRX mechanism has a fixed length of sleep duration, which increases the energy consumption and delay. Hence, LTE-DRX is not suitable for low latency communications in 5G networks.

Authors in [3] propose DRX for 5G network, which requires UE to search for best beam pairs after completion of each sleep cycle in order to serve the packets. The additional beam searching operation after each sleep cycle enhances the energy consumption of UE. The work in [10] suggests beam aware DRX approach for 5G enabled machine to machine communications, in which UE has prior information about best beam pairs. Similarly, beam aware DRX mechanism for energy saving in the 5G network is proposed in [11]. The beam aware mechanism may not perform well in case of UE mobility and beam misalignment [12]. Kwon et al. [12] show that the probability of beam misalignment increases with an increase in UE velocity. The probability of misalignment is 0.1 and 0.38 if a UE moves with the velocity of 30 Km/h and 60 Km/h, respectively.

The DRX with built-in state of beamforming is proposed by Liu et al. [13]. Authors present the concept of DRX for multiple beam communications and utilize the semi-Markov model to design eleven-states of DRX. Their approach considers the beam training process only in case of beam misalignment and after completion of long sleep duration. Hence, authors save the energy of UE while minimizing the delay for 5G services. However, the fixed duration of sleep cycles in their approach may cause more energy consumption.

On the other hand, Recurrent Neural Network (RNN) in AI has shown incredible results to predict the upcoming value of a given sequence [14]. Long Short-Term Memory (LSTM) is a popular type of RNN that is specially designed to learn long-term dependencies of a sequence for predicting the upcoming value of a sequence [15]. The term long-term dependency refers to the sequence, whose desired/current output values (prediction results) depend on long-sequence of previous input values rather than the only single previous input value.

Motivated by the success of RNN to learn and predict long-term dependent sequence values in various applications [11,14,16], we use RNN to extract the pattern of packet arrival time from real wireless traffic traces and to predict the values of the next packet arrival time. Based on the prediction results, we propose an AI-DRX algorithm that works on a ten-state DRX model to enable energy saving. AI-DRX for multiple beam communications in 5G network saves the UE energy by enabling dynamic short and long sleep cycles, respectively. To be more specific, the following are our key contributions to save UE energy in multiple beam communications scenario of 5G networks.

- We perform the training of the LSTM neural network on wireless traffic. During training, LSTM network extracts the packet arrival time pattern from the wireless traffic trace. The prediction results show that the trained model predicts with minimum RMSE of 5 ms on random test set from trace 1 and 6 ms on random test set from trace 2, respectively.
- We devise DRX as a ten-state model.
- We propose an artificial intelligence based DRX mechanism for multiple beam communications in 5G networks. We suggest AI-DRX algorithm using ten state model to enable dynamic short or long sleep cycles, depending on the prediction results.

- We evaluate the performance of AI-DRX in terms of energy efficiency and mean delay. AI-DRX achieves the energy efficiency of 59% on trace 1 and 95% on trace 2, respectively; while considering the mean delay requirements of different services.

The remaining paper is organized as follows. In Section 2 we present an overview of existing DRX mechanism and introduce the RNN with focus on LSTM neural network. Section 3 proposes AI-DRX algorithm with a ten-state DRX model, to enable dynamic short or long sleep cycles in multiple beams communications of the 5G network. Section 4 presents the performance analysis of AI-DRX in terms of energy efficiency and mean delay. Finally, we conclude our work in Section 5.

2. Related Work

2.1. DRX in LTE

LTE networks use DRX mechanism to minimize the power consumption of UE. The energy expenditure of UE can be reduced by switching off radio components during the unavailability of incoming data packets [17]. LTE-DRX is configured and controlled by evolved node B (eNB). The eNB informs the UE to turn off its radio components in case of no data in the buffer. The LTE-DRX can be regulated by radio resource control (RRC) layer at eNB. RRC sends the packets' information to UE via a physical downlink control channel (PDCCH). RRC operates in two modes: (1) RRC_connected mode; (2) RRC_idle mode, after a UE is turned on [18]. The RRC_idle mode is only responsible for paging operations, UE neither receives nor transmits the data, but only monitors the paging signals during the paging occasion. Whereas, all the data exchange between UE and eNB takes place during RRC_connected mode. Since all transmissions take place during the RRC_connected mode and this mode is responsible for more energy consumption of UE, our work focuses on the DRX in RRC_connected mode. Figure 1 shows the LTE-DRX in RRC_connected mode that works on two types of states. These states are:

- Active State
- Sleep State

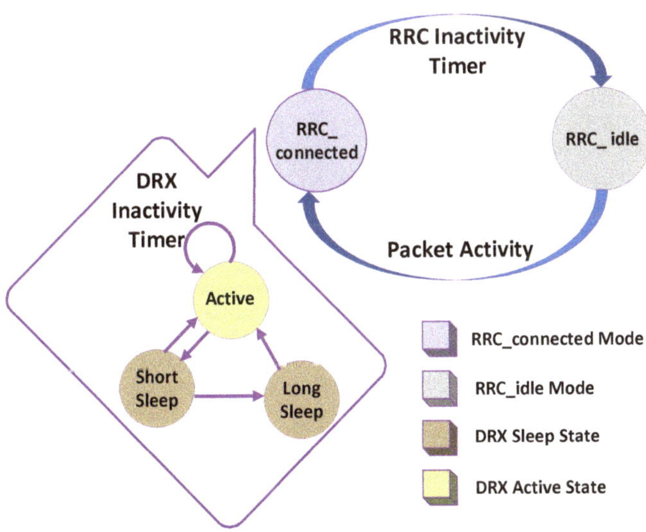

Figure 1. LTE-DRX in RRC_connected mode.

Based on the above two states, different DRX parameters are configured in RRC_connected mode while considering various services' delay requirements.

2.1.1. Active State

A UE sends or receives the data packets during DRX active state. This state does not allow UE to save the energy because UE has to receive the data packets during this state. Moreover, a UE has to turn on the radio circuitry throughout the active state. The active state of DRX can be controlled by two parameters.

- Inactivity timer
- Active time

All the data packets are transmitted/received during the active time. The inactivity timer is a countdown timer in DRX active state. This timer re-starts every time a new data packet arrives at eNB and then eNB serves the received packets to UE. In case of no new packet arrival, the inactivity timer gets expired and the UE switches to sleep state for a certain time period.

2.1.2. Sleep State

A UE turns off the radio components during sleep state in order to save power. UE cannot receive or send the data packets during DRX sleep state but can only monitor the PDCCH for any incoming data. DRX in a sleep state is controlled by the following parameters:

- ON time
- Short sleep cycle
- Short sleep timer
- Long sleep cycle
- Long sleep timer

During ON time, a UE monitors the PDCCH. ON time always starts after completion of each sleep cycle. A short sleep cycle is a small duration of time that saves the UE energy by switching off the transceiver part. A short sleep cycle is repeated up to N_{sc} number of short sleep cycles. After the expiry of N_{sc}, a UE transits to the long sleep cycle. The long sleep cycle and long sleep timer are similar to short sleep cycle and short sleep timer but have a longer time period than the counterpart, respectively.

Figure 2 delineates the timing diagram of LTE-DRX. As shown in Figure 2, all the data transmission and reception take place during the active state. After reception of each new packet, inactivity timer restarts. In the case that no new packet reaches and inactivity timer finishes the countdown timer, then UE switches to short sleep cycle. After every short sleep cycle, a UE monitors the PDCCH for any incoming packet during the ON time. If no new packet arrives before the completion of short sleep timer, the UE switches to the long sleep cycle and remains there until intimation of the new packet is received. UE transits from a long sleep to idle state if no new packets arrive and the long sleep timer expires.

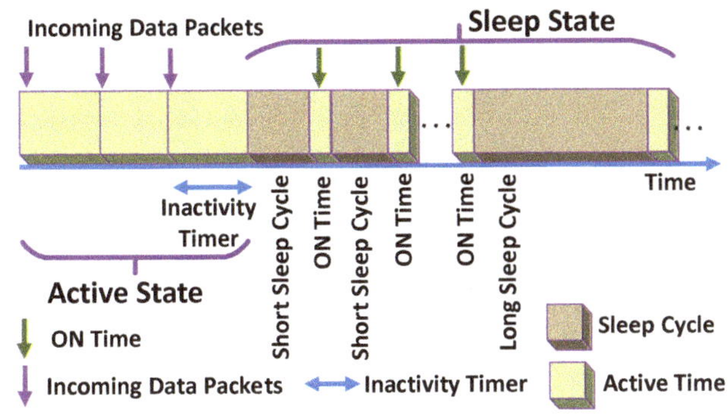

Figure 2. LTE-DRX timing diagram.

2.2. Non-Compatibility of LTE-DRX for 5G Networks

The LTE-DRX mechanism in its existing form is not suitable for 5G networks. The main reason for non-suitability is the directional communications in 5G networks, which require beamforming operation prior to data transmission [4]. Whereas, LTE-DRX mechanism does not include beam searching operation [19,20]. Figure 3 shows the concept of directional communication with multiple beams in 5G network. The gNB transmits K number of beams and a UE has L number of beams. A UE needs to search and align the best beam pair form $K \times L$ beam pairs, prior to the start of communications [4]. The process of beam searching and beam alignment in 5G is not included in LTE-DRX. Moreover, the process of beam searching in 5G network causes a UE to remain in the active state more than that of the LTE-DRX, which also increases the power consumption of UE [21]. Furthermore, the beamforming process in 5G requires additional time for beam searching and aligning best beam pairs [22]. The additional beam searching and beam alignment time cause more delay than the delay in LTE-DRX. Hence, directional communications in 5G is one of the reasons, which makes LTE-DRX mechanism in its existing form, not a suitable solution for power saving in 5G networks [21,22].

The second main reason for non-compatibility of existing LTE-DRX mechanism with 5G networks is the fixed length of sleep cycles in LTE-DRX [11]. LTE-DRX uses fixed length short and long sleep cycles to economize the power consumption of UE. Whereas, 5G network is expected to deal with different types of services, simultaneously [4]. These services may have different size of packet lengths, variable packet arrival time and variable transmission time interval (TTI) [23]. The fixed length of sleep cycles in LTE-DRX mechanism may not be suitable for 5G services as these cycles may under-utilize or over-utilize TTI. Moreover, fixed length sleep cycles in 5G may be responsible for least power savings in 5G enabled devices [23]. However, DRX design can still be used for 5G [9,24,25].

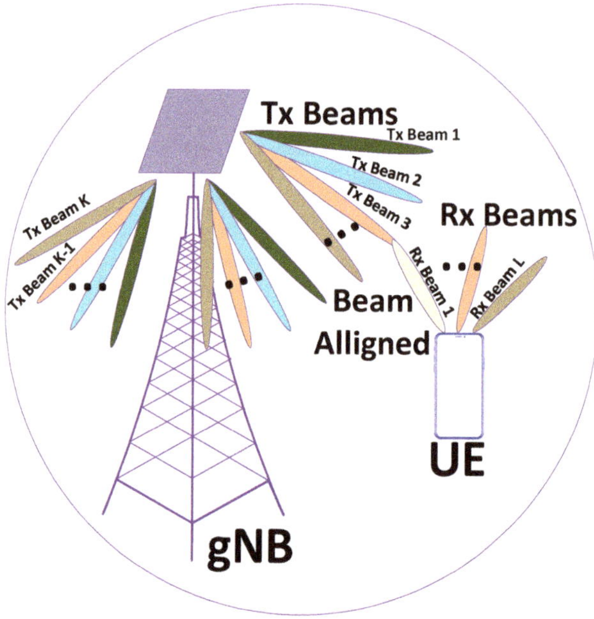

Figure 3. Directional communication with multiple beams in 5G network.

2.3. Recurrent Neural Network for Predicting Sequential Data

Recurrent Neural Network (RNN) is a part of AI that is widely used for predictions over sequential data [26]. Long short-term memory (LSTM) network is one of the popular kinds of RNN that performs well while predicting the long-term dependent time sequences [15]. LSTM is different from other kinds of RNN due to its gated structure and internal cell state in every single unit, as shown in Figure 4. LSTM neural network utilizes various previous inputs to decide the values of internal gates (forget gate, input gate, input modulation gate, output modulation gate), which contributes to cell state. Cell state helps the LSTM to remember or forget the impact of past inputs while deciding the prediction results. This is the main reason LSTM has better performance for predicting the long-term dependent sequences [14,15]. Table 1 shows the notations used in our work.

LSTM unit includes forget gate F_T that can be mathematically shown by Equation (1). The output of forget gate varies from 0 to 1, due to sigmoid $Sig = \frac{1}{1+e^{-T}}$ function. The letters w_x and w_h in the Equation (1) represent the weights associated with the current input x_T and previous output h_{T-1}, respectively. The term b_F is the bias for the forget gate. Similarly, the input gate (I_T) and input modulation gate (G_T) can be calculated by Equations (2) and (3), respectively. The hyperbolic tangent function in input and output modulation gates can be mathematically written as $\tanh(T) = 2Sig(2T) - 1$ and ranges from -1 to $+1$. Moreover, b_I and b_C are biases for input gate and cell state, respectively.

$$F_T = Sig(w_x x_T + w_h h_{T-1} + b_F) \quad (1)$$

$$I_T = Sig(w_x x_T + w_h h_{T-1} + b_I) \quad (2)$$

$$G_T = \tanh(w_x x_T + w_h h_{T-1} + b_C) \quad (3)$$

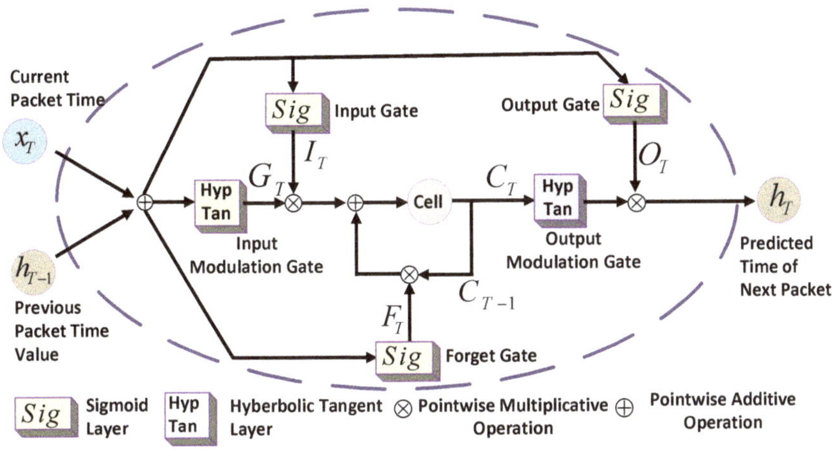

Figure 4. Gated structure of single long short-term memor (LSTM) unit.

Table 1. List of abbreviations.

Abbreviation	Notation	Abbreviation	Notation
Forget gate	F_T	DRX active time	T_{Ac}
Input modulation gate	G_T	DRX ON time	T_{ON}
Input gate	I_T	DRX Inactivity timer	T_{IN}
Output layer	O_T	Packet inter-arrival time	λ_κ
Cell state	C_T	DRX Dynamic sleep time	T_{DY}
Output modulation gate	h_T	Weight corresponding to each gate	w_i

Cell state C_T and the output layer O_T can be calculated by Equations (4) and (5), respectively. The final prediction results can be observed by output modulation gate (h_T) using Equation (6).

$$C_T = F_T \otimes C_{T-1} + I_T \otimes G_T \tag{4}$$

$$O_T = Sig(w_x x_T + w_h h_{T-1} + b_O) \tag{5}$$

$$h_T = O_T \otimes \tanh(C_T) \tag{6}$$

During the training process, LSTM learns to extract the relationship between the input and desired output (upcoming value of sequence) by adjusting various weight values in Equations (1)–(3) and (5). Once the model is trained with the least error, the learned weight values can be used in Equations (1)–(6) to calculate the output values of a sequence (prediction). The error between prediction result and observed (actual) value can be computed by the root mean square error (RMSE) and is given as:

$$\text{RMSE} = \sqrt{\frac{\sum_{T=1}^{N}(Predicted_T - Observed_T)^2}{N}} \tag{7}$$

where the notation T represents the number of samples and N shows the maximum number of samples.

3. AI-DRX for Multiple Beam Communications

3.1. System Model

In 5G directional air interface, beam information is a very crucial operation as a UE wakes up to receive data or control packet [3]. Hence, our work considers DRX for multiple beam communications [13], having dynamic long and short sleep cycles instead of fixed length sleep cycles. Dynamic sleep cycles have different values of sleep time in each cycle rather than one static time value in each sleep cycle. The random values of dynamic sleep time can be predicted based on previous values of packet arrival time.

We model DRX as a ten-state model. These states range from S_0 to S_9 and are shown in Figure 5. Let us elaborate on the states below.

- S_0 is the active time. A UE sends and receives the data during S_0 state. UE predicts the dynamic time T_{DY} during S_0 state.
- S_1 state is threshold comparison state. S_1 compares T_{DY} with both threshold values (Th_{Min} and Th_{Max}) to decide whether to remain in the active time or switch to the dynamic sleep cycle.
- S_2 state is dynamic short sleep cycle. It saves the UE power for a short period of time up to T_{DY}. The predicted value of T_{DY} in the dynamic sleep cycle is always lower than that of value in S_3 state.
- S_3 state highlights the dynamic long sleep cycle. It is similar to S_2 state but has a larger sleep period than S_2. The predicted value of T_{DY} always has larger value of dynamic long sleep cycle than that of the dynamic short sleep cycle.
- S_4 state shows beam training during active time. This state allows UE to train and search for best available beam pairs between gNB and UE.
- S_5 state represents feedback after beam training operation in active time.
- S_6 state shows active time after best beam pairs are aligned. S_6 is different from S_0 as S_6 always occurs after beam training and feedback process. Moreover, S_6 enables UE to search beam pairs in active time until best beam pairs are found.
- S_7 state delineates beam training process after the execution of dynamic long or short sleep cycle.
- S_8 state is feedback operation that occurs after beam training and after the execution of dynamic long or short sleep cycle, respectively.
- S_9 state is the ON period, a UE only monitors PDCCH for an incoming packet. UE could not send or receive data packets during ON time.

3.2. Proposed AI-DRX Algorithm

We propose artificial intelligence based DRX (AI-DRX) algorithm for multiple beams communications in 5G network. Our proposed algorithm enables DRX to achieve dynamic long or short sleep cycles and to reduce the power consumption of UE. AI-DRX algorithm works on ten-state model of DRX as shown in Figure 5. Algorithm 1 demonstrates AI-DRX mechanism and is elaborated below:

- AI-DRX algorithm takes ON timer T_{ON}, minimum threshold value Th_{Min} and maximum threshold value Th_{Max} as input (Line 1).
- AI-DRX examines the buffer for any incoming data packets (Line 2).
- If any packet is received in the buffer, the packets will be served and the value of dynamic sleep time T_{DY} will be predicted, simultaneously (Line 4).
- During the active time, if no beam misalignment occurs between UE and gNB, AI-DRX compares the predicted value of dynamic sleep time T_{DY} with threshold values Th_{Min} and Th_{Max}.
- If the predicted value of dynamic sleep time T_{DY} is less than the minimum threshold value Th_{Min} (Line 8), UE continues to remain in the active time (Line 9).

- If the predicted value of dynamic sleep time T_{DY} is greater than or equal to the minimum threshold value Th_{Min} and less than maximum threshold value Th_{Max} (Line 11). UE observes dynamic short sleep cycle (Line 12).
- In case of any beam misalignment after UE wakes up from dynamic short sleep cycle (Line 13), UE starts the beam training process (Line 14) and Feedback (Line 15) followed by ON time (Line 16).
- During ON time, if beam pairs are still misaligned (Line 17), UE searches beam pairs again (Line 18).
- During ON time, if a new packet arrives (Line 21), UE switches to active mode and start receiving data packets (Line 22).
- After completion of ON time, if no new packet arrives, UE continues to sleep for previous predicted time value T_{DY} of the long sleep cycle or short sleep cycle. respectively (Line 24).
- If T_{DY} is greater than Th_{Max} (Line 28) the UE will go to dynamic long sleep cycle (Line 29).
- UE will perform beam training and feedback after completion of each dynamic long sleep cycle (Line 30).
- In the case of beam misalignment during an active time (Line 35), UE performs beam training (Line 36) and feedback (Line 37) and then re-enter active time (Line 38).
- During the active time (Line 38), if UE still find beam misalignment (Line 39), AI-DRX executes beam training process again (Line 40).
- After the alignment of beam pairs, the UE transits to active time (Line 42).

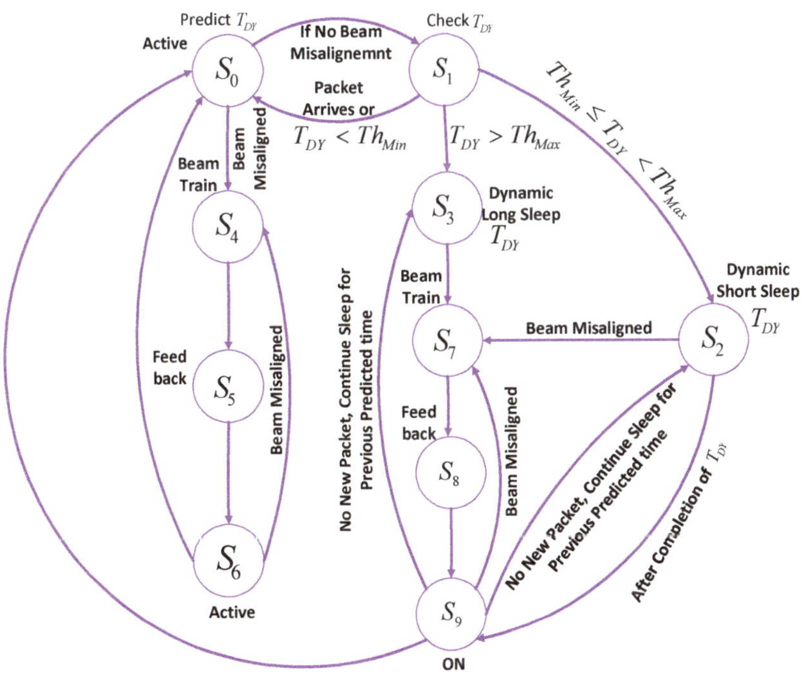

Figure 5. State diagram for AI-DRX.

Algorithm 1 AI based DRX for Multiple Beam Communications.

1: Input T_{ON}, Th_{Min}, Th_{Max}
2: Examine buffer for incoming packets
3: **if** Packets in buffer > 0 **then**
4: Serve the packets and Predict T_{DY} (S_1)
5: **else**
6: **if** No Beam Misalignment **then**
7: Check T_{DY} (S_1)
8: **if** $T_{DY} < Th_{Min}$ **then**
9: Go to Step 4 (S_0)
10: **else**
11: **if** $Th_{Min} \leq T_{DY} < Th_{Max}$ **then**
12: Go to Dynamic Short Sleep up to T_{DY} (S_2)
13: **if** Beam Misalignment **then**
14: Execute Beam Training (S_7)
15: Feedback (S_8)
16: ON (S_9)
17: **if** No Beam Aligned **then**
18: Go to Step 14 (S_7)
19: **end if**
20: **else**
21: **if** New packet arrives before completion of ON Time T_{DY} **then**
22: Go to Step 4 (S_9 to S_0)
23: **else**
24: Go to Dynamic sleep for previous predicted time T_{DY} (S_2/S_3)
25: **end if**
26: **end if**
27: **else**
28: **if** $T_{DY} > Th_{Max}$ **then**
29: Go to Dynamic Long sleep (S_3)
30: Go to Step 14 (S_7)
31: **end if**
32: **end if**
33: **end if**
34: **else**
35: **if** Beam Misalignment **then**
36: Execute Beam Training (S_4)
37: Feedback (S_5)
38: Active after Beam Training (S_6)
39: **if** No Beam Aligned **then**
40: Go to Step 36 (S_4)
41: **else**
42: Go to Step 4 (S_0)
43: **end if**
44: **end if**
45: **end if**
46: **end if**

3.3. *AI-DRX for Enabling Dynamic Long and Short Sleep Cycles*

Algorithm 1 (AI-DRX) demonstrates the use of artificial intelligence in the implementation of DRX for 5G networks. AI-DRX makes dynamic short and long sleep cycles in DRX. AI-DRX utilizes trained LSTM model to predict the upcoming value of packet arrival time and subsequently to enable dynamic sleep cycles in DRX. The training process is conducted offline on two traces of real wireless traffic acquired from the University of Massachusetts (UMass) trace repository [27] and Crawdad data set repository [28]. Training process learns the packet arrival time pattern of increasing sequence values from both traces. Once the LSTM network is trained offline with least prediction error, the trained model predicts the upcoming packet time of real wireless traffic. AI-DRX algorithm calculates the dynamic sleep cycles by using prediction results of upcoming packet arrival time value.

AI-DRX enables a UE to reduce power consumption in multiple beam communications scenario of 5G networks. AI-DRX saves energy by enabling dynamic short and long sleep cycles, respectively. AI-DRX enables dynamic short sleep cycles or dynamic long sleep cycles based on two threshold values Th_{Min} and Th_{Max}. The values of Th_{Min} and Th_{Max} can be varied according to delay requirements of various services. We discuss three cases of AI-DRX: (1) dynamic short sleep cycles; (2) dynamic long sleep cycles; (3) dynamic inactivity timer.

During an active time of AI-DRX, the packets are served to UE. The trained LSTM model predicts the value of upcoming packet arrival time (T_{DY}) while serving the packets. UE checks for new packets in the buffer via PDCCH during ON state of AI-DRX. If no new packet is observed in the buffer, UE continues to sleep up to T_{DY}. However, if a new packet is observed in the ON state, UE transits to active time and receives the packet(s). Meanwhile, inactivity timer in the active mode is restarted on the reception of every new packet. Furthermore, AI-DRX also considers the case of the active state having empty buffer, the UE counts down for dynamic inactivity timer (third case below) to complete and then transits to sleep cycle for predicted sleep time (T_{DY}).

The first case enables dynamic short sleep cycle if the value of T_{DY} is greater than or equal to Th_{Min} and less than Th_{Max}. The condition of dynamic short sleep cycle gets satisfied and UE sleeps for the predicted time value of T_{DY}. There are very small chances of beam misalignment after a short sleep period [12,13]. Hence, AI-DRX considers the beam training process after dynamic short sleep cycles only if beams are misaligned. Figure 6 shows the timing diagram of dynamic short sleep cycle using AI-DRX algorithm.

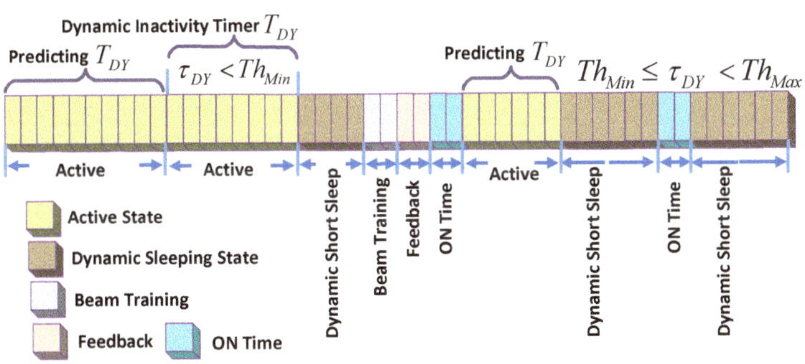

Figure 6. Timing diagram of AI-DRX based dynamic short sleep cycle.

The second case considers the dynamic long sleep cycle up to the predicted value of T_{DY}, if T_{DY} is greater than Th_{Max}. The concept of the dynamic long sleep cycle is depicted in Figure 7. The dynamic long sleep cycle saves more energy than that of short sleep cycle. Moreover, there are more chances of beam misalignment after the dynamic long sleep cycles. Hence, AI-DRX performs beam training and feedback after completion of each dynamic long sleep cycle.

The third case deals with dynamic inactivity timer. If the predicted value of T_{DY} is less than Th_{Min}, the UE will remain active until the period T_{DY} or any new packet arrives. The concept of dynamic inactivity timer can be seen in Figures 6 and 7. AI-DRX also addresses the problem of beam misalignment during active time. AI-DRX performs beam training and feedback in case of any beam misalignment during active time.

It may be noted that AI-DRX utilizes dynamic short and long sleep cycles instead of static fixed time sleep cycles. Moreover, the inactivity timer value is also dynamic according to the prediction results (using trained LSTM model). Our proposed algorithm keeps updating the predicted time value based on most recent received packets and their packet arrival time.

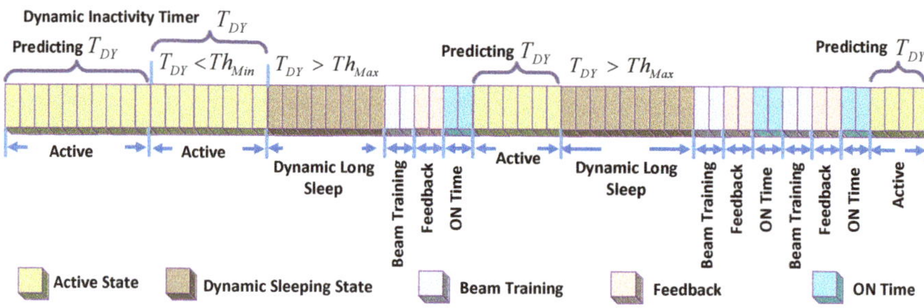

Figure 7. Timing diagram of AI-DRX based dynamic long sleep cycle.

DRX saves the power of UE at the cost of delay. Hence, we considered two performance parameters: energy efficiency (EE) and mean delay. The performance parameter; EE of the UE is the ratio of dynamic short sleep time and dynamic long sleep time to the sum of active time (T_{AC}), ON time (T_{ON}), beam training time, feedback time, dynamic short sleep (T_{DY}) time and dynamic long sleep time (T_{DY}). EE can be calculated by Equation (8). The beam training and feedback processes are considered during active time (T_{AC}) and dynamic short and long sleep cycles take place during (T_{DY}).

$$EE = \frac{T_{DY}}{T_{AC} + T_{ON} + T_{DY}} \tag{8}$$

Similarly, for packet inter-arrival time λ_κ, the mean holding period of active state can be calculated as [5]:

$$T_{AC} = \frac{1 - e^{-\lambda_\kappa T_{IN}}}{e^{-\lambda_\kappa T_{IN}}(1 - e^{-\lambda_\kappa})} \tag{9}$$

Furthermore, the packets arrived during the sleep state and ON duration are stored in the buffer until next ON period. The packet arrival events are the random observer to the sleep period and ON state. Therefore, the mean delay is defined as the sum of mean sleep time and ON duration and is given as:

$$\text{Mean Delay} = \frac{T_{DY}}{2} + \frac{T_{ON}}{2} \tag{10}$$

4. Performance Analysis

We use MATLAB 2019a for training and testing of RNN on two different traces (data sets). These traces of burst traffic type are taken from the Crawdad dataset repository (trace 1) [28] and UMass trace repository (trace 2) [27]. The trace 2 shows the traffic pattern of HTTP and video streaming applications. The video streaming trace is used as it is expected by 2024 over three-quarters of mobile data traffic will be video traffic [1]. The trace 1 and trace 2 include seven parameters as shown in [11]. These parameters are: (1) serial number; (2) packet arrival time in seconds; (3) source address of the packets; (4) destination address of the packet; (5) protocol used; (6) length of packets in bytes; and (7) additional information. We have utilized the time parameter from both traces to train the LSTM network. Time parameter contains the information of packet arrival time. Our purpose is to train the LSTM network until it learns the time interval pattern of packet arrival time from a given sequence. Data pre-processing makes the training process simple and avoid training from the divergence [11,29]. Hence, we standardize the data with zero value of mean and unit value of the variance of the training set during the training process. Moreover, we also standardize the test set during the prediction time.

Hyper-parameters in LSTM network are selected manually to make the training process more efficient. These parameters include the number of hidden units, learning rate drop factor, the maximum number of epochs, initial learning rate, and the optimizer used. The best selection of hyper-parameters may result in the learning with least prediction error. The training process performance can be measured in terms of RMSE. The less the value of RMSE, the better is the prediction results of the trained model on unseen data (test set). We have selected 200 and 125 hidden units in LSTM network during training process over trace 1 and trace 2, respectively. The value of the learning rate drop factor remains 0.2 for both traces. Maximum number of epochs during the training process of trace 1 are considered to be 600 while for trace 2 are 1000 epochs. Initial learning rate during the training process of both traces remains 0.004. Furthermore, the optimizer used during the training process for both traces is Adam optimizer. The above mentioned hyper-parameters reduce the RMSE during training and testing processes.

We have considered the total length up to 130,820 sample values (323.886 s) of trace 1, while trace 2 has 77,470 values of time samples (417.642 s) [11]. We have divided both traces into 10% of samples as the training set and a number of different test sets randomly selected from the remaining 90% of both traces. Each test set has an equal number of samples. Figure 8a shows the RMSE for the initial test set of trace 1 that is as small as 12 ms. Whereas, Figure 8b shows the RMSE value for trace 2 is 10 ms for the first random test set.

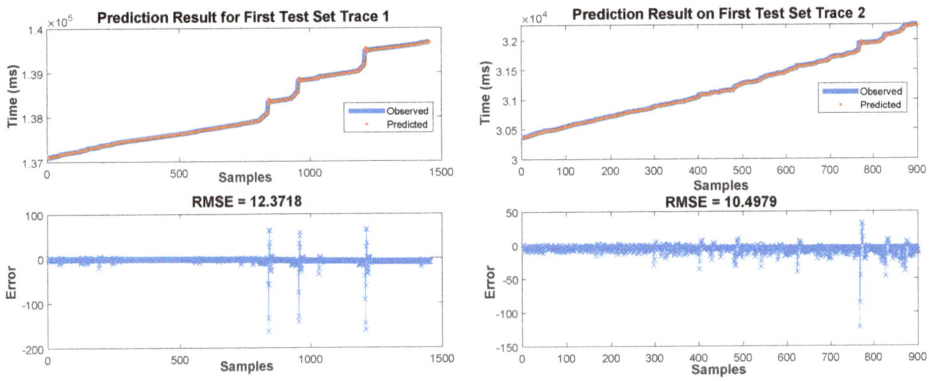

(**a**) Trace 1. (**b**) Trace 2.

Figure 8. Prediction result and root mean square error (RMSE) for first test set.

The RMSE values for random test sets from the remaining 90% samples of trace 1 and trace 2 are shown in Figure 9. It can be seen from Figure 9a,c that the minimum value of RMSE is least at 6 ms and 5 ms on random test sets from samples 70,500 to 71,953 and 90,700 to 92,153 of trace 1, respectively. Whereas, Figure 9b,d show the RMSE value of 6 ms and 8 ms from 10,000 to 10,899 and 65,000 to 65,899 samples of trace 2, respectively.

AI-DRX can be implemented at the gNB of 5G network. During the execution of AI-DRX, we consider packet generation event, active event, dynamic long sleep event, dynamic short sleep event, ON duration event, beam searching event and feedback event. Moreover, our approach enables dynamic sleep cycles in multiple beam communications scenario for 5G networks. The packet generation event in our simulation scans the time column of both traces and produces the data packets of the identical length on the same instant in the respective trace. The generated data packets are collected in the buffer and served to the UE during the active event. UE checks the buffer during ON period, in case of any packet in buffer, the UE switches to the active event, or else continues to sleep. During an active event, the buffered packets are served to UE after getting the beam pairs alignment between UE and gNB. At the same time, packet arrival time is inputted to the trained LSTM model to predict the upcoming packet arrival time. We can obtain the dynamic sleep duration by subtracting the previous packet arrival time value from the predicted dynamic time value T_{DY} (upcoming packet arrival time).

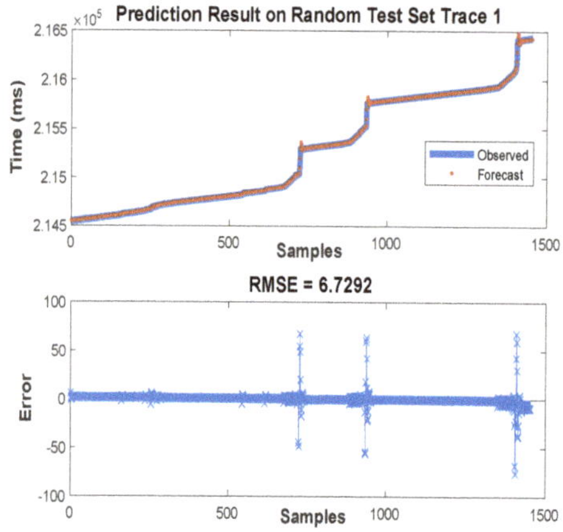

(a) Test set from trace 1 (samples: 70,500 to 71,953).

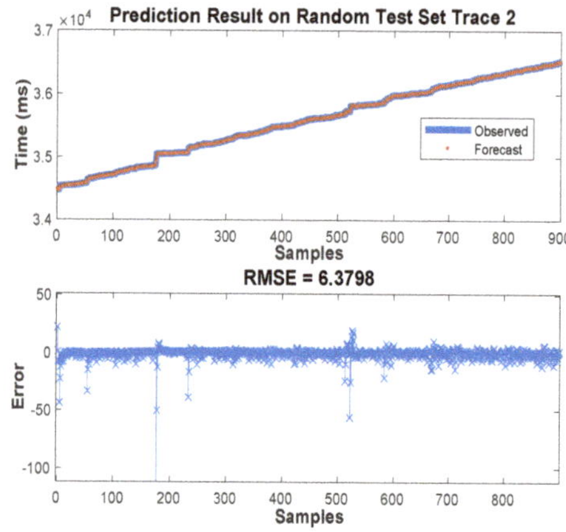

(b) Test set from trace 2 (samples: 10,000 to 10,899).

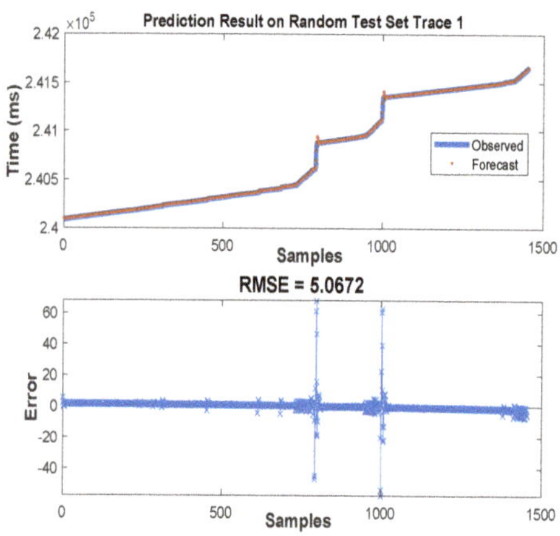

(c) Test set from trace 1 (samples: 90,700 to 92,153).

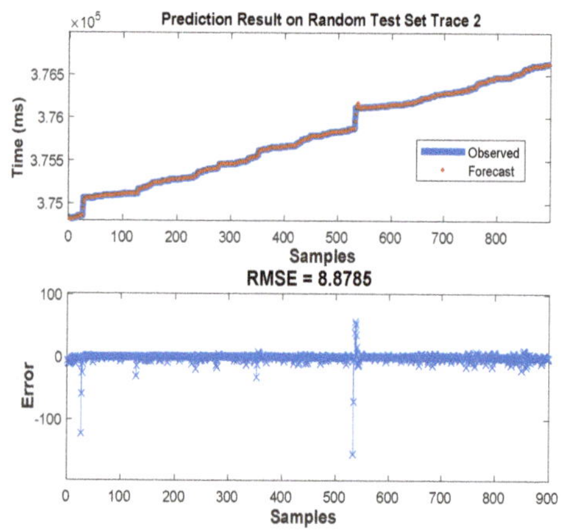

(d) Test set from trace 2 (samples: 65,000 to 65,899).

Figure 9. Prediction results and RMSE for random test sets.

Figure 10a,b, demonstrate the energy efficiency and mean delay with varying ON Period (T_{ON}) for AI-DRX and LTE-DRX over trace 1 & trace 2, respectively. In Figure 10, T_{ON} is varied from 1 ms to 160 ms while the values for Th_{Min} = 20 & Th_{Max} = 100 are considered. Figure 10a highlights the decrease in energy efficiency with increase in T_{ON}. The reason for the drop in energy efficiency is, UE waits for longer period in ON state prior switching to active state (to serve packets). It can be seen from Figure 10b, the mean delay for trace 1 ranges from 150 ms to 194 ms with an increase in ON period. The reason for rise in mean delay lies in the feet that with an increase in ON duration, the time spend by UE in ON state will be higher, which results in higher delay. Moreover, by selecting the optimum value of T_{ON}, delay observed by UE can be minimized. From Figure 10a, it is noticed that the energy efficiency of trace 2 is higher than that of trace 1 due to higher arrival rate of trace 1 as compared to trace 2.

We have compared the performance of AI-DRX with LTE-DRX. To implement the LTE-DRX, we select the value of short sleep cycle to Th_{Min}, the value of long sleep cycle to Th_{Max} and fed trained model with trace 1 and trace 2. It is observed in Figure 10a AI-DRX energy efficiency for trace 1 is 69% higher than that of LTE-DRX, at the cost of higher delay. The reason arises from the fact that AI-DRX calculates sleep time form real wireless traffic trace based on arrival rate, while LTE-DRX uses Th_{Min} and Th_{Max} to set the short and long sleep time. For AI-DRX, the small value of Th_{Min} and Th_{Max} achieves higher energy efficiency as UE easily transits to long sleep time, which results in a larger delay. Whereas, for small values of Th_{Min} and Th_{Max} in LTE-DRX, a UE sleeps for a short period that results in less energy efficiency and smaller mean delay. AI-DRX achieves 55% higher energy efficiency as compared to that of LTE-DRX using trace 2.

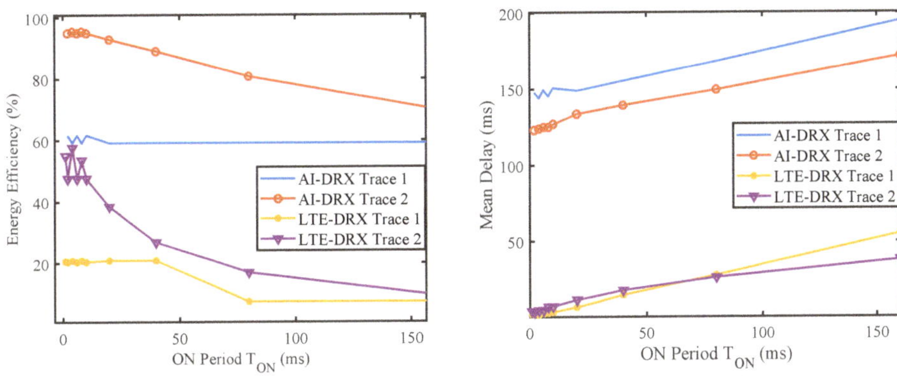

(**a**) Energy Efficiency. (**b**) Mean Delay.

Figure 10. AI-DRX energy efficiency and mean delay with varying T_{ON} ($Th_{Min} = 20, Th_{Max} = 100$).

Figures 11 and 12 present the energy efficiency and mean delay with varying ON period T_{ON} for $Th_{Min} = 200$ & $Th_{Max} = 1000$ and $Th_{Min} = 300$ & $Th_{Max} = 1600$, respectively. It is observed from Figures 11 and 12 that the energy efficiency and mean delay of LTE-DRX increase with an increase in Th_{Min} & Th_{Max}, as UE sleep longer. The energy efficiency of AI-DRX decreases with an increase in Th_{Min} & Th_{Max}, as UE will not able to transit to sleep state if $T_{DY} < Th_{Min}$. With a small value of sleep time T_{DY} or higher value of Th_{Min} & Th_{Max}, UE remains in the active state, which results in less energy efficiency.

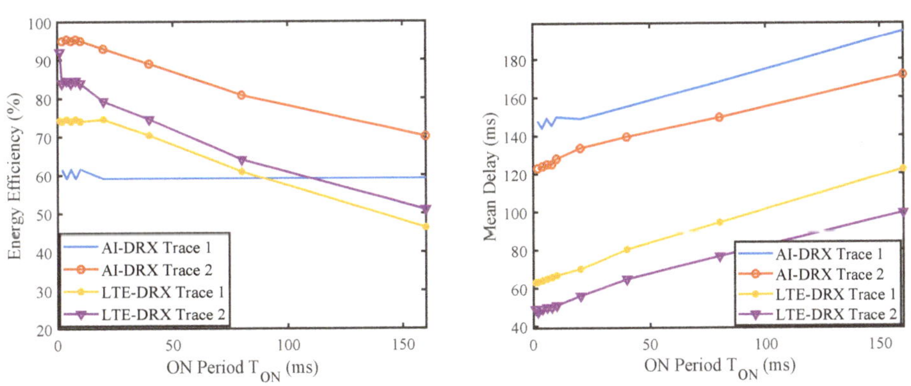

(**a**) Energy Efficiency. (**b**) Mean Delay.

Figure 11. AI-DRX energy efficiency and mean delay with varying T_{ON} ($Th_{Min} = 200, Th_{Max} = 1000$).

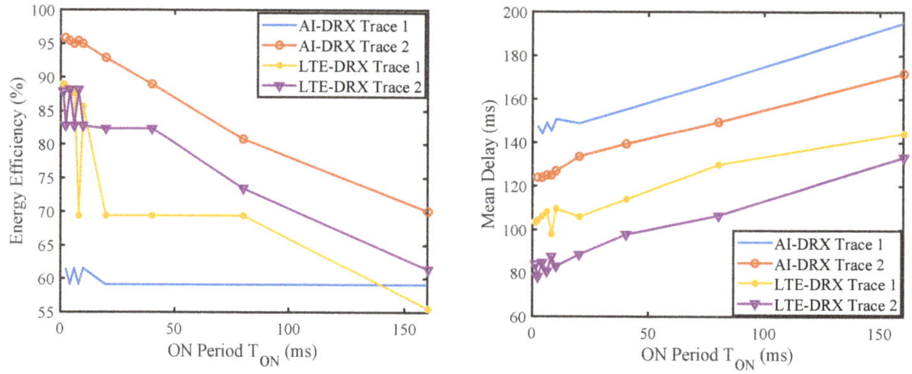

(**a**) Energy Efficiency. (**b**) Mean Delay.

Figure 12. AI-DRX energy efficiency and mean delay with varying T_{ON} ($Th_{Min} = 300, Th_{Max} = 1600$).

To validate our proposal, we have compared our work to traditional Poisson arrival model. We generated three data sets considering Poisson arrival rate (λ) with mean value of $\lambda = 1/20$, $\lambda = 1/10$, and $\lambda = 1$. We have trained model using Poisson arrival rate. The generated traces are fed to AI-DRX algorithm to analyze the energy efficiency and mean delay. Figure 13a,b shows the energy efficiency and mean delay for AI-DRX trace 2 and Poisson arrival with varying T_{ON}. The energy efficiency of AI-DRX is 70% higher than Poisson arrival rate ($\lambda = 1/20$). The corresponding mean delay of AI-DRX is 100 ms (on an average) higher as compared to Poisson arrival for $\lambda = 1/20$. The gain in energy efficiency is achieved as AI-DRX considers real traffic arrival rate for selection of sleep cycles and inactivity timer, while Poisson arrival considers mean arrival rate. The energy efficiency and mean delay both are zero for higher Poisson arrival rate $\lambda = 1$. The reason lies in the fact that for higher arrival rate, UE could not transit to sleep state to save the power.

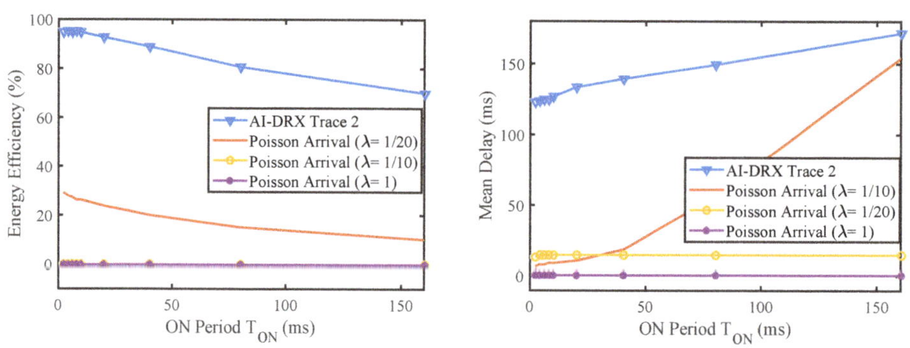

(**a**) Energy Efficiency. (**b**) Mean Delay.

Figure 13. AI-DRX energy efficiency and mean delay comparison with Poisson Arrival with varying T_{ON} ($Th_{Min} = 300, Th_{Max} = 1600$).

Various services in a wireless network can tolerate different delay levels while not compromising quality of service (QoS). QoS class identifier (QCI) is a metric that is used to identify the characteristics of traffic. QCI measures the QoS with two parameters; (1) packet loss rate (PLR) and (2) packet delay budget (PDB). PDB can be defined as maximum tolerable waiting time by a packet during its delivery from eNB to UE. Standardized QCI characteristics are shown in Table 2 [6,30,31]. In various kinds of non-real-time services like email, web browsing, after a certain time period, a UE does not require to monitor PDCCH continuously [6,32]. Hence, these types of services can tolerate higher delays up to 300 ms [6,30,31]. These types of services require DRX with higher values of sleep time ($T_{DY} > Th_{Min}$ & $T_{DY} > Th_{Max}$) and smaller values of ON timer for better energy efficiency. Whereas, the real-time services like voice and live video streaming cannot tolerate delay [33]. Therefore, the delay should be a higher priority than energy saving. If we limit the mean delay to 125 ms, the energy efficiency will be 95.02%. For this energy efficiency we have to select $Th_{Min} = 200$, $Th_{Max} = 1000, T_{ON} = 6$ ms. If we increase the T_{ON} the energy efficiency of UE decreases as UE remains in active state. The mean delay also increases with increase in T_{ON} as UE does not receive the data during ON period but only monitors the Physical Downlink Control Channel (PDCCH). The mean delay observed by UE increases to 150 ms when $T_{ON} = 80$, with energy efficiency of 80.2%. The network can maximize the energy efficiency of UE by selecting optimum value of Th_{Min}, Th_{Max}, T_{ON} depending on QCI value of different services.

Table 2. Standardized QoS class identifier (QCI) characteristics in LTE/LTE-A [6,30,31].

Type	Qos Class Identifier	Packet Loss Rate	Packet Delay Budget	Examples
GBR	1	10^{-2}	100 ms	Voice services
GBR	2	10^{-3}	150 ms	Live streaming services
GBR	3	10^{-3}	50 ms	Real time gaming services
GBR	4	10^{-6}	300 ms	Buffered streaming services
Non-GBR	5	10^{-6}	100 ms	IMS Signaling services
Non-GBR	6	10^{-6}	300 ms	TCP based application services
Non-GBR	7	10^{-3}	100 ms	Interactive gaming services
Non-GBR	8	10^{-3}	300 ms	TCP based video services
Non-GBR	9	10^{-6}	300 ms	TCP based video services

5. Conclusions

In this work, we have suggested an AI-based DRX mechanism for energy saving in multiple beams communications scenario. We have modeled DRX as a ten-state model and suggested AI-DRX algorithm depending on these 10 states. AI-DRX algorithm enables dynamic short and long sleep cycles for energy efficiency of UE in the 5G network. We have trained LSTM network, a popular type of RNN, to extract the packet arrival time pattern from real wireless traffic traces. Later, we have utilized the learned model in AI-DRX algorithm for energy saving in 5G enabled devices. AI-DRX economizes power consumption of a UE by enabling dynamic short and long sleep cycles. Extensive training with selected hyper-parameters achieves the least RMSE of 5 ms on a random test set from trace 1 and 6 ms on a random test set from trace 2, respectively. The energy efficiency obtained with AI-DRX is approximately 60% and 95% for trace 1 and trace 2, respectively. AI-DRX achieves 69% higher energy efficiency on trace 1 and 55% more energy efficiency on trace 2 as compared to LTE-DRX, respectively. We also validated the performance of AI-DRX with traditional Poisson packet arrival model. AI-DRX attains 70% more energy efficiency on trace 2 as compared to Poisson packet arrival rate for $\lambda = 1/20$.

Author Contributions: Conceptualization, M.L.M. and M.K.M.; methodology, A.R.; investigation, M.L.M. and M.K.M.; resources, M.L.M.; data curation, M.L.M.; writing—original draft preparation, M.L.M. and M.K.M.; writing—review and editing, N.S. and A.R.; supervision, N.S., A.R. and D.R.S.; project administration, D.R.S.; funding acquisition, D.R.S.

Funding: This research received no external funding.

Acknowledgments: This research was supported by the Basic Science Research Program through the National Research Foundation of Korea (NRF) funded by the Ministry of Education (NRF-2016R1D1A1B03935633).

Conflicts of Interest: The authors declare no conflict of interest.

References

1. Ericsson Mobility Report June 2019. Available online: https://www.ericsson.com/assets/local/mobility-report/documents/2019/ericsson-mobility-report-june-2019.pdf (accessed on 12 June 2019).
2. Fowler, S.; Shahidullah, A.O.; Osman, M.; Karlsson, J.M.; Yuan, D. Analytical evaluation of extended DRX with additional active cycles for light traffic. *Comput. Netw.* **2015**, *77*, 90–102. [CrossRef]
3. Maheshwari, M.K.; Agiwal, M.; Saxena, N.; Roy, A. Directional Discontinuous Reception (DDRX) for mmWave enabled 5G Communications. *IEEE Trans. Mobile Comput.* **2018**. [CrossRef]
4. Agiwal, M.; Roy, A.; Saxena, N. Next generation 5G wireless networks: A comprehensive surve. *IEEE Commun. Surv. Tut.* **2016**, *18*, 1617–1655. [CrossRef]
5. Zhou, K.; Nikaein, N.; Spyropoulos, T. LTE/LTE-A discontinuous reception modeling for machine type communications. *IEEE Wireless Commun. Lett.* **2013**, *2*, 102–105. [CrossRef]
6. Maheshwari, M.K.; Roy, A.; Saxena, N. DRX over LAA-LTE—A New Design and Analysis based on Semi-Markov Model. *IEEE Trans. Mob. Comput.* **2018**, *18*, 276–289. [CrossRef]
7. Ferng, H.-W.; Tseng, Y.-T.; Wang, T.-H. Parameter-Adaptive Dynamic and Adjustable DRX Schemes for LTE/LTE-A. *IEEE Trans. Green Commun. Netw.* **2019**. [CrossRef]
8. Gódor, I. Revitalizing DRX for enhanced mobile sleep modes. *Comput. Commun.* **2019**, *138*, 98–105. [CrossRef]
9. 3GPP TSG-RAN WG2 #94 Tdoc R2-164023 Std., 2016. Available online: https://www.3gpp.org/ftp/tsg_ran/WG2_RL2/TSGR2_94/Docs/R2-164023.zip (accessed on 4 April 2019).
10. Ho, C.H.; Huang, A.; Hsieh, H.-J.; Wei, H.-Y. Energy-Efficient Millimeter-Wave M2M 5G Systems with Beam-Aware DRX Mechanism. In Proceedings of the 2017 IEEE 86th Vehicular Technology Conference (VTC-Fall), Toronto, ON, Canada, 24–27 September 2017.
11. Mudasar, L.M.; Maheshwari, M.K.; Shin, D.R.; Roy, A. Deep-DRX: A framework for deep learning–based discontinuous reception in 5G wireless networks. *Trans. Emerg. Telecommun. Technol.* **2019**, *30*, e3579.
12. Kwon, S.-W.; Hwang, J.; Agiwal, A.; Kang, H. Performance analysis of DRX mechanism considering analogue beamforming in millimeter-wave mobile broadband system. In Proceedings of the 2014 IEEE Globecom Workshops (GC Wkshps), Austin, TX, USA, 8–12 December 2014.
13. Liu, D.; Wang, C.; Rasmussen, L.K. Discontinuous Reception for Multiple-Beam Communication. *IEEE Access* **2019**, *7*, 46931–46946. [CrossRef]
14. Wang, J.; Tang, J.; Xu, Z.; Wang, Y.; Xue, G.; Zhang, X.; Yang, D. Spatiotemporal modeling and prediction in cellular networks: A big data enabled deep learning approach. In Proceedings of the IEEE INFOCOM 2017—IEEE Conference on Computer Communications, Atlanta, GA, USA, 1–4 May 2017.
15. Gers, A.F.; Schmidhuber, J.; Cummins, F.A. Learning to forget: Continual prediction with LSTM. *Neural Comput.* **1999**, *12*, 850–855.
16. Google. Google AI Blog. Available online: https://ai.googleblog.com/2018/05/duplex-ai-system-for-natural-conversation.html (accessed on 1 June 2019).
17. Bontu, C.S.; Illidge, E. DRX mechanism for power saving in LTE. *IEEE Commun. Mag.* **2009**, *47*, 48–55. [CrossRef]
18. Evolved universal terrestrial radio access (E-UTRA); Medium access control (MAC) protocol specification. Available online: https://portal.3gpp.org/desktopmodules/Specifications/SpecificationDetails.aspx?specificationId=2437 (accessed on 2 March 2018)
19. Chang, H.-L.; Tsai, M.-H. Optimistic DRX for Machine-Type Communications in LTE-A Network. *IEEE Access* **2018**, *6*, 9887–9897. [CrossRef]
20. Chang, H.-L.; Lu, S.-L.; Chuang, T.-H.; Lin, C.-Y.; Tsai, M.-H.; Sou, S.-I.Optimistic DRX for machine-type communications. In Proceedings of the 2016 IEEE International Conference on Communications (ICC), Kuala Lumpur, Malaysia , 22–27 May 2016 .
21. Agiwal, M.; Maheshwari, M.K.; Saxena, N.; Roy, A. Directional-DRX for 5G wireless communications. *Electron. Lett.* **2016**, *52*, 1816–1818. [CrossRef]
22. Maheshwari, M.K.; Agiwal M.; Saxena, N.; Roy, A. Hybrid directional discontinuous reception (HD-DRX) for 5G communication. *IEEE Commun Lett.* **2017**, *21*, 1421–1424. [CrossRef]

23. Maheshwari, M.K.; Roy, A.; Saxena, N. Analytical modeling of DRX with flexible TTI for 5G communications. *Trans. Emerg. Telecommun. Technol.* **2018**, *29*, e3275. [CrossRef]
24. Study on new radio access technology Physical layer aspects. Available online: https://portal.3gpp.org/desktopmodules/Specifications/SpecificationDetails.aspx?specificationId=3066 (accessed on 2 March 2018)
25. Study on new radio access technology Radio interface protocol aspects. Available online: https://portal.3gpp.org/desktopmodules/Specifications/SpecificationDetails.aspx?specificationId=3070 (accessed on 2 March 2018).
26. Rumelhart, D.E.; Hinton, G.E.; Williams, R.J. *Learning Internal Representations by Error Propagation*; Technical Report; Institute for Cognitive Science, University of California: San Diego, CA, USA, 1985.
27. UMass Trace Repository. ACM MMSys conference dataset archive. Available online: http://traces.cs.umass.edu/index.php/Mmsys/Mmsys (accessed on2 March 2018).
28. CRAWDAD Data Set Cambridge/Inmotion. Available online: https://crawdad.org/cambridge/inmotion/20051001 (accessed on 2 March 2018).
29. Baughman, M.; Haas, C.; Wolski, R.; Foster, I.; Chard, K. Predicting Amazon spot prices with LSTM networks. In Proceedings of the 9th Workshop on Scientific Cloud Computing, Tempe, AZ, USA, 11 June 2018.
30. Policy and charging control architecture. Available online: https://portal.3gpp.org/desktopmodules/Specifications/SpecificationDetails.aspx?specificationId=810 (accessed on 2 March 2018).
31. Liang, J.M.; Chen, J.-J.; Cheng, H.-H.; Tseng, Y.-C. An energy-efficient sleep scheduling with QoS consideration in 3GPP LTE-advanced networks for internet of things. *IEEE J. Emerg. Sel. Top. Circ. Syst.* **2013**, *3*, 13–22. [CrossRef]
32. Tseng, C.-C.; Wang, H.-C.; Kuo, F.-C.; Ting, K.-C.; Chen, H.-H.; Chen, G.-Y. Delay and power consumption in LTE/LTE-A DRX mechanism with mixed short and long cycles. *IEEE Trans. Veh. Technol.* **2015**, *65*, 1721–1734. [CrossRef]
33. Huang, B.; Tian, H.; Chen, L.; Zhu, J. DRX-aware scheduling method for delay-sensitive traffic. *IEEE Commun. Lett.* **2010**, *14*, 1113–1115.

electronics

MDPI

Article

Planar Array Diagnostic Tool for Millimeter-Wave Wireless Communication Systems

Oluwole J. Famoriji [1,*], Zhongxiang Zhang [2], Akinwale Fadamiro [1], Rabiu Zakariyya [1] and Fujiang Lin [1]

1 Micro-/Nano Electronic System Integration R & D Centre (MESIC), University of Science and Technology of China (USTC), Hefei 230026, China; aofadamiro@mail.ustc.edu.cn (A.F.); rabiu123@mail.ustc.edu.cn (R.Z.); fujiang_lin@ieee.org (F.L.)

2 Applied Electromagnetic Field Group, Microwave and Radio Frequency Laboratory, Hefei Normal University, Hefei 230601, China; zhzhx@mail.ustc.edu.cn

* Correspondence: famoriji@mail.ustc.edu.cn

Received: 8 October 2018; Accepted: 23 November 2018; Published: 3 December 2018

Abstract: In this paper, a diagnostic tool or procedure based on Bayesian compressive sensing (BCS) is proposed for identification of failed element(s) which manifest in millimeter-wave planar antenna arrays. With adequate a priori knowledge of the reference antenna array radiation pattern, a diagnostic problem of faulty elements was formulated. Sparse recovery algorithms, including total variation (TV), mixed ℓ_1/ℓ_2 norm, and minimization of the ℓ_1, are readily available in the literature, and were used to diagnose the array under test (AUT) from measurement points, consequently providing faster and better diagnostic schemes than the traditional mechanisms, such as the back propagation algorithm, matrix method algorithm, etc. However, these approaches exhibit some drawbacks in terms of effectiveness and reliability in noisy data, and a large number of measurement data points. To overcome these problems, a methodology based on BCS was adapted in this paper. From far-field radiation pattern samples, planar array diagnosis was formulated as a sparse signal recovery problem where BCS was applied to recover the locations of the faults using relevance vector machine (RVM). The resulted BCS approach was validated through simulations and experiments to provide suitable guidelines for users, as well as insight into the features and potential of the proposed procedure. A *Ka*-band (28.9 GHz) 10×10 rectangular microstrip patch antenna array that emulates failure with zero excitation was designed for far-field measurements in an anechoic chamber. Both simulated and measured far-field samples were used to test the proposed approach. The proposed technique is demonstrated to detect diagnostic problems with fewer measurements provided the prior knowledge of the array radiation pattern is known, and the number of faults is relatively smaller than the array size. The effectiveness and reliability of the technique is verified experimentally and via simulation. In addition to a faster diagnosis and better reconstruction accuracy, the BCS-based technique shows more robustness to additive noisy data compared to other compressive sensing methods. The proposed procedure can be applied to next-generation transceivers, aerospace systems, radar systems, and other communication systems.

Keywords: far-field; antenna array; diagnosis procedure; noisy data; BCS; millimeter-wave

1. Introduction

Antenna array is a key technology component in various communication systems such as radar, radio-astronomy, remote sensing, satellite communications, and next-generation (fifth generation, 5G) wireless communications [1], where a very large number (in the hundreds) of radiating elements are particularly used to meet the increasing demands of high radiation performance and reconfigurability [2]. Conversely, the higher the number of radiating elements in the beam-forming configuration, the higher the probability of failed element(s) will be. This causes abrupt field variations across the aperture of the array, and distortion in the radiation features (e.g., beamwidth, peak sidelobe, and boresight). Therefore, the availability of reliable and effective diagnosis tools for large arrays remains an asset, because manual dismantling and replacement operations consume excessive time and cost, and are even unfeasible in satellite-borne installations. Currently, failure identification in antenna arrays is a theoretical and practical important research domain. Detection of faulty elements in antenna arrays is of great interest in both military and civilian markets. Upcoming technologies adopt active or passive antenna arrays with a large number of elements [1–5]. For instance, millimeter-wave transceivers implement multiple-input multiple-output (MIMO) features and beamforming for future 5G applications, as shown in Figure 1. The block diagram shows the location of the AWMF-0108 in a 5G MIMO system [6]. The integrated circuit (IC) contents in the circle are the gain and phase control blocks with amplification and RX/TX switching. The first industrial and commercial millimeter-wave quad-core IC transceiver for 5G applications is the AWMF-0108 [6]. Many compactable antennas were designed for that purpose. Thus, some communication systems will evolve for 5G technology, even before full deployment, which is not expected until 2020. The large number of elements in the planar antenna required by the transceiver must function optimally. Failure in the element(s) causes far-field degradation of antenna systems. The detection of failed elements from field measurements taken from a suitable observation point is very important to re-calibrate the feeding network and to reinstall the needed radiation characteristics by reconfiguring the excitations of the healthy elements [1,3]. Testing of antennas is then a necessity when a certain number of elements exhibit fault. Therefore, the fast diagnosis of complex antenna structures is always a fundamental need.

Far-field measurements are a very powerful approach for antenna array testing. The measurement data are sequentially presented for probable failure identification in the array under test (AUT). The matrix method algorithm (MMA) and back propagation algorithm (BPA) are the most commonly used mechanisms to detect the number and corresponding positions of defective elements using a reference antenna (healthy) and the AUT (defective). BPA [7] was established using the Fourier relationship between radiated far-field and the field situated on the array aperture, and it is applicable to planar antenna arrays. A generalized form of BPA is MMA [8,9]. MMA uses linear algebra standard tools to stabilize the inversion matrix, which relates the array aperture field to radiated far-field. However, MMA and BPA demand a large number of measurements, thus causing long post-processing of large arrays. One approach to mitigate the problem is the use of a priori knowledge of the array without failure; consequently, only the defective array elements are identified. The modeled diagnostic problem is solved by employing available customs whose computation time is a little longer than standard methodologies. At this point, it is evident that the total time taken to get the antenna array diagnosed greatly depends on measurement time, with post-processing times having a higher order of magnitude. This is why sparse recovery-based methods require fewer measurement numbers and provide faster antenna array diagnosis.

Recently, compressive sensing emerged rapidly as a potential technique for solving sparse recovery problems [10–20]. Within this context, the appropriateness of compressive sensing in addressing the array diagnosis problem was examined in References [3,12,14,16,18,21]. Evidently, faulty element distribution in array configurations in practice were found to be highly sparse because it accounts for small non-null entries in the excitation vector of the transmit/receive modules. Beginning from that hypothesis, ℓ_1-norm minimization mechanism was applied successfully to detect failures in planar arrays using a small number of near-field [9] or far-field [14] measurements. Conversely, deterministic compressive sparse techniques require a "measurement matrix" to comply with the restricted isometric property (RIP) condition, for which the estimation of large matrices remains an open challenge [3,14,21]. An alternative is the probabilistic compressive sensing approach reported in Reference [21] to diagnose linear arrays from far-field measurements. However, most of these techniques were not tested experimentally.

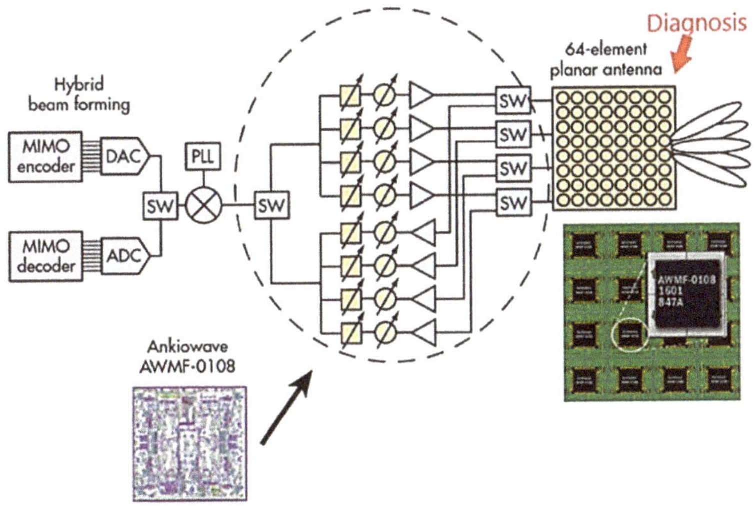

Figure 1. *Ka*-band active antenna array formation which integrates multiple-input multiple-output (MIMO) and beamforming.

In this work, the problem of antenna element excitation level was not examined; however, we estimated field distribution on the array aperture. This helps us identify the modifications of aperture field distribution as a result of factors that cannot be quantified by simple failure of elements, such as different reflections of the array and its feed. The problem faced in getting more information about the AUT is the larger number of unknowns. However, in sparse recovery methods, the required number of measurements increases slowly and logarithmically with the number of unknowns [10–13]. Hence, the field reconstruction scheme benefits more in sparse recovery-based mechanisms. Different sparse recovery algorithms used to conduct antenna array diagnosis were unveiled [14] and compared to the traditional BPA and MMA. In particular, total variation (TV) norm, mixed ℓ_1/ℓ_2 norm, and minimization of the ℓ_1 norm were used to proffer solutions to the resulting inversion issues. From the field reconstructed on the antenna aperture, Fuchs et al. [20] acquired a good antenna diagnosis. The approach was applied to far-field simulation data generated from a 100-element antenna array. The performance of the diagnosis was evaluated and compared to the two standard techniques (BPA and MMA) under different conditions. The approaches were also applied to far-field measurement data of an antenna array with failure to justify the practical applicability of the proposed schemes. Although there were many more works on sparse recovery methods in applied electromagnetics and microwaves involving the diagnosis processes of antenna arrays [13,14], experimental data, which are fundamental for testing any procedure, were reported in few of them.

In References [15–20], differential scenarios with sparse recovery algorithms were employed to perform antenna diagnosis and retrieve element excitations. Reference [21] proposed a joint scheme for adaptive diagnosis of antenna arrays using communication signal fusion (radar-communication scheme) and the echoes of probe signals received at the same antenna. This method equally solved the antenna diagnosis problem at low signal-to-noise ratios (SNRs) to ensure optimal performance of smart sensors in wireless sensor networks. Also, Reference [22] attempted array diagnosis in millimeter waves using compressive sensing. This work considered both full and partial blockage, which occurs from a plethora of particles (such as ice, water droplets, salt, and dirt) and the technique jointly computed the locations of the blocked elements, and the induced phase-shifts and induced attenuation provided the prior knowledge of the angles of departure/arrival. Reference [23] proposed a deterministic sampling strategy for failure detection in uniform linear arrays via compressed sensing or a sparse recovery approach. This is an extension of the Weyl formula which is basically used for prime numbers. The strategy obtained was good for nonprime number (i.e., valid for any number of array elements). This sampling approach is good for sparse electromagnetic (EM) problems encompassing Fourier matrices. Reference [24] gives a review of different capacities of sparse recovery by analyzing how compressive sensing can be applied to antenna array synthesis, diagnosis, and processing. Illustrations of a set of applicative examples were given, including direction-of-arrival estimation, along with present challenges and current trends in compressive sensing applications to the solution of innovative and traditional antenna array challenges. In general, compressive sensing generates few unknown numbers; however, it needs a comprehensive array model with exact knowledge of the radiating element patterns to produce useful results. The technique is sparse with respect to the whole array structure, and requires a priori information to recast it as the function of minimization of ℓ_1 norm. However, these techniques are applicable only if the relationships between the data and the unknowns satisfy the restricted isometry property (RIP). To overcome this challenge, the Bayesian compressive sensing (BCS) approach is adopted. This technique was explored in many electromagnetic problems, such as antenna design and synthesis [25,26], microwave imaging [27–30], and direction-of-arrival estimation [31–33]. It is employed in this paper to estimate the number, magnitude, and location of failures in antenna arrays from far-field measurements. The BCS approach was attempted to diagnose large linear arrays [11], and more recently, planar array configurations [34]; however, no experimental validation was reported. Hence, there is a need for a more reliable procedure tested experimentally and via simulation, because experiments are fundamental tests of any given procedure.

Specifically, this work is an extension of that described in References [15,16]. The BCS method is applied to both the simulated and measured far-field data of a millimeter-wave 100-element microstrip patch antenna array in which failures were added intentionally. A new regularization technique was unveiled and applied to field distribution in order to enhance the efficiency and reliability of antenna array diagnosis. The proposed BCS-based approach is a better choice due to its fast nature and robustness under different noise conditions. The key contributions of this paper are summarized as follows:

1. The BCS technique was applied to diagnose a millimeter-wave planar array, and the result was compared with other approaches reported in the literature.
2. The procedure shows high effectiveness and reliability with fewer measurement points compared to the other methods, and is highly robust to additive noisy data. This was validated experimentally and via simulations.

However, some boundary conditions were observed. The BCS-based approach detects diagnostic problems with few measurements, provided prior knowledge of the reference array radiation pattern, and the number of faults is relatively smaller than the array size. The remainder of this paper is arranged as follows: Section 2 contains the problem formulation of antenna array diagnosis. Section 3 presents compressed sparse recovery methods. Resolution via the BCS-based approach is given in Section 4. The numerical simulations are presented in Section 5. Diagnoses from experimental data are presented and discussed in Section 6. Finally, some conclusions are drawn in Section 7.

2. Antenna Array Diagnosis Problem Formulation

Consider an antenna array in space (Figure 2a). The antenna radiated far-field is usually quantified by phase and amplitude. The AUT is depicted in Figure 2b. All the parameters associated with the AUT are marked with superscript "u". Specifically, $\boldsymbol{E}^u(x,y)$ is the tangential field situated on the antenna aperture, i.e.,

$$\boldsymbol{E}^u(x,y) = E_x^u(x,y)\hat{x} + E_y^u(x,y)\hat{y}, \tag{1}$$

where $E_x^u(x,y)\hat{x}$ and $E_y^u(x,y)\hat{y}$ are the x and y planes of the aperture's electric field, respectively. Far-field $F^u(r,\theta,\varnothing)$ is the measured field on part of the hemispherical surface $(0 \le \theta \le \pi/2, 0 \le \phi \le 2\pi)$ at radius r from the phase center of the AUT, and $r > 2D^2/\lambda$, where D is the diameter. Also, the amplitude and phase of a reference array (RA; array without failures) shown in Figure 2b are assumed to be available. Associated quantities are marked with superscript "o". $E^o(x,y)$ is the field on the aperture Σ of the reference array (RA) and $F^o(r,\theta,\varnothing)$ represents the far-field radiation. For the differential antenna (DA) shown in Figure 2c, the tangential distribution $E(x,y)$ on the aperture Σ is equal to the difference between the field distributions of the reference array and the antenna under test, and the corresponding far-field $F(r,\theta,\phi)$ is expressed as the difference between the fields of reference array (RA) and AUT as

$$E(x,y) = E^u(x,y) - E^o(x,y), \tag{2}$$

$$F(r,\theta,\phi) = F^u(r,\theta,\phi) - F^o(r,\theta,\phi). \tag{3}$$

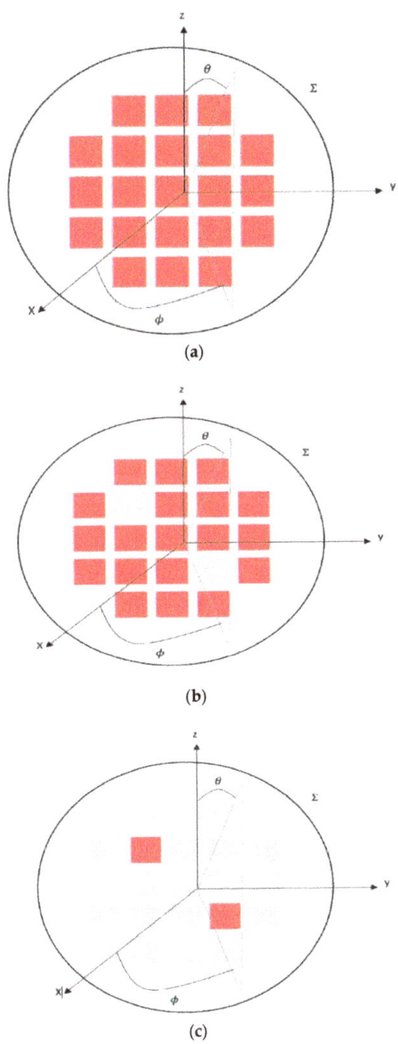

Figure 2. Antenna array: (**a**) reference antenna without failures; (**b**) antenna under test (AUT); (**c**) differential antenna (DA). The number of failures is 2 within the total element number $N = 21$.

The differential antenna gives a resulting problem in which only the corresponding area to the field modification radiates as a result of failure. By visually monitoring the field distribution on the DA, the identification of faulty elements in the AUT can be observed.

2.1. Number of Far-Field Measurement Points Required

BPA and MMA approaches require a large number of measurement points. In the differential antenna, we assumed the field was localized, i.e., the unknown we wanted to retrieve was very sparse, as shown in Figure 2c. In practice, there are a fewer number of failures than the overall elements N. Sparse recovery algorithms estimate $\mathbf{x}$ from a number of measurement points smaller than the number of measurements required by the standard mechanisms. Hence, it is possible to theoretically get a reduction in the number of measurement points. Fuch et al. [20] demonstrated this using total variation (TV), mixed ℓ_1/ℓ_2 norm, and minimization of ℓ_1. However, better methods/algorithms that require fewer numbers of far-field measurement points for faster array diagnosis are still in demand.

2.2. Signal-to-Noise Ratios (SNRs)

Total variation (TV), mixed ℓ_1/ℓ_2 norm, and minimization of ℓ_1 techniques are the leading compressive techniques, and they exhibit low efficiency and reliability in antenna diagnosis for low SNRs. This drawback fosters the need for a more robust diagnosis procedure in the presence of noisy data.

3. Compressed Sparse Recovery Methods

The essence of matrix inversion regularization is to initiate a priori facts within the inversion. An adequate approach is needed to get this regularization by approximately reducing the selected norm q of $\mathbf{x}$ solution. Then, the optimization to be solved is

$$\min_{\mathrm{x}} \|X\|_q \quad \text{subject to } \|\boldsymbol{y} - \boldsymbol{A}\boldsymbol{X}\|_2 \leq \gamma, \tag{4}$$

where $\| * \|_q$ represents l_q norm, and γ is a function of noise and factors influencing the data. There are various routines available to effectively solve the convex optimization problem of Equation (4) such as References [25–27]. The three norms l_q, selected based on a priori knowledge of the differential antenna set-up with the diagnosis problem, can now be described for regularization of the inversion. We applied them to conduct diagnosis of both the simulated and measured radiating antennas.

3.1. Total Variation (TV) Norm

Based on a priori knowledge that solution X has small discontinuities as a result of failures present, in addition to the failures, we expect field X to be leveled and almost zero. Hence TV norm is a smooth function to regularize $\boldsymbol{X}$ [27]. Thus, minimizing TV norm is minimizing its gradient, which is the effect of smoothing. Consider a two-dimensional complex dataset $\boldsymbol{X} \in \mathbb{C}^{M \times N}$; TV norm gives

$$\|\boldsymbol{X}\|_{TV} = \sum_{m,n} |\boldsymbol{X}_{m+1,n} - \boldsymbol{X}_{m,n}| + |\boldsymbol{X}_{m,n+1} - \boldsymbol{X}_{m,n}| \tag{5}$$

$$\|vec(\nabla_x \boldsymbol{X})\|_1 + \|vec(\boldsymbol{X}\nabla_y)\|_1.$$

Vec($\boldsymbol{X}$) generates vector $N \times M$ which holds the columns of $\boldsymbol{X}$ stacked beneath each other. Gradient matrices ∇_x and ∇_y are of $M \times M$ and $N \times N$ size, respectively, which are expressed as

$$\nabla_x = \begin{bmatrix} 1 & & 1 & 0 \\ & \ddots & \ddots & \\ 0 & & -1 & 1 \end{bmatrix}, \text{and}$$

$$\nabla_y = \begin{bmatrix} -1 & & 0 \\ 1 & \ddots & \\ & \ddots & -1 \\ 0 & & 1 \end{bmatrix}.$$

Then, the optimization problem in Equation (4) transforms to

$$\min_{\boldsymbol{X}} \|\boldsymbol{X}\|_{TV} \quad \text{subjects to } \|\boldsymbol{y} - \boldsymbol{A}vec(\boldsymbol{X})\|_2 \le \epsilon. \tag{6}$$

3.2. The ℓ_1 Norm

Since there is a sparse solution X; then, a space of search could be drastically reduced by the introduction of a priori knowledge in inversion. Specifically, the ℓ_1-norm ($\|X\|_1 = \sum_k |x_k|$) is the leading convex surrogate of an acceptable estimate sparsity of the vector (quasi-norm ℓ_0 that calculates nonzero occurrences of a given vector). As a result, ℓ_1 norm is an efficient approach to enhance sparse solutions [2,5,10,19]. The regularization problem is then

$$\min_{X} \|X\|_1 \quad \text{subject to } \|\boldsymbol{y} - \boldsymbol{A}X\|_2 \le \epsilon. \tag{7}$$

Minimizing ℓ_1-norm imposes the pointwise sparsity of solution per sample x_k of the field on the aperture of the DA.

3.3. Mixed ℓ_1/ℓ_2 Norm

The radiating aperture's position and dimensions can also be taken. The solution X is grouped into G groups X^g, which corresponds to the individual radiating element's aperture g. For a faulty element, all regions of discretization x_k^g in the aperture will be faulty (nonzero). Let vector X of dimension $M \times N$ be divided into G non-overlapping groups depicted X^g of size N_g, such as $\sum_{g=1}^{G} N_g = MN$. Hence, the mixed ℓ_1/ℓ_2 norm is given as

$$\|X\|_{1,2} = \sum_{g=1}^{G} \|X^g\|_2 = \sum_{g=1}^{G} \sqrt{\left|x_1^g\right|^2 + \ldots + \left|x_{N_g}^g\right|^2}. \tag{8}$$

The mixed ℓ_1/ℓ_2 norm has similar behavior to ℓ_1 norm on vector $\|X^1\|_2, \ldots, \|X^g\|_2, \ldots, \|X^G\|_2$; it, therefore, induces group sparsity at the radiating aperture level. The regularized inversion problem is then expressed as

$$\min_{X} \|\boldsymbol{X}\|_{1,2} \quad \text{subject to } \|\boldsymbol{y} - \boldsymbol{A}\boldsymbol{X}\|_2 \le \epsilon. \tag{9}$$

4. Resolution via Bayesian Compressive Sensing

For a planar antenna configuration of N elements positioned at coordinates $(x_n y_n)$, $n = 1, \ldots, N$, with error-free excitations α_n, $n = 1, \ldots, N$, beaming a familiar field $E(u, v)$, $(where\ u = sin\theta cos\varphi\ and\ v = sin\theta sin\varphi)$, referencing a noisy case with element failure, the estimated far-field radiation pattern of (AUT) is expressed as

$$\widetilde{E}(u_l, v_l) = \sum_{n=1}^{N} \beta_n e^{j\frac{2\pi}{\lambda}(x_n u_l + y_n v_l)} + v_l, \tag{10}$$

where (u_l, v_l) for $l = 1, \ldots, L$ is the angular location of the *l-th* angular sample, and v_l is the noise effect considered as Gaussian-distributed with zero mean and variance σ^2. β_n, $n = 1, \ldots, N$, is the failed excitations vector, expressed as

$$\beta_n = \begin{cases} h\alpha_n & \text{with probability } \Phi \\ \alpha_n & \text{otherwise} \end{cases}, n = 1, \ldots, N. \tag{11}$$

$h \in (0,1)$ is the failure factor, while Φ is the rate of failure, and α_n is the weighting coefficient. From knowledge of the difference in field pattern, $W(u_l, v_l) = E(u_l, v_l) - \widetilde{E}(u_l, v_l),\ l = 1, \ldots, L$, array failures can be estimated by determining the minimum $\ell_0 - norm$ vector

$$\underline{Y} = \{Y_n = \alpha_n - \beta_n; n = 1, \ldots, N\}, \tag{12}$$

which satisfies

$$\underline{W} - \underline{\underline{\Psi}}\underline{Y} = \underline{v}. \tag{13}$$

The aim is to determine the entries of the "failure" vector $\underline{W} = \{W(u_l, v_l); l = 1, \ldots, L\}, \underline{v} = \{v_l; l = 1, \ldots, L\}$ from the prior knowledge of the difference between the field samples in Equation (10) on the AUT and those of the *golden* antenna with coefficients α_n $n = 1, \ldots, N$. Against the deterministic approaches aimed at retrieving the vector $\underline{\gamma}$ with minimum $\ell_0 - norm$ satisfying the condition of Equation (13), $\underline{\underline{\Psi}}$ is the $L \times N$ radiation measurement matrix expressed as

$$\underline{\underline{\Psi}} = \begin{bmatrix} e^{[j2\pi(x_1 u_1 + y_1 v_1)]} & \cdots & e^{[j2\pi(x_N u_1 + y_N v_1)]} \\ \vdots & \ddots & \vdots \\ e^{[j2\pi(x_1 u_L + y_1 v_L)]} & \cdots & e^{[j2\pi(x_N u_L + y_N v_L)]} \end{bmatrix}. \tag{14}$$

Hence, the BCS technique (summarized in Figure 2) can be employed to determine the sparsest solution $\hat{\underline{Y}}$ to the problem

$$\hat{\underline{Y}} = arg\left\{\max_{\underline{Y}}[P(\underline{Y}|\underline{W})]\right\}, \tag{15}$$

which gives

$$\hat{\underline{Y}} = \frac{1}{\sigma_{NP}}\left[\frac{\underline{\underline{\Psi}}^T\underline{\underline{\Psi}}}{\sigma_{NP}} + diag\left(\underline{f}_{NP}\right)\right], \tag{16}$$

where T is the transpose operator, and σ_{NP} and $\underline{f}_{NP}$ are the figures that are used to maximize the likelihood function

$$L\left(\sigma, \underline{f}\right) = -\frac{1}{2}\left[Nlog2\pi + log\left|\underline{\underline{C}}\right| + \underline{W}^T\,\underline{\underline{C}}^{-1}\,\underline{W}\right]. \tag{17}$$

Equation (11) is computed using RVM [29], with $\underline{\underline{C}} = \sigma + \underline{\underline{\Psi}}\underline{\underline{F}}^{-1}\,\underline{\underline{\Psi}}^T$, where $\underline{\underline{F}} = diag(\underline{f})$.

The implementation of the BCS technique (as shown in Figure 3) is summarized in Algorithm 1.

Algorithm 1. Proposed diagnostic procedure

a. Step 1—*Array parameter selection and definition of problem*: For each element ($n = 1, \ldots N$), accumulate the array field at N measurement points (reference antenna) and randomly sample the AUT field. Appropriate the noise variance and threshold for the maximization of $L\left(\sigma, \underline{f}\right)$.

b. Step 2—*Radiation pattern measurement matrix* $\underline{\underline{\Psi}}$ *definition*. Input the parameters of $\underline{\underline{\Psi}}$.

c. Step 3—*Posterior mode estimation*. Maximize Equation (10) iteratively to estimate σ *and* $\underline{f}$ using the RVM procedure [29].

d. Step 4—*Source difference reconstruction*. Determine $W(u_l, v_l)$ for $l = 1, \ldots, L$.

e. Step 5—Failed field excitation reconstruction. Determine the vector of failed excitations β_n ($n = 1, \ldots, N$) using Equation (11).

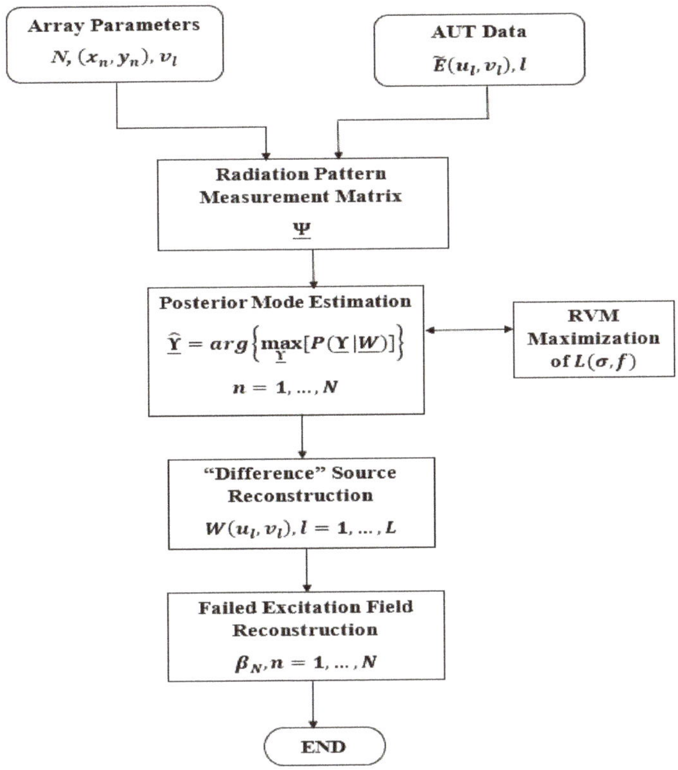

Figure 3. Flowchart of the proposed Bayesian compressive sensing (BCS) procedure.

5. Simulations and Analysis

Assessing the performance of the BCS algorithm, we consider an RA with $N = 316$ with Taylor taper and peak sidelobe level of −25 dB. Assuming a complete failure ($h = 0$), the given percentage of failed elements is $\Phi = 4\%$. To determine the detection error numerically, the index of detection can be expressed as

$$\zeta = 100 \times \frac{\sum_{n=1}^{N} \left| \mathrm{Y}_n - \hat{\mathrm{Y}}_n \right|^2}{\sum_{n=1}^{N} \left| \mathrm{Y}_n \right|^2}, \tag{18}$$

where Y_n and $\hat{\mathrm{Y}}_n$, $n = 1, \ldots, N$, are the real and predicted failure entries of vector $\underline{\mathrm{Y}}$. The uniformly sampled ($k = 316$ samples) far-field radiation pattern is within (u, v) space, with the signal-to-noise ratio $SNR = 30$ dB. The configuration of the AUT is presented in Figure 4, while the excitation coefficients β_n of the failed array are equally shown. Figure 5 shows $\hat{\mathrm{Y}}_n$ of the failure vector estimated via the proposed technique. As expected, there is good correlation between the location and the number of the real and predicted failed elements. Accuracy of the estimation was ascertained by a very small index of detection figure of $\zeta = 3.83 \times 10^{-3}$.

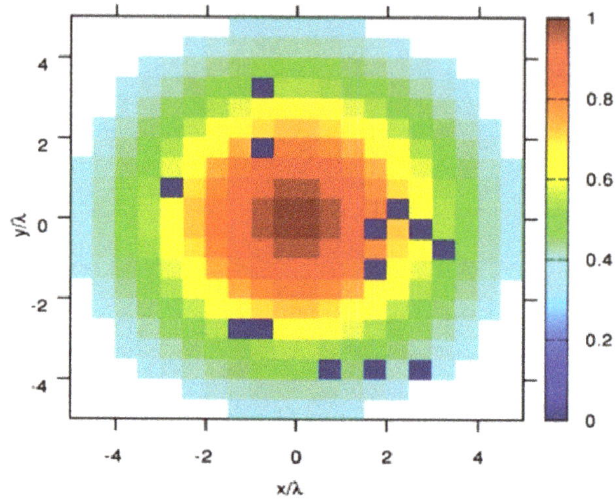

Figure 4. Normalized array excitation of the array under test (AUT).

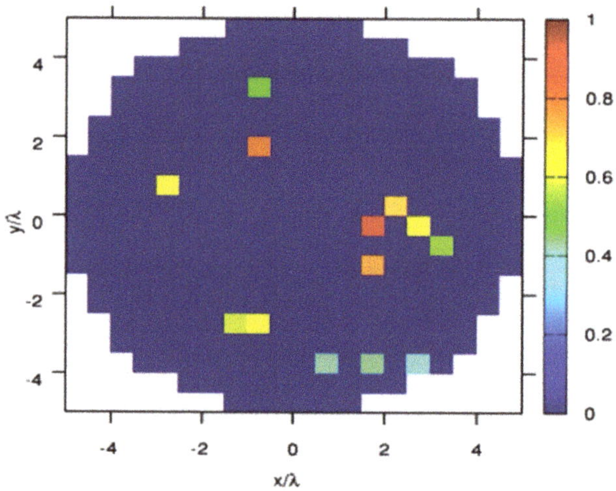

Figure 5. Normalized array excitation with estimated failure vector $\underline{\hat{Y}}$ and location and amplitude of entries.

To analyze the impact on performance metrics of the technique of the noise on far-field patterns, we ranged the SNR between 0 dB and 100 dB. Figure 6 shows the obtained result. The estimation error ζ is high for low SNRs irrespective of failed element percentage. Also, for higher SNR, the robustness of the approach increases, which shows that the best performances are attained at $\Phi = 3\%$. The impact of the percentage of failed elements on the performance of the method proposed was also assessed by varying the percentage of the failed elements from $\Phi = 2\%$ to $\Phi = 20\%$ (see Figure 7). Expectedly, the performance metrics of the approach reduced for a higher percentage of failed elements, even for very low noise levels. Conversely, the approach achieved a degree acceptable accuracy until about 10% of damaged elements at any SNR. This result validates the efficiency of the BCS technique in the diagnosis of sparse failures in arrays.

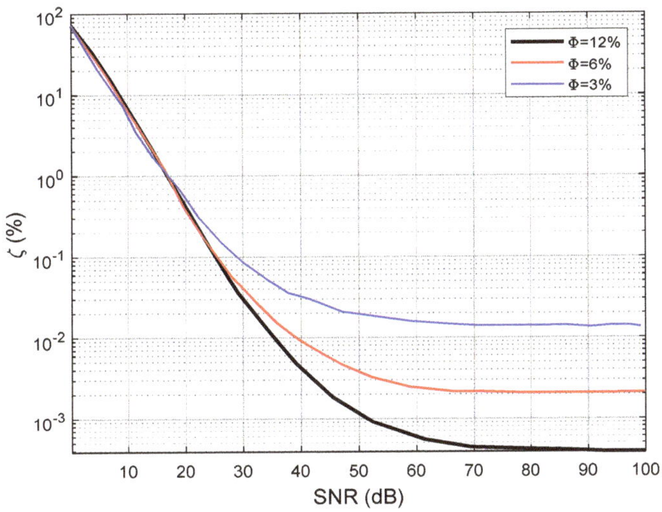

Figure 6. Detection index against signal-to-noise ratio (SNR) for different failure percentages.

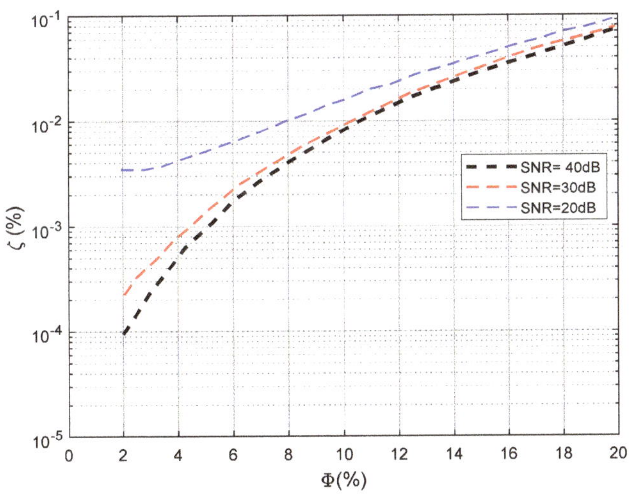

Figure 7. Detection index against failure rate for various SNRs.

Example Using Full-Wave Simulation Set-Up

A 10×10 microstrip patch antenna array with an aperture size of 31×31 mm^2 operating at 28.9 GHz was designed, as shown in Figure 8. We designed and computed the radiation pattern and S-parameter (Figure 9) of the antenna using full-wave three-dimensional (3D) EM software Ansys HFSS v. 17. The elements were uniformly spaced along the x and y directions. Each element had an excitation port. Practically, measurements are made impure by noise; hence, a Gaussian noise $\boldsymbol{n}$ was added to the data on both radiation patterns as $\boldsymbol{y}_n^q = \boldsymbol{y}^q + \boldsymbol{n}^q$, with $q = \{r, d\}$. The noise level was determined by signal-to-noise ratio (SNR) defined from the maximum received field magnitude fitting the dynamic measurement range. The noise was estimated as

$$n^q = \frac{\mathcal{N}(0,1) + j\mathcal{N}(0,\,1)}{\sqrt{2}}\; max|\boldsymbol{y}^q| \,\times\, 10^{-SNR_{dB}/20}, \tag{19}$$

where $\mathcal{N}(0, 1)$ is a Gaussian random vector of mean 0 and standard deviation 1.

The faulty elements cause low gain, high sidelobe level, wider beamwidth, lower front-to-back ratio, and a boresight pointing error. The effects are shown in Figure 9b. A Rogers 5880 dielectric substrate with 20 mm thickness and 3.48 dielectric constant was used for the antenna design because it has low signal loss, low dielectric loss, low outgassing (which is good for space applications), and cheap circuit fabrication. The total number of elements in the array was 100. The length and width of a single patch were estimated to be 7.4 mm and 9.5 mm, respectively. Elements were uniformly spaced by 8.947 mm and 6.847 mm along x and y, respectively.

All elements were fed with the same excitation value which equal to one to emulate the array without failure (reference array). Then K failures (in this case, the estimated percentage of failure rate) were also initiated intentionally by making the excitation equal to zero in order to model the AUT effectively. At first, we considered the reference array, i.e., the array without failures. The excitation coefficients are depicted in Figure 10, and the reconstructed excitation coefficients are presented in Figure 11. Also, for quantitative knowledge of the error estimated, the computed excitation error in dB is shown in Figure 12. The result indicates an exact reconstruction in the case of the reference array, and shows a low probability of a false alarm.

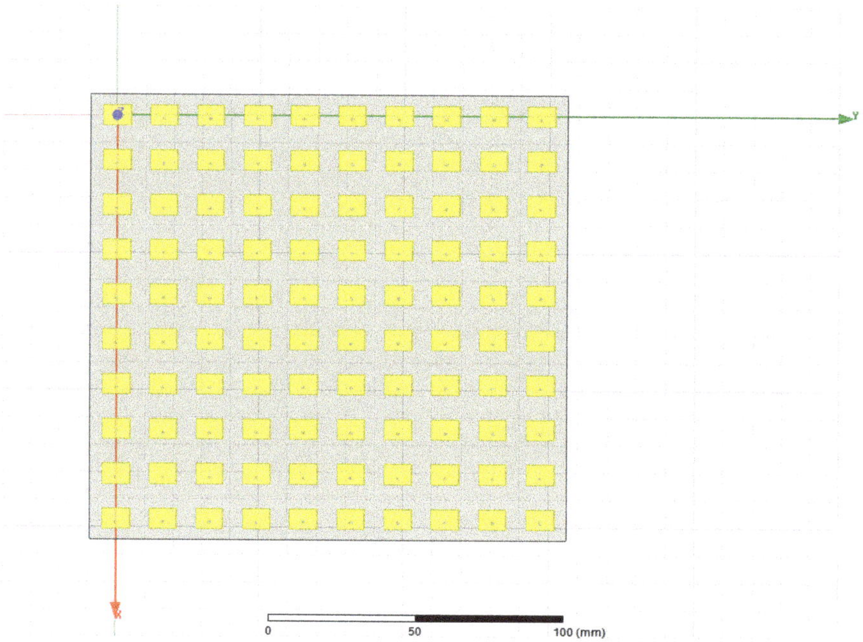

Figure 8. Designed microstrip patch antenna array in Ansys HFSS for diagnosis.

Also, considering an AUT with $K = 5$ element failures ($\Phi = 5\%$) (elements with zero excitation) because failed elements are usually of small number in practice, the resulted excitation coefficients are presented in Figure 13, and the estimated excitations by 30 random noisy measurement points are presented in Figure 14, while the dB excitation error is depicted in Figure 15. The result is an indication of good estimation of RA excitations and the locations of the faulty elements.

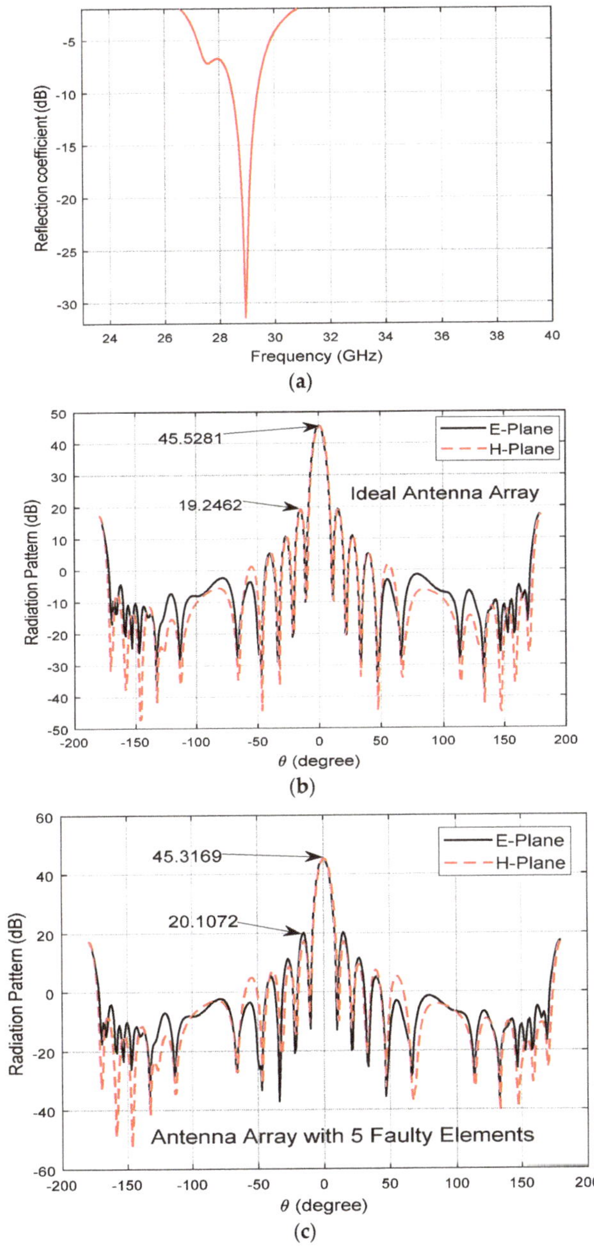

Figure 9. The antenna array for diagnostic purposes: (**a**) S-parameter; (**b**) RA radiation pattern; (**c**) AUT radiation pattern.

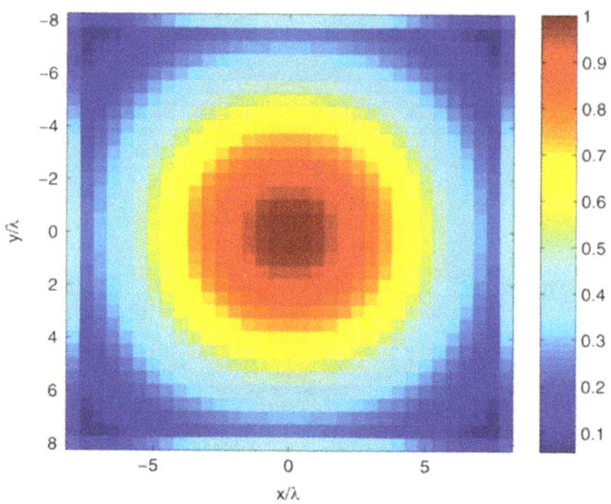

Figure 10. Excitation field of reference array.

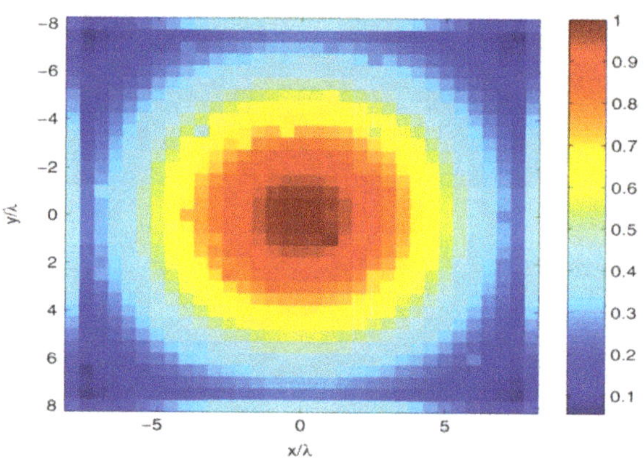

Figure 11. Reconstructed excitation field of reference array employing 30 random noisy measurement points.

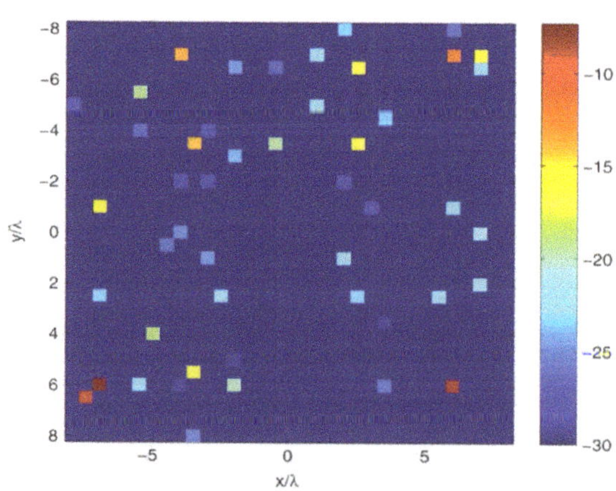

Figure 12. Reference array reconstructed excitation error in dB by 30 random noisy measurement points.

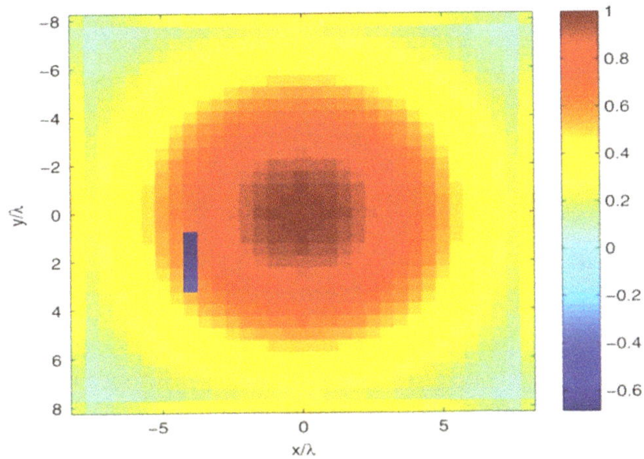

Figure 13. Reference excitation field of AUT with $\Phi = 5\%$.

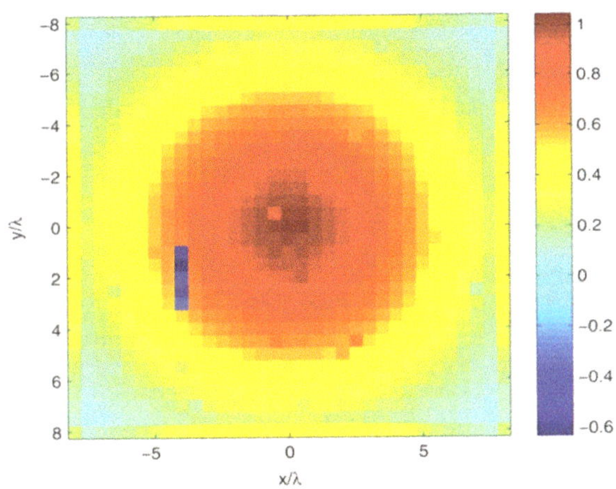

Figure 14. AUT with $K = 5$ failures $\Phi = 5\%$: reconstructed excitation error field by 30 random noisy measurement points.

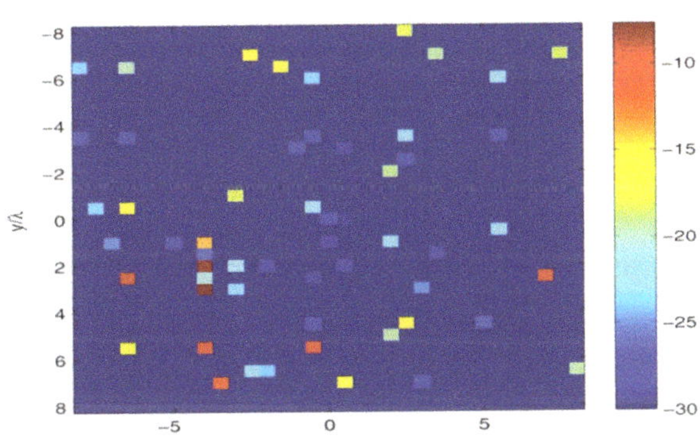

Figure 15. AUT with $K = 5$ failures $\Phi = 5\%$: reconstructed excitation error in dB by 30 random noisy measurement points.

6. Antenna Array Diagnosis from Measured Data

6.1. Measurement Set-Up

The proposed diagnostic technique was subjected to an experimental test, as presented in Figure 16. Although, this was a controlled environment (anechoic chamber), a more practical condition (uncontrolled environment) was also possible using the same set-up in Figure 16, without the chamber. The AUT was a 10 × 10 microstrip patch antenna array (see Figure 17a), reradiating a signal tilted in the two planes. Figure 17b,c show the measured radiation patterns for an ideal antenna and defected antenna, respectively. It can be see that the failure causes higher sidelobe level, reduced gain, lower front-to-back ratio (FBR), wider beamwidth, and a boresight pointing error.

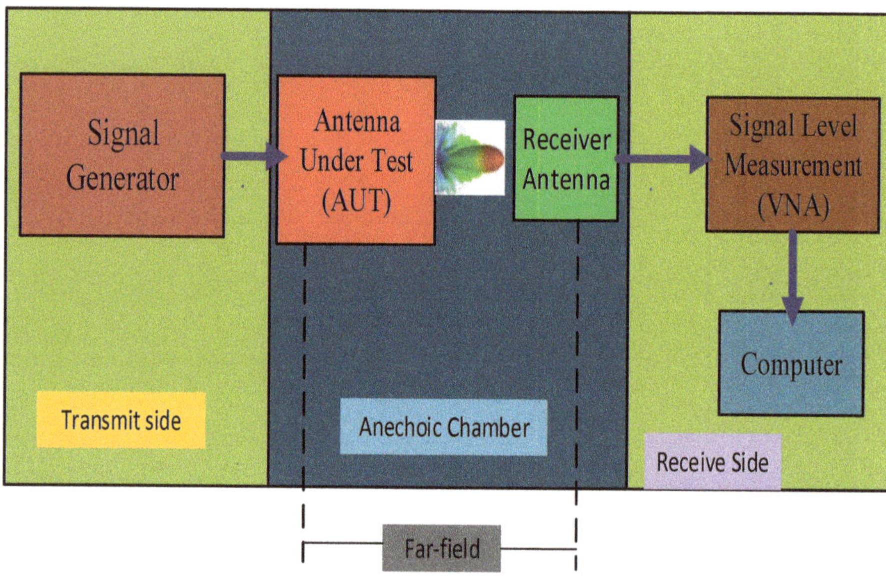

Figure 16. Schematic of experimental set-up.

The antenna was particularly designed and fabricated for this purpose, and each element had its feeding port which was excited using a power divider. Five radiating elements in the array were not excited (zero excitation) to successfully emulate the failure of elements. The AUT set-up is depicted in Figure 18. About 1000 co-polar and cross-polar measurements were taken on the far-field half-sphere at 28.9 GHz in an anechoic chamber (see Figure 18).

Radiation pattern measurements obtained from the array with five faulty elements were fed into the proposed algorithm for post-processing. The reconstructed excitation error which identified the specific faulty elements is depicted in Figure 19, and the corresponding dB equivalence is shown in Figure 20. Moreover, the performance metrics of the BCS-based approach were experimentally tested, and are shown in Figure 21. Figure 21a shows the obtained reconstruction error versus the measurement number at different degrees of failure. The error decreased as the number of measurement points increased irrespective of failure percentage. The reconstruction error profile increased with increased failure rate. In Figure 21b, it is demonstrated how the reconstruction error profile changes with SNR for different failure rates. The reconstruction error degraded exponentially with increased SNR independent of failure rate. The reconstruction error increased with increased failure rate. Figure 21c depicts reconstruction error versus different levels of failure for various SNRs. It can be observed that the error increased with increased failure rate. Also, the reconstruction error decreased as SNR increased.

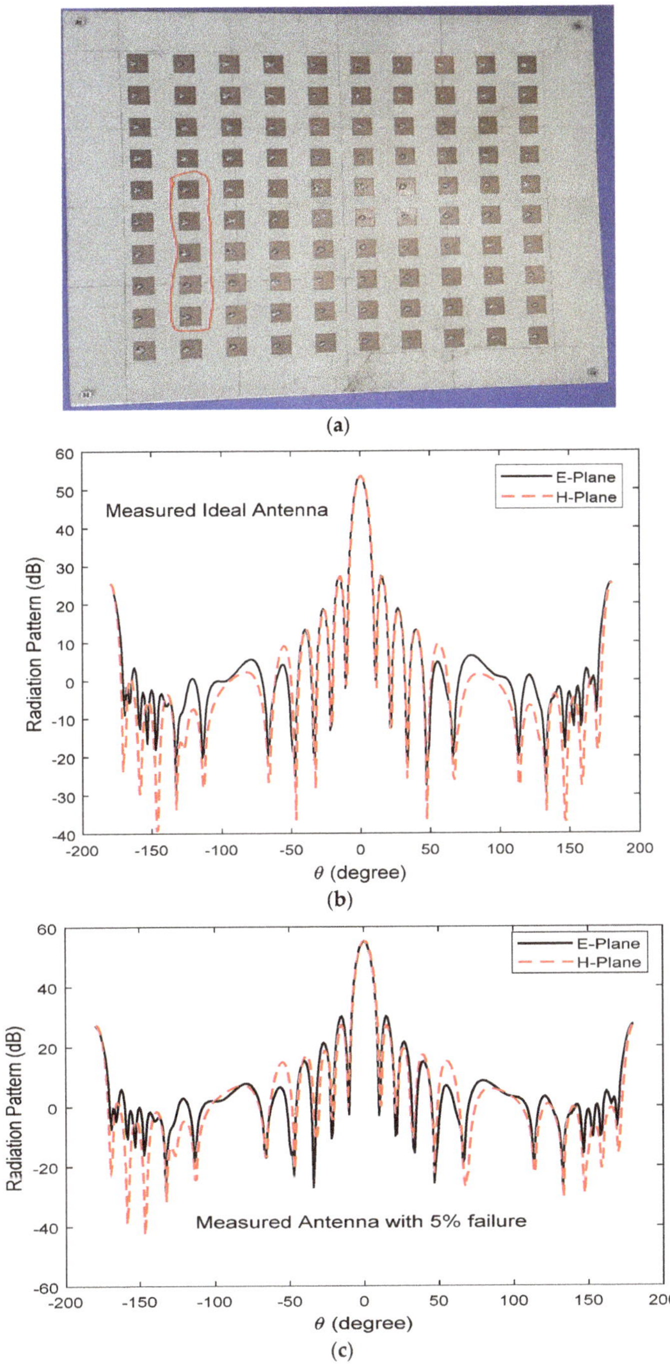

Figure 17. Photograph of (**a**) the fabricated antenna array, (**b**) measured radiation pattern without fault, and (**c**) measured radiation pattern with emulated failures.

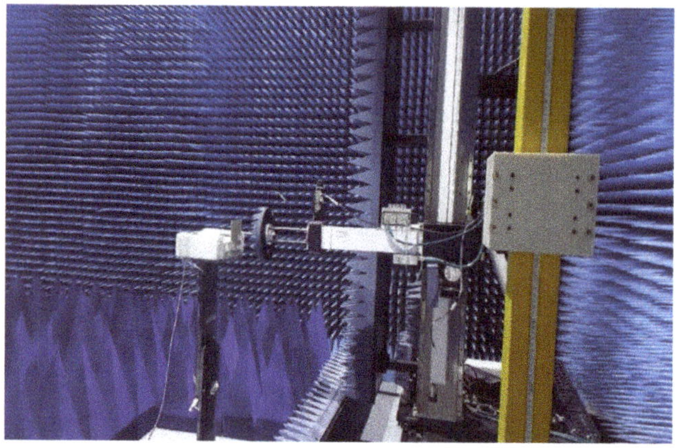

Figure 18. Measurement set-up of AUT for diagnostic purpose.

The imperfection of the curves (compared to the simulation) could be attributed to measurement errors, and errors due to experimental set-ups which provide different conditions from those in the simulations. The experiment was conducted in an anechoic chamber, which is a controlled environment. Hence, the results presented here may show a little variation if the experiment is conducted in a more practical environment (i.e., uncontrolled environment). Moreover, the BCS-based procedure presented here can be trusted to effectively and reliably address sparse recovery problems, particularly the detection of faulty radiators in planar arrays for next-generation 5G wireless communications. Once the suitable data are collected, and used to diagnose the array, then the array feeding network can be recalibrated to restore the needed radiation features via excitation reconfiguration of the healthy antenna elements. However, prior knowledge of the *golden* array must be provided, and the failure rate is relatively smaller than the array size. Therefore, from the simulation results verified by the experiment, the BCS-based approach is adequate and reliable for noisy data. This technique overcomes the shortcomings of BPA, MMA, etc., demanding off-line phase training to form accurate mapping between the response of the AUT and the failure location. Hence, the proposed procedure will be highly useful for millimeter-wave planar array optimal performance.

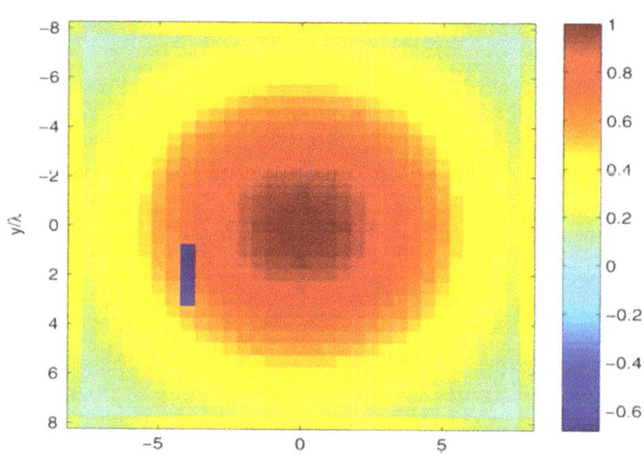

Figure 19. Reconstructed excitation error field by 30 random noisy measurement points with 20 dB signal-to-noise-ratio.

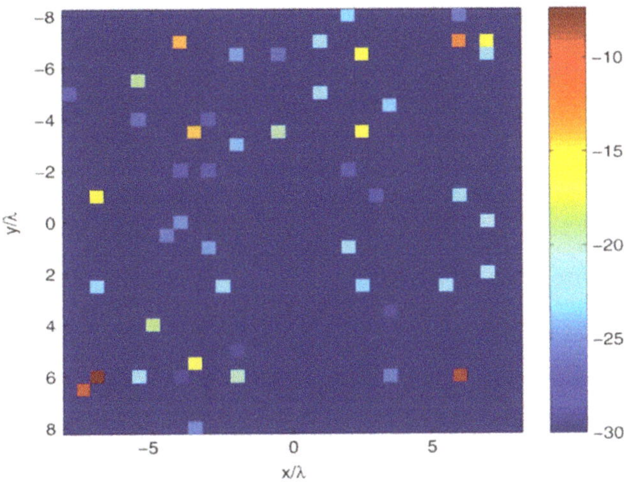

Figure 20. Reconstructed excitation error in dB by 30 random noisy measurement points.

6.2. Antenna Array Diagnosis from Simulated and Measured Far-Field Radiation Patterns

There are differences between the simulated and measured antenna patterns due to measurements errors, uncontrollable array fabrication errors, and experimental set-ups that give different conditions from the simulations. For example, in our design, a finite flange was employed to feed the ground plane. Hence, the induced current on flange rescattered and redistributed somehow against the very large ground plane which was used for the simulations. Antenna array diagnosis procedures based on simulated radiation pattern (such as References [3,16–18,27,35]) may not be reliable and accurate, except when tested with the corresponding measured data. Although, in this work, simulation and measurement data exhibit little difference in field intensity of the identified faulty elements (in Figures 14 and 19, respectively) caused by the electromechanical coupling effect. In general, the proposed BCS-based approach shows good reliability and accuracy against both simulated and measured far-field radiation patterns.

6.3. Number of Measurement Points versus Noisy Measured Data

Measurement points affect field reconstruction fidelity and, hence, the scheme of diagnosis. The proposed BCS-based procedure performed well despite significantly reduced measurement data due to added sparse information. According to the study and experiments performed by Fuch et al. [14], total variation (TV), mixed ℓ_1/ℓ_2 norm, and minimization of ℓ_1 techniques require 64 measurement points for accurate reconstruction and diagnostics, compared to the BCS technique that requires 30 or less measurement points. The proposed method significantly reduces the number of measurements needed for diagnosis as compared to those three approaches. Since the speed of diagnosis is inversely related to the required number of measurement points, the BCS approach enables a faster diagnosis of antenna arrays.

Also, according to the diagnostic procedures proposed by Fuch et al. [14], total variation (TV), mixed ℓ_1/ℓ_2 norm, and minimization of ℓ_1 techniques can only accommodate measured data with the lowest SNR of 40 dB. A lower SNR results in bad diagnostics for the three procedures. However, there are practical measurements that exhibit SNRs lower than 40 dB which require diagnosis. To this advantage, the proposed BCS approach was theoretically and practically used to diagnose antenna arrays from measurement data with 20 dB SNR. It can equally adapt to measured data with lower SNR. The comparison is summarized in Table 1. The BCS approach requires a few seconds more computational time; however, this is very small with respect to the measurement time cost. Therefore, the proposed method shows its robustness to noisy measured data, and a reliable diagnosis was obtained for low SNRs.

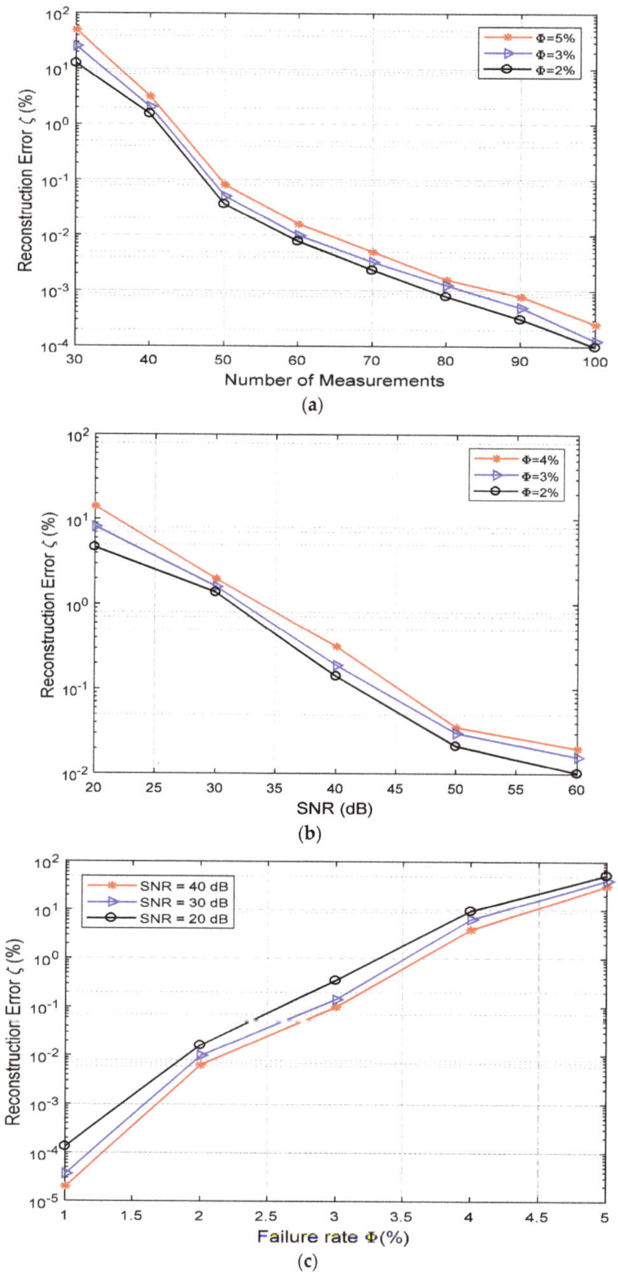

Figure 21. Experimental performance assessment of the proposed BCS-based diagnostic procedure. (**a**) Reconstruction error versus number of measurements, (**b**) Reconstruction error against SNR, (**c**) Reconstruction error with different failure rate

Table 1. Comparison between Bayesian compressive sensing (BCS)-based approach and previous compressive sensing techniques. SNR—signal-to-noise ratio.

Approach	This Work (BCS-Based)	Minimization of ℓ_1 [14,36]	Total Variation (TV) [14,37]	Mixed ℓ_1/ℓ_2 Norm [14,38]
Required signal characteristics	Amplitude	Amplitude or Phase	Amplitude or Phase	Amplitude or Phase
Minimum SNR (dB)	20	40	40	40
Nature	More complex	Simpler	Simple	Complex
Required measurement points	30	64	64	64
Post-processing time (s)	7.6	0.4	2.0	4.9

7. Conclusions

A faster and robust antenna array diagnosis procedure from far-field radiation pattern measurement points using Bayesian compressive sensing (BCS) approach was proposed in this paper. Previous compressive sensing procedures exhibit shortcomings based on reliability with noisy data, and require a large number of far-field measurement points. The proposed method solves these problems by formulating planar array diagnosis within the concept of the BCS framework, resolved using fast relevance machine (RVM). We are not the first to apply the BCS approach to antenna array diagnosis. It was applied only to linear configurations in References [15,16] without practical measurements, which are fundamental for testing any procedure. To the best knowledge of the authors, this is the first attempt to apply the BCS approach to planar antenna array diagnosis from far-field measurement points, validated with experimental measurements. Diagnoses from simulated and measured far-field points from a designed microstrip patch antenna array show the method's robustness to additive data noise, as well as its reconstruction accuracy and faster diagnosis speed, which is desired in practical applications. Hence, the proposed method is a better practical choice whenever an efficient, faster, and more reliable antenna array diagnosis (testing) is needed.

Also, it is important to comment on the choice of sampling strategy. We considered a random selection of measurement points from a uniform lattice. The choice of the sampling technique is not critical because it affects all the techniques in the same manner at the far-field. However, it was pointed out, from a non-uniform near-field lattice, that proper non-uniform random sampling (NURS) using a priori information on the problem provides meaningful reduction in the cardinality of the set of measured data compared to uniform random sampling and random sampling from a $\lambda/2$ equispaced dataset [39]. Moreover, the BCS-based technique was compared to other methods using the data reported in the literature. In the future, we will compare different techniques using experimental data from controlled and uncontrolled environments, and the same parameters in order to quantify the error affecting the result of different techniques. For example, we will determine what happens if we use 30 measurements instead of 64 measurements in the experimental data using ℓ_1 minimization, i.e., the same number of data used by the BCS, as well as the error compared to BCS. A complete comparison among the techniques using real data is still absent in the present literature.

Author Contributions: Conceptualization, methodology, writing-review and editing, original draft preparation, O.J.F.; resources, Z.Z.; visualization, software, A.F. and R.Z.; supervision, F.L.

Funding: This research received no external funding. The APC was funded by University of Science and Technology of China.

Acknowledgments: The majority of the work was conducted at MESIC (a joint lab of USTC and IMECAS), and partially carried out at the USTC Center for Micro and Nanoscale Research and Fabrication. The authors would like to thank the Information Science Laboratory Center of USTC for software and hardware services. The authors acknowledge the Applied Electromagnetics and Microwave Engineering Group of Hefei Normal University for the provision of the anechoic chamber. The support of the Chinese Academy of Science and the World Academy of Science (CAS-TWAS) is appreciated. The authors appreciate Zhang [40] and the CVX research group [41] for making useful codes accessible online.

Conflicts of Interest: The authors declare no conflict of interest.

References

1. Oliveri, G.; Donelli, M.; Massa, A. Linear array thinning exploiting almost difference sets. *IEEE Trans. Antennas Propag.* **2009**, *57*, 3800–3812. [CrossRef]
2. Oliveri, G.; Manica, L.; Massa, A. ADS-based guidelines for thinned planar arrays. *IEEE Trans. Antennas Propag.* **2010**, *58*, 1935–1948. [CrossRef]
3. Carlin, M.; Oliveri, G.; Massa, A. On the robustness to element failures of linear ADS-thinned arrays. *IEEE Trans. Antennas Propag.* **2011**, *59*, 4849–4853. [CrossRef]
4. 5G Vision White Paper, DMC R&D Center. Samsung Electronics Co., Ltd., 2015. Available online: http://www.samsung.com/global/business-images/insights/2015/Samsung-5G-Vision-0.pdf (accessed on 4 October 2018).
5. 5G: A Technological Vision, HUAWEI white Paper. Huawei Technologies Co.: Shenzhen. Available online: http://www.huawei.com/5gwhitepaper/ (accessed on 4 October 2018).
6. Frenzel, L. Ka-Band Transceiver Implements MIMO and Beamforming for 5G. Available online: Electronicdesign.com (accessed on 8 October 2018).
7. Lee, J.J.; Ferrer, E.M.; Woollen, D.P.; Lee, K.M. Near-field probe used as a diagnostic tool to locate defective elements in an array antenna. *IEEE Trans. Antennas Propag.* **1988**, *36*, 884–889. [CrossRef]
8. Gattoufi, L.; Picard, D.; Samii, Y.R.; Bolomey, J.C. Matrix method for near-field diagnostic techniques of phased array antennas. In Proceedings of the IEEE International Symposium on Phased Array Systems and Technology, Boston, MA, USA, 15–18 October 1996; pp. 52–57.
9. Bucci, O.M.; Migliore, M.D.; Panariello, G.; Sgambato, G. Accurate diagnosis of conformal arrays from near-field data using the matrix method. *IEEE Trans. Antennas Propag.* **2005**, *53*, 1114–1120. [CrossRef]
10. Candes, E.J.; Romberg, J.; Tao, T. Robust uncertainty principles: Exact signal reconstruction from highly incomplete frequency information. *IEEE Trans. Inf. Theory* **2006**, *52*, 489–509. [CrossRef]
11. Donoho, D.L.; Tanner, J. Precise undersampling theorems. *Proc. IEEE* **2010**, *98*, 913–923. [CrossRef]
12. Migliore, M.D.; Fuchs, B.; le Coq, L.; Ferro-Famil, L. Compressed sensing approach for reflectarray diagnostic from far field measurements. In Proceedings of the European Microwave Conference, Paris, France, 9–11 September 2015; pp. 289–292.
13. Massa, A.; Rocca, P.; Oliveri, G. Compressive sensing in electromagnetics—A review. *IEEE Antennas Propag. Mag.* **2015**, *57*, 224–238. [CrossRef]
14. Fuchs, B.; Coq, L.L.; Migiliore, M.D. Fast antenna array diagnosis from a small number of far-field measurements. *IEEE Trans. Antennas Propag.* **2016**, *64*, 2227–2235. [CrossRef]
15. Migliore, M.D. A compressed sensing approach for array diagnosis from a small set of near-field measurements. *IEEE Trans. Antennas Propag.* **2011**, *59*, 2127–2133. [CrossRef]
16. Oliveri, G.; Rocca, P.; Massa, A. Reliable diagnosis of large linear arrays—A Bayesian compressive sensing approach. *IEEE Trans. Antennas Propag.* **2012**, *60*, 4627–4636. [CrossRef]
17. Migliore, M.D. Array diagnosis from far-field data using the theory of random partial Fourier matrices. *IEEE Trans. Antennas Wirel. Propag. Lett.* **2013**, *12*, 745–748. [CrossRef]
18. Fuchs, B.; Migliore, M.D. Accurate array diagnosis from near-field measurements using reweighted minimization. In Proceedings of the IEEE Antennas Propagation Symposium, Orlando, FL, USA, 7–13 July 2013; pp. 2255–2256.
19. Fuchs, B.; le Coq, L.; Ferro-Famil, L.; Migliore, M.D. Comparison of methods for reflectarray diagnostic from far field measurements. In Proceedings of the IEEE International Symposium on Antennas Propagation, Boston, MA, USA, 19–25 July 2015; pp. 398–399.
20. Donoho, D.L. Compressive sensing. *IEEE Trans. Inf. Theory* **2006**, *52*, 1289–1306. [CrossRef]
21. Li, Q.; Zhang, Y.; Dai, K.; Huang, L.; Pan, C.; Song, J. An adaptive diagnosis scheme for integrated radar-communication antenna array with huge noise. *IEEE Access* **2018**, *6*, 25785–25796. [CrossRef]
22. Eltayeb, M.E.; Al-Naffouri, T.; Heath, R.W., Jr. Compressive sensing for millimeter wave antenna array diagnosis. *IEEE Trans. Commun.* **2018**, *66*, 2708–2721. [CrossRef]
23. Li, W.; Deng, W.; Migiliore, M.D. A deterministic far-field sampling strategy for array diagnosis using sparse recovery. *IEEE Antennas Wirel. Propag. Lett.* **2018**, *17*, 1261–1265. [CrossRef]

24. Massa, A.; Bertolli, M.; Gottardi, G.; Hannan, A.; Marcantonio, D.; Oliver, G.; Polo, A.; Robol, F.; Rocca, P.; Viani, F. Compressive sensing as applied to antenna arrays: Synthesis, diagnosis, and processing. In Proceedings of the IEEE International Symposium on Circuits and Systems (ISCAS), Florence, Italy, 27–30 May 2018.
25. Oliveri, G.; Massa, A. Bayesian compressive sampling for pattern synthesis with maximally sparse non-uniform linear arrays. *IEEE Trans. Antennas Propag.* **2011**, *59*, 467–481. [CrossRef]
26. Oliveri, G.; Carlin, M.; Massa, A. Complex-weight sparse linear array synthesis by Bayesian Compressive Sampling. *IEEE Trans. Antennas Propag.* **2012**, *60*, 2309–2326. [CrossRef]
27. Jaehyun, P.; Lee, J.; Chun, J.; Kwak, S. Robust monopulse beam synthesis with sparse elements in linear and planar arrays with element failure detection. *IET Radar Sonar Navig.* **2017**, *11*, 1251–1258.
28. Poli, L.; Oliveri, G.; Massa, A. Microwave imaging within the first-order Born approximation by means of the contrast-field Bayesian Compressive Sensing. *IEEE Trans. Antennas Propag.* **2012**, *60*, 2865–2879. [CrossRef]
29. Poli, L.; Oliveri, G.; Rocca, P.; Massa, A. Bayesian compressive sensing approaches for the reconstruction of two-dimensional sparse scatterers under TE illuminations. *IEEE Trans. Geosci. Remote Sens.* **2013**, *51*, 2920–2936. [CrossRef]
30. Poli, L.; Oliveri, G.; Viani, F.; Massa, A. MT-BCS-based microwave imaging approach through minimum-norm current expansion. *IEEE Trans. Antennas Propag.* **2013**, *61*, 4722–4732. [CrossRef]
31. Carlin, M.; Rocca, P. A Bayesian compressive sensing strategy for direction-of-arrival estimation. In Proceedings of the 6th European Conference on Antennas Propag. (EuCAP 2012), Prague, Czech Republic, 26–30 March 2012; pp. 1508–1509.
32. Carlin, M.; Rocca, P.; Oliveri, G.; Viani, F.; Massa, A. Directions-of-arrival estimation through Bayesian compressive sensing strategies. *IEEE Trans. Antennas Propag.* **2013**, *61*, 3828–3838. [CrossRef]
33. Carlin, M.; Rocca, P.; Oliveri, G.; Massa, A. Bayesian compressive sensing as applied to directions-of-arrival estimation in planar arrays. *J. Elect. Comput. Eng.* **2013**, *2013*, 1–12. [CrossRef]
34. Salucci, M.; Gelmini, A.; Oliver, G.; Massa, A. Planar array diagnosis by means of an advanced Bayesian compressive processing. *IEEE Trans. Antenna Propag.* **2018**, *66*, 5892–5906. [CrossRef]
35. Oliveri, G.; Rocca, P.; Massa, A. A Bayesian-compressive-sampling-based inversion for imaging sparse scatters. *IEEE Trans. Geosci. Remote Sens.* **2011**, *49*, 3993–4006. [CrossRef]
36. Candès, E.J.; Wakin, M.B.; Boyd, S.P. Enhancing sparsity by reweighted minimization. *J. Fourier Anal. Appl.* **2008**, *14*, 877–905. [CrossRef]
37. Rudin, L.; Osher, S.; Fatemi, E. Nonlinear total variation based noise removal algorithms. *Phys. D* **1992**, *60*, 259–268. [CrossRef]
38. Bach, F.; Jenatton, R.; Mairal, J.; Obozinski, G. Optimization with sparsity-inducing penalties. *Found. Trends Mach.* **2012**, *4*, 1–106. [CrossRef]
39. Migliore, M.D. On the sampling of the electromagnetic field radiated by sparse sources. *IEEE Trans. Antennas Propag.* **2014**, *63*, 553–564. [CrossRef]
40. Zhang, Y.; Yang, J.; Yin, W. YALL1: Your ALgorithms for L1. Available online: Yall1.blogs.rice.edu (accessed on 8 October 2018).
41. CVX Research, Inc. CVX: Matlab Software for Disciplined Convex Programming, Version 2.0 Beta. 2012. Available online: http://cvxr.com/cvx (accessed on 8 October 2018).

electronics

MDPI

Article

Mobile-Phone Antenna Array with Diamond-Ring Slot Elements for 5G Massive MIMO Systems

Naser Ojaroudi Parchin [1,*], Haleh Jahanbakhsh Basherlou [2], Mohammad Alibakhshikenari [3], Yasser Ojaroudi Parchin [4], Yasir I. A. Al-Yasir [1], Raed A. Abd-Alhameed [1] and Ernesto Limiti [3]

1 Faculty of Engineering and Informatics, University of Bradford, Bradford BD7 1DP, UK; Y.I.A.Al-Yasir@bradford.ac.uk (Y.I.A.A.-Y.); R.A.A.Abd@bradford.ac.uk (R.A.A.-A.)
2 Bradford College, Bradford BD7 1AY, UK; Hale.Jahanbakhsh@gmail.com
3 Electronic Engineering Department, University of Rome Tor Vergata, Via del Politecnico 1, 00133 Rome, Italy; Alibakhshikenari@ing.uniroma2.it (M.A.); Limiti@ing.uniroma2.it (E.L.)
4 Young Researchers and Elite Club, Germi Branch, Islamic Azad University, Germi 63764-56517, Iran; y.ojaroudi@gmail.com
* Correspondence: N.OjaroudiParchin@Bradford.ac.uk; Tel.: +44-7341436156

Received: 31 March 2019; Accepted: 7 May 2019; Published: 10 May 2019

Abstract: A design of mobile-phone antenna array with diamond-ring slot elements is proposed for fifth generation (5G) massive multiple-input/multiple-output (MIMO) systems. The configuration of the design consists of four double-fed diamond-ring slot antenna elements placed at different corners of the mobile-phone printed circuit board (PCB). A low-cost FR-4 dielectric with an overall dimension of 75 × 150 mm^2 is used as the design substrate. The antenna elements are fed by 50-Ohm L-shaped microstrip-lines. Due to the orthogonal placement of microstrip feed lines, the diamond-ring slot elements can exhibit the polarization and radiation pattern diversity characteristic. A good impedance bandwidth ($S_{11} \leq -10$ dB) of 3.2–4 GHz has been achieved for each antenna radiator. However, for $S_{11} \leq -6$ dB, this value is 3–4.2 GHz. The proposed design provides the required radiation coverage of 5G smartphones. The performance of the proposed MIMO antenna design is examined using both simulation and experiment. High isolation, high efficiency and sufficient gain-level characteristics have been obtained for the proposed MIMO smartphone antenna. In addition, the calculated total active reflection coefficient (TARC) and envelope correlation coefficient (ECC) of the antenna elements are very low over the whole band of interest which verify the capability of the proposed multi-antenna systems for massive MIMO and diversity applications. Furthermore, the properties of the design in Data-mode/Talk-mode are investigated and presented.

Keywords: 5G; diamond-ring slot; dual-polarized antenna; massive MIMO; mobile-phone antenna; pattern diversity

1. Introduction

Nowadays, there is an increased interest in research on MIMO systems in wireless communication [1,2]. It has incomparable advantages in improving the wireless link transmission capacity and reliability. In MIMO systems, multiple antennas are deployed at both transmitter and receiver sides [3]. This technology is a key component and probably the most established to truly reach the promised transfer data rates of fifth generation (5G) communication systems [4,5]. MIMO antennas are important for increasing channel capacity and link reliability [6,7]. Standard MIMO networks tend to use two or four antennas in a single physical package. However, massive MIMO is a MIMO system with an especially high number of antennas [8]. A 2 × 2 MIMO system has been successfully applied for fourth generation (4G) mobile communications and it is expected that the massive MIMO system with a large number of MIMO antennas is very promising for 5G wireless communications [9]. The greater number of elements in a network will make it more resistant to interference and intentional jamming [10].

Among the antennas which are used for MIMO applications, printed antennas are more appropriate due to their low cost, easy fabrication and their capability of easily being integrated to small terminal devices [11]. However, placing multiple antennas in the limited space of a transceiver poses a significant challenge in the incorporation of the MIMO technique. According to the requirement of cellular communications, compact, wideband and high isolation MIMO antenna is an urgent demand in the future mobile terminal and the portable applications [12–17]. Recently, several techniques have been introduced to design massive MIMO antennas for 4G and sub-6 GHz 5G mobile terminals [18–27].

We propose here a new design of Eight-port mobile-phone antenna with compact dual-polarized radiation elements providing wide impedance bandwidths for 5G applications. Eight-element MIMO smartphone antenna can achieve the channel capacity of 37 bps/Hz which is close to eight times that of a single antenna for single-input/single-output operation. With such a channel capacity and a wide frequency spectrum (200 MHz, at least), the data rate can be much higher than 1 Gbps. The antenna is designed to operate at 3.6 GHz, a candidate frequency band for sub-6-GHz 5G cellular networks, proposed by Ofcom, UK [28]. The design configuration contains four elements of double-fed/dual-polarized slot-ring antennas placed at corners of the printed circuit board (PCB). The antenna elements exhibit wide impedance bandwidth with low mutual coupling function providing pattern and polarization diversity characteristics at different sides of the mobile-phone PCB. As a result, the design not only can provide full radiation coverage but also it can support different polarizations. The computer simulation technology (CST) software was used to investigate antenna characteristics [29]. Fundamental properties of the single-element and its MIMO design in terms of S parameters, efficiency, radiation pattern, envelope correlation coefficient (ECC), total active reflection coefficient (TARC) are investigated. In addition, the performance of the designed mobile-phone antenna in Data-Mode/Talk-Mode are studied.

2. Single-Element Diamond-Ring Slot Antenna

Figure 1 depicts the configuration of the dual-polarized diamond-ring antenna. The antenna is designed on an FR-4 substrate (h = 1.6 mm, $\varepsilon = 4.4$, and $\delta = 0.025$).

Its configuration contains a diamond-ring slot radiator with a pair of L-shaped microstrip feed lines. Parameter values of the designed antenna and its MIMO configuration are specified in Table 1. The motive behind the presented design is to achieve a dual-polarized wideband antenna with compact-size and capability of integration onto smartphone PCB. This has been achieved by using the diamond-ring slot antenna with L-shaped microstrip feed lines. The slot antenna is one type of printed antenna that has been investigated extensively for different wireless systems for several decades because of its attractive features including light-weight, compactness and ease of integration with radio frequency (RF) circuit [30]. The ring-slot antenna can excite two orthogonal polarization if it is fed differently [31]. This makes the printed-ring-slot antennas attractive. The resonant frequency of the antenna is mainly determined by the circumference length of the employed diamond-ring slot. Therefore, the circumference length of the diamond-ring slot needs to satisfy the dielectric wavelength at the corresponding frequency point, where $W_x/2 + g = \lambda$. However, the length of feed-line ($L_f + L_1$) also has a little effect on the frequency point and impedance-bandwidth of the design.

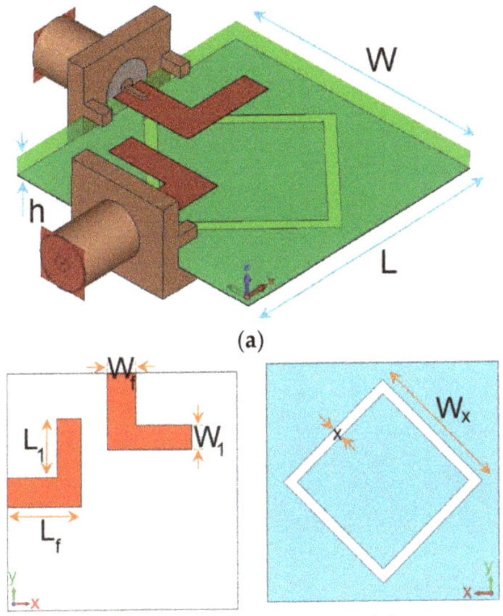

Figure 1. The antenna schematic, (**a**) side view, (**b**) top and (**c**) bottom layers.

Table 1. Parameter values of the single-element antenna and its MIMO array design.

Parameter	Value (mm)	Parameter	Value (mm)	Parameter	Value (mm)	Parameter	Value (mm)
W_S	75	L_S	150	x	1	W	24
L	24	W_f	3	L_f	5.25	W_1	2.5
L_1	3.9	W_2	8.9	L_2	9.9	W_x	7.25

Configurations and S parameters of the square-ring slot with rectangular feed line, a diamond-ring slot with rectangular feed line, and the proposed diamond-ring slot radiator with L-shaped feed line are illustrated and compared in Figure 2a–c, respectively. It can be observed that by using the proposed design (Figure 2c), the antenna not only provides wider impedance bandwidth but it also exhibits high isolation with low mutual coupling characteristic (less than −20 dB) at the desired operation band. As shown, the operation frequency of the slot radiator with L-shaped feed lines spans from 3.2 to 4 GHz (800 MHz bandwidth). For $S_{11} \leq -6$ dB, this value is 3–4.2 GHz.

S_{11} characteristics of varying design parameters including W_x, L_1, x and W_1 are illustrated in Figure 3. Figure 3a depicts the effects of diamond-ring size (W_x) on the resonance frequency: when its size decreases from 8 to 6 mm, the antenna resonance varies from 3.2 to 4.6 GHz. The frequency resonance of the antenna is also affected by the size of the L-shaped feed line arm (L_1). As shown in Figure 3b, the antenna operation frequency tunes to lower frequencies (without any changes on its bandwidth or isolation). Figure 3c illustrates the S_{11} results for various values of x (width of the diamond-ring slot-line). As shown, it mainly affects the impedance bandwidth of the antenna: when its size changes from 0.5 to 2 mm, the antenna operation bandwidth varies from 0.3 to 1.2 GHz. Another important parameter of the dual-polarized diamond-ring slot antenna design is the length of the feed line arm (L_1) which tunes the isolation characteristic. As can be observed from the results shown in Figure 3d, the antenna reflection coefficient tunes from −18 to less than −40 dB.

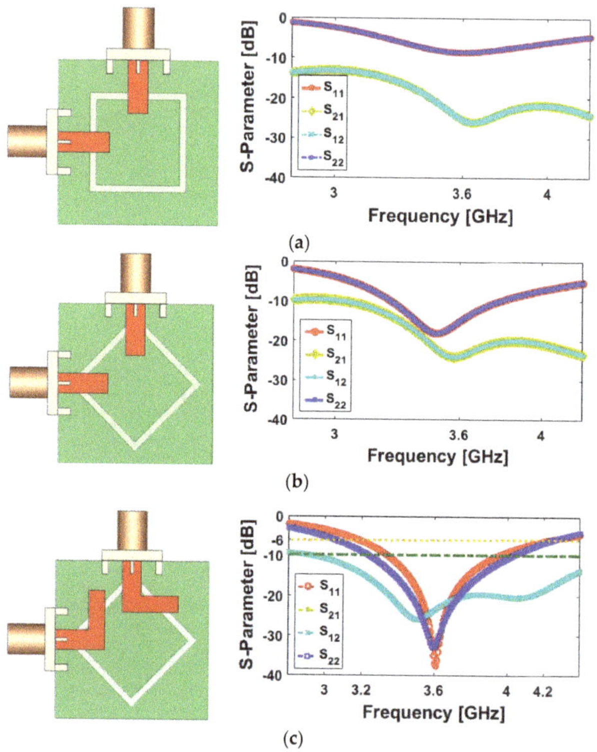

Figure 2. Various structures and S parameter results of: (**a**) square-ring slot with rectangular feed line, (**b**) diamond-ring slot with rectangular feed line, and (**c**) the proposed diamond-ring slot radiator with L-shaped feed line.

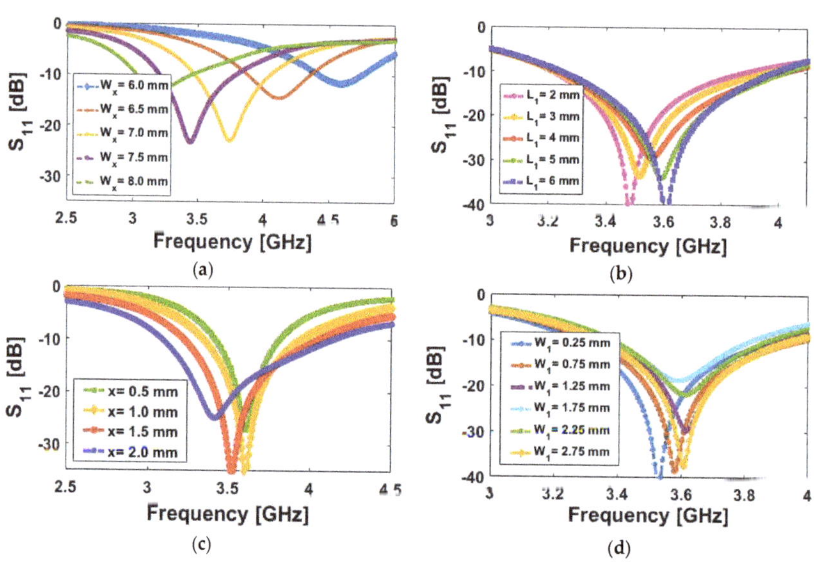

Figure 3. S_{11} results of the diamond-ring antenna for various values of (**a**) W_x, (**b**) L_1, (**c**) x, and (**d**) W_1.

Figure 4 shows the surface current distributions in the ground plane of the antenna at 3.6 GHz. As shown, the surface currents are mainly distributed around the diamond-ring slot radiator. In addition, for the different feeding ports of the antenna, the currents densities are equal and opposite due to the polarization diversity function [32–34]. Figure 5 illustrates the 3D radiation patterns of the antenna when it is fed differently (Port 1 and Port 2). As seen, the antenna exhibits similar radiation patterns with different orthogonal polarizations and more than 3 dB realized-gain. Radiation characteristics of the dual-polarized diamond-ring slot antenna in terms of radiation efficiency, total efficiency, and maximum gain are illustrated in Figure 6. As seen, the antenna provides high efficiencies. More than 80% radiation and total efficiency properties are obtained over the entire operation band (3.2–4 GHz). It can be observed that the antenna exhibits almost similar radiation and total efficiency. In addition, the antenna has around 2.5 dBi directivity.

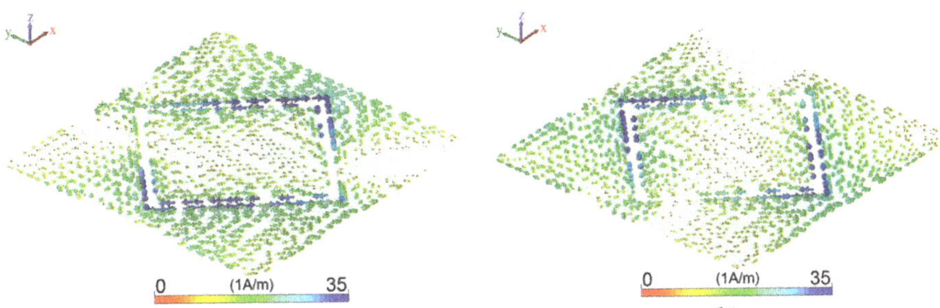

(a) (b)

Figure 4. Simulated current densities at 3.6 GHz for (**a**) 1st feeding port and (**b**) 2nd feeding port.

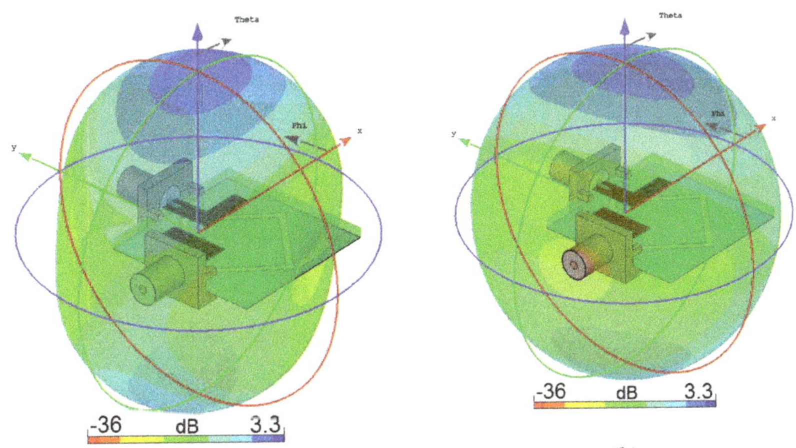

(a) (b)

Figure 5. 3D views of the dual-polarized radiation patterns from (**a**) feeding port 1 and (**b**) feeding port 2.

A prototype of the design was fabricated and its S parameters were tested. Figure 7 shows a photograph of the fabricated prototype in the measurement setup. Figure 8 illustrates the measured and simulated S parameter results of the fabricated antenna. It is observed that the fabricated antenna works properly at the desired frequency range.

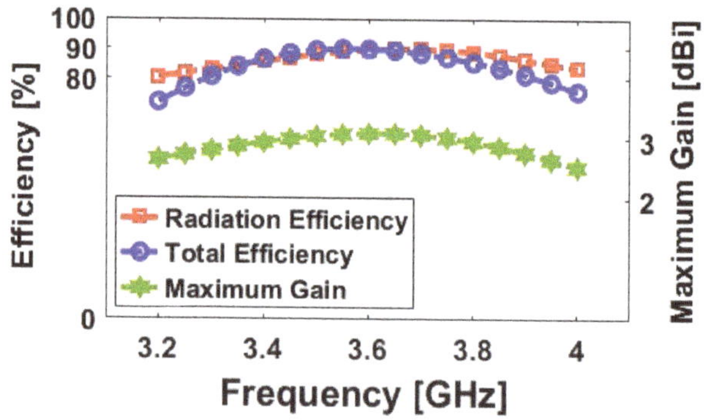

Figure 6. Radiation, total efficiencies, maximum gain of the diamond-ring slot antenna.

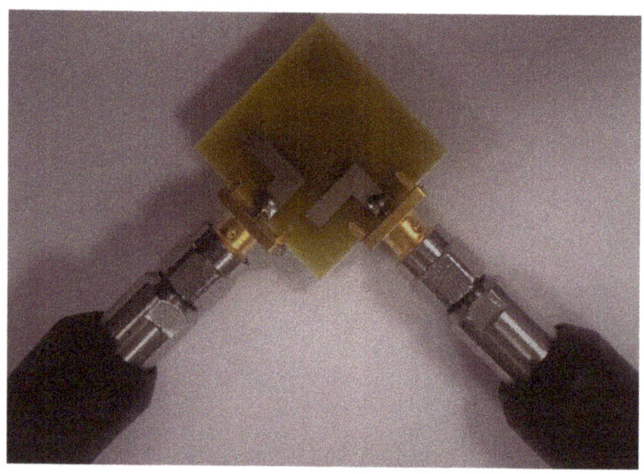

Figure 7. Photograph of the fabricated prototype in the measurement setup.

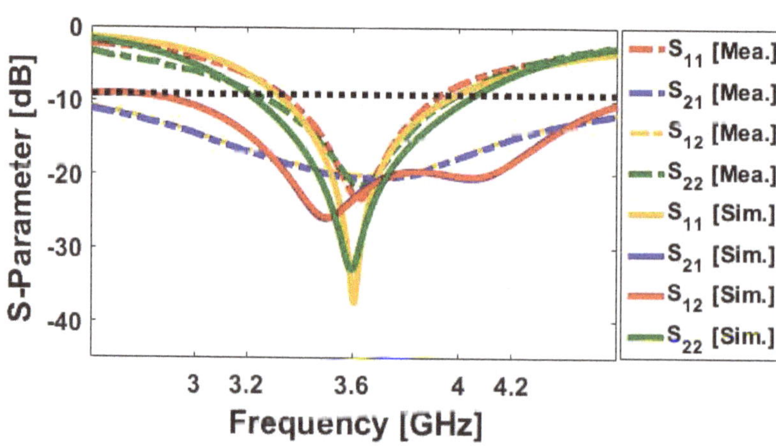

Figure 8. Measured and simulated S parameter results of the fabricated antenna.

3. Mobile-Phone Antenna Design

The simulated design layout of the mobile-phone antenna design is shown in Figure 9. It was implemented on an FR4 substrate with an overall dimension of 75 × 150 mm^2. Four elements of the dual-polarized diamond-ring slot radiators are employed at the corner of the mobile-phone PCB. As can be observed, due to the compact size of the employed radiator, there is enough spaces in the configuration of smartphone antenna PCB to add other antennas covering different frequencies of 3G/4G mobile terminals. Figure 10 illustrates the simulated S parameters (including S_{nn} and S_{mn}) of the design over its operation band. It is evident that the proposed mobile-phone antenna exhibits good S parameters with wide bandwidth and low mutual coupling characteristics. As mentioned above, the dual-polarized radiation elements provide similar performances. 3D radiation patterns of antennas 1 and 2 at 3.6 GHz is represented in Figure 11. As seen, the antenna elements have quasi-omnidirectional radiation patterns covering the top and bottom portions of the mobile-phone PCB.

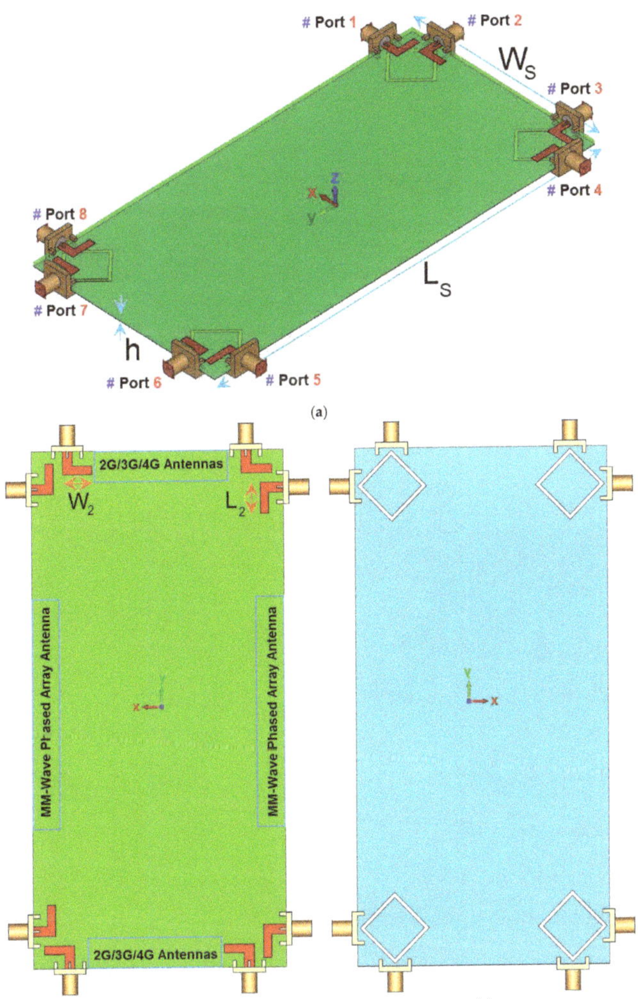

Figure 9. Designed mobile-phone antenna (**a**) transparent view, (**b**) top-layer and (**c**) bottom layer.

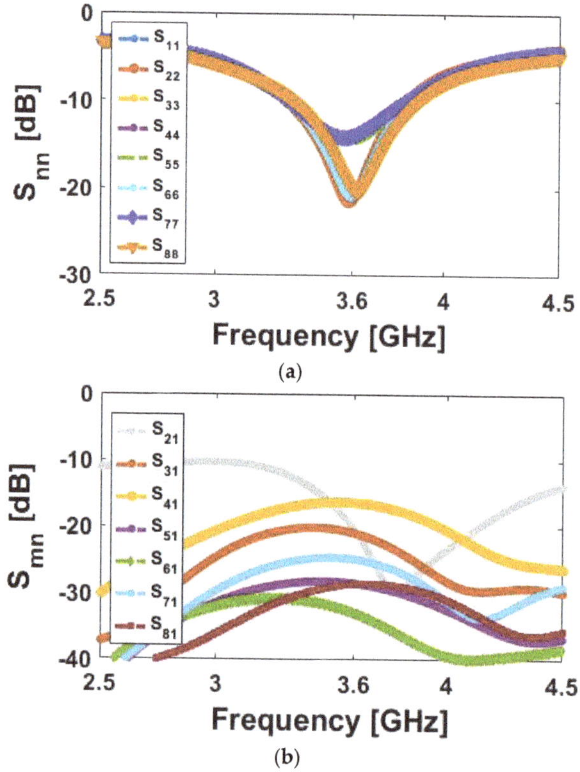

Figure 10. (**a**) S_{nn} and (**b**) S_{mn} results of the mobile-phone antenna.

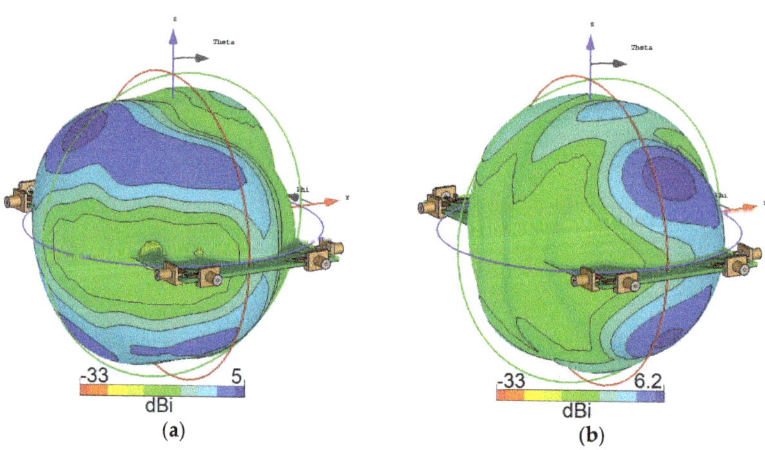

Figure 11. 3D transparent views of the radiation patterns for (**a**) feeding port 1 and (**b**) feeding port 2.

Top-views of the radiation patterns for the proposed mobile-phone antenna design are displayed in Figure 12. It can be seen that each side of the mobile-phone PCB has been covered with differently polarized radiation patterns. Thus, the MIMO antenna exhibited good radiation coverage and can support different polarizations which make it more suitable to be used in future smartphones. Furthermore, the antenna provides high radiation and total efficiencies over the operation band, as illustrated in Figure 13: more than 70% radiation and total efficiencies were obtained for the radiation elements at 3.6 GHz.

The proposed mobile-phone antenna was fabricated and its characteristics were tested in the Antenna Laboratory at the University of Bradford. Top and bottom views of the prototype are shown in Figure 14a,b, respectively. The mobile-phone antenna is constructed on a cheap FR4 dielectric with an overall dimension of 75 × 150 × 1.6 mm^3. During the measurement process, 50-Ω RF loads are employed for the elements not under test to avoid their effects, as shown in Figure 14c. The measured and simulated S parameters (S_{nn}: S_{11}–S_{88} and S_{mn}: S_{21}–S_{81}) of the fabricated design are illustrated in Figure 15. As illustrated, the diamond-ring slot resonators achieve good S parameters with sufficient impedance bandwidth and low mutual coupling characteristic in the desired frequency range. Some deviations from the measurements arise from the errors in fabrication, feeding and experiment processes.

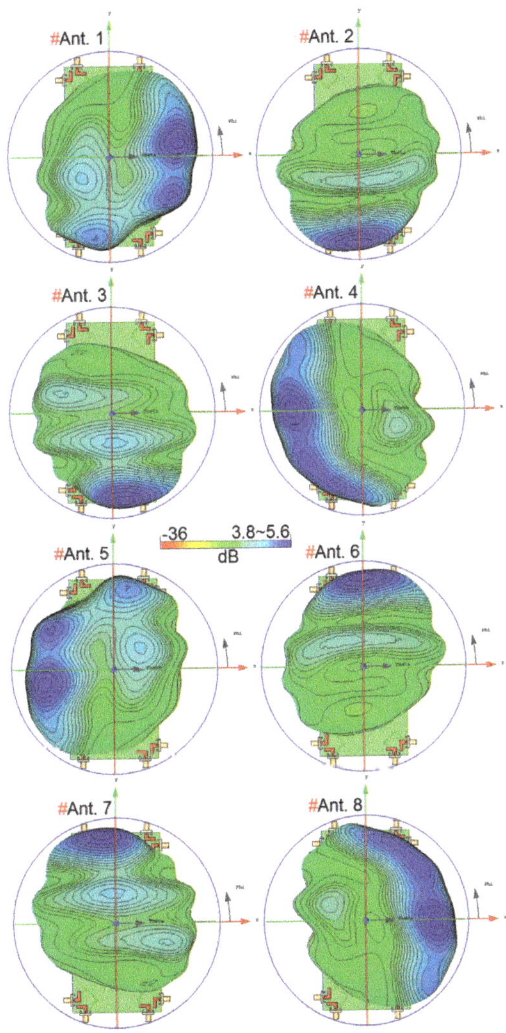

Figure 12. 3D radiation patterns of the fifth generation (5G) mobile-phone antenna.

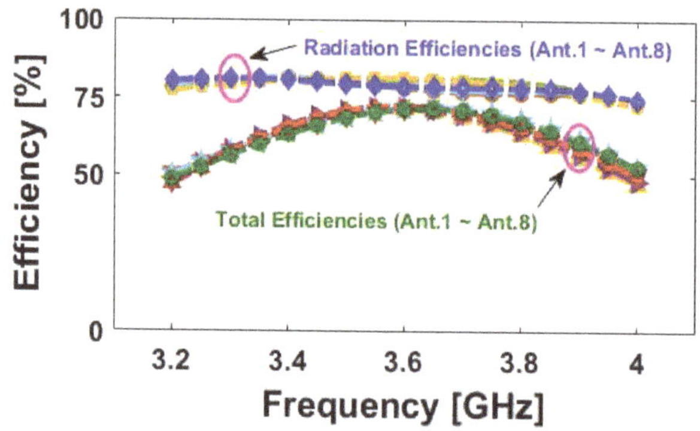

Figure 13. Efficiencies of the antenna elements (Ant. 1–Ant. 8) for the proposed design.

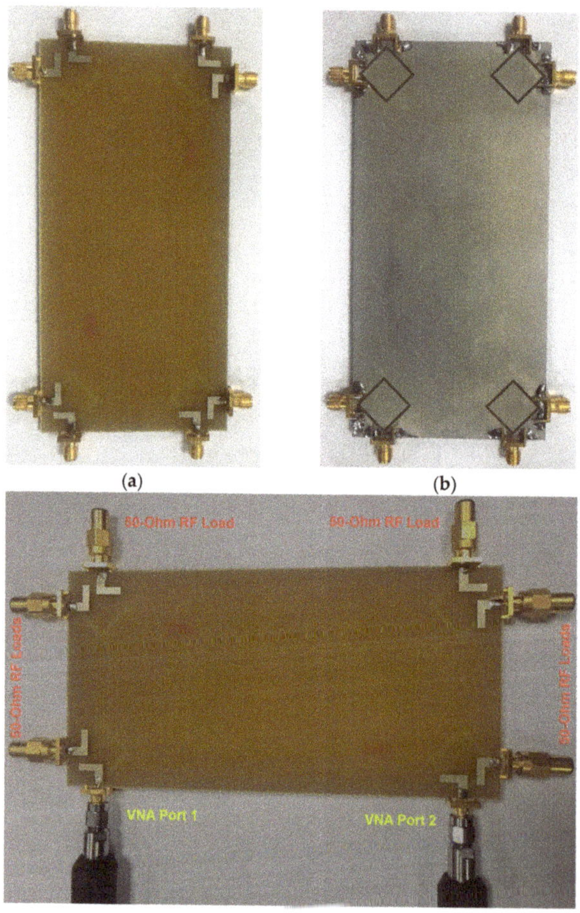

Figure 14. (**a**) Top, (**b**) bottom views of the fabricated design and (**c**) the prototype connected to the cables and 50-Ohm loads.

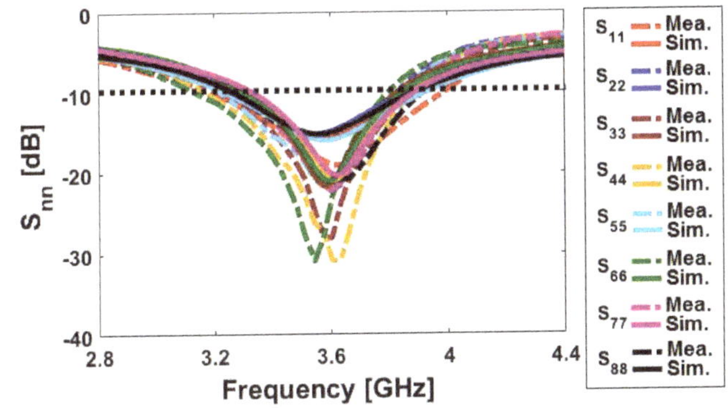

(a)

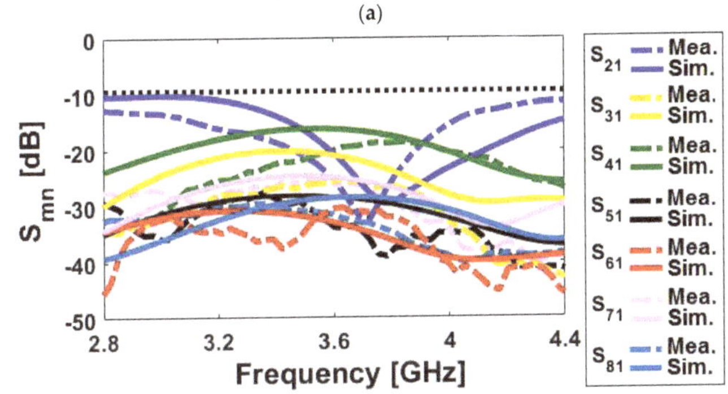

(b)

Figure 15. Measured and simulated (**a**) S_{nn} (S_{11}–S_{88}) and (**b**) S_{mn} (S_{21}–S_{81}) of the fabricated prototype.

According to the point that the radiation elements with the same placements and polarizations provide similar radiation patterns, 2D polar radiation patterns of the adjacent resonators (including Ant. 1 and Ant. 2) were measured at center operating frequency (3.6 GHz) and illustrated in Figure 16. As shown, the sample prototype exhibits good radiation patterns and provides acceptable agreement with simulations. In addition, the antenna elements with different polarizations provide sufficient gain values at the center frequency of the operation band.

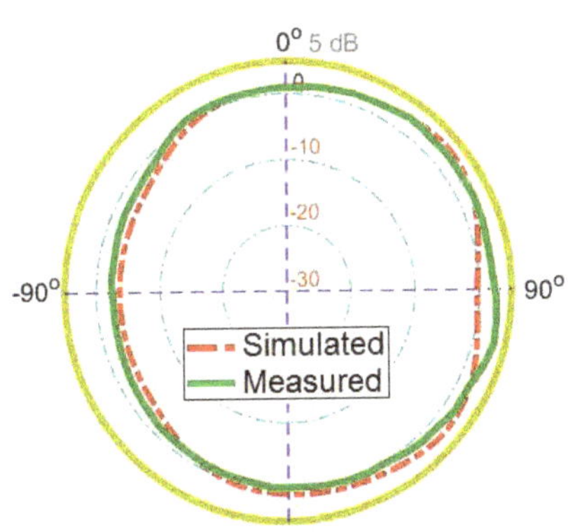

(a)

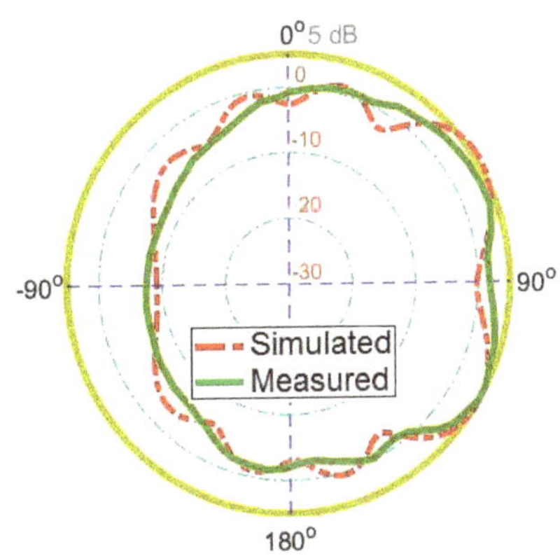

(b)

Figure 16. Measured and simulated 2D radiation patterns for (**a**) Ant.1 and (**b**) Ant.2.

In order to ensure that the MIMO antenna can work properly, ECC and TARC characteristics are two important parameters which should be considered in MIMO antennas [35,36]. The ECC and TARC of two elements can be calculated from the S parameters using the formula described as:

$$ECC = \frac{\left|S_{mm}^{*}S_{nm} + S_{mn}^{*}S_{nn}\right|^{2}}{\left(1 - |S_{mm}|^{2} - |S_{mn}|^{2}\right)\left(1 - |S_{nm}|^{2} - |S_{nn}|^{2}\right)^{*}}, \tag{1}$$

$$TARC = -\sqrt{\frac{(S_{mm} + S_{mn})^{2} + (S_{nm} + S_{nn})^{2}}{2}}. \tag{2}$$

Figure 17 shows the calculated ECC and TARC results from simulated and measured S parameters of the mobile-phone antenna design. As evident from figures, the calculated ECC and TARC results are very low over the whole band of interest. The design provides less than 0.01 ECC over the entire operating band and proves that two adjacent antenna elements are irrelevant. In addition, its TARC value is less than −30 at 3.6 GHz. Table 2 summarizes and compares the fundamental characteristics of the presented mobile-phone antenna with the recently reported 5G mobile-phone antenna designs [16–25]. It can be observed that the proposed design can provide better performances in terms of efficiency, isolation and ECC. In addition, it exhibits wider bandwidth with pattern and polarization diversity characteristics to cover different sides of the mobile-phone PCB.

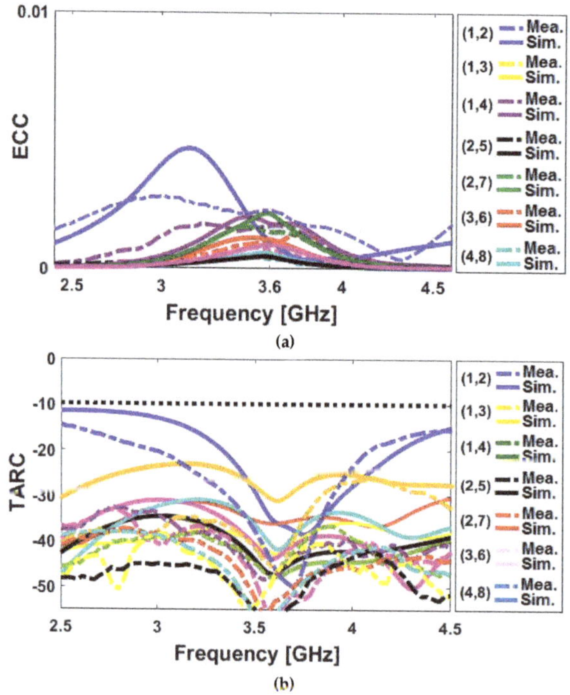

Figure 17. Calculated (**a**) envelope correlation coefficient (ECC) and (**b**) total active reflection coefficient (TARC) from measured S parameters.

Table 2. Comparison between the presented and recently reported 5G mobile-phone antennas.

Reference	Bandwidth (GHz)	Efficiency (%)	Size (mm^2)	Isolation (dB)	ECC
[18]	3.4–3.6	55–60	100 × 50	10	-
[19]	3.4–3.6	-	136 × 68	15	-
[20]	3.55–3.65	52–76	150 × 75	11	-
[21]	3.4–3.6	35–50	150 × 75	11	<0.40
[22]	3.4–3.6	30–50	145 × 70	15	<0.2
[23]	3.4–3.6	40–60	136 × 68	14	<0.2
[24]	3.4–3.6	60–75	150 × 80	17	<0.05
[25]	3.4–3.6	50–70	150× 73	17	<0.07
[26]	3.4–3.6	50–80	150 × 75	15	<0.2
[27]	3.4–3.6	60–70	150 × 75	18	<0.015
Proposed	3.3–3.9	60–80	150 × 75	17	<0.01

4. User-Hand/User-Head Impacts on the Mobile-Phone Antenna Performance

The impact of human-hand/human-head on the characteristics of the design in terms of total efficiency and antenna realized-gain were investigated in this section [37–39]. As illustrated in Figure 18, different scenarios including right-hand and left-hand modes for top-layer and back-layer of the design are studied in simulations. According to the obtained results, the mobile-phone antenna design and its radiators exhibit good performances and provide sufficient total efficiencies in the presence of the human hand. Due to the symmetric configuration of the proposed design, it performs almost similarly for different hand scenarios. The maximum reductions of the radiation properties are observed for the radiation elements partially covered by the user hand which is due to the nature of hand tissue properties which can highly absorb the radiation power. As can be observed, the antenna elements provide 25–55% total efficiencies over the operation band of 3.2–4 GHz.

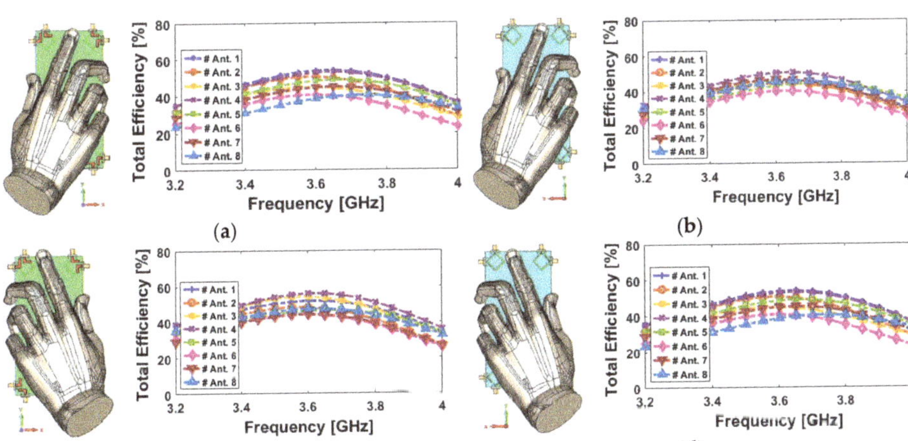

Figure 18. Placement and total efficiencies of the design for different user-hand scenarios (**a**) right-hand/top-layer, (**b**) right-hand/back-layer, (**c**) left-hand/top-layer and (**d**) left-hand/back-layer.

3D radiation patterns of the mobile-phone antenna in Talk-Mode at 3.6 GHz are illustrated in Figure 19a. It should be noted that the radiation performance of each antenna element mainly depends on its locations in the Talk-Mode scenario. As shown, the realized gain characteristic of the design varies from 3.2 to 4.9 dB. Compared with the radiation patterns in free space (Figure 12), due to the existence of the user's head and hand, radiation patterns are a bit distorted and become weaker. One can see that antenna elements are touched by different parts of the hand and head phantoms in the presented Talk-Mode.

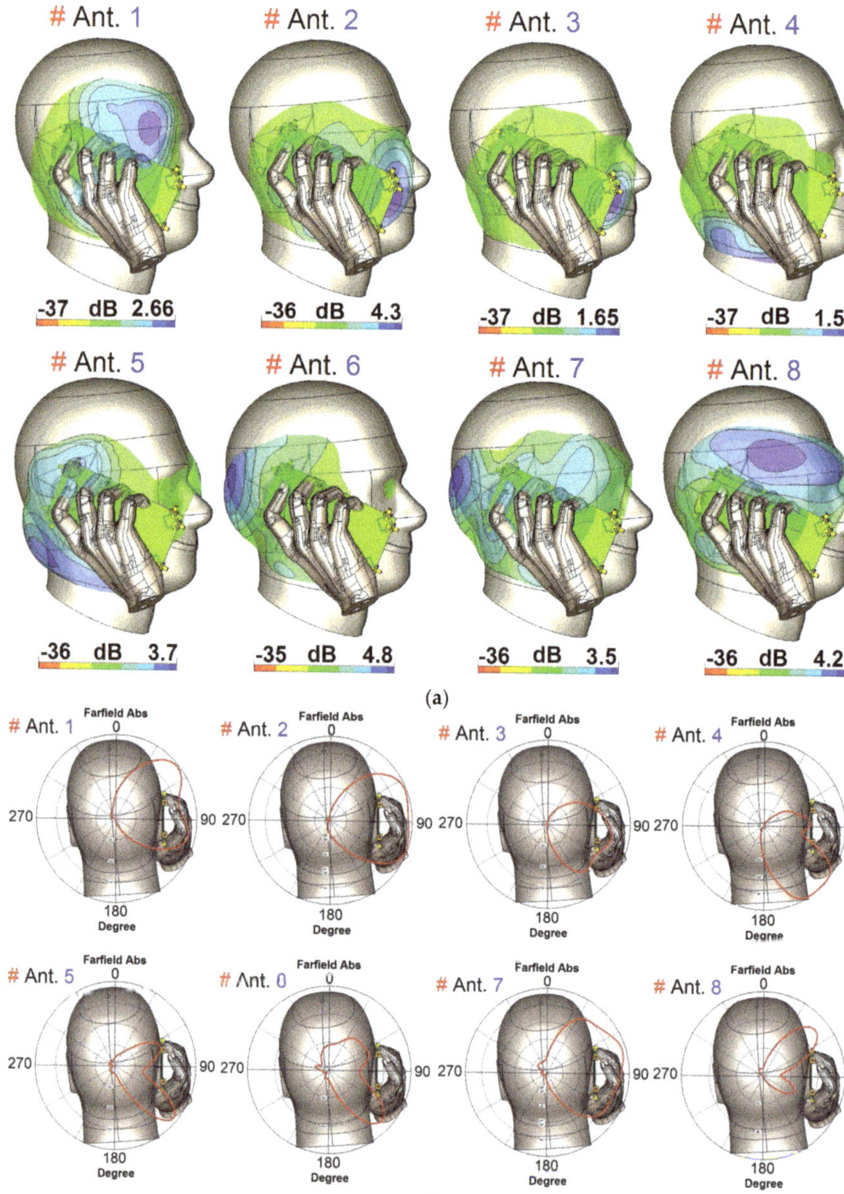

Figure 19. (**a**) 3D and (**b**) 2D linearly-scaled radiation patterns of the mobile-phone antenna in Talk-Mode scenario.

The maximum reductions of the radiation properties are observed for the elements that are located near to user-head (Ant. 3 and Ant. 4) [40]. However, the difference is not very significant. The 2D-polar (linear-scaling) radiation patterns of the design are illustrated in Figure 19b. As can be observed, the directivity of the antenna radiation pattern is maximum in the opposite direction of the head, which is most important part of the body to protect from the radiation. The main lobe of each single-element directs most of the power while the other lobes should be negligible.

Figure 20 depicts the total efficiencies and reflection coefficient (S_{nn}) of the antenna elements in the presence of the user-head and user-hand in Talk-Mode scenario. As seen, the diversion of the S_{nn} characteristic of the design is not significant. In addition, the proposed MIMO design exhibit sufficient efficiency in its operation bandwidth. Based on the above analysis, we can conclude the MIMO design provides sufficient efficiency, radiation coverage and gain levels for diamond-ring slot radiators.

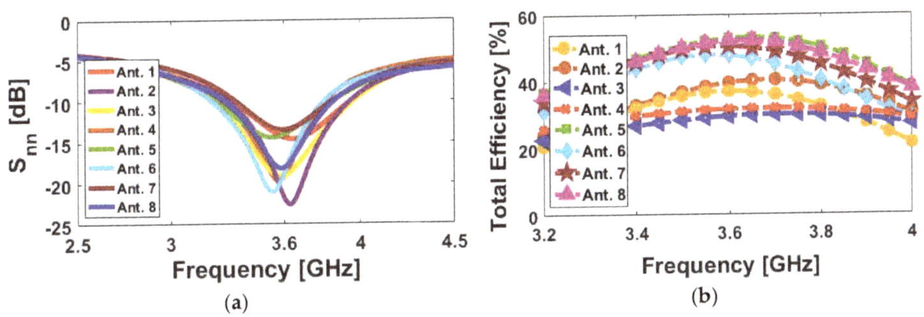

Figure 20. (**a**) S_{nn} and (**b**) total efficiencies of the proposed smartphone antenna in Talk-Mode scenario.

5. Conclusions

A mobile-phone antenna design with dual-polarized radiators is proposed for 5G massive MIMO communications. The antenna configuration contains eight-port/four elements of diamond-rings slot radiators with L-shaped microstrip feed lines deployed at four corners of the PCB. The antenna elements exhibit wide bandwidth with the center frequency of 3.6 GHz. S parameters, radiation patterns, efficiency, ECC and TARC results of the design are studied and sufficient results are achieved. In addition, a prototype of the mobile-phone antenna was fabricated and measured. Moreover, the performances of the antenna in Hand-Mode and Talk-Mode scenarios are investigated. The obtained results demonstrated that the proposed smartphone antenna provides good characteristics and meets the requirements for use in future mobile handsets.

Author Contributions: Writing—original draft preparation, N.O.P., H.J.B., M.A., Y.O.P., Y.I.A.A.-Y., R.A.A.-A., and E.L.; writing—review and editing, N.O.P. and R.A.A.-A.; investigation, N.O.P., H.J.B., M.A., Y.O.P., Y.I.A.A.-Y.; resources, N.O.P., R.A.A.-A., E.L. and; For other cases, all authors have participated.

Funding: This project has received funding from the European Union's Horizon 2020 research and innovation programme under grant agreement H2020-MSCA-ITN-2016 SECRET-722424.

Acknowledgments: The authors wish to express their thanks to the support provided by the innovation programme under grant agreement H2020-MSCA-ITN-2016 SECRET-722424.

Conflicts of Interest: The authors declare no conflict of interest.

References

1. Jensen, M.; Wallace, J. A review of antennas and propagation for MIMO wireless communications. *IEEE Trans. Antennas Propag.* **2004**, *52*, 2810–2824. [CrossRef]
2. Gesbert, D.; Shafi, M.; Shiu, D.S.; Smith, P.J.; Naguib, A. From theory to practice: An overview of MIMO space-time coded wireless systems. *IEEE J. Sel. Areas Commun.* **2003**, *21*, 281–302. [CrossRef]
3. Votis, C.; Christofilakis, V.; Raptis, V.; Tatsis, G.; Chronopoulos, S.K.; Kostarakis, P. Design and analysis of a multiple-output transmitter based on DDS architecture for modern wireless communications. *AIP Conf. Proc.* **2010**, *1203*, 421–426.

4. Kammoun, A.; Debbah, M.; Alouini, M.S. Design of 5G full dimension massive MIMO systems. *IEEE Trans. Commun.* **2018**, *66*, 726–740.
5. Osseiran, A.; Boccardi, F.; Braun, V.; Kusume, K.; Marsch, P.; Maternia, M.; Queseth, O.; Schellmann, M.; Schotten, H.; Taoka, H.; et al. Scenarios for 5G mobile and wireless communications: The vision of the METIS project. *IEEE Commun. Mag.* **2014**, *52*, 26–35. [CrossRef]
6. Angelis, C.T.; Chronopoulos, S.K. System performance of an LTE MIMO downlink in various fading environments. In Proceedings of the Ambient Media and Systems, Berlin, Germany, 27–28 January 2011; pp. 36–43.
7. Ojaroudiparchin, N.; Shen, M.; Pedersen, G.F. Multi-layer 5G mobile phone antenna for multi-user MIMO communications. In Proceedings of the 23rd Telecommunications Forum (TELFOR), Belgrade, Serbia, 24–26 November 2015; pp. 559–562.
8. Hassan, N.; Fernando, X. Massive MIMO Wireless Networks: An Overview. *Electronics* **2017**, *6*, 63. [CrossRef]
9. Larsson, E.; Edfors, O.; Tufvesson, F.; Marzetta, T. Massive MIMO for next generation wireless Systems. *IEEE Commun. Mag.* **2014**, *52*, 186–195. [CrossRef]
10. Yang, H.H.; Quek, T.Q.S. *Massive MIMO Meet Small Cell*; Springer Briefs in Electrical and Computer Engineering; Springer International Publishing: Basel, Switzerland, 2017.
11. Sharawi, M.S. *Printed MIMO Antenna Engineering*; Artech House: Norwood, MA, USA, 2014.
12. Parchin, N.O.; Basherlou, H.J.; Al-Yasir, Y.I.A.; Abd-Alhameed, R.A.; Abdulkhaleq, A.M.; Noras, J.M. Recent developments of reconfigurable antennas for current and future wireless communication systems. *Electronics* **2019**, *8*, 128. [CrossRef]
13. Parchin, N.O.; Al-Yasir, Y.; Abdulkhaleq, A.M.; Elfergani, I.; Rayit, A.; Noras, J.M.; Rodriguez, J.; Abd-Alhameed, R.A. Frequency reconfigurable antenna array for mm-Wave 5G mobile handsets. In Proceedings of the 9th International Conference on Broadband Communications, Networks, and Systems, Faro, Portugal, 19–20 September 2018.
14. Al-Yasir, Y.; Abdullah, A.; Ojaroudi Parchin, N.; Abd-Alhameed, R.; Noras, J. A new polarization-reconfigurable antenna for 5G applications. *Electronics* **2018**, *7*, 293. [CrossRef]
15. Hussain, R.; Alreshaid, A.T.; Podilchak, S.K.; Sharawi, M.S. Compact 4G MIMO antenna integrated with a 5G array for current and future mobile handsets. *IET Microw. Antennas Propag.* **2017**, *11*, 271–279. [CrossRef]
16. Chen, Q.; Lin, H.; Wang, J.; Ge, L.; Li, Y.; Pei, T. Single ring slot based antennas for metal-rimmed 4G/5G smartphones. *IEEE Trans. Antennas Propag.* **2019**, *67*, 1476–1487. [CrossRef]
17. Ojaroudi Parchin, N.; Abd-Alhameed, R.A. A compact Vivaldi antenna array for 5G channel sounding applications. In Proceedings of the 12th European Conference on Antennas and Propagation, London, UK, 9–13 April 2018.
18. Al-Hadi, A.A.; Ilvonen, J.; Valkonen, R.; Viikan, V. Eight-element antenna array for diversity and MIMO mobile terminal in LTE 3500MHz band. *Microw. Opt. Technol. Lett.* **2014**, *56*, 1323–1327. [CrossRef]
19. Abdullah, M. Compact 4-Port MIMO antenna system for 5G mobile terminal. In Proceedings of the International Applied Computational Electromagnetics Society Symposium, Florence, Italy, 26–30 March 2017.
20. Ojaroudi Parchin, N.; Al-Yasir, Y.I.A.; Abd-Alhameed, R.A.; Noras, J.M. Dual-polarized MIMO antenna array design using miniaturized self-complementary structures for 5G smartphone applications. In Proceedings of the EuCAP Conference, Krakow, Poland, 31 March–5 April 2019. accepted.
21. Wong, K.L.; Lu, J.Y.; Chen, L.Y.; Li, W.Y.; Ban, Y.L. 8-antenna and 16-antenna arrays using the quad-antenna linear array as a building block for the 3.5-GHz LTE MIMO operation in the smartphone. *Microw. Opt. Technol. Lett.* **2016**, *58*, 174–181. [CrossRef]
22. Rao, L.-Y.; Tsai, C.-J. 8-loop antenna array in the 5 inches size smartphone for 5G communication the 3.4 GHz–3.6 GHz band MIMO operation. In Proceedings of the Progress in Electromagnetics Research Symposium (PIERS), Toyama, Japan, 1–4 August 2018.
23. Abdullah, M.; Ban, Y.L.; Kang, K.; Li, M.Y.; Amin, M. Eight-element antenna array at 3.5GHz for MIMO wireless application. *Prog. Electromagn. Res. C* **2017**, *78*, 209–217. [CrossRef]
24. Li, Y.; Luo, Y.; Yang, G. High-isolation 3.5-GHz 8-antenna MIMO array using balanced open slot antenna element for 5G smartphones. *IEEE Trans. Antennas Propag.* **2019**. [CrossRef]
25. Sun, L.; Feng, H.; Li, Y.; Zhang, Z. Compact 5G MIMO mobile phone antennas with tightly arranged orthogonal-mode pairs. *IEEE Trans. Antennas Propag.* **2018**, *66*, 6364–6369. [CrossRef]

26. Li, M.Y.; Ban, Y.L.; Xu, Z.Q.; Guo, J.; Yu, Z.F. Tri-polarized 12-antenna MIMO array for future 5G smartphone applications. *IEEE Access* **2018**, *6*, 6160–6170. [CrossRef]
27. Zhao, A.; Ren, Z. Size reduction of self-isolated MIMO antenna system for 5G mobile phone applications. *IEEE Antennas Wirel. Propag. Lett.* **2019**, *18*, 152–156. [CrossRef]
28. Statement: Improving Consumer Access to Mobile Services at 3.6 GHz to 3.8 GHz. Available online: https://www.ofcom.org.uk/consultations-and-statements/category-1/future-use-at-3.6-3.8-ghz (accessed on 21 October 2018).
29. *CST Microwave Studio*; ver. 2018; CST: Framingham, MA, USA, 2018.
30. Chang, K. *Microwave Ring Circuits and Antennas*; Wiley: New York, NY, USA, 1996.
31. Batchelor, J.C.; Langley, R.J. Microstrip annular ring slot antennas for mobile applications. *Electron. Lett.* **2019**, *32*, 1635–1636. [CrossRef]
32. Yoshimura, Y. A microstripline slot antenna. *IEEE Trans. Microw. Theory Tech.* **1972**, *20*, 760–762. [CrossRef]
33. Elfergani, I.; Hussaini, A.S.; Rodriguez, J.; Abd-Alhameed, R. *Antenna Fundamentals for Legacy Mobile Applications and Beyond*; Springer Nature: Basingstoke, UK, 2017; pp. 1–659.
34. Al-Nuaimi, M.K.T.; Whittow, W.G. Performance investigation of a dual element IFA array at 3 GHz for MIMO terminals. In Proceedings of the Antennas and Propagation Conference (LAPC), Loughborough, UK, 14–15 November 2011; pp. 1–5.
35. Nguyen, Q.D.; Le, T.T.; Le, D.T.; Tran, X.N.; Yamada, Y. A compact MIMO ultra-wide band antenna with low mutual coupling. *Appl. Comput. Electromagn. Soc. J.* **2018**, *31*, 252–260.
36. Parchin, N.O.; Al-Yasir, Y.I.; Ali, A.H.; Elfergani, I.; Noras, J.M.; Rodriguez, J.; Abd-Alhameed, R.A. Eight-element dual-polarized MIMO slot antenna system for 5G smartphone applications. *IEEE Access* **2019**, *9*, 15612–15622. [CrossRef]
37. Syrytsin, I.; Zhang, S.; Pedersen, G.F. Performance investigation of a mobile terminal phased array with user effects at 3.5 GHz for LTE advanced. *IEEE Antennas Wirel. Propag. Lett.* **2017**, *16*, 1847–1850. [CrossRef]
38. Sharawi, M.S. Printed multi-band MIMO antenna systems and their performance metrics [wireless corner]. *IEEE Antennas Propag. Mag.* **2013**, *55*, 218–232. [CrossRef]
39. Ojaroudi Parchin, N.; Alibakhshikenari, M.; Jahanbakhsh Basherlou, H.; AAbd-Alhameed, R.; Rodriguez, J.; Limiti, E. mm-Wave phased array quasi-Yagi antenna for the upcoming 5G cellular communications. *Appl. Sci.* **2019**, *9*, 978. [CrossRef]
40. Basar, M.R.; Malek, M.F.; Juni, K.M.; Saleh, M.I.; Idris, M.S.; Mohamed, L.; Saudin, N.; Mohd Affendi, N.A.; Ali, A. The use of a human body model to determine the variation of path losses in the human body channel in wireless capsule endoscopy. *Prog. Electromagn. Res.* **2013**, *133*, 495–513. [CrossRef]

electronics

MDPI

Article

High-Isolation Leaky-Wave Array Antenna Based on CRLH-Metamaterial Implemented on SIW with ±30° Frequency Beam-Scanning Capability at Millimetre-Waves

Mohammad Alibakhshikenari [1,*], Bal Singh Virdee [2], Chan H. See [3,4], Raed A. Abd-Alhameed [5], Francisco Falcone [6] and Ernesto Limiti [1]

1 Electronic Engineering Department, University of Rome "Tor Vergata", Via del Politecnico 1, 00133 Rome, Italy; limiti@ing.uniroma2.it
2 Center for Communications Technology, School of Computing & Digital Media, London Metropolitan University, London N7 8DB, UK; b.virdee@londonmet.ac.uk
3 School of Engineering & the Built Environment, Edinburgh Napier University, 10 Colinton Rd.; Edinburgh, EH10 5DT, UK; C.See@napier.ac.uk
4 School of Engineering, University of Bolton, Deane Road, Bolton BL3 5AB, UK
5 Faculty of Engineering & Informatics, University of Bradford, Bradford, West Yorkshire BD7 1DP, UK; R.A.A.Abd@Bradford.ac.uk
6 Electrical and Electronic Engineering Department, Public University of Navarre, 31006 Pamplona, Spain; francisco.falcone@unavarra.es
* Correspondence: alibakhshikenari@ing.uniroma2.it

Received: 22 April 2019; Accepted: 5 June 2019; Published: 6 June 2019

Abstract: The paper presents a feasibility study on the design of a new metamaterial leaky-wave antenna (MTM-LWA) used in the construction of a 1 × 2 array which is implemented using substrate-integrated waveguide (SIW) technology for millimetre-wave beamforming applications. The proposed 1 × 2 array antenna consists of two LWAs with metamaterial unit-cells etched on the top surface of the SIW. The metamaterial unit-cell, which is an E-shaped transverse slot, causes leakage loss and interrupts current flow over SIW to enhance the array's performance. The dimensions of the LWA are 40 × 10 × 0.75 mm^3. Mutual-coupling between the array elements is suppressed by incorporating a metamaterial shield (MTM-shield) between the two antennas in the array. The LWA operates over a frequency range of 55–65 GHz, which is corresponding to 16.66% fractional bandwidth. The array is shown to exhibit beam-scanning of ±30° over its operating frequency range. Radiation gain in the backward (−30°), broadside (0°), and forward (+30°) directions are 8.5 dBi, 10.1 dBi, and 9.5 dBi, respectively. The decoupling slab is shown to have minimal effect on the array's performance in terms of impedance bandwidth and radiation specifications. The MTM-shield is shown to suppress the mutual coupling by ~25 dB and to improve the radiation gain and efficiency by ~1 dBi and ~13% on average, respectively.

Keywords: Metamaterials (MTM); leaky-wave antenna (LWA); antenna arrays; substrate integrated waveguide (SIW); transverse slots; beam-scanning; mutual coupling isolation; millimetre-wave; composite right/left-handed transmission line (CRLH-TL)

1. Introduction

Leaky-wave antennas (LWAs) are travelling wave antennas with electrically large radiating aperture [1,2]. Such antennas can provide high gain directive beam without using a complex feeding network [3,4]. The advantage of LWA over a conventional array antenna is a simple feed structure with low loss [5–7]. Conventional planar LWAs radiate higher order modes in the forward direction [8]. However, the periodic structure-based LWA can radiate in both forward and backward directions. It has been shown that metamaterial-based LWA designs can achieve a continuous main beam scanning from the backward to the forward direction as a function of frequency [9]. Various LWA designs based on metamaterial structures have been considered in the past. Such designs include (1) an LWA with a composite right/left-hand (CRLH)-folded substrate-integrated waveguide (SIW) structure that is shown to provide beam scanning from of −58° to 65° with a gain of 1 dBi [10]; (2) an interdigital-shaped slotted-SIW-based LWA that is reported to achieve a scanning angle of −60° to 70° with gain of around 8 dBi [11]; (3) a CRLH LWA based on a rectangular waveguide structure that has been demonstrated for a continuous main beam scanning range from −70° to 70° with gain of 8.64 dBi [12]; and (4) a planar slotted SIW LWA that provides a scanning range of −66° to 78° with consistent gain [13].

In this paper, a new type of LWA in array configuration is proposed based on SIW with metamaterial inclusions for millimetre-wave applications. Mutual coupling between the closely-spaced antennas in the array can undermine the array's performance. To circumvent this, a metamaterial shield is embedded between the two LWAs. With this approach, mutual coupling is shown to reduce by an average of ~25 dB over the array's operating frequency range. Engraved on the upper layer of the SIW LWA are several metamaterial unit-cells comprising of transverse E-shaped slots. Dimensions of the LWAs were modified for optimum array performance.

2. Design Process of The Proposed Mtm-Lwa Array Based On Siw

The proposed 1 × 2 array antenna based on MTM-LWA implemented on SIW technology is designed on a RO3003 dielectric substrate with ε_r = 3.0, tangδ = 0.0010 and thickness of 0.75 mm. Figure 1 displays the layout of the proposed array structure that is constructed with two MTM-LWA on SIW. Engraved on the top of each SIW antenna are several metamaterial unit-cells consisting of transverse E-shaped slots. Leakage loss at the slots interrupts the current flow over SIW, which is shown to enhance the array's impedance bandwidth and beamwidth that scans as a function of frequency. In the structure, the transverse slots behave as series left-handed capacitance and the grounded via holes acts as shunt left-handed inductors.

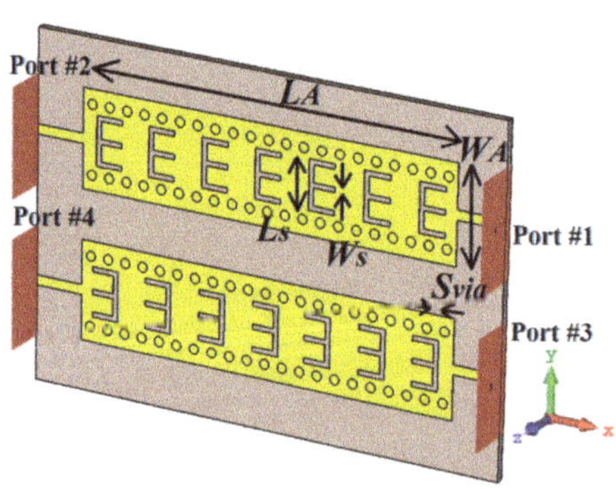

(a)

Figure 1. *Cont.*

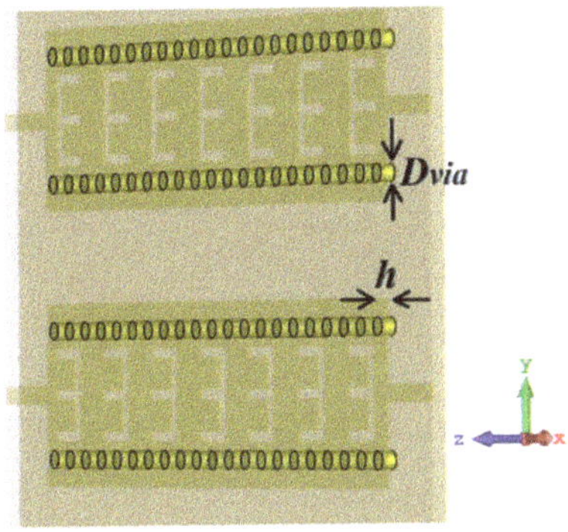

(**b**)

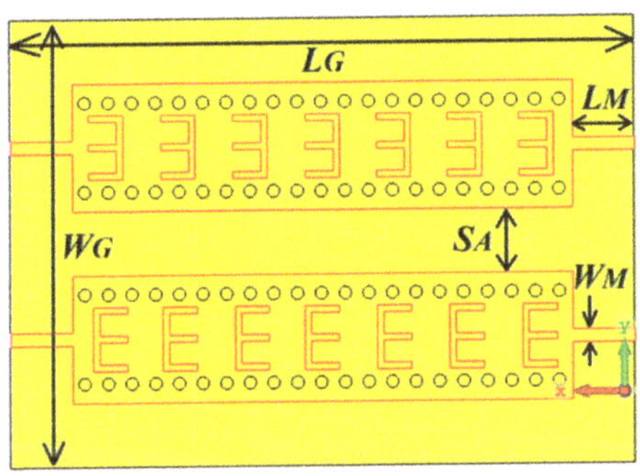

(**c**)

Figure 1. Proposed 1 × 2 array antenna based on MTM-LWA using SIW technology. (**a**) Top view; (**b**) view to show the Substrate integrated waveguide slots; (**c**) back side to show the ground plane.

The structural parameters of the MTM-LWA array are summarized in Table 1. Each antenna has dimensions of 40 × 10 × 0.75 mm^3. The overall ground plane dimensions are 50 × 35 × 0.75 mm^3. The S-parameter responses of the proposed array antenna are exhibited in Figure 2, which shows it operates throughout the frequency range of 55–65 GHz, which corresponds to 16.66% fractional bandwidth.

Table 1. Structural parameters of the array.

Structural Parameters	Dimensions (Units in mm)
$L_A L_A$	40
$W_A W_A$	10
$L_G L_G$	50
$W_G W_G$	35
$L_S L_S$	5
$W_S W_S$	0.5
DviaDvia	0.5
SviaSvia	0.5
$S_A S_A$	5
$L_M L_M$	5
$W_M W_M$	1
hh	0.75

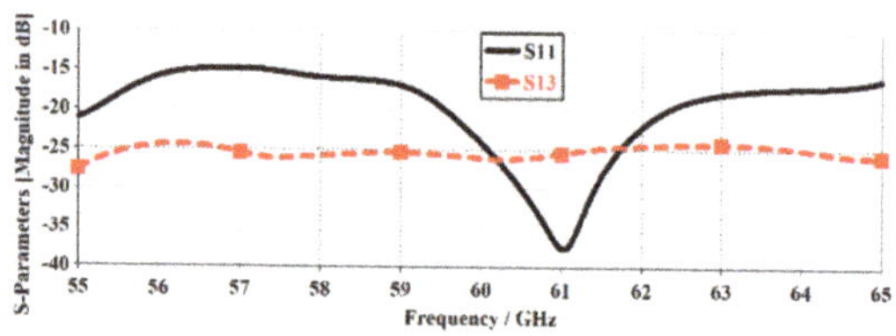

Figure 2. S-parameter responses of the proposed array antennas. Since the structure is symmetrical, we have not plotted all curves.

Radiation gain patterns of the proposed array antenna at three spot frequencies within its operating frequency range are plotted in Figure 3. It is evident that the array antenna is capable of beam-scanning from −30° to +30° with backward radiation at −30°, broadside radiation at 0°, and forward radiation at +30°. In backward, broadside, and forward directions, the gain is 8.5, 10.1, and 9.5 dBi, respectively.

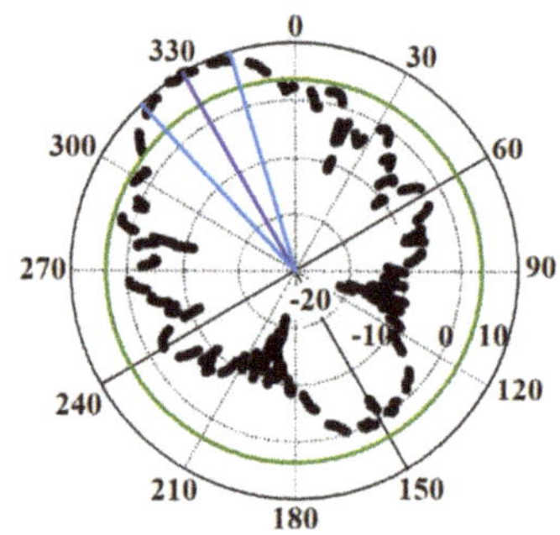

(a)

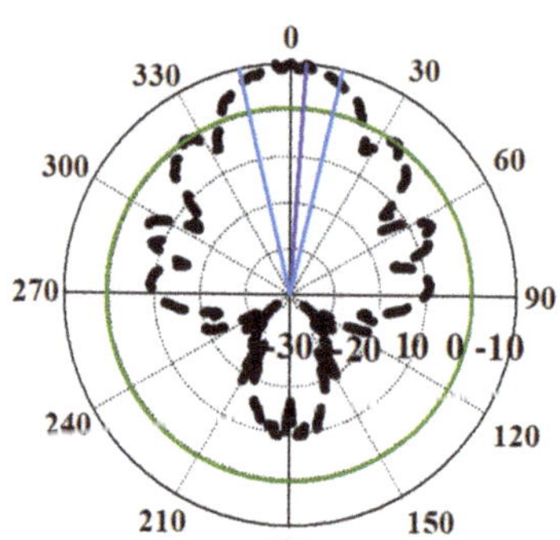

(b)

Figure 3. *Cont.*

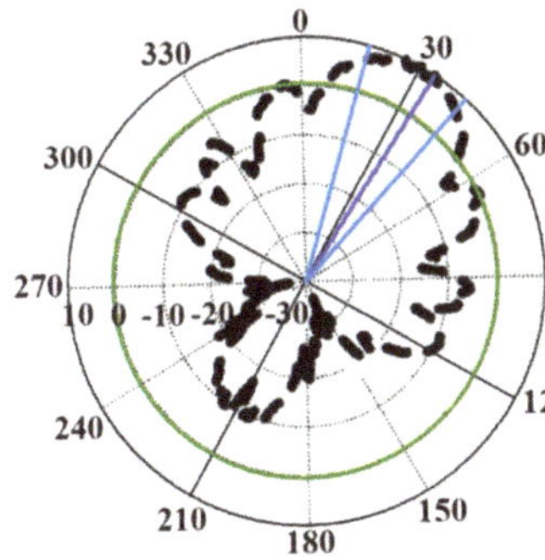

(c)

Figure 3. Radiation characteristics of the proposed 1 × 2 MTM-LWA array at 55 GHz, 60 GHz, and 65 GHz. (**a**) Backward-radiation at 55 GHz; (**b**) broadside-radiation at 60 GHz; (**c**) forward-radiation at 65 GHz.

3. Suppress the Mutual Coupling Between the Closely Spaced Mtm-Lwa Arrays

Mutual coupling between closely-spaced radiation elements can severely undermine the array's radiation performances. Here the isolation between the two MTM-LWAs is increased by introducing a metamaterial shield which is based on the SIW structure, as indicated in Figure 4. It comprises of transverse slots that are tapered. The slots have a width of 0.5 mm and essentially play the role of the series left-handed capacitances (C_L), where the metallic via-holes with diameter of 0.25 mm act as shunt left-handed inductances (L_L). The MTM-shield suppresses surface waves created by the LWAs to increase isolation between the two antennas in the array. The overall dimensions of the shield are 40×4 mm^2.

Figure 5 shows the S-parameter response before and after applying the MTM-shield. After applying MTM-shield, the minimum, average and maximum suppression observed are 8 dB, ~25 dB, and 42.5 dB, respectively. This shows the effectiveness of the proposed isolation technique. The shield has no influence on the reflection coefficient response which is $S_{11} \leq -10$ dB.

Radiation patterns of the proposed antenna arrays with no and with MTM-shield through its operational bandwidth at spot frequencies of 55 GHz, 60 GHz & 65 GHz are plotted in Figure 6. It is clear from this figure that with the shield the cross-polarized radiation over its operating range is substantially reduced. The average gain of the co-polarized radiation is only marginally affected. All details are tabulated in Table 2.

In addition, the radiation gain and efficiency curves over frequency bandwidth for both antennas without and with the proposed shield are shown in Figure 7. Obviously, after realizing the metamaterial shield based on SIW, the radiation gain and efficiency performances improved by ~1dBi and ~13% on average, respectively.

Surface current density distributions without and with the MTM-shield are exhibited in Figure 8. This figure shows that the MTM-shield is an effective EM band-gap structure to remarkably block surface currents from EM interacting with adjacent radiation elements in the antenna array. Destructive influences of surface currents in the antenna are dramatically suppressed from effecting the far-field of the array antennas.

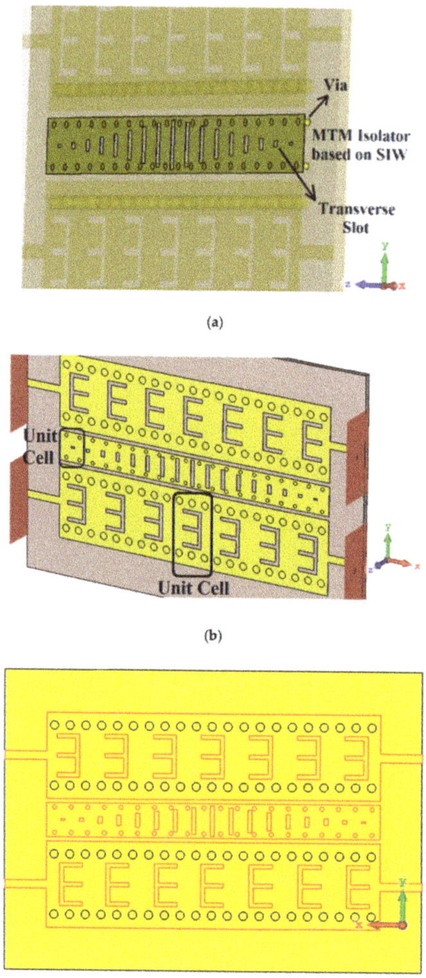

Figure 4. Proposed SIW-based leaky-wave antenna array with MTM-shield. (**a**) Proposed MTM-SIW shield located between the array antennas; (**b**) top-view of the leaky-wave array antennas; (**c**) back-side to show ground plane.

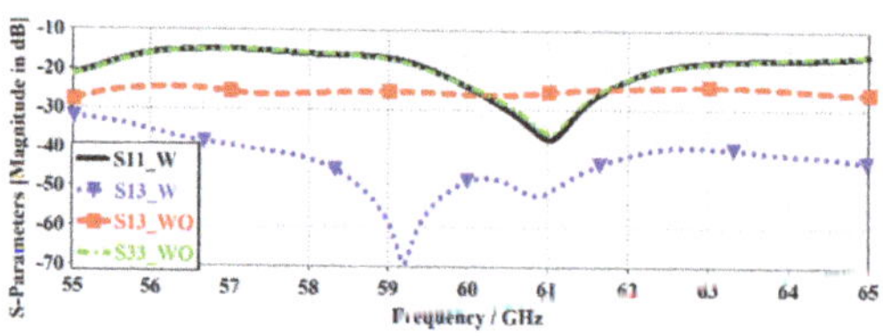

Figure 5. Reflection and transition coefficients of the proposed antenna array before and after applying the MTM-shield. Since the structure is symmetrical, we have not plotted all curves.

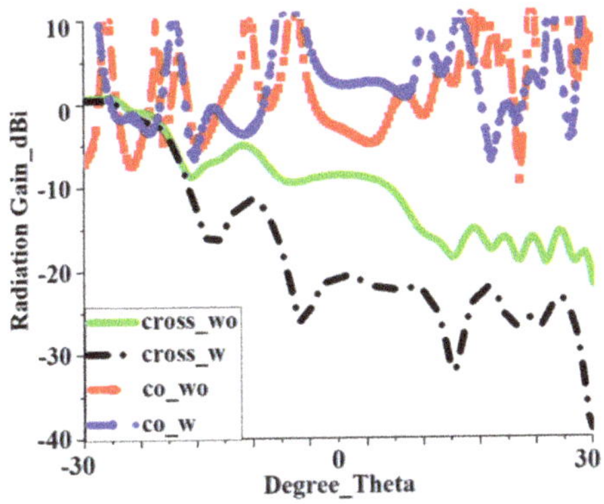

@ 55 GHz

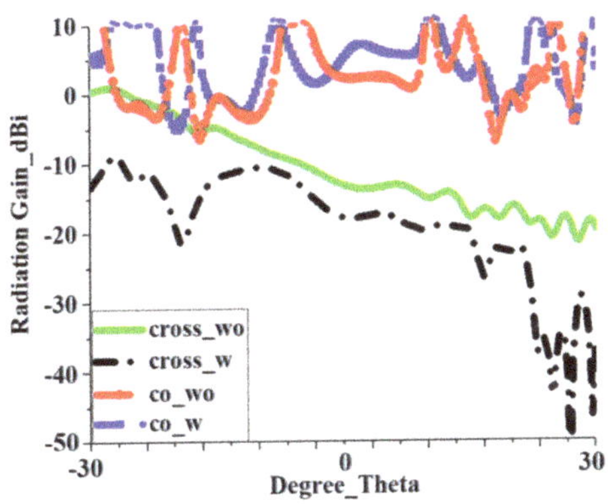

@ 60 GHz

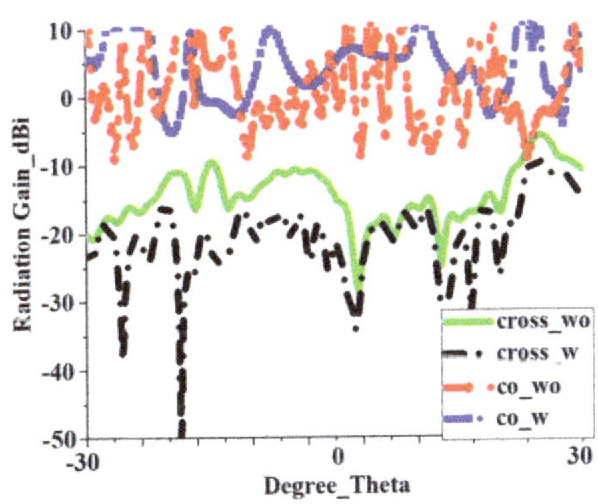

@ 65 GHz

Figure 6. Co- and cross-polarized radiation gain patterns of the proposed structure without (WO) and with (W) metamaterial-shield.

Table 2. Radiation properties.

Radiation Gain		
	Without MTM	**With MTM**
Minimum	8.69 dBi	9.6 dBi
Maximum	9.11 dBi	10.16 dBi
Average	8.90 dBi	9.86 dBi
Improvement in Average	~1 dBi	
Radiation Efficiency		
	Without MTM	With MTM
Minimum	62.53%	73.44%
Maximum	65.21%	78.12%
Average	63.5%	76.38
Improvement in Average	~13%	

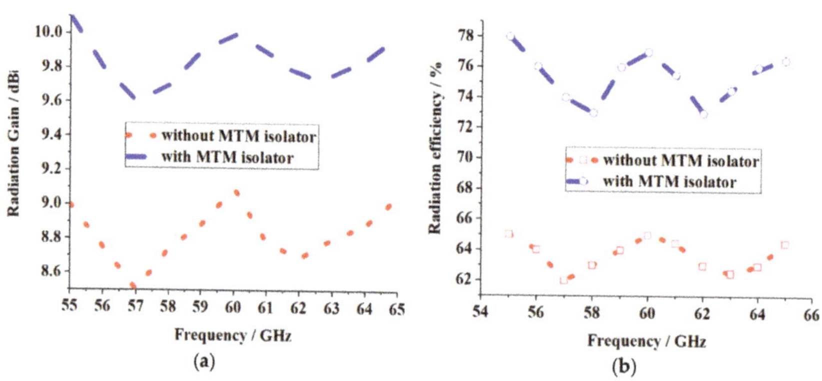

Figure 7. Radiation gain and efficiency curves over frequency band for both cases with no and with MTM shield. (**a**) Radiation-gain; (**b**) Radiation-efficiency.

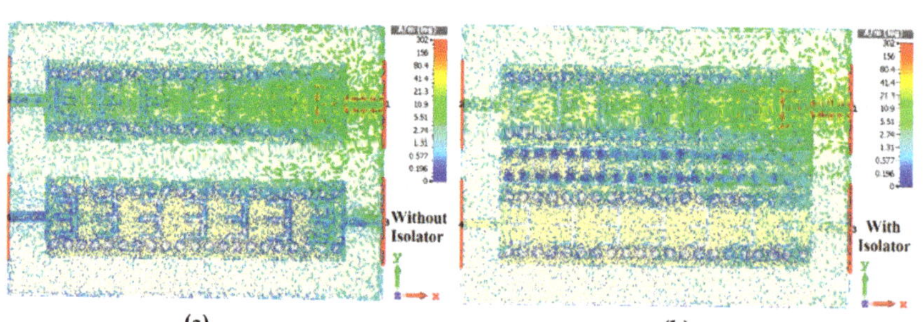

Figure 8. Surface current density distributions without and with the metamaterial shield at 60 GHz. (**a**) without the metamaterial shield; (**b**) with the metamaterial shield.

4. Circuit Model of The Proposed MTM-Lwa Array and its Dispersion Phenomenon

One way to explain the metamaterials (MTMs) is the transmission-line theory in terms of the circuit models. The concept of the composite right/left-handed metamaterial transmission lines (CRLH-MTM TLs) is investigated and realized based on this approach. This solution has been broadly recognized and adopted as a powerful analysis tool for the understanding and modelling of MTM devices. By considering the right-handed (RH) effects within a purely left-handed (LH) circuit, it demonstrates a general configuration of a practical MTM-TL. The circuit model of a generic symmetrical CRLH transmission-line unit-cell has exhibited in Figure 9 where the loss is neglected for simplicity. The series capacitance (C_L) and the shunt inductance (L_L), which have been realized by the slots and via-holes, respectively, contribute to the left-handedness while the series inductance (L_R) and the shunt capacitance (C_R), which have been implemented by the unwanted currents flowing on the patches and the gap space between the patches and ground plane, respectively, actualize its right-handed (RH) dual counterpart. The one indicated in Figure 9a is called *T*-type model with the LH capacitance placed at the two ends. The mushroom unit cell belongs to this type [14]. The circuit exhibited in Figure 9b is called π-type model with the LH capacitance in the centre. One example is the CRLH-SIW unit cell [15–17]. Therefore, each unit cell of the proposed leaky wave array antennas is based on the π-type model, which has been identified in Figure 4b.

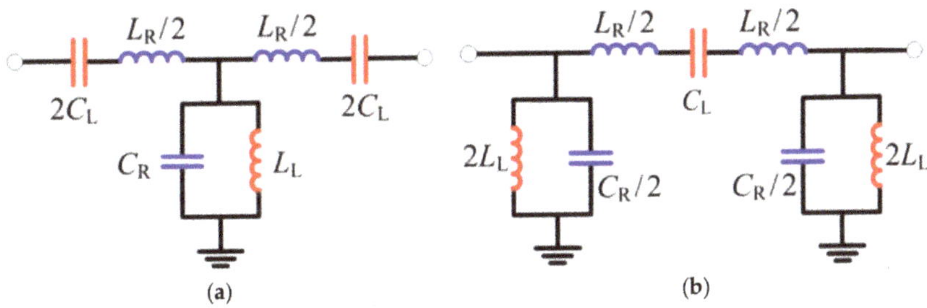

Figure 9. Equivalent circuit models for the symmetrical CRLH-metamaterial unit-cells. (**a**) *T*-type circuit model. (**b**) π-type circuit model.

By using the periodic boundary conditions corresponding to the Bloch–Floquet theorem, these two CRLH transmission line unit-cells basically become equal and their dispersion relevance is determined to be [9]:

$$\beta(\omega) = \frac{1}{p} cos^{-1}(1 - \frac{1}{2}(\frac{\omega_L^2}{\omega^2} + \frac{\omega^2}{\omega_R^2} - \frac{\omega_L^2}{\omega_{se}^2} - \frac{\omega_L^2}{\omega_{sh}^2})) \tag{1}$$

where $p = 5\ mm$ is the length of the unit-cell and

$$\omega_L = \frac{1}{\sqrt{C_L L_L}} \tag{2}$$

$$\omega_R = \frac{1}{\sqrt{C_R L_R}} \tag{3}$$

$$\omega_{se} = \frac{1}{\sqrt{C_L L_R}} \tag{4}$$

$$\omega_{sh} = \frac{1}{\sqrt{C_R L_L}} \tag{5}$$

Seemingly, there are two frequency spots referred to as the infinite-wavelength points ($\beta = 0$) with a bandgap in between. In the balanced case ($\omega_{se} = \omega_{sh}$), the bandgap vanishes. Generally, just one particular zeroth-order resonance will be excited that depends on the boundary conditions and the circuit values. For the short-ended resonator, it is determined by ω_{se}, while for the open-ended case, it is represented by ω_{sh} [9]. Multiplex resonances containing the negative-, zeroth-, and positive- order resonances can be produced by cascading more than one unit-cell. Those resonance frequencies of different order modes for an M-stage CRLH-transmission line can be discovered on the dispersion diagram when the following condition is satisfied [9]:

$$\theta_M = \beta p_M = \beta M_p = n\pi \tag{6}$$

$$\beta p = \frac{n\pi}{M}\begin{cases} n = 0, \pm 1, \ldots, \pm\{M-1, \ for\ T\ type\ unice\ cell \\ n = 0, \pm 1, \ldots, \pm M, \ for\ \pi\ type\ unice\ cell\,. \end{cases} \tag{7}$$

The proposed leaky wave antenna (LWA) array applying the CRLH-SIW unit-cell as shown in Figures 1 and 4 has an π-type circuit model with the two ends terminated by the LH inductances (L_L) realized by the metallic via-holes. The E-shaped transverse and the tapered transverse slots have presented within the circuit model as the LH capacitances (C_L). RH contribution comes from the distributed shunt capacitor (C_R) and the series inductor (L_R). Figure 10 shows the dispersion curves of the proposed leaky wave array antennas based on the CRLH-SIW unit-cell that has been extracted by the CST Microwave Studio package and the equivalent circuit model exhibited in Figure 9b. It has been observed that the results achieved from CST Microwave Studio package and the circuit elements are in an excellent coherence with each other, and also it illustrates the dispersion relation of the unit-cell very well. There is one zeroth-order resonance frequency occurring at 61 GHz that defines the upper and lower edges of the stop-band. This LW antenna array falls into the short-ended case because it is operated below the original waveguide cutoff-frequency and the metallic via-holes offer the short-ended condition. According to Equations (6) and (7), there is one resonance that can be excited comprising the zeroth-order resonance at f_{se} = 61 GHz. Figure 5 shows the reflection coefficient which has verified the proposed model. Notice that this resonance frequency can be simply controlled by engineering the dispersion diagram. Magnitudes of the equivalent circuit parameters, which were determined from full-wave EM simulation using CST Microwave Studio package, are L_L = 6.45 nH, C_L = 8.69 pF, L_R = 1.53 nH, and C_R = 4.12 pF. This data was then utilized to determine the equivalent circuit-model of the antenna, displayed in Figure 9b, which was verified applying ADS (RF-circuit solver).

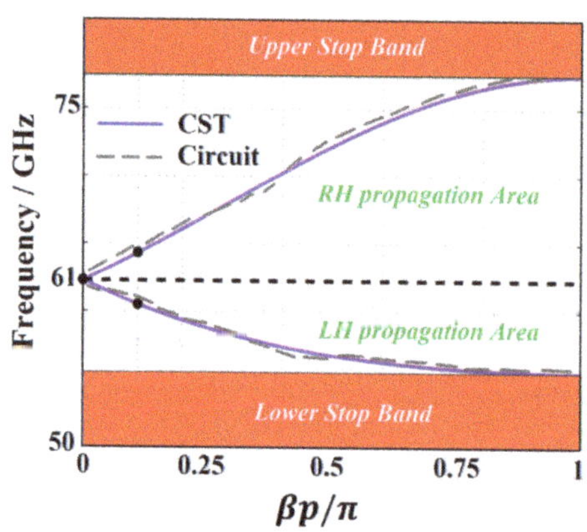

Figure 10. Dispersion diagrams for the proposed LWA array based on SIW-MTM extracted by CST Microwave Studio package and the corresponding equivalent circuit shown in Figure 9b.

The proposed LWA is exhibited as a traveling-wave antenna, where the current propagates along a guiding structure. Since that, the perturbations are introduced along the structure by implementing the E-shaped transverse and the traveling-wave leaves the structure and radiates into free-space. Therefore, in the ideal case, no energy reaches the end of the structure. In a practical scenario, any energy that reaches the end is absorbed by a matching load. Usually, LWA is designed, in which at the least 90% of the power at the structures leaks away before the traveling-waves reach the end of the antenna. Leaky-wave phenomenon is demonstrated with fast propagating waves only. The propagating wave number K_p is defined by [18,19]

$$K_p = \sqrt{K_0^2 - K_z^2} \tag{8}$$

In this case, $K_z = -j\alpha$ is for surface-wave or slow-wave, or $K_z = \beta$ is for leaky-wave or fast-wave, where K_z is the longitudinal wave-number and K_0 refers to the free-space propagating wave-number. The complexity of radiation K_z is given by

$$K_z = \beta - j\alpha \tag{9}$$

where α and β are the attenuation and phase constants respectively. Supposing that there is a standard free-space wave equation for the above wave, the waves outside the leaking-structure are given by:

$$\Psi(r) = \Psi_0 e^{-j(K_p p + \beta_z)} e^{\alpha_z} \tag{10}$$

If $\beta < K_p$ i.e., if the phase velocity is smaller than the free-space velocity of light, $V_p < C$, then it is a slow-wave and K_p is imaginary. The wave decays exponentially in amplitude along the length of the structure and it is a bounded wave. If, $\beta > K_p$ i.e., if the phase velocity is greater than the free space velocity of light, $V_p > C$, then it is a fast-wave and K_p is purely real; therefore, the real power at an angle is radiated with respect to the normal defined by [20]:

$$\sin(\theta) = sin^{-1}\left(\frac{\beta}{K_0}\right) = sin^{-1}\left(\frac{c\beta}{\omega}\right) \tag{11}$$

Since all of the abovementioned terms are functions of the angular frequency, the angle changes with frequency; hence, this shows frequency scanning behaviour. The main beam-width is

$$\Delta\theta_0 = \frac{0.91}{\left(\frac{l}{\lambda}\right)cos\theta_0} \tag{12}$$

If the above equation is applied for large antenna lengths, high directivity can be specified as

$$D = \frac{4\pi A_e}{\lambda_0} \tag{13}$$

However, the effect of enhancing directivity is negligible if there is no power left near the end of the structures. To specify this parameter, the attenuation/leakage constant is determined as [20]:

$$\alpha_z = \frac{e_r^{\frac{A^2(z)}{2}}}{\int_0^l A^2(z)dz - e_r \int_0^z A^2(z)dz} \tag{14}$$

Thus, if α is sufficiently small so that $(1 - e^{-2\alpha l} > 0)$, the improvement of directivity is perceptible as length l enhances.

5. Comparison between This Work and The Literature

Performance parameters of the proposed 1 × 2 MTM-LWA array antenna based on SIW is compared with the recent works employing various mutual coupling suppression techniques. The comparison in Table 3 is for array antennas composed of two radiation elements. Most of the arrays listed in Table 3 exhibit narrow band performance, and to increase isolation between the radiation elements they employ defected ground structures (DGS) to complement their suppression technique. The proposed array antenna presented here has the advantage of (i) symmetry; (ii) very wide bandwidth from 55

GHz to 65 GHz; (iii) simple design; (iv) improved radiation patterns; (v) enhanced radiation gain; (vi) low cross-polarization levels; and (vii) mutual coupling suppression on average of ~25dB over its operating band.

Table 3. Performance parameters of the proposed lw array antennas in comparison with the recent papers.

Ref.	Method	Max. Isolation Improvement	Bandwidth (FBW)	Rad. Gain Pattern Deterioration	No. of Elements	Application of DGS	Symmetry and Simplicity
[21]	SCSRR	10 dB	Narrow	Yes	2	Yes	No
[22]	SCSSRR	14.6 dB	Narrow	Yes	2	Yes	No
[23]	Compact EBG	17 dB	Narrow	Yes	2	Yes	No
[24]	Fractal MTM-EMBG	37 dB	Wide (~15%)	No	4	No	Yes
[25]	Meander Line	10 dB	Narrow	No	2	Yes	No
[26]	EBG	8.8 dB	Narrow	-		Yes	No
[27]	EBG	13 dB	Wide (~12%)	Yes	2	Yes	No
[28]	Substrate Integrated Waveguide (SIW) with Transverse Slots	15 dB	Wide (22.22%)	-	1	No	Yes
[29]	Substrate Integrated Waveguide (SIW) and Slots	20 dB	Wide (16.32%)	-	1	No	Yes
[30]	Periodic Space-Time Modulation	-	Wide (58.82%)	-	1	No	Yes
[31]	Space Time Modulation (External Linear Momentum)	-	-	-	1	No	Yes
[32]	MTM-DS	57 dB	Wide	No	2	No	Yes
[33]	Fractal load + DGS	16 dB	Narrow (2.5%)	No	2	Yes	No
[34]	Slotted Meander-Line	16 dB	Narrow	Yes	2	Yes	No
[35]	Slots	>26 dB	Wide	No	4	No	Yes
[36]	I-Shaped Resonator	30dB	Narrow	Yes	2	Yes	No
[37]	W/g MTM	18 dB	Narrow	No	2	Yes	No
[38]	MSWI	13.5 dB	Wide	No	2	No	Yes
[39]	Metamaterial Superstrate	25	Narrow	No	2	No	Yes
[40]	Slot Combined Complementary Split Ring Resonator	19	Narrow	Yes	2	No	No
[41]	Metamaterial	40	Wide (8.8%)	No	4	No	Yes
[42]	Fractal Load	37	Wide	No	2	No	Yes
This work	Metamaterials and Substrate Integrated Waveguide	42.5 dB	Wide (16.66%)	No	2	No	Yes

6. Conclusion

A feasibility study of a new metamaterial leaky-wave array antenna based on substrate-integrated waveguide (SIW) technology with transverse slots and metallic via-holes for operation over 55 GHz to 65 GHz was proposed and investigated. The array antenna provides beam-scanning capability of ±30° with the gain of 8.5, 10.1, and 9.5 dBi at backward (-30°), broadside (0°), and forward (+30°) directions, respectively. To increase the isolation between the array's elements, a metamaterial shield based on SIW was introduced between the antennas, which has reduced the mutual coupling by an average value of ~25 dB. In addition, the proposed MTM shield has increased the radiation gain and efficiency by an average value of ~1dBi and ~13%, respectively.

Author Contributions: Conceptualization, M.A.; B.S.V., F.F.; E.L.; methodology, M.A.; C.H.S.; F.F.; E.L.; software, M.A.; B.S.V.; C.H.S.; validation, M.A.; B.S.V.; C.H.S.; R.A.A.-A.; F.F.; E.L.; formal analysis, M.A.; B.S.V.; F.F.; E.L.; investigation, M.A.; B.S.V.; C.H.S.; R.A.A.-A.; F.F.; E.L.; resources, M.A.; C.H.S.; R.A.A.-A.; E.L.; data curation, M.A.; C.H.S.; R.A.A.-A.; F.F.; writing—original draft preparation, M.A.; writing—review and editing, B.S.V.; C.H.S.; R.A.A.-A.; F.F.; E.L.; visualization, M.A.; B.S.V.; C.H.S.; F.F.; E.L.; supervision, E.L.; project administration, R.A.A.-A.; F.F.; E.L.; funding acquisition, R.A.A.-A.; F.F.; E.L..

Funding: This work is partially supported by grant agreement H2020-MSCA-ITN-2016 SECRET-722424 and the UK EPSRC under grant EP/E022936/1.

Conflicts of Interest: The authors declare no conflict of interest. The funders had no role in the design of the study; in the collection, analyses, or interpretation of data; in the writing of the manuscript, or in the decision to publish the results.

References

1. Jackson, D.R.; Oliner, A.A. Leaky-wave antennas. In *Modern Antenna Handbook*; John Wiley & Sons: New York, NY, USA, 2008; pp. 325–367.
2. Sievenpiper, D.F. Forward and backward leaky wave radiation with large effective aperture from an electronically tunable textured surface. *IEEE Trans. Antennas Propag.* **2005**, *53*, 236–247. [CrossRef]
3. Jackson, D.R.; Caloz, C.; Itoh, T. Leaky-wave antennas. *Proc. IEEE* **2012**, *100*, 2194–2206. [CrossRef]
4. Lim, S.; Caloz, C.; Itoh, T. Metamaterial-based electronically controlled transmission-line structure as a novel leaky-wave antenna with tunable radiation angle and beamwidth. *IEEE Trans. Microw. Theory Tech.* **2004**, *52*, 2678–2690. [CrossRef]
5. Caloz, C.; Itoh, T.; Rennings, A. CRLH metamaterial leaky-wave and resonant antennas. *IEEE Antennas Propag. Mag.* **2008**, *50*, 25–39. [CrossRef]
6. Paulotto, S.; Baccarelli, P.; Frezza, F.; Jackson, D.R. A novel technique for open-stopband suppression in 1-D periodic printed leaky-wave antennas. *IEEE Trans. Antennas Propag.* **2009**, *57*, 1894–1906. [CrossRef]
7. Chen, Z.N.; Qing, X. Slotted SIW Leaky-Wave Antenna with Improved Backward Scanning Bandwidth and Consistent Gain. In Proceedings of the 2017 11th European Conference on Antennas and Propagation (EUCAP), Paris, France, 19–24 March 2017; pp. 752–755.
8. Qian, Y.; Chang, B.C.C.; Itoh, T.; Chen, K.C.; Tzuang, C.K.C. High efficiency and broadband excitation of leaky mode in microstrip structures. In Proceedings of the 1999 IEEE MTT-S International Microwave Symposium Digest (Cat. No.99CH36282), Anaheim, CA, USA, 13–19 June 1999; pp. 1419–1422.
9. Caloz, C.; Itoh, T. *Electromagnetic Metamaterials: Transmission Line Theory and Microwave Applications*; Wiley-IEEE Press: New York, NY, USA, 2005; 376p.
10. Yang, T.; Chi, P.L.; Xu, R.M. Novel composite right/left-handed leaky-wave antennas based on the folded substrate-integrated waveguide structures. *Prog. Electromagn. Res.* **2012**, *29*, 235–248. [CrossRef]
11. Dong, Y.; Itoh, T. Composite right/left-handed substrate integrated waveguide and half mode substrate integrated waveguide leaky-wave structures. *IEEE Trans. Antennas Propag.* **2011**, *59*, 767–775. [CrossRef]
12. Weitsch, Y.; Eibert, T.F. Analysis and design of a composite left/right-handed leaky wave antenna based on the H10 rectangular waveguide mode. *Adv. Radio Sci.* **2008**, *6*, 49–54. [CrossRef]
13. Chen, Z.N.; Qing, X. Multilayered composite right/left-handed leaky-wave antenna with consistent gain. *IEEE Trans. Antennas Propag.* **2012**, *60*, 5056–5062.
14. Lai, A.; Leong, K.M.; Itoh, T. Infinite wavelength resonant antennas with monopolar radiation pattern based on periodic structures. *IEEE Trans. Antennas Propag.* **2007**, *55*, 868–876. [CrossRef]
15. Dong, Y.; Itoh, T. Miniaturized substrate integrated waveguide slot antennas based on negative order resonance. *IEEE Trans. Antennas Propag.* **2010**, *58*, 3856–3864. [CrossRef]
16. Dong, Y.; Itoh, T. Substrate integrated waveguide negative order resonances and their applications. *IEEE Trans. Antennas Propag.* **2010**, *4*, 1081–1091. [CrossRef]
17. Dong, Y.; Itoh, T. Metamaterial-Based Antennas. *Proc. IEEE* **2012**, *100*, 2271–2285. [CrossRef]
18. Johnson, R.C.; Jasik, H. *Antenna Engineering Handbook*, 4th ed.; McGraw-Hill Education: New York, NY, USA, 2007.
19. Mohsen, M.K.; Isa, M.S.M.; Isa, A.A.M.; Zin, M.S.I.M.; Saat, S.; Zakaria, Z.; Ibrahim, I.M.; Abu, M.; Ahmad, A.; Abdulhameed, M.K. The Fundamental of Leaky Wave Antenna. *J. Telecommun. Electron. Comput. Eng.* **2018**, *10*, 19–127.
20. Frank, B.G. *Frontiers in Antennas: Next Generation Design & Engineering*; McGraw Hill Professional: New York, NY, USA, 2010; 512p.
21. Bait-Suwailam, M.M.; Siddiqui, O.F.; Ramahi, O.M. Mutual coupling reduction between microstrip patch antennas using slotted-complementary split-ring resonators. *IEEE Antennas Wirel. Prop. Lett.* **2010**, *9*, 876–878. [CrossRef]

22. Shafique, M.F.; Qamar, Z.; Riaz, L.; Saleem, R.; Khan, S.A. Coupling suppression in densely packed microstrip arrays using metamaterial structure. *Microw. Opt. Tech. Lett.* **2015**, *57*, 759–763. [CrossRef]
23. Islam, M.T.; Alam, M.S. Compact EBG Structure for Alleviating Mutual Coupling between Patch Antenna Array Elements. *Prog. Electromagn. Res.* **2013**, *137*, 425–438. [CrossRef]
24. Alibakhshikenari, M.; Virdee, B.S.; See, C.H.; Abd-Alhameed, R.; Ali, A.H.; Falcone, F.; Limiti, E. Study on Isolation Improvement Between Closely Packed Patch Antenna Arrays Based on Fractal Metamaterial Electromagnetic Bandgap Structures. *IET Microw. Ant. Propag.* **2018**, *12*, 2241–2247. [CrossRef]
25. Ghosh, T.; Ghosal, S.; Mitra, D.; Bhadra Chaudhuri, S.R. Mutual Coupling Reduction Between Closely Placed Microstrip Patch Antenna Using Meander Line Resonator. *Prog. Electromag. Res. Lett.* **2016**, *59*, 115–122. [CrossRef]
26. Yang, F.; Rahmat-Samii, Y. Microstrip Antennas Integrated with Elctromagnetic Band-Gap (EBG) Structures: A Low Mutual Coupling Design for Array Applications. *IEEE Trans. Antennas Propag.* **2003**, *51*, 2936–2946. [CrossRef]
27. Mu'ath, J.; Denidni, T.A.; Sebak, A.R. Millimeter Wave Compact EBG Structure for Mutual Coupling Reduction Applications. *IEEE Trans. Antennas Propag.* **2015**, *63*, 823–828.
28. Liu, J.; Jackson, D.R.; Long, Y. Substrate integrated waveguide (SIW) leaky-wave antenna with transverse slots. *IEEE Trans. Antennas Propag.* **2012**, *60*, 20–29. [CrossRef]
29. Ettorre, M.; Sauleau, R.; Le Coq, L. Multi-beam multi-layer leaky-wave SIW pillbox antenna for millimeter-wave applications. *IEEE Trans. Antennas Propag.* **2011**, *59*, 1093–1100. [CrossRef]
30. Taravati, S.; Caloz, C. Mixer-duplexer-antenna leaky-wave system based on periodic space-time modulation. *IEEE Trans. Antennas Propag.* **2017**, *65*, 442–452. [CrossRef]
31. Taravati, S.; Caloz, C. Space-time modulated nonreciprocal mixing, amplifying and scanning leaky-wave antenna system. In Proceedings of the 2015 IEEE International Symposium on Antennas and Propagation & USNC/URSI National Radio Science Meeting, Vancouver, BC, Canada, 19–24 July 2015; pp. 639–640.
32. Alibakhshikenari, M.; Virdee, B.S.; Shukla, P.; See, C.H.; Abd-Alhameed, R.; Khalily, M.; Falcone, F.; Limiti, E. Interaction Between Closely Packed Array Antenna Elements Using Metasurface for Applications Such as MIMO Systems and Synthetic Aperture Radars. *Radio Sci.* **2018**, *53*, 1368–1381. [CrossRef]
33. Yang, X.; Liu, Y.; Xu, Y.X.; Gong, S.X. Isolation Enhancement in Patch Antenna Array with fractal UC-EBG Structure and Cross Slot. *IEEE Antennas Wirel. Propag. Lett.* **2017**, *16*, 2175–2178. [CrossRef]
34. Alsath, M.G.N.; Kanagasabai, M.; Balasubramanian, B. Implementation of Slotted Meander Line Resonators for Isolation Enhancement in Microstrip Patch Antenna Arrays. *IEEE Antennas Wirel. Propag. Lett.* **2013**, *12*, 15–18. [CrossRef]
35. Alibakhshikenari, M.; Virdee, B.; Shukla, P.; See, C.; Abd-Alhameed, R.; Khalily, M.; Falcone, F.; Limiti, E. Antenna Mutual Coupling Suppression Over Wideband Using Embedded Periphery Slot for Antenna Arrays. *Electronics* **2018**, *7*, 198. [CrossRef]
36. Ghosh, C.K.; Parui, S.K. Reduction of Mutual Coupling Between E-Shaped Microstrip Antennas by Using a Simple Microstrip I-Section. *Microw. Opt. Tech. Lett.* **2013**, *55*, 2544–2549. [CrossRef]
37. Qamar, Z.; Park, H.C. Compact Waveguided Metamaterials for Suppression of Mutual Coupling in Microstrip Array. *Prog. Electromagn. Res.* **2014**, *149*, 183–192. [CrossRef]
38. Alibakhshikenari, M.; See, C.H.; Virdee, B.S.; Abd-Alhameed, R.A. Meta-surface Wall Suppression of Mutual Coupling between Microstrip Patch Antenna Arrays for THz-band Applications. *Prog. Electromagn. Res. Lett.* **2018**, *75*, 105–111. [CrossRef]
39. Qamar, Z.; Naeem, U.; Khan, S.A. Mutual coupling reduction for high performance densely packed patch antenna arrays on finite substrate. *IEEE Trans. Antennas Propag.* **2016**, *64*, 1653–1660. [CrossRef]
40. Qamar, Z.; Riaz, L.; Chongcheawchamnan, M.; Khan, S.A.; Shafique, M.F. Slot combined complementary split ring resonators for mutual coupling suppression in microstrip phased arrays. *IET Microw. Antenna Propag.* **2014**, *8*, 1261–1267. [CrossRef]

41. Alibakhshikenari, M.; Khalily, M.; Virdee, B.S.; See, C.H.; Abd-Alhameed, R.A.; Limiti, E. Mutual-Coupling Isolation Using Embedded Metamaterial EM Bandgap Decoupling Slab for Densely Packed Array Antennas. *IEEE Access* **2019**, *7*, 5182–51840. [CrossRef]
42. Alibakhshikenari, M.; Khalily, M.; Virdee, B.S.; See, C.H.; Abd-Alhameed, R.A.; Limiti, E. Mutual Coupling Suppression Between Two Closely Placed Microstrip Patches Using EM-Bandgap Metamaterial Fractal Loading. *IEEE Access* **2019**, *7*, 23606–23614. [CrossRef]

MDPI
St. Alban-Anlage 66
4052 Basel
Switzerland
Tel. +41 61 683 77 34
Fax +41 61 302 89 18
www.mdpi.com

Electronics Editorial Office
E-mail: electronics@mdpi.com
www.mdpi.com/journal/electronics

www.ingramcontent.com/pod-product-compliance
Lightning Source LLC
Chambersburg PA
CBHW061616220926
51850CB00005B/523
9783039360161